Landmarks of Botanical History

LANDMARKS OF BOTANICAL HISTORY

Edward Lee Greene

EDITED BY FRANK N. EGERTON

With contributions by Robert P. McIntosh and Rogers McVaugh

PART I

A Publication of the Hunt Institute for Botanical Documentation
Carnegie-Mellon University

STANFORD UNIVERSITY PRESS
Stanford, California 1983

The text of Part I, with sparse notes by Greene, was published as a book by the Smithsonian Institution in 1909 but has long been out of print. The text of Part II is based on a typescript, now in the Hunt Institute, prepared at the Smithsonian in 1936 from a handwritten manuscript found among Greene's effects at the time of his death in 1915. Part II is published here for the first time, by permission of the University of Notre Dame.

Work on the present edition began at the Hunt Institute in 1965. It was interrupted in 1970 for lack of funding, and was finally resumed in 1977. Minor emendations in the text of both Parts have been made in the interest of clarity and consistency.

STANFORD UNIVERSITY PRESS, STANFORD, CALIFORNIA

Printed in the United States of America ISBN 0-8047-1075-9 LC 79-66057

Acknowledgments

Many members of the Hunt Institute staff have contributed to the production of this edition, and to them I am deeply grateful. The late George H. M. Lawrence planned this edition, commissioned the studies on Greene by Robert P. McIntosh and Rogers McVaugh, introduced the project to me, and provided me with the opportunity to work on it during the four years in which I served as his assistant. McIntosh deserves credit not only for having written the biography of Greene included here, but also for having suggested that the Hunt Institute undertake this publication.

Greene's manuscript and the posthumous typescript of Part II of these *Landmarks* are the property of the University of Notre Dame Archives. Both were on loan for over fifteen years to the Hunt Institute while this edition was being prepared. On behalf of the Institute, I thank the University of Notre Dame for allowing us to hold these materials for so long and for permission to edit and publish them.

The original expectation was that I would complete the editing within three years while I was at the Hunt Institute. When that period proved to be insufficient, I stayed on for another year (altogether, from 1966 to 1970). While there, I especially appreciated the advice and friendship of John V. Brindle, Curator of Art. Günther Buchheim, both then and after I left, provided expert opinions on the identities of plants being illustrated or discussed. Jeanne Gina Gazzola, my faithful secretary at the Hunt Institute, occasionally translated Italian quotations for my notes.

My departure from the Hunt Institute coincided with a change in its administration and with budgetary constraints I had never considered when I planned this edition. I was given the opportunity to find a new publisher, but during the following few years pressing new teaching responsibilities prevented my making a sustained effort to do so. This edition might not have been published at all had not the Institute's third director, Robert W. Kiger, taken a strong interest in it and arranged in 1977 for its publication for the Institute by Stanford University Press.

In those intervening years, the manuscript I had produced was read first by Jerry Stannard, and later by Karen Reeds. They deserve the deep gratitude both of myself and of you, the reader. Stannard is one of the very few scholars, perhaps the only one, whose expertise covers the entire period of this work. I began seeking his advice shortly after I started editing, and have benefited greatly from his scholarship all along the way. Reeds expressed an interest in the manuscript at a time when I had no prospects for its publication. She nevertheless provided additional advice and corrections (over five years having elapsed since Stannard made his comments).

Meanwhile, the Committee on Research and Creative Activity at the University of Wisconsin–Parkside furthered my efforts to complete the project by providing funds to purchase essential reference works not otherwise available to me here.

Assisted also by new funds from the Hunt Institute, I devoted my summer in 1978 to the completion of this edition. Those three months were not enough for the task, but they did enable me to regain sufficient momentum to continue until completion. Elizabeth A. Mosimann worked with me during this final period, answering questions from volumes in the Hunt Botanical Library and completing the acquisition of reproducible illustrations. Prints had already been made of the large majority of illustrations, which were taken from volumes in the Hunt Botanical Library; what remained were mostly high hopes and indefinite possibilities. After a year she had ascertained what was possible: a good deal more of what I had wanted than seemed likely when she started.

On 1 June 1978 I wrote to John Scarborough to ascertain the publication date of his paper on Theophrastian medical botany. Before long, he had agreed to read all of my revised notes on ancient Greek and medieval Byzantine botany. This resulted in his rewriting my notes on Galen as well as a few others. His abundant advice far exceeded what I could compress into my notes. He has also very kindly proofread galleys for me on the ancient Greek and Byzantine notes.

When I began in 1969 to work on Mesopotamian and Islamic contributions to botany for Appendixes A and B, I consulted Martin Levey, who generously provided useful advice and references. No doubt he would also have read the final draft of these discussions had I completed them before his death. The unfortunate loss of his advice on the final copy has been very well compensated for by Emilie Savage-Smith, who very kindly read my final account of Islamic botany and provided valuable corrections and guidance on both transliterations and references.

Originally, I had hoped to provide a summary account on Chinese botany in some greater detail than is found in Appendixes A and B. In 1968, my wife and I spent a memorable week as guests of Joseph Needham, at Gonville and Caius College, Cambridge University. Then I had the opportunity to read his long, still unpublished, volume on Chinese botany. It was a fine opportunity, but there were limits to what I could absorb and transmit in the time that I could afford for this material. I have therefore settled for only brief indications of the richness of Chinese botany, comforted by the knowledge that the full story is to appear in Needham's *Science and civilization in China*.

My research has depended on the resources and cooperation of many libraries, to which I wish to extend my thanks. My greatest indebtedness is to those I used most extensively. In Pittsburgh, of foremost value to me was the Hunt Botanical Library, which contains an outstanding research collection and, for the period before 1700, is elegantly described by Jane Quinby, *Catalogue of botanical books* (1958). Its history-of-science holdings, especially in the history of botany, have increased steadily since the Institute was established at Carnegie-Mellon University in 1960. Also important for my work were the Carnegie Public Library of Pittsburgh and the University of Pittsburgh Libraries.

When I came to the University of Wisconsin–Parkside in 1970, its library had only recently been established and was very largely furnished with duplicate volumes from the Madison campus. It has quickly grown into a nationally recognized undergraduate library. I had the advantage of suggesting many purchases in the history of science before inflation made serious inroads into the library budget. Yet, our holdings could never suffice for the research necessary to complete the editing of this work. I have depended heavily on the interlibrary loan service at Parkside, which made the great collections at the Madison campus very largely available to me here throughout the course of my work. When the works I needed were not in Madison, this loan service extended its search to the rest of the country.

The research at other libraries was always limited in extent by time constraints. Certainly some of my unanswered questions could have been answered had I been at them for longer periods. But my gratitude for their help is no less for the brevity of my visits, and in each case I gained some of the answers I sought. The list includes some of the great libraries of the world: U.S. Library of Congress; Library of Medical History in the National Library of Medicine; the Rare Book Room and other libraries at the University of Wisconsin–Madison; Houghton, Widener, Museum of Comparative Zoology, and other libraries of Harvard University; The British Library (formerly the British Museum); The Wellcome Historical Medical Library; and the Cambridge University Libraries. The last of the libraries I visited was that at the University of Notre Dame, to check references in Greene's own copies of works he cited that I had been unable to find elsewhere. This was a great pleasure, since questions that had remained unanswered for over a decade could finally be resolved. During the week in which I used Greene's library in the Special Collections and Rare Book Department, I was graciously assisted by the head of that department, Anton Masin, and also by Robert P. McIntosh, whose familiarity with Greene's library has developed through using it as a resource in his course on the history of biology. I am grateful to both these men.

Next to myself, Jean McIntosh, at Stanford University Press, has spent more time than anyone else in preparing this work for publication. She raised questions that I had considered either inadequately or not at all, and the published work is much more useful for my having struggled to answer them.

Robert W. Kiger was not only responsible for reviving the plans for publication, he has read the entire manuscript with the close scrutiny to which he subjects all the works published by the Hunt Institute. It has been gratifying for me to work closely with both him and Jean McIntosh. Many difficult editorial problems seemed easily solved after they began working on them. When my editorial ideas seemed right, they found the best way to implement them. Conversely, they also found alternatives to my awkward ideas. To the extent that clarity, accuracy, and conciseness are achieved, they deserve much of the credit. They also have my deepest appreciation for it.

My wife, Andrea Lynn Egerton, has never been directly involved in any aspect of this project. Yet, for the entire time, she has been my sounding board and moral support, and I wish to thank her for having had more patience and optimism through it all than I could manage myself.

F.N.E.

Contents

PART I

PART II

Landmarks of Botanical History

INTRODUCTORY MATTER

Editor's Introduction

FRANK N. EGERTON

The history of botany is obviously a part of the history of science; and the history of science was just becoming an academic discipline at about the time Greene ceased writing his *Landmarks*. How his work appears from the vantage point of that now-mature discipline, and, more narrowly, how it appears from the narrower vantage point of the history of botany, merit a careful look.

The History of Science

Science is so important an activity that there is a general interest among both scientists and non-scientists in how it works, what it has accomplished, and what its implications are for civilization. Many episodes in the history of science are inspiring and exciting, because they are success stories with lasting implications. Various writers have attempted to satisfy the demand for clear accounts of the nature of science, which they often have explained as a commonsense kind of activity that requires mainly diligence and hard work, with a dash of ingenuity. Those who have adopted this view have recognized such factors as education, the interaction between scientists, and the general intellectual climate of the age as important influences on the progress of science, but influences that are nevertheless separable from scientific method. And they have explained scientific method by a simple formulation that has been widely accepted. Consequently, the history of science is often seen as only a purely descriptive branch of cultural history that does not contribute to the understanding of how science worked, since that is already understood. However, the close study of the history of science has exposed this view of science as an oversimplification. The advancement of science is in reality a complex process that can be understood adequately only when the explanations of science are based on a good understanding of the historical records of science. Thus, it has gradually become clear that the history of science, knowledgeably written, describes not only the steps that science has taken, but how it took them.

Since the time of Greene, many important contributions have been made to both the knowledge and the methodology of the history of science. Most historians of science have come to feel that the effective command of their subject requires knowledge of the content and the methodology of both a science and a part of history. Both descriptive and analytical approaches are essential and complementary. For practical reasons they are sometimes presented separately, but ideally they should be combined. Descriptive studies often reveal an inadequately discriminating point of view, and analytical writings are sometimes based on insufficient data.

Another difference in approach turns on whether one studies the history of a particular science or the general history of science. Scientists are often interested in the history of their specialty and choose the former approach; historians of science often choose the latter. The differences between these points of view are illustrated by the differences between the accounts of Theophrastus by Greene, given in this volume, and those of some later historians of science. Greene asked what were the methods that Theophrastus used in his botany, and the point of comparison primarily was later botany. Historians of science have later asked what scientific methods are revealed in the botanical writings of Theophrastus, and their point of comparison has been Greek science before and after Theophrastus. The differences that these contrasting approaches reveal do not constitute disagreements about the way ancient science ought to be approached. Since the contributions of scholars with different orientations can be useful, this difference in perspectives need not be stressed. Greene wrote the present work from a botanist's point of view, and this editor's notes reflect that of a historian of science.

One may ask if, by later standards, Greene's *Landmarks*, written over 65 years ago, is not obsolete. Evidence that it was not came to me in a letter from a great contemporary historian of science, Professor Joseph Needham, written to me on 26 April 1968:

> We were most interested to get your letter of 8 April last and to learn that the classical work of E. L. Greene, *Landmarks of Botanical History*, is going to be re-issued in full. I have always counted myself very fortunate in that I inherited a copy of the first edition of the first volume from that great historian and philosopher of botany, Agnes Arber, whom we knew for many years. Of all the histories of botany which have helped us in our long chapter on Chinese botany in *Science and Civilization in China*, this has been by far the most precious.

Progress in writing history has been slower in botany than in some other sciences. It might be argued that Part I of Greene's *Landmarks* has not been superseded because those interested in its subject found that it met their needs, but this argument could not apply to the unpublished Part II, which has not been superseded either. The recent reprinting of Meyer's *Geschichte der Botanik* (1854–57) is further indication of the slow pace of scholarship in the history of botany.

During the last half-century there have been many studies of more limited scope on the early history of botany. These have advanced our knowledge of this subject, and a major contribution of this edition may be to direct attention to these studies. Some are restricted in scope; others, e.g., Arber's fine study of herbals, give overviews rather than detailed assessments.

The History of Botany Before 1700

The history of botany before 1700 displays a pattern that will surprise anyone unfamiliar with the brilliant achievements of Greek science, but it is a pattern similar to that of several other sciences. It can be accounted for primarily by general scientific and cultural factors rather than by any special characteristics of the particular botanical subject matter.

Fragmentary evidence survives of activities before Aristotle that can be considered protobotany. Early cultures developed a considerable practical knowledge of plants, and much of this was passed on to the Greeks, who incorporated it into scientific treatises. But Greek biology undoubtedly owed a greater debt to Greek philosophy. There are two great Grecian treatises on botany, both attributed to Theophrastus, a member of the Aristotelian philosophical and scientific school. The Theophrastian treatises organized the science of botany and were great achievements, being solidly based on earlier practical and philosophical knowledge.

The next important theoretical advance was the clarification of the sexuality of plants, which did not occur until the eighteenth century! The many important advances in botany during the intervening centuries raised questions that were not resolved until later. The failure of botany to reach maturity earlier can be attributed in part to the failure of students of science to use the Theophrastian treatises as a paradigm. In his brilliant analysis of scientific advancement, Thomas S. Kuhn has argued that every mature science has a paradigm[1] that provides those practicing that science with a common orientation.

Though Ruel, Gesner, and Cesalpino used Theophrastus's works as a paradigm in the sixteenth century, and other scholars certainly read Theophrastus and were influenced by him, most of the scholarly works on plants written before 1700 were in the area of medical or rather pharmacological botany. The great paradigm for this discipline was Dioscorides' *Materia medica*, and most of the history of botany before 1700 was really the history of pharmacy. Plants were the major source of medicines. Yet to write the early history of botany and include only the works of Theophrastus, Ruel, Gesner, and Cesalpino and a few other theoretical discussions would be a distortion, because botany in the eighteenth century was built on pharmaceutical as well as theoretical botany.

In modern science important advances usually stimulate further activity, but in antiquity a different response was very likely. If accounts written on a subject were good, as were those by Euclid on geometry, Ptolemy on astronomy, Aristotle on zoology, and Theophrastus on botany, then that subject was considered to be understood and students of science could turn to something else. The situation was different in medicine. Although the Hippocratic corpus might be satisfying theoretically, physicians were continually forced to face their lack of success in helping patients. Consequently, medical areas of biology continued to be studied and advanced long after there had ceased to be much attempt to further those other areas of biology that had been treated by Aristotle and Theophrastus. The short treatise on medicinal plants that forms Book IX of Theophrastus's *Historia plantarum* could not satisfy the demand.

Before the first century A.D. there was a succession of works on medical botany that have been lost except for the two on antidotes by Nicander of Colophon. In the first century the Roman encyclopedists Celsus and Pliny both wrote works that included discussions of medicinal plants, and at about the same time Dioscorides produced his *Materia medica*, a synthesis that rendered the earlier works obsolete and was the most authoritative treatise on the subject during the next 15 centuries. Had printing existed during this period, it is probable that Dioscorides' influence would have been so great that the later writings on the subject would have been confined almost entirely to glosses on his *Materia medica*. As it was, most of them were very heavily indebted to him. But the continuing need for medicines stimulated others to continue the quest; Galen and several Arabic and Latin authors produced new works of varying degrees of originality.

Agriculture would seem to have been more promising as a source of botanical knowledge and advancement than medicine, as the treatises produced during the Roman era tend to confirm. Cato, Varro, and Columella wrote important treatises on agriculture, based partly on Roman experience and partly on earlier Greek works, some of which have disappeared. However, though medical schools existed in antiquity, not until the nineteenth century did agricultural education and research make significant progress. Medicine has always been a profession, whereas farming until recently has been on a level with trade or business. Because of this, until the present day many more highly trained and creative thinkers have attempted to advance medical than agricultural science. When educated men wrote on agriculture, it was often to report the results of others who had less education, rather than to lead the way themselves. Therefore, before 1700, medicine provided the main motivation for studying and advancing botany, and the contributions from agriculture were considerably more modest.

During the medieval period the scientific writings were inferior to the best achievements of antiquity. Albertus Magnus (1193–1280) wrote the only important botanical treatise within the Aristotelian tradition. There were, however, empirical innovations in pharmaceutical botany, more extensive in Arabic than in Latin scientific literature. In the eleventh century, scholars began translating scientific works from Arabic to Latin, and this practice continued for about three centuries. Not only Arabic but also many Greek works were first available to the Latin West in these translations.

This movement to acquire Arabic and Greek science was a step toward outgrowing the authoritarianism to which science had succumbed during the Middle Ages. At first, the authority of Latin authors was merely supplemented by that of Greek and Arabic authors. But as the Greek and Arabic works became more and more important in science education, the situation changed. Greek authors usually inspired a more critical spirit than did the Romans. Good texts and translations of Arabic and Greek works into Latin were in demand, and beginning in the 1460's this demand encouraged the printing of science books.

The transition from the veneration and mastery of ancient and Arabic science to the general realization that scientific knowledge could be broadened and improved was long and arduous. It began certainly by 1492, if not earlier, for in that year Columbus exposed a world unknown to the ancients and Leonicenus revealed the fallibility of so illustrious an ancient authority as Pliny. Yet the hold of ancient authority was so great that it was far from being completely destroyed a century and a half later. Brunfels's *Herbarum vivae eicones* (1530–36) was intended to clarify ancient knowledge and make it more easily available. The additions to knowledge that he, Fuchs, Bock, and Valerius Cordus made could be viewed as accidents or special cases. By 1700 naturalists still had not become fully aware of how different the New World flora was from that known to the ancients. Nevertheless, from about 1550 onward some naturalists came to realize that new species could be found by those who sought them. By the time that Tournefort published his *Institutiones rei herbariae* in 1700, there were so many species of plants known that the means of ordering botanical knowledge had become a serious problem. The old Dioscoridean paradigm was not being challenged, but a major problem had arisen that demanded the development of a supplementary paradigm. This demand was successfully met by Linnæus in the eighteenth century, but that story lies beyond the limits of this history.

Historians of Botany Before Greene

An active interest in the history of botany seems to have arisen with the emergence of botany as a scientific discipline. Tournefort, in including a comprehensive summary of botanical history in his *Institutiones rei herbariae* (1700), was indebted to the scholarship of a number of predecessors.[2] Greene depended on two important eighteenth-century works for his own researches: the "Préface istorike" of *Familles des plantes* (1763) by Michel Adanson (1727–1806) and the *Bibliotheca botanica* (1771–72) by Albrecht von Haller (1708–77).

Adanson's "Préface istorike" was not published as a history of botany. He was concerned with the history of classification, and he discussed earlier writings from that perspective.[3] His example of evaluating the past from the perspective of his botanical concerns may have inspired Greene to undertake historical research in a similar vein. In contrast to Adanson, whose ideas were not very influential, Haller was one of the most eminent botanists of his day. Perhaps for that reason the annotations in his bibliography do not appear to have been made to support any particular perspective. Nevertheless, Greene frequently found reason to quarrel with Haller's judgments.[4]

In his recent bibliography of the life sciences, Smit comments on a dozen histories and bibliographies from the nineteenth century.[5] Greene was not acquainted with most of them, using only four and occasionally citing a fifth.

A botanist and physician at Halle University, Kurt Polycarp Joachim Sprengel (1766–1833), wrote not only many botanical works, but also a five-volume history of medicine (1792–1803). His *Historia rei herbariae* (1807–8; German ed., 1817–18) was of broader scope than Adanson's "Préface istorike" and more detailed than Tournefort's "Isagoge." Sprengel was primarily interested in the discovery, naming, and describing of plants, and his accounts are full of long lists of species.[6] Greene nevertheless used his work far more than Winckler's more concise *Geschichte der Botanik* (1854), which was eclipsed as soon as it appeared.

A detailed survey of the history of botany is the *Geschichte der Botanik* (1854–57) by Ernst H. F. Meyer (1791–1858), professor of botany and director of the Royal Botanic Garden at Königsberg.[7] Meyer's *Geschichte* begins with the ancient Greeks, includes the Romans, Byzantines, Arabs, and medieval Europeans, and comes down to 1600 A.D. For all its detail, it is far more a biobibliography of botanists than a history of botanical science. Greene relied most heavily on it, but he took every possible occasion to disagree with Meyer's judgments.

Of great bibliographical use to Greene, as indeed to any subsequent historian of botany, was the *Thesaurus* (1871–77) by Georg August Pritzel (1815–74), who was heavily indebted to Meyer for references to the early literature, as he gladly acknowledged.[8]

The last of the histories Greene used was the *Geschichte der Botanik* (1875; English transl., 1890) by Julius Sachs (1832–97). Sachs was the greatest plant physiologist of the nineteenth century and a great science teacher as well. Nevertheless, as Greene discovered, Sachs was biased toward the German herbalists, and his judgments of Cesalpino and other non-German botanists were unsympathetic.[9]

Greene as Historian of Botany

Greene wrote his *Landmarks* at a time when few Americans had written anything on the history of science and fewer still had made significant contributions to the subject of his work. Considering this it is remarkable that he had the inclination and the capacity to make an important addition to the literature on the history of botany. But the major conditions of his life, which when viewed separately seem rather unfavorable to scholarship, coalesced in a way that enabled him to write incisively on the subject.

An understanding of Greene as a historian can best be reached with some knowledge of his previous life. The studies of him in this volume by Robert P. McIntosh and Rogers McVaugh provide useful insights, and the following assessment is indebted in part to their studies.

Greene's independent temperament inclined him to make his own decisions, and this inclination led him to original sources in botany. His theological training provided him with linguistic competence in Greek and Latin later used for a fresh study of the early botanical authors, and his theological education and reading undoubtedly contributed to the development of his often graceful, sometimes pompous, but always precise command of English. The solitude in which he so often lived gave him both the motivation and the opportunity to read and write. His long career as a taxonomic botanist provided him with a knowledge of plants and an insight into taxonomic problems so essential to his historical studies. Botany as a unified science did not exist in the seventeenth century, and the herbals of that date and earlier emphasize the characteristics of individual species much more than they do the functions and features we associate with botany. For this reason many historians found it difficult to say much about them. This difficulty did not bother Greene. His long experience as a taxonomist gave him the capacity to react meaningfully to the catalogues of botanical descriptions that make up most of the contents of the herbals.

Nevertheless, even he did not attempt to comment on each account in a herbal. He was content to discuss certain samples, which were extensive enough to provide good insights into the methods and capabilities of the author. By his close analysis he exposed implicit procedures and ideas that others had overlooked. He was alert to the meanings of terms and used them to trace subtle shifts in ideas. His linguistic competence gave him the background needed to discuss the Renaissance texts and translations of ancient authors, subjects of his Chapters 10–15.

Greene had the capacity to put himself in the position of the man under discussion. He did not consistently do this, but it occurs often enough to give special strength to his work, and it often saves the modern reader from the tiresome witnessing of the awarding of merits and demerits for statements judged by modern standards. For some of the men he discussed, his accounts are still the most important available.

As a historian Greene had his limitations. He was willing to discuss a broad range of topics, but his personal interests were strongly committed to those few narrow topics that dominated his discussions. Unfortunately, 65 years after the dust has settled over his personal controversies, these topics seem too trivial to have warranted the amount of attention he gave them. Although he admitted that the pre-Linnæan students of botany had not shared his concern for proper nomenclature, he nevertheless looked upon their work as a series of steps, however faltering, toward his position. On more than one occasion he suggested that the writer under discussion secretly believed in Greene's "correct" principles of nomenclature but was inhibited by either convention or practicality from open subscription to them.

Reading his own image into history was only one aspect of Greene's bias as a historian. Another, less serious but more disagreeable, was his tactic of using history as a club to batter his fellow taxonomists. Having failed in his botanical writings to win general support for his ideas, he turned to a more subtle approach. The implicit theme in his *Landmarks* is that botanical taxonomy has gone astray ever since Linnæus, that this has happened because botanists have neglected the pre-Linnæan literature and its nomenclature, and that the situation could be corrected if taxonomists would heed his admonition and follow his lead in basing the necessary nomenclatural changes upon a knowledge of the early literature. This is ironic, because his contemporaries were accusing him of an inadequate usage of *modern* botanical literature!

His quarrel with fellow botanists is responsible for the polemical and moralistic tone in Greene's *Landmarks*. Charles Singer, the author of extensive and widely read works on the history of biology, lamented that "the late E. L. Greene exhausted himself and his readers and turned a good piece of work into a weary wilderness of words by his determined invective against the errors of his predecessors. . . ."[10] Although Singer may have underestimated the stamina of Greene's readers, he certainly described a major flaw. Even for those who enjoy polemics, Greene's position was unwise. He usually assumed that errors were due either to negligence or to some untenable bias, and on neither account was he charitable. This position can be defended effectively only by those who can themselves achieve perfection. His worst mistake, as Singer went on to point out, was "to suppose that absolute truth is more attainable in the History of Science than in Science itself." To write history from this point of view not only discourages readers who do not enjoy polemics but also renders ironic all the uncharitable condemnations of the mistakes of others. For Greene, there was the further consequence that his attitude led him to underestimate the contributions of earlier historians. He seldom mentioned Meyer, Sprengel, or Sachs except to criticize them, and yet his debt to them, especially to Meyer, was enormous.

Greene's strength lay in his textual analyses, and his errors are most conspicuous in his biographical accounts. In the former, he wrote from primary sources; in the latter, mainly from secondary sources. Even so, he apparently remained unaware of many important secondary sources published in Europe after 1860. The inadequacies of his biographical studies make it clear that he chose wisely when he decided to write a series of "landmarks" rather than to attempt, as had Meyer, a general survey of botany. In his third chapter, on late antiquity, Greene lost sight of his objective and wrote instead a superficial survey that is conspicuously weaker than the others. The first chapter, on the rhizotomi, and the eighth, on Euricius Cordus, are also briefer than seems desirable, but they are more limited in scope than is the third chapter.

Given Greene's "landmark" approach, the omission of the Middle Ages was legitimate, though, as he realized, it would not have been inappropriate to have included a chapter on Albertus Magnus. The five "German fathers of botany" whom Greene chose for Part I nicely balanced the ancient material, but it is a giant step from Galen to Brunfels, and it is difficult for the unprepared reader to take such a step and adequately appreciate the essential prerequisites to the achievements of these Germans. In Part II Greene bridges this gap by starting not in 1562, where the first ends, but in 1440. His Preface to Part II gives an *ex post facto* justification for not following chronological order, but I advise the reader to take it as Greene's procedural justification rather than as advice on the sequence for reading the chapters. For the sake of a better chronological sequence, several of the chapters in Part II are rearranged from the order in which Greene left them.

The study of botany during the Italian Renaissance may sound like a promising theme for a half-dozen or more chapters, but a closer look could occasion second thoughts. Despite its being a period of exciting cultural developments, not all of these developments are as easily accessible to a modern audience as a painting by Michelangelo or a poem by Dante. The botanical advances hinged on the proper translation of ancient terms and names and the identification of the plants indicated by them. It is thus a story that might well interest a philologist more than a botanist or historian of science. Nevertheless, Greene succeeded quite well in keeping the story directed toward the latter two groups. He did so partly by using a biographical approach, and by pointing up controversy among the scholars; and he drew to good effect on the ancient botanical tradition that provided the material for the controversy and the prospect for a new botanical science to come.

The Renaissance translations were really the culmination of a great period of medieval translations, and the Renaissance universities were culminations of another aspect of medieval learning, the growth of universities. That growth, among other things, resulted in the attachment of herbal gardens to the medical schools by the mid-sixteenth century. Those who wrote on plants were often professors who taught specialized courses in medical botany, using the botanical garden to give their students firsthand knowledge of the plants discussed.

The sixteenth century witnessed the beginnings of both the scientific revolution and the exploration of the world by Europeans, some of whom were naturalists. These two exciting movements, scientific and geographical, influenced the botanical writings of that century. The immediate influence of geographical discovery on botany was in the description of new species, which led in turn to active searching for still more new species. Of course the naturalists realized that it was not only foreign species that were unknown to science, but also still others from various corners of Europe and the Mediterranean region. The explorations Greene mentions, from Gesner to Tournefort, were entirely the reexamination of the native continent. Appendix C introduces prominent naturalists of the seventeenth century who published important accounts of plants found in lands beyond Europe and the Mediterranean.

An inevitable result of this ongoing discovery and description was confusion over how best to organize the ever-increasing body of knowledge. It was more of a logical than a biological problem, and Gesner's pioneering effort to classify animals was borrowed from earlier classifications of the branches of knowledge (see Figs. 191, 192). Nevertheless, the traits that could be classified were biological, and Cesalpino's attempt to classify plants was based on Aristotelian biology. The thrust of Gesner's and Cesalpino's efforts was continued throughout the later sixteenth century and all of the seventeenth century. Greene tells parts of the story, and references to other parts of it are contained in Appendix C.

Greene's approach freed him from the necessity of discussing every significant figure in the history of botany before 1700, but by his own standards there are at least a dozen men who are conspicuously absent from his treatment. Among those are Christobal Acosta, Prospero Alpino, Kaspar Bauhin, Johann Bauhin, Joachim Camerarius the Younger, Charles de l'Ecluse, Nehemiah Grew, Joachim Jung, Marcello Malpighi, Nicolas Monardes, Garcia de Orta, John Ray, and Guillaume Rondelet. He had hoped partially to remedy this situation with a Part III of *Landmarks*, to be devoted to British botanists, but he did not live to write it.

Even if Greene had lived to publish Part II, however, it seems unlikely that he would have written his Part III, since two substantial works on the history of British botany appeared in 1913 and 1914, one edited by F. W. Oliver and the other written by J. Reynolds Green.[11]

Contributions of This Edition

Greene's *Landmarks* can stand alone, but to reprint Part I in 1983 without editorial additions would be a disservice to readers who will want to know what has been published since Greene's death on the history of this period of botany. A major editorial objective is to emphasize the very large quantity of work that has been done since Greene's time. The "landmark" approach does not provide a skeleton on which everything can be hung, but where it has seemed feasible, references are added to supplement Greene's discussion and to give an indication of the presently available knowledge. In some respects the footnotes in this edition provide a broad bibliographical guide for the history of botany before 1700, yet the coverage of the notes is restricted by the scope of the text, and the reader should not assume that all important gaps have been filled. However, general sources have been cited that can guide the reader to areas of omission. There is a paradox here: on the one hand, the history of botany has made little progress since Greene wrote; on the other, important recent studies are cited in the editor's notes. However, recent studies have not always exerted as much influence as they should have, because they tend to be scattered in diverse journals, and few synthetic works of much detail have brought their results before a general audience. This edition attempts to remedy that deficiency.

Greene assumed that his readers would have a good knowledge of the characteristics of European plant species; hence he often discusses a species without describing it adequately. This makes it difficult for non-taxonomists to appreciate fully some of his points. Since he often reacted as much to the illustrations in the herbals as to the text, this difficulty is largely eliminated in this edition by supplying pertinent illustrations from the herbals. With a few obvious exceptions, the illustrations reproduced here are those discussed by him. Others are chosen to illustrate or to provide commentaries on the points that he made. Greene frequently used technical terminology in his discussions, but most of these terms are treated in standard dictionaries.

The literature of the history of botany contains a type of study that Greene utilized only rarely, but which is too important to ignore: the literature that traces man's use of or ideas about certain species. These studies provide various insights that otherwise would be missed. Therefore, such studies are cited in the editor's notes where relevant, but no attempt is made to provide a complete guide to this literature.

Because it is hoped that this edition will interest readers of a variety of backgrounds, the notes contain information that will seem common knowledge to some, wholly new ground to others. It is only with this approach that the history of science can serve as a bridge between disciplines. In a few instances my attempt to meet the needs of different audiences might appear in conflict with the maintenance of scholarly standards. The most important example of this is my decision to cite Goodyer's English translation of Dioscorides, despite Jerry Stannard's objections. I have done so simply because there is no other English translation for the whole *Materia medica*, though in this case, by way of compromise, I have also added references to Berendes' German translation and to the authoritative Greek text by Wellmann.[12]

Too many editorial annotations will drown a text in footnotes. The guideline followed here is one of what is appropriate for the character of the text. If Greene's objective had been to provide a good overview of the trends in the history of botany, then perhaps an editor's notes should be confined to brief comments on errors. However, Greene attempted to write a detailed analysis of a select group of works in the history of botany, and even if stripped bare of all notes including his own, his *Landmarks* are unsuitable for skimming. His objective was a firm understanding of the history of botany, and the editorial footnotes have been added to strengthen this objective.

These notes and additions to Greene's notes are placed in square brackets. None of Greene's notes has been omitted, but in many cases the exact form of his note has been altered to make it more comprehensible. For a few of his notes it has been necessary, because of the addition of editor's notes, to change his point of footnote citation within the paragraph, but at the same time care has been taken that such adjustment does not distort his meaning.

The corrections and expansion of Greene's observations and documentation made in the editor's notes depend very heavily on a multitude of relevant studies that have been published since he wrote. In addition, I have had the advantage of consulting experts on various matters, whereas Greene had to rely on his own erudition and the reference works then at the Smithsonian Institution and the Surgeon General's office. Even so, questions occurred to me or to my consultants that I lacked the time, resources, or expertise to answer. These questions I gladly leave to others. It has been fifteen years since I undertook the editing of this work, and it now seems preferable to publish it with a few minor questions still unresolved than to take any more years to try to answer them.

This edition reprints Part I, with minor corrections of punctuation and spelling and a renumbering and format change in the notes, from the text that the Smithsonian Institution published in 1909. Greene never completed the text for Part II, but except for his incomplete chapter on Mattioli, there is a copy in cursive, which was made from his original draft. Surviving portions of that draft indicate some of the topics he planned to discuss in his chapter on Mattioli. Most of the original draft has disappeared, probably discarded by Greene after the cursive copy was made. After his death, the Smithsonian Institution made a typescript from the cursive copy in 1936. The extant portions of the draft, the cursive copy, and the Smithsonian typescript were available to me at the Hunt Institute. I edited the typescript, and when it seemed questionable, I checked it against the cursive copy. If the question was still unresolved, then I checked the draft, if the page in question was extant. In a few cases, when I attempted to check the typescript against the cursive copy, the cursive pages were missing, which means they were lost after the Smithsonian typescript was made.

Although Greene may have felt that the chapters he completed and had recopied in cursive were ready to submit to a publisher, they were still some distance away from being ready to publish. There were repetitious passages, unclear sentences, contradictory statements, incomplete statements, and inconsistent usages of names, italics, capitalizations, and so forth. I edited the

Smithsonian typescript and produced a Hunt typescript, which was read and marked for corrections by Jerry Stannard and Karen Reeds. When Stanford University Press agreed to publish this edition, all of the versions were sent there for further editing. The publisher's editing was undertaken in consultation with me and Robert W. Kiger, Director of the Hunt Institute.

The subtitle Greene used for his *Landmarks* is not retained in this edition: *A study of certain epochs in the development of the science of botany.* His title for Part I, "Prior to 1562 A.D.," is accurate, for it does not claim to include all botanical landmarks before 1562. His title for Part II, "From the Italian Renaissance to the time of Linnæus," is not so accurate; his last chapter discusses Tournefort, who died when Linnæus was one year old, and "the time of Linnæus" implies the period when Linnæus was making his own contributions to botany. Although Greene wrote a short book on Linnæus (1912, 91 pp.), there is no indication that he wished to reprint it as a final chapter to his *Landmarks.*

It is difficult to establish a consistent policy on the forms of the names for authors before 1700 who are discussed in the text. Meyer's *Geschichte der Botanik* (1854–57) exhibits certain tendencies, but not absolute consistency. He does consistently use the transliterated Greek names for ancient Greek and Byzantine authors, e.g., Theophrastos, Stephanos Magnetes; he consistently uses the Latin name for ancient Roman authors, *e.g.*, Plinius, Virgilius. However, his consistency breaks down for the medieval and Renaissance periods, where he uses Latinized forms in some cases, *e.g.*, Bartholomäeus Anglicus, Joannes Manardus, but vernacular forms in other cases, *e.g.*, Albert der Grosse, Otto Brunfels.

Since almost all of the "landmark" books that Greene used were in Latin, he felt it appropriate to use Latinized names for their authors, as given on the title pages of their books: Brunfelsius rather than Brunfels, Fuchsius rather than Fuchs. But he was not consistent in this policy. Paradoxically, he preferred the Anglicized names Virgil and Pliny for authors whose official names are the Latinized forms, Virgilius and Plinius. In the case of Anguillara, he used as the first name Aloysius, even though it is given as Luigi in Anguillara's only book. For the last two authors he discussed, Lobel and Tournefort, Greene used the vernacular forms, since these authors preferred them on the title pages of their books, even when their books were in Latin.

For this edition of Greene's *Landmarks*, the names are generally given according to the predominant current usage in English (exceptions are Fabius Columna and Hieronymus Tragus). Usually this means using the native-language form rather than the Latinized form, *e.g.*, Jean Ruel rather than Joannes Ruellius. But current English usage is still Virgil and Pliny rather than Virgilius and Plinius. The medieval authors whom Greene occasionally mentions are given in Latinized form, but in the editor's Appendix B, both the Latinized and a vernacular form are used.

Another editorial change is in citations of scientific names assigned to Linnæus. The abbreviation "Linn.," which Greene used, is now obsolete, and it is changed to "L." In other respects, spelling and punctuation follow the usage in Part I. Greene frequently referred, using cross-references, from discussions in Part II to discussions in Part I and earlier in Part II. All of these cross-references are adjusted to refer to the pertinent discussions in this edition. All of Greene's note numbers are also changed, to form a consecutive sequence with mine.

Greene's citations in his notes sometimes include full titles, but usually he gave a short title. There is no indication that he contemplated adding a bibliography, but with the much larger number of works cited in this edition than in his manuscript, a bibliography is essential. Since full titles are given in the Bibliography, most of the works cited in the notes are identified there with short titles and dates of publication. In cases where more than one edition is cited, the title is given only once, the other editions being indicated with dates and pages or chapter numbers. For example, when both the Latin translation of Theophrastus's *Historia plantarum* (1644) and the English translation (1916) are cited, the note would read: Theophrastus, *Historia plantarum* (1644), bk. 1, ch. 7; (1916), bk. 1, ch. 7, sec. 3. If this is a note in which Greene cited the 1644 translation and I cite the 1916 translation, then the portion following the semicolon is placed in brackets. At times Greene's notes are expanded to include additional information that he had not bothered to include. For example, if he left off the date of a work, when it is clear which edition he used, that date is added to the citation. In those few cases where the edition he used is unknown, no date is added to his portion of the note, but it might be added to the bracketed editor's portion, when I have located the citation in a particular edition.

Notes

Full citations of the sources cited below are given in the Bibliography, in Part II of this work.

1. Kuhn, *Structure of scientific revolutions* (1962), p. 10.
2. Tournefort's summary history, Isagoge in rem herbariam (1700), is translated by Becker into French (1957).
3. For a discussion of Adanson's Préface istorike (1763), see Stafleu, Adanson (1963), pp. 145–203. Other studies are Burtt, Adanson (1965); Guédès, Méthode taxonomique d'Adanson (1967); Jacobs, Adanson (1966); Lawrence, *Adanson* (1963–64); Nicolas, Adanson (1970); Stafleu, *Taxonomic literature* (1967), pp. 1–2, 236–37.
4. Hintzsche, Haller (1972); Stafleu, *Taxonomic literature* (1967), p. 190; Zoller, Hallers Pflanzensammlungen (1958).
5. Smit, *History of life sciences* (1974), cols. 46–47, 657–66.
6. Risse, Sprengel (1975); Stafleu, *Taxonomic literature* (1967), pp. 454–57.
7. Verdoorn, Introductory essay (1965).

8. Pritzel, Praefatio, in *Thesaurus* (1871–77); Stafleu, *Taxonomic literature* (1967), pp. 367–68.

9. Bopp, Sachs (1975); Larson, *Natural order . . . Linné* (1971), pp. 6–7.

10. Singer, Review of Harvey-Gibson, *History of botany* (1920).

11. Oliver, *Makers of British botany* (1913); Green, *Botany in the United Kingdom* (1914). These are supplemented by two fine bibliographies: Desmond, *British and Irish botanists* (1977); Henrey, *British botanical . . . literature* (1975).

12. Jerry Stannard has himself provided four authoritative English translations of chapters from Dioscorides, *Materia medica*, on *moly* (*Allium sp.*), *sampsychon* (sweet marjoram, *Majorana hortensis* Moench), *melanthion* (nigella, *Nigella sativa* L.), and *skilla* (squill, *Urginea maritima* (L.) Baker), found respectively in his papers: Plant called moly (1962), p. 260; Dioscorides and Renaissance (1966b), p. 3; Herbal as medical document (1969a), p. 215; Squill (1974c), pp. 693–94. Compare to Goodyer transl. (1934), pp. 290, 283, 326, 213.

Edward Lee Greene: The Man

ROBERT P. McINTOSH

On 2 December 1913 the Botanical Society of Washington, D.C., held a banquet and special program to honor Edward Lee Greene on his seventieth birthday. Various of his students and associates expressed high appreciation of his botanical work. Professor Aven Nelson praised Greene's "facile pen, the classical clearness, brevity and exactness, of his diction, the rapier-like thrusts of his criticism" and "his splendid achievements in systematic, historical and philosophical botany."[1] The affair, attended by many prominent botanists, was acknowledged by Greene "as attesting an appreciation and a cordial good will on your part, my fellow botanists, which I had no reason that I know of to look for and which I wished I had better deserved,"[2] and as an indication that "you appreciate the zeal and the success with which, through now long years, I have wrought for the advancement of our science."[3] Appreciation and good will were, no doubt, the feelings of those present, but not all of Greene's contemporaries had such kindly feelings toward him as those expressed at the banquet. J. M. Coulter, a distinguished botanist and university president, said of him, "If he could only quietly pass away in one of his apoplectic fits, how much better for American botany!"[4] The irascible M. E. Jones wrote, "There have been several notable deaths in the botanical world. . . . Greene, the pest of systematic botany, has gone and relieved us from his botanical drivel. They say that the good that men do lives after them, but the evil is interred with their bones. I suspect that his grave must have been a big one to hold it all."[5]

Jones reversed the aphorism but, good or evil, Greene's impact on botany is still felt a half-century after his death, however diverse the opinions of his worth. Greene is one of the striking and controversial figures in the turbulent, formative period of botany in America. He made major contributions, particularly to the systematic botany of the West. Ewan notes that his arrival in California "rather initiated a botanical revival."[6] Notwithstanding the zeal he brought to his botanical work, Greene's standing among botanists was limited in two important respects. He was widely regarded as a proponent of fixed species,[7] though he did not oppose in print the concept of organic evolution. He was a vigorous and, at times, vitriolic warrior in the "movement for correct nomenclature,"[8] which was rejected in the universal acceptance of Linnæus' *Species plantarum* (1753) as a starting point for plant nomenclature. Correct nomenclature, in Greene's view, required the acceptance of plant names from the pre-Linnæan literature, with which he was more familiar than most of his contemporaries. In spite of Greene's intransigent and unsuccessful stand in the latter area, H. H. Bartlett (1916), in a posthumous review of Greene's botanical work, wrote, "Certain it is, however, that few, if any, exerted a greater influence than he in the development of systematic botany in America." It is certain also that Greene remains one of the most colorful and controversial personalities in the history of American botany.

Edward Lee Greene was born 20 August 1843 at Hopkinton, Rhode Island. The Greenes were living at the time on the property of his mother's ancestor Zaccheus Reynolds. In a fragment of an unpublished autobiography Greene records that his interest in botany was foreshadowed by "distinct and vivid impressions of two potted plants in the window of the old ancestral dwelling remembered from the age of one and one half to last through life."[9] His mother, Abby Maria (Crandall) Greene, told him years later that the "plants were a soothing panacea to the infant [Greene] when in peevish and fretful mood; whose sorrows were easily dispelled when held up to a near view of the flowers." He also recalled, from the age of four, a decadent garden on the property as "the first earthly paradise that I knew." Greene wrote that the persistence of these childhood memories of plants, enabling him "after the lapse of sixty years to catalogue them from memory, is a fact that ought to be recorded." He believed that these early experiences stimulated his lifelong interest in plants.

As a child, barely six, he was inspired spiritually and instructed intellectually in botany by a book in his mother's library, Mrs. Lincoln's *Lectures on botany* (1842), an early botanical primer with a strongly religious tone. His mother, a lover of plants and a skilled gardener, one year won four first prizes and two silver cups for her bouquets and roses at the Wisconsin State Fair. Throughout her lifetime, Greene continued to send her rhizomes, bulbs, and seeds of plants, and he attributed his early love of plants to her influence. In 1855, when his father, William Maxson Greene, moved the family to Illinois, Edward Lee, aged eleven, was already "an altogether infatuated searcher after plants."[10]

After briefly staying near the Sangamon River, Illinois, the family moved to southern Wisconsin, settling in the town of Albion. Near Albion, on a hill overlooking Lake Koshkonong, lived Thure Kumlien, a Swedish naturalist. Kumlien was a graduate of the University of Uppsala, one of Europe's great universities, and a student of the famous biologist Elias Fries. Known primarily as an ornithologist, Kumlien was a broadly trained naturalist with a thorough knowledge of plants who, according to his admirer Greene, "in the most polished society would have been recognized as an intelligent, refined and most courtly gentleman."[11] Greene came under his influence about 1860, their friendship continuing until Kumlien's death in 1888. In 1872 he wrote to the young Greene, "You see there is between the true votaries of science a singular bond of attachment that Nature has kindly put there, if possible to make their work more pleasant still."[12] That a powerful bond developed between the seventeen-year-old boy and Kumlien, whom Greene describes as "my earliest teacher and companion in botanical studies," is apparent in their long correspondence.[13] It seems likely that Greene's interest in, and later command of, classical and modern languages and philology was stimulated by Kumlien, whom he praised as "a finished collegian . . . whose pen could indite Ciceronian Latin and whose tongue could address any foreigner in, I believe, any one of the languages of Europe spoken between Spain and Sweden."[14] Greene's own facility in languages, later an important asset to his historical and botanical studies, was evidenced early. As a boy in high school he was employed during the summers by a merchant in nearby Milton, Wisconsin. His command of the Norwegian language, gained in three summers of contact with Norwegian immigrants, amazed new arrivals who could not believe he was not also an immigrant.

In 1859 Greene entered Albion Academy, a coeducational school of 200–300 students which, at the time, compared favorably with the young University of Wisconsin.[15] It was founded by Seventh Day Baptists and had a strong non-sectarian religious emphasis.[16] The curriculum was classical with a considerable admixture of natural science. Greene's second-year course included philosophy, botany, geology, zoology, and mineralogy. The daily regimen at Albion was spartan, from the rising bell at 5:00 A.M., followed by an hour of study before breakfast and a full day of classes and study, to 9:30 P.M. bedtime. The Academy prided itself upon making education available to the poor. Tuition was $6–$8 per term plus twenty-five cents for incidentals. Room rent was $2.75 per semester, and board was $2.00 per week. Men and women attended classes together, although social life was rather restricted. Tobacco, gambling, drinking, swearing, and visiting other rooms during study hours were taboo. The school met three terms of fourteen weeks each. Groups of students from Albion visited Thure Kumlien at his home nearby and took trips in the field with him. Although Kumlien's name appears on an 1865 diploma before Greene's graduation, and Kumlien was on the faculty in the 1870's, it does not appear that Greene had classes with him at the Academy.

In his junior year at Albion, Greene, his father, and two of his four younger brothers joined the Union Army. He served nearly three years (21 August 1862 to 13 July 1865) as a private in the 13th Wisconsin Infantry.

His army service, largely in garrison and patrol duties, took him through Tennessee, Kentucky, and Alabama, where he broadened his knowledge of plants.[17] His military career, however, was not without its hazards; he recorded that his unit surprised an encampment of Confederate cavalry eating supper and engaged them in brief conflict before they escaped. Later, his own company was routed by a surprise attack, and Greene must have made an unsoldierly appearance as he retreated, for he wrote Kumlien on 7 January 1865:

> I had well nigh lost all my botanical and other little collections on my last march. We were ordered away from Claysville . . . and were obliged to burn up a good many valuable things to keep them from falling into rebellious hands after our departure. I managed to crowd most of my plants into my knapsack by carrying some of my clothing in another manner and have now preserved the most of them.

His failure to rise from the rank of private may well have been due to his devotion to botanical pursuits throughout his military career and, to judge from his correspondence, he seemed to have been unconcerned about the military, political, or slavery aspects of the war. He carried in his knapsack Alphonso Wood's *Classbook of botany*, wrote long letters to Kumlien telling him of plants seen and of his religious concerns, and sent specimens and seeds to his mother and Kumlien. Greene's extant war letters refer to little else. After one fifteen-mile march he wrote Kumlien (cf. Main, p. 150): "being in a new place, I, of course, took a tramp as soon as I was rested." Strange behavior for an infantryman!

Like many spiritually and intellectually inclined youths in armies everywhere, he found that "A great change has taken place in the young men with whom I was acquainted. They are losing all taste for intellectual enjoyments and spend their hours in gambling and indulging their vile appetites." (Main, p. 151.) In May of 1863 he wrote Kumlien, "We begin to hope for an end of the war very soon" (p. 157). However, the bright prospects for an early end to the war were dimmed shortly afterwards by Union reverses, and it was two years before Private Greene returned to Albion Academy, finished his studies, and graduated in 1866 with a degree of Bachelor of Philosophy.

His own driving interest in natural history, his intimate association with Thure Kumlien, and his academic work had given him an excellent start for his career in botany.

From 1867 to 1870 he taught in rural schools at Monticello, Cerro Gordo, and Pleasant Grove, Illinois. These were raw midwestern towns, Cerro Gordo being just two years old. Greene described it as "about like Edgerton, Wisconsin though the young are far less thoroughly drilled in evil. . . . There are no grog shops in this place. . . ."[18] Writing Kumlien 1 May 1867 (pp. 1 and 6), he said that he was happily enough situated with "a fine prospect of making new acquaintances in Flora's domain" and "no difficulty to get from fifty to eighty dollars a month." In addition to pursuing his botanical interests he practiced taxidermy, spurred on by Thure Kumlien. He developed sufficient skill so that a cat ate his stuffed birds and we hear no more of taxidermy.

Greene was no ordinary rural schoolmaster. He continued his interest in music after his melodeon had been shipped to him from Albion. He studied Latin and, in his letter of 13 May 1867 (p. 3), wrote to his friend Rasmus Anderson, later a distinguished student of Scandinavian languages, "I have a desire to become a thorough Latin scholar," adding, "About the Greek I am not so particular as I have decided not to enter the ministry." This decision concerning the ministry was to be reversed a few years later, and the comment suggests Greene's early difficulties in fixing his religious views. By boarding with a Pennsylvania Dutch family, he learned German well enough to be hired at Albion Academy as a professor of German. However, he resigned his post in a very short time and returned to Illinois after an unpleasant controversy involving his friend Rasmus Anderson and the principal of Albion, Professor Cornwall.

Though Greene was a serious-minded and religiously inclined young man, active in the Methodist Church at this time, he was not a recluse. He wrote to Anderson on 29 October 1869 (p. 2):

> A week ago tonight we had a merry ingathering of young friends who occupied the evening and even several of the small hours in "keeping time to musical chime." I, of course, enjoy being present at such scenes, as you well know, although we teachers must not indulge too liberally in these things if we would keep in favor with a good many of our staid patrons.

The staid patrons of Pleasant Grove might not have approved of his spending several nights sitting up in a haunted house listening for ghosts, which much intrigued him as, perhaps, did "an intelligent and, by the way, pretty daughter of some eighteen years"[19] who also resided in the haunted house. He confided to Anderson in a later letter, of 28 May 1870, that after the owners of the house moved away leaving it vacant he would not spend the night there alone when he could not persuade a friend to go with him.

During this period, as reported by Main (p. 164) from a letter of 2 May 1865, Greene considered returning to school, perhaps to Illinois University or Northwestern, but instead acted on an earlier plan of his army days to "make my home somewhere in that trans-Mississippian region." The beginning of the year 1870 found him ready to spend the coming summer in the mountains of Colorado. His interest in the West had been awakened by reading, with Thure Kumlien, the botany of the Pacific Railway Survey Reports. Kumlien prophesied "that his boy friend would go there and do much for the development of the botany of those new lands." This prophecy was to be amply fulfilled. In preparation for his journey, Greene wrote to the famous American botanists Asa Gray at Harvard and George Engelmann at St. Louis, with whom he had already been corresponding, inquiring about books and papers useful in his field researches. Both offered cordial assistance and requested that he send them specimens of his western collections.

Greene left from Decatur, Illinois, 4 April 1870 on the new transcontinental railway and arrived in Evans, Colorado, in a spring snowstorm. The next day he went by stagecoach to Denver. Early the following morning, like many another neophyte, he learned how deceptive distances can be in mountainous areas. Setting off for the near foothills he found "the long day was consumed in reaching what I had believed would be the easy goal of a not very long morning walk."[20]

Greene spent his first summer in Colorado botanizing in the valley of Clear Creek, the plains between there and Denver, the foothills around Golden City, and the alpine areas. The strong ties to family and friends which persisted throughout his lifetime are evidenced in a letter Greene wrote to Thure Kumlien on 12 May 1870, after a month in Colorado:

> Your letter is very, very welcome. It does me much good to hear from home and friends. I am not altogether gone crazy over the world of beauty and novelty that now surrounds me. I am among strangers of the human kind and do really at times feel that I could appreciate as I never did before the society of those whom I have long known. . . . This afternoon I took five letters from the office and it almost made me cry! One from my mother and two from my brothers. What a blessed thing our mail system is! (Main, p. 170.)

His joy in his new venture is, however, evident in this next sentence, "Well, as you would suppose, I am having really a glorious time; all the flowers I can possibly manage now." He still had Wisconsin ties, for on 30 April 1870 he wrote to Rasmus Anderson (p. 3): "Albion is my home, or at least I intend it shall be by and by." However, in the autumn of 1871 Greene's family moved to Colorado, where they bought a farm near Denver, and he never returned to Wisconsin to live.

Greene was, particularly for his day, a large man, six feet tall. His earliest known photograph, taken before his first years in Colorado, shows him clean shaven, with a heavy shock of long, straight, black hair parted on the left and hanging down over a broad forehead. He had strong features and a large mouth. Later descriptions frequently mention his fine carriage and it seems evident that he made a striking appearance. One hot day, a group of botanists, including Greene, came to a swimming hole and took time out to plunge in. One later recalled he was the most perfect specimen of manhood he had ever seen and that others in the group were also struck by Greene's physical appearance.[21]

That Greene was a fearless, active, and powerful man is evident from his arduous field excursions. He often walked twenty to forty miles in a day, and he "always went alone, afoot, unarmed, carrying but a portfolio, a notebook and a dull knife for digging up the roots of plants."[22] He wrote to George Engelmann, "I am a good climber and a strong man."[23] Many of his journeys carried him into little-known and even unexplored territory. The Indians were by no means completely pacified in the 1870's, and Stephen T. Olney, another of his botanical correspondents, wrote him, "God preserve you and your scalp in your wanderings among the Rocky Mountains."[24] Greene was one of a dedicated group of early botanical collectors in the West of this period who were, like the early mountain men, courageous, hardy, and self-sufficient. Other settlers considered him too reckless in some of his solitary meanderings. After the Apaches had massacred a group of miners in the Mogollon Mountains, New Mexico, where he was collecting, he and everyone else stayed in Silver City; but on 3 June 1880 he wrote to Engelmann (p. 5), "week after next I will try it again."

Once, while collecting in an isolated region, he found a party of 100 mounted Indians coming along the trail behind him. Rather than appear to be fleeing, he moved back towards them. As he approached them, Colorado, the Ute chief, pulled his horse across the trail forcing him to halt. The chief examined his portfolio of plants and decided he was a medicine man. Several of the younger men had dismounted and were gathered around Greene, who records, "For a more impressive make-belief of fearlessness, I challenged a pleasant faced one of these stalwarts to a wrestle and presently I threw him."[25] He then wrestled another who in turn threw him. Later, according to Greene, this same chief murdered the agent and one or more of the assistants on the reservation.

In the autumn of his first year in Colorado, Greene called on Bishop Randall of the Episcopal Church in Denver, and shortly thereafter he was confirmed in that denomination. Although Greene had been baptised only a few years before, he had long been interested in religion and was a devout believer. He stated that as a child he had received no religious instruction, and that at the age of thirteen he had never heard of Good Friday or Easter, and that even Christmas had no religious significance for him.[26] There is corroboration from his mother in a letter dated 17 October 1870, in which she expressed regret that her children had not had early training in Christianity.[27] Nevertheless, he had been raised in a Baptist household, and he had attended church meetings regularly. His passion, even as a child, for natural things and the heavily religious overtones of Mrs. Lincoln's *Lectures on botany* probably contributed to Greene's early spiritual development. His letters to Thure Kumlien during the Civil War reveal Greene's piety. While in the army, he visited churches of various denominations in quest of religious security. During his years as a teacher in Illinois, though having many religious doubts, Greene served as a licentiate preacher in the Methodist denomination. One of his avowed reasons for going to Colorado was to make a choice between the Roman Catholic Church and the Episcopal, a change he had been considering privately for two or three years. He wished to avoid disturbing his family and friends and to make his choice freely, far from any

FIG. 1. E. L. Greene in clerical robes, age 28. A photograph sent to Rasmus B. Anderson with the inscription "Your friend E. L. Greene. Admitted to the ministry of the Church of God on the 13th Sunday after Trinity [3 September] 1871 at Denver Col. Ter." (Courtesy State Historical Society of Wisconsin)

outside influences. He wrote Rasmus Anderson (21 March 1871, p. 2), "I have kept the fact of my having left the Methodists a secret from my Wisconsin friends." Foreshadowing his future religious development he assured Anderson that he was no low churchman.

Having elected the Episcopal Church, he was persuaded by Bishop Randall to enter Jarvis Hall, the newly founded seminary at Golden, Colorado, as a candidate for holy orders and also as teacher of botany. The following year (22 July 1871) he wrote Thure Kumlien (Main, p. 176), "I have all the spring been applying myself so closely to divinity studies that until within a few weeks I have quite neglected botany. During the past week I have received a note from Prof. Gray and one from Dr. Engelmann, both inquiring what has become of me." On 3 September 1871, Greene was admitted to the Sacred Order of Deaconry and later that month he moved to Greeley, Colorado, to take charge of the congregation there. He was ordained a priest of the Episcopal Church on 31 January 1873 and assumed the pastorate of St. Peter's Episcopal Church in Pueblo, Colorado. During the next several years he successfully served mission churches in Colorado and northern California. He described his situation in a letter of 3 September 1874 to his friend Ludwig Kumlien, Thure Kumlien's son:

I have had a fine time, also have worked my way into the pulpit so as to have no trouble about the wherewithal to pay my expenses. I have a large congregation and good salary, but with all that, so much pastoral work, that my scientific studies are interfered with not a little. I have also the prospects of being even called to a large San Francisco church on three thousand dollars a year to which, if I go, I fear I shall turn out either a poor shepherd for so large a flock or else give up more still of my favorite studies. However, I am thankful for so much happiness and success as seems to attend me in several ways. (Main, p. 181.)

Apparently rumors of Greene's impending or actual marriage reached Wisconsin during this period. But there is no indication, then or later, that he ever had a serious romantic interest, and he remained a bachelor. On Easter 1872 he wrote to Thure Kumlien:

And so you fancy you see me with a wife—a mistaken fancy I assure you. I am not married, and to me it seems now quite probable that I never shall be. I shall soon be twenty-nine years of age and am not engaged or in love. It seems with my preaching and botanizing I have enough to keep me busy and happy. Happy as any mortal ought to expect to be. I am getting that old bachelor notion that a wife would be in the way. So, I guess I am bound to know the whole of single blessedness! I at least do not ask to be happier in the world than I have been since I was ordained a minister. (Main, p. 178.)

Although later in life Greene urged at least two of his friends not to marry, the tone of his reply to Kumlien does not sound like that of a misogynist, and, to judge by his correspondence, his relations with women students and professional women associates were generally cordial, if not intimate.

Greene had been an ardent naturalist and student of botany since childhood, but he first became known as a serious botanist during his residence in Colorado. He had published several short botanical observations in the *American naturalist* during his stay in Illinois (1867–70), but these were hardly serious efforts. He had also studied a low form of *Tradescantia* in the midwest and sent a description of it to the *American naturalist*. This was returned to him with a letter from Asa Gray telling him that there were more *Tradescantia* published already than there ought to be. Many years later, Greene wrote that this experience "taught me very effectively that at that time no new proposition in botany could get printed in the United States in a public journal until it had been submitted to and obtained the approval of Asa Gray."[28] Greene, as a novice botanist, regularly sent his Colorado collections to Gray, and initiated correspondence and exchanges with many well-known botanists. His ability as a keen-eyed collector was quickly recognized by Gray and Engelmann. On 6 August 1872 Gray wrote rather doubtfully to him shortly after his arrival in Denver, "I hope you may find some new things; but you will be sharp if you do."[29] Three years later Gray wrote, "*There*. You always find something good and new wherever you may be. Bully for you!"[30]

When Gray planned a visit to Colorado he asked Greene to join him in an ascent of Mt. Gray. Greene was ecstatic and at Easter 1872 wrote to Thure Kumlien, "Don't you think I was almost crazy with delight when that letter came? You know that since I have been in Colorado I have had many letters from him [Gray] and have sent him specimens of all my collections, and he has named in my honor, at least two of my species." (Main, p. 177.) Greene met Gray at Empire City, Colorado, on 10 August 1872. On the twelfth he accompanied Gray, C. C. Parry, and a number of "unscientific gentlemen" on a trip to the top of Parry's Peak. Two days later the party climbed Gray's Peak. Gray spent a week in Empire, and after he left Greene described him in a letter to Thure Kumlien as "one of the most delightful men I ever saw." (Main, p. 179.) Greene had also begun his association with Charles Christopher Parry, a well-known plant collector, who became a long-time friend, field companion, and correspondent.

Greene was a somewhat peripatetic preacher and botanist during his early years in the West. Apart from his local rambles in the vicinity of the several parishes he served in Colorado, he made use of the Union Pacific Railway for side trips into Wyoming Territory in winter, spring, and summer. In May, June, July, and November of 1872 he visited the plains and hills in the vicinity of Cheyenne, Wyoming. In February of 1874 he moved from Pueblo, Colorado, to a parish at Vallejo, near San Francisco, California. His mother's death in the fall of that year and his father's poor health caused him to resign his post at Vallejo and return to Colorado in April 1875 as rector of Grace Episcopal Church at Georgetown. That fall he severed his connection at Georgetown, intending to return to California. Instead, he visited some of his midwestern friends and returned to Georgetown in November. However, in March of 1876 he left Colorado and moved to Yreka, in extreme northern California, 40 miles from the base of Mt. Shasta. He wrote to Asa Gray:

> My address is to be *Yreka, Siskiyou Co*; away up between Mt. Shasta & Klamath River!! I can hardly sleep of nights since I have secured my appointment to that field of Missionary labor, so delighted am I. I have now a pretty ample supply of sermons on hand: don't mean to compose a single new one all next spring, summer and fall, but just herborize to my heart's content.[31]

He did not remain long in Yreka. In the spring of 1877 he moved to Silver City, New Mexico. This area, especially the Mogollon Mountains, caught his fancy, for he returned to Silver City in the spring of 1880. He remained as a missionary in Silver City until May of 1881 and botanized in the Pinos Altos, San Francisco, and Mogollon Mountains.

He traveled extensively, on one trip walking 1,400 miles through the desert and Apache country. The Reverend Greene was, according to the Yuma, Arizona, newspaper, "no chicken eating bummer."[32] His success as a missionary on the rough-and-ready western frontier may have been due to his own rigorous life, and he was sufficiently unorthodox to drink a tumbler of mescal at Fort Bowie. He comments that it did not intoxicate him, although he did get lost that day for the only time in his long journey. Once while walking the stagecoach route, in a canyon with a bad reputation for robbers, Greene, being thirsty, stopped two Methodist missionaries and asked for a drink. He noticed that they seemed disturbed and found that they had been warned to "look out for a crazy man . . . carrying nothing but an umbrella and a big book [Greene's plant press]."[33] Greene was hardy and adaptable; he traveled without camping gear, slept in the open when necessary, and shared the simple fare of miners and Mexican peasants in between the comforts of the rude outposts of civilization on the frontier. By this time, Greene had changed in appearance in one respect at least. He records, "I was several times designated as 'that good old man' before I was well past the age of thirty . . . for at thirty years my hair was much too nearly white to be described as gray."[34]

Greene was a prolific collector and a dutiful apprentice botanist in the decade following his arrival in Colorado (1870–80). His botanical publications during this period were few and entirely descriptive, *e.g.*, "Rambles of a botanist in Wyoming Territory."[35] However, as his botanical experience grew he became restive under the gentle yoke of Asa Gray, and no doubt the hero worship of his earlier years diminished. In 1876 he criticized Gray's application of the specific epithet "carneum" as inadequate, one of the earliest implications that he had begun to regard Gray as a closet botanist. Greene commented, "Name it what you please, if it needs a name but do not think that it ever, in growing, shows such a color as *the dried specimens* display."[36]

By 1880 Greene was anxious to stand on his own botanical feet and publish the many new species he was discovering according to his own botanical insights. Gray wrote Greene in 1880, "Now if you wish to publish your discoveries yourself you will receive cordial help from me and Mr. Watson and you may be a help and we shall all love you. But if you rush new species into print til we have seen them you will bother and we shall all get to hate you."[37] Shortly after, probably in anticipation of his own absence on a trip to Europe, Gray wrote, "I advise that you confer with Watson before publishing. He will respect all your names sacredly, and may often help you to a diagnostic description of the species—tho' in that he is not so precise as I am. I hate descriptions that are not diagnostic."[38] Gray, anxious to complete his projected *Synoptical flora of North America*, wished to see and name all the American collections. If his associate, Sereno Watson, failed to meet his standards of precision in describing new species, how could Greene and the other botanists from the western provinces do so?

Rodgers in his *American botany* (1944a) noted that Greene's controversy with Asa Gray was one of the most celebrated in the history of American botany. Each had his supporters and, in a day when invective was less proscribed by editors, tart comments were published on both sides, many letters indicating the resentment that was engendered. A. Hunter Dupree, Gray's biographer, intimates that Greene rather underhandedly seized the oppor-

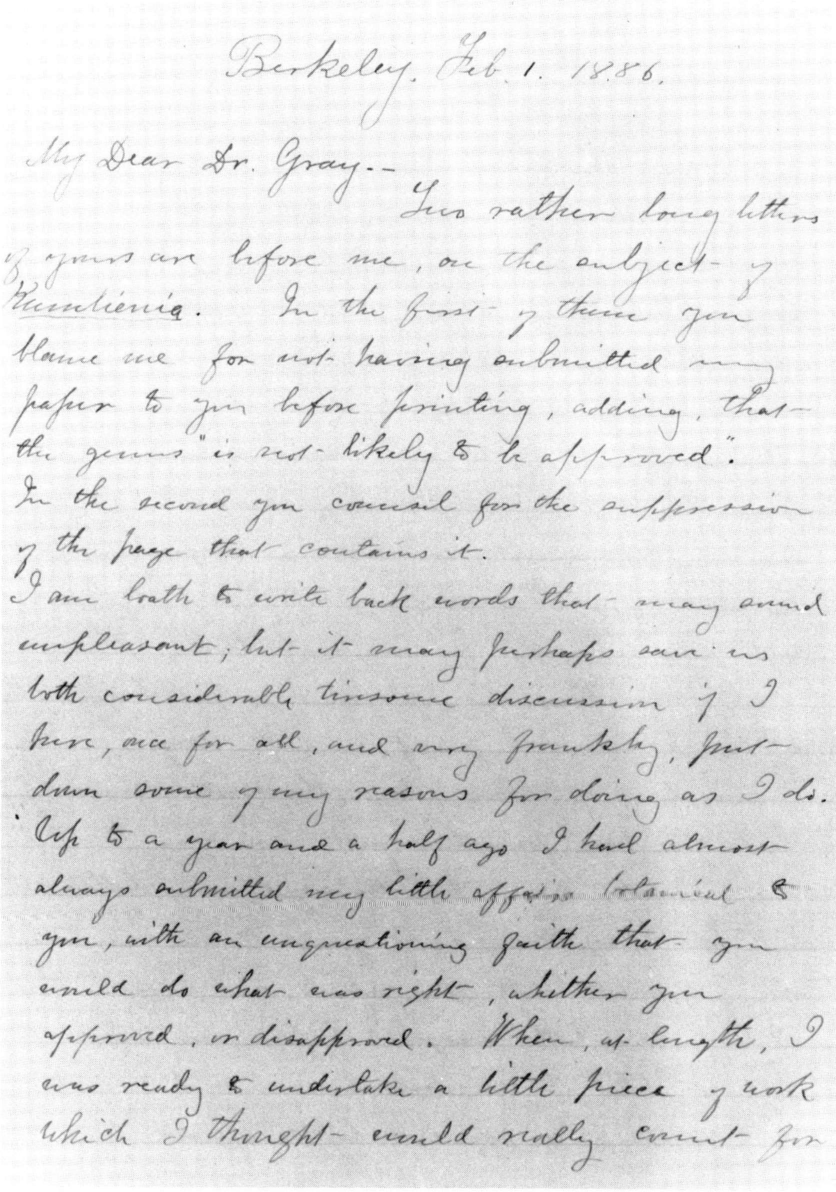

Berkeley. Feb. 1. 1886.

My Dear Dr. Gray.—

Two rather long letters of yours are before me, on the subject of [illegible]. In the first of these you blame me for not having submitted my paper to you before printing, adding, that the genus "is not likely to be approved". In the second you counsel for the suppression of the page that contains it.

I am loath to write back words that may sound unpleasant; but it may perhaps save us both considerable tiresome discussion if I here, once for all, and very frankly, put down some of my reasons for doing as I do.

Up to a year and a half ago I had almost always submitted my little affairs botanical to you, with an unquestioning faith that you would do what was right, whether you approved, or disapproved. When, at length, I was ready to undertake a little piece of work which I thought would really count for

FIG. 2. Part of a letter from Greene to Asa Gray, 1 February 1886, in which Greene declined to submit his taxonomic decisions to Gray for approval. (Permission of the Director of the Gray Herbarium)

tunity of Gray's absence in Europe in 1881 to declare his independence and publish his own new species (*Asa Gray*, p. 396). Since Greene's first new species, *Asclepias uncialis*, was published in the *Botanical gazette* of June 1880, well before Gray's departure, this allegation is unwarranted. Subsequent publications of new species of *Asclepias* and other genera were the subject of correspondence between Gray and Greene, who wrote, "You need not fear after all the scoldings, which I am so happy as to possess in your handwriting that I shall *print* n.[ew] sp.[ecies] without obtaining your approval first."[39] He was, however, an ambitious and increasingly independent young botanist who was becoming critical of his acknowledged masters, Gray and Engelmann. One can imagine his reaction to Asa Gray's comment, "Watson writes me that you are out with your *Asclepias*. . . . Well, I could wish you had waited. It is a great bother to take care of these scattered lost sheep."[40] Although Greene was later to accept the authority of a Pope in Rome, he is said to have asserted, "There is no Pope in botany."[41]

In spite of Greene's denying a wish to publish new species, by 1882 the editor of the *Botanical gazette* (7:89. 1892) commented, "Mr. Greene is coming rapidly to the fore as a publisher of new species and his work among our western plants has been invaluable." Other collectors submitted their specimens to Greene for determination, and Gray, who was normally the arbiter of all matters botanical in America, wrote to Greene, "Do you wish to stop the flow [of specimens] altogether or do you mean to allow it after you have screened it for new species?"[42] Greene assured Gray that he, at least, had no wish to be reckoned among the botanical authors of the West Coast and commented that the reasons why "many of our supposedly new or doubtful species should be submitted to Cambridge Herbarium are manifest enough." Greene continued:

> Of the several scores of n.[ew] sp.[ecies] I have published within a year the type specimens (in several instances the only ones extant) are in *your* herbal at Cambridge. My work upon them was interesting and instructive to myself but it was chiefly your own and Mr. Watson's long absence that tempted me to do what I have done in that line. My *Senecio Howellii* is I think the only species which I printed this year without waiting for your "imprimatur."[43]

During this period Greene was involved in religious as well as botanical controversy. In 1881 he was called to his last mission, at Berkeley, California. Asa Gray commented, "Well Berkeley will be a change for you but I had rather have you in some vast wilderness."[44] While he was there, as rector of St. Mark's Episcopal Church, his conversion to Roman Catholicism was completed. He had long protested that he would not become a Roman Catholic, but he changed his mind so completely that, as he wrote, "Two years passed, during which I tried with indifferent success to teach Catholic doctrine to a congregation that was ultra-Protestant."[45] As one might expect, this aroused strenuous objections among the members of his congregation.[46]

On 27 June 1883, Bishop Kip, the Episcopal Bishop of California, wrote to him that the dissensions at St. Mark's parish required action, and that Greene must resign by a certain date or be fired. Half of the congregation seceded, but Greene remained at St. Mark's until All Saints' Day, 1884. Jepson recorded that at one point during the strife the vestry locked Greene out of the church, "and we have preserved for us the picture of the Reverend Mr. Greene, in surplice, passing down Bancroft Way, an axe over his shoulder and the faithful of his flock behind him, beating down the doors of St. Mark's. . . ."[47] This controversy eventually led to court action. The episode has been described in an article by Rideout entitled "Priests, Pistols and Polemics."[48]

Shortly afterwards Greene resigned his Episcopal ministry and in 1884 he became a lay Roman Catholic. Charles C. Parry, his long-time friend and associate in botany, wrote him, "How is that? We had heard of your high church notions but did not suppose you would go as far as Rome. Well, as we have historical proof that Catholicism and botany are not incompatible we shall cling to you botanically and leave you to enjoy the repose of the mother church."[49]

Greene continued his botanizing and publishing and was active in the California Academy of Sciences, where he became curator of the herbarium. His relations with Asa Gray became increasingly strained. He wrote to Gray, "If you cannot recognize difference of habit put the *four* [species] into one. But a California schoolmaster might teach children to distinguish each of the four blindfolded by odor of herbage alone."[50] Gray responded, "As to smell you have me there. We can't tell on the dry. But a species that has not structural characteristics must be shaky indeed."[51]

Greene was forty-two when he finally abandoned the ministry in 1885 and was appointed the first full-time instructor in botany in the Department of Agriculture of the University of California at Berkeley at $75 per month. This was his first position as a professional botanist. Other than his apprenticeship with Thure Kumlien, he was self-trained in botany, as were many of the prominent American botanists of his day. He wrote to Asa Gray the following February:

> I am going to request of you that you try to think of me as a man of two and forty years, a worshipper of plants and trees from my babyhood and one whose veneration for every great name in botanical science long kept him from venturing to set forth any opinions of his own. I make some mistakes, mistakes that I am greatly ashamed of, and you in telling me of them, tho you do it harshly and with the tone of a master to a small school boy, will be better to me than if you praised me where I deserve no praise.[52]

It is not clear what precipitated this pronouncement and Gray did not reply, as the letter is annotated, "Not answered." It does suggest Greene's impatience at being an apprentice over long years and marks a definite change in the relations between the two men, as Greene assumed the role of professional rather than amateur botanist.

In six years at Berkeley Greene progressed from instructor to full professor. In 1890–91 a separate Botany Department was established, with Greene as chairman. By 1892–93 the department offered three advanced courses in botany: Systematic and Economic Botany, Advanced Systematic Botany, and Medical Botany. Later, courses were added in the History of Botany and the Phanerogamic Natural Orders. The department grew during his tenure to one of 70–80 students and three staff members. A botanical garden of native plants flourished in spite of an attempt by the tennis club to usurp some of the ground allotted to it. Greene was a stimulating, effective, and apparently well-liked teacher. Numerous students were drawn to him by his enthusiasm for his subject. Some became botanists, notably Victor Chesnut and Willis Linn Jepson. One, Frederic Bioletti, described in a rather tongue-in-cheek article the zestful air of the botany department under Greene. He wrote that the students used botany as a substitute for football games, "Not as a laboratory but as a playground."[53]

Jepson was to become the most famous of Greene's botanical students, and he gives us the best contemporary description of Greene. According to Jepson (1918), Greene was ruddy-faced, with cottony, white hair, and was always well and carefully dressed. Jepson records, and himself illustrates, Greene's marked ability to make warm and strongly partisan friends. The admiration and affection with which Greene was regarded by many of his associates are clearly expressed by Jepson, in striking contrast to the acrid comments of some of Greene's critics, such as Katharine Brandegee and Marcus Jones. To Jepson and many others, however, Greene was genial and approachable, "often with a mischievous smile and radiating good humored sallies." His manner was courtly, his conversation elevated and elegant. Jepson particularly admired the deftness with which he managed the difficult task of carving a leg of lamb (*cf.* Jepson, 1943).

Greene was a friend indeed to many of his relatives, friends, and botanical associates. He gave or lent money freely from his limited resources. He assisted other botanists often with criticism and, on occasion, with bank drafts to support their work. As his reputation grew, he developed an extensive botanical correspondence. He wrote to Jepson at one point that he was completely worn out with answering letters, but throughout his life he gave his time and energies generously to the many botanists who solicited his help (*cf.* Lange, 1955).

During the fourteen years at Berkeley (1881–95), Greene established himself as the outstanding botanist of the West Coast. He published prolifically on the botany of the West, particularly of California. Bartlett (1916) noted that Greene's papers "Studies in the botany of California and parts adjacent" "were the first noteworthy contributions to western botany published in the West by a western botanist." His major efforts were the incomplete *Flora franciscana, an attempt to classify and describe the vascular plants of middle California* (1891) and the *Manual of the botany of the region of San Francisco Bay* (1894).

He also developed during that time his views on the nature of species and genera and of priority in plant nomenclature, views that perpetuated his conflict with Gray and subsequent botanists at Cambridge. Greene was constantly publishing new genera and species, most of which Gray did not approve. Greene was reported by Hopkins (1942, p. 199) to be "as great a splitter as American systematic botany has yet encountered." He was, it is true, inclined to give specific or generic rank to differences which were not acceptable to Gray and like-minded botanists. Greene's keen eyes detected differences that were distinctive, although many of his contemporaries and successors would not accept them as of specific value. Of one of Greene's efforts Gray commented, "But the plant you chose without flower or fruit; on what ground then could you think it a new genus?"[54] The story is told of descriptions of two of Greene's species that were carefully compared, point by point. The only difference that could be found was that one had a certain indescribable grace which the other lacked.[55] Years later Greene asserted that very probably he could have found new species in Asa Gray's backyard,[56] and in one large notebook in his effects is a blank page headed, "Errors of Asa Gray." He wrote to Gray in July 1885:

> It is probably because I have been for the most part a field student and have had ten or a dozen years of almost weekly detecting the errors of learned men of the herbarium and library that I dissent strongly from your view that Benth.[am] and Hook.[er] should have influence with you as they do in this case.[57]

Greene placed great emphasis upon attributes of the living plant not always apparent in the herbarium specimen. Referring to a species of *Matricaria* he said:

> It is a good species without a "character" standing on its habit and properties and I like such species. Every one such stands as an emphatic protest against the artificiality of our accepted system of classification—"natural" we call it, and perfectly natural it may be when here and there genera as well as species shall be freely admitted in despite of their lacking what we call a "character."[58]

Gray commented at one point, "About your now well developed views in botany and my ideas about them it were better to say nothing."[59] Other botanists were more astringent. Katharine Brandegee and Marcus Jones wrote to Jepson that Greene was insane.[60] John M. Coulter wrote that Greene's *Pittonia*, No. 3, "reads like the work of a crazy man."[61]

Greene's rebellion against the dominance of Gray and the Cambridge botanists was one of the turning points of his career, and he was not without sympathizers. M. S. Bebb wrote to Greene, "Dr. Gray's services to botany cannot be overestimated. . . . But Dr. Gray is characteristically, I might say, a closet botanist—circumstances make him such—and his present position makes his influence one which operates both for good and ill. Like a noble tyrant who loves his constituents he strengthens the power of the central government and makes it respected."[62] Parry wrote, "Dr. Gray complains of your jumping to conclusions! If you only jump right *keep on jumping*."[63] Parry, a friend of both men, was in the awkward position of maintaining good relations with both in the midst of the controversy. In 1886, T. C. Porter wrote Greene, "Allow me to express my sincere pleasure at the departure you have made in the botany of the Pacific slope. You have rightly taken an independent position and it is not relished by Cambridge, but what of that? Her botanists have done a good deal of their work too rapidly."[64] Greene wrote to N. L. Britton, head of the New York Botanical Garden and his chief ally in the struggle over botanical nomenclature, "Cambridge must lose prestige gradually and it is our religious duty to pull it down as fast as we discreetly can."[65] A few months later Britton wrote Greene, "Keep up the fire my glorious colleague, we will bring them all down after a little!" He signed the letter, "Exultingly."[66] With such exhortation and his own faith in his abilities it is hardly surprising that Greene challenged the Cambridge powers and earned their dislike. Professor M. L. Fernald, in his course in the History and Techniques of Taxonomy, at Harvard University, related that when he came to the Gray Herbarium as a boy of seventeen his instructions were that if E. L. Greene were ever to appear, he was to slam the door in his face and lock it.[67]

Not all of the several thousands of species that Greene named as new were accepted by his contemporaries or by subsequent generations of botanists. Yet it is quite wrong to say, as did Katharine Brandegee, in *Zoe* (4:64. 1893), that one in fifteen or twenty was the proportion of acceptable species, or, as did Marcus Jones (1912b), that 95 percent have no foundation, or, as did A. Hunter Dupree (1959), that few "have won the approval of botanists." As Rogers McVaugh shows in the following study, Greene's average of currently accepted species for the years 1881–86 is a quite respectable 70 percent. Most generally agreed with Jepson (1918) that Greene's best systematic work was done during his years in California.

The motives for Greene's rebellion are complex. Andrew D. Rodgers suggests, in his *American botany* (p. 151), "the fire of egotism" as one. Dupree (p. 399) ascribes it to "sources peculiar to himself," including his disbelief in evolution, dislike of Europeans, belief in absolute priority, and his religious conflicts. Dupree suggests that Greene disregarded such ecological influences as climate and soil. It seems most unlikely that Greene, who wrote of the ecological effects on plant form in his *Landmarks of botanical history*, would have been unaware of these effects. That he named many species on relatively subtle differences is true, but a large fraction of these remain valid today. It is not clear what effect his religious controversy had upon his botanical work, but it is of note that it was simultaneous with his best systematic work. Greene's insistence on priority hardly, as Dupree asserts (p. 401), created a revolution in settled practice of nomenclature, for the problems and controversies arose because much was unsettled at this time and general acceptance of rules of nomenclature was yet to be achieved.

No doubt Greene was confident of his own growing abilities to the point of egotism, and he was certainly sensitive to adverse criticism, of which he received good measure. Nevertheless, it is fair to say with Rodgers, in his *American botany* (p. 256), "Greene must be remembered as a keen observer and one who stood fearlessly in defense of the right to practice science as one saw truth to be." Greene wrote to Gray:

> I am sorry you will be displeased with my treating of *Brodiaea* and *Hookera*, really sorry, for I dislike to give annoyance to you who have labored so long so zealously and so fruitfully in this great field. But what am I to do? Keep silence, let all that I *know* [Greene's underscoring] about these things remain untold and let somebody else who has not a tithe of my knowledge of the plants by and by do something.[68]

Charles C. Parry wrote of Greene, "I am on the whole rather interested in Mr. Greene's enthusiasm for descriptive work, as I know him to be sharp sighted and cautious though sometimes stubborn."[69] While Greene's personal idiosyncracies no doubt contributed to the caustic tone of his rebellion against Gray and the Cambridge botanists, it must be admitted that the rebellion itself was a struggle for the truth as he saw it. Jepson, a student and partisan of Greene's, noted (1931), "Greene's assertion, however, of the right of a botanist to publish his results without first submitting specimens or manuscript to Asa Gray may well prove in historical retrospect his most significant effort in behalf of North American botany." Charles E. Bessey, himself a notable figure in American botany, wrote to Greene in later years, "I have a great leaning towards any man who has something of heresy in him. . . . You have dared to be original and that is what pleasures me immensely."[70]

Despite, or perhaps because of, his botanical heresies Greene occupied an increasingly important place in American botany. Following the death in 1888 of Asa Gray, none stood higher in taxonomic botany than Edward Lee Greene. In 1892 he was one of three American members of the International Committee on Botanical Nomenclature. In 1893 he was elected President, by acclamation, of the Botanical Congress held at Madison, Wisconsin, where Lucian Underwood, a younger contemporary, assured him, "throughout the meetings every movement that you made resulted in a spontaneous feeling of many of the younger men that you were the true leader. Your stock unconsciously to yourself received a genuine boom."[71] In 1894, he was awarded an honorary LL.D. by the University of Notre Dame.

Greene's tenure at Berkeley was not without its conflicts. There were echoes of strains with President Davis in 1889 in a letter to N. L. Britton wherein he asserted that he was in peril of losing his position.[72] Britton hurriedly wrote the Board of Trustees that Greene was the outstanding western botanist and was irreplaceable.[73] Later there were conflicts over the Botany Department and the role of systematic botany in it. In 1894 Greene, apparently tiring of the conflict, was persuaded to accept the professorship of botany at the Catholic University of America, in Washington, D.C.

At that time Greene made his only trip to Europe. He sailed from New York on 25 August 1894.[74] He told Jepson (1894), his former student and colleague, that he planned to spend two months at the herbaria at Kew and the British Museum and later to visit the herbaria at Paris, Berlin, and Geneva. His purpose was to make a critical study of the type specimens of Pacific Coast plants, many of which were deposited in these herbaria. Jepson reported further that this task awaited a master's hand and clearly Greene was the man for the job. Greene's name appears in the visitors book at the Kew Herbarium regularly from 7 September 1894 to 23 November 1894. Since he returned to the United States on 3 December 1894,[75] it does not appear that he did visit the Continental herbaria. He makes little reference to his work in Britain, and how much it may have influenced his systematic work, if at all, is not clear. He was impressed, for he wrote to Jepson that he must save his eyes for all he must see in leaves of this "great herbarium"[76] and commented ironically that at Kew the pages of his own journal *Erythea* were not cut until the binder cut them.[77] On the last day out of New York, on his trip home, he wrote Jepson, "I am quite exempt from seasickness and some of the companions of the voyage I shall be loath to part with. I had meant to do a lot of botanical writing on the way across but have failed utterly, having given myself up wholly to conversations and pastimes of several sorts with some half dozen fellow voyagers."[78] If in his later years he became something of the "devoted hermit," as he described himself,[79] at this time he was apparently affable and companionable.

Greene's departure from the California scene was described in the *San Francisco Call* (1895) as "Berkeley's great loss." The departure of a scholar from a university was apparently more newsworthy in 1895 than today, when only the movements of athletic coaches receive much public notice. Leaving the West in which his botanical skills had matured over a period of twenty-five years no doubt caused him some misgiving, and he wrote that he found "some little hardship in severing connection here with so many friends and so interesting a department of university work."[80] H. H. Bartlett wrote in 1916 that Greene's impress on the University of California still remained. For Greene established on the Pacific Coast a center of systematic research that has held its own through decades of changing emphasis in botany.

Greene had been brought to the attention of Bishop Keane, the Rector of the Catholic University, by Father John Zahm, himself a noted scientist. Bishop Keane met Greene while on a visit to California and persuaded him to accept the chair in botany at Catholic University. Later, Greene asserted that he had not sought this appointment, but indeed had taken it at some personal sacrifice, and that promises of freedom to write and publish without students to absorb his time had not been kept. A friend of his Washington days, Mrs. Margaret Downing, wrote in *Catholic world* that the period at Catholic University was neither fruitful nor entirely happy. However, the nine years there (1895–1904) saw some 178 publications from his pen, so they were hardly unproductive, even though his great period of systematic work lay behind him.

Kistler's bibliography of E. L. Greene's writings includes 565 titles, nearly half of which were published before he left Berkeley. While head of the Botany Department there, he had been instrumental in founding the journal *Erythea*. Many of Greene's publications appeared in journals of his own creation. The first of these was *Pittonia*, whose first number appeared in February 1887. Two volumes were published in California and three in Washington, D.C. He explained the name *Pittonia* as commemorating Joseph Pitton de Tournefort (1656–1708), commenting, "All this is well known to the few of our botanists who do not ignore the fact that there existed a botanical nomenclature before Linné."[81] His second privately published journal, *Leaflets of botanical observation and criticism*, first appeared in November 1903 in a limited edition of 150 copies. Two volumes of this journal were published in Washington, D.C. Greene's last journal, *Cybele columbiana: A series of studies in botany chiefly North American*, was begun in 1914, the year before he died. Only the first number, of December 1914, was completed. He was also influential in the founding of the *American midland naturalist*, to which he contributed numerous articles.

It is probable that he established his own journals because of disagreements with the editors of the standard journals of the day, notably *Botanical gazette* and *Bulletin of the Torrey Botanical Club*. Marcus Jones, a vitriolic critic of Greene's botanical work, asserts that these two journals had a compact to publish no new species until the types had been deposited in an eastern herbarium, and that Gray attempted to suppress publication of new species by Greene. Greene probably had similar beliefs and found publishing his own journals easier than dealing with eastern and sometimes antagonistic editors. John M. Coulter, editor of *Botanical gazette* in the mid-nineties, is described by his biographer, Andrew D. Rodgers, as unmerciful toward Greene. Though Greene was a fluent writer and many of his contemporaries commented favorably on his style, most modern editors would rigorously edit what would seem to them a rather florid prose. Yet Greene was much concerned with language and style, and his comment in a lecture on the anniversary of the University of California has a surprisingly modern ring. He pointed out that the days of great English prose and rhetoric preceded the presence of English as a subject in the curriculum, but "Never did colleges in America send forth so large a proportion as now of graduates unable to clothe their thoughts in passably decent English."[82]

During his tenure at Catholic University, Greene was still immersed in the conflict over botanical nomenclature. He was very critical of the proposal to establish an arbitrary starting point for plant names beginning with those in publications by the Swedish botanist Linnæus, proposing instead to extend priority back to writings of medieval and even classical Latin and Greek authors. He wrote to Watson, "I believe in the dark ages and always did: am not nor ever was well disposed toward revolutionists like Luther and Linné. . . ."[83] Although usually a proponent of absolute priority, he apparently considered a compromise referring to "botanists who really obey the rule of priority and go to Tournefort."[84] Greene's strenuous efforts in this conflict were largely unsuccessful and he gradually gave up the struggle. However, even late in life he apparently harbored some hope of stemming the Linnæan tide, for he wrote to his student Father Nieuwland of a paper he was contemplating, "It will be the only really important document on nomenclature that there has been since the beginning of the 20th century."[85]

Greene was a stern taskmaster to others, doubtless because he demanded so much of himself. He was unsparing and often caustic in his criticism and could be devastating, even to his own protégés. He wrote of one effort of Father Nieuwland, "Every statement and every paragraph is untrue! As a tissue of errors I think it cannot be matched in scientific journalism."[86] No doubt his acrid commentaries on the work of other botanists earned him their dislike. In one review he managed a double-barreled comment, hitting the author and the Cambridge botanists in the same sentence, "We might have expected much of bibliographical laxity and inaccuracy in any author, who could speak of Watson's Index as being a book 'remarkably exact'."[87] He was candid about his failings in this respect, writing to Britton, "I know I must repress all sarcasm and ridicule, but it is hard for me."[88]

In those days, when there was less editorial proscription of polemics, he received as good as he gave. His work on *Eschscholzia* and *Apocynum* was described by Jones (1912b) as differing "in no essential particular from a typhoid fever patient in the last stages picking at the bedding." One of his most outspoken critics was Katharine Brandegee, a contemporary California botanist, who wrote of his work (April 1893a, p. 63), "It seems to suit his convenience, where ever there is the slightest ground for difference, to at once describe a new species as vaguely as possible, both as to character and station, and leave to others the unhappy task of finding out whether it is admissible or not." In the same paper (p. 65) she also said of his botanical views, "This kind of botany was taught, probably, in the Middle Ages to which Mr. Greene properly belongs." In 1942 Joseph A. Ewan suggested that Mrs. Brandegee's tart criticisms of Greene's botanical work were due to unrequited love; earlier, while a widow named Mary K. Curran, she had been associated with Greene at the California Academy of Sciences for several years. Ewan stated (1955) that she was curator of the Academy herbarium, but, according to Greene, he was curator and she was assistant.[89]

Ewan's idea as to the reason for Mrs. Brandegee's attacks upon Greene was vigorously challenged by Albert W. C. T. Herre (1960) as a reflection on Mrs. Brandegee's character. She had not yet met her next husband, Townshend S. Brandegee, and she may or may not have had designs upon the handsome and eligible pastor. Herre's privately printed pamphlet defends Mrs. Brandegee of the charge of being in love with Greene by asserting that he was a homosexual whom she could only despise. This charge of homosexuality has been widely circulated about Greene. Proof or refutation of such allegations is difficult at any time, let alone 65 years after the death of the subject. Marcus Jones (1929b) claimed that Greene was "kicked out of the pulpit because of sexual vices," but the letter of deposition from the Bishop of California explicitly excluded such cause.[90]

Both Barry (1950) and Hogan (1949) reported that Catholic University, during Greene's tenure (1895–1904), was in a state of administrative confusion and in dire financial straits. The School of Biological Sciences consisted of the Department of Botany, with Professor Greene its sole member. As early as 1897, rumors of his impending resignation were denied by Greene. Intrafaculty strife added to his difficulties. Hogan reports (p. 134) that one faculty member wrote, "The professors have their hand in and are willing to bounce the drones from the nest as soon as they can. Look out for the heads of Professors Viridis [Greene]. . . ."

In 1901 Greene was called on to answer a series of questions concerning the Botany Department. According to his responses, three courses were taught, two having one student, and one having two students. Greene devoted nine hours per week to instruction. At the same time, he was "deeply engaged in the general Systematic Botany of North America publishing each year (and at my own expense) much more in the way of contribution to knowledge of the Plant World than is issued from any other University in America, or even in Europe, I doubt not."[91] He called for an additional instructor in plant anatomy and histology, an efficient curator for the herbarium, and two microscopes in addition to the one then possessed by the department. This request was promptly rejected by the administration, and in 1902 an economy drive resulted in a recommendation to cut Greene's salary by one thousand dollars, the saving to be used to hire a Professor of Pedagogy.[92] Greene agreed to this, but shortly thereafter he was called on to defend the utility of retaining the Department of Botany. At this, he threatened to recall his assent to giving up one-third of his salary, and he defended the Department of Botany as "the one brilliant success, which the gentlemen of the board of trustees can point to in attestation of the possible success of this whole enterprise of a Catholic University in and of the United States. This success, under God, I have made and that unaided."[93]

Greene had few students at Catholic University, but among them were Theodore Holm and Father Julius Nieuwland. Nieuwland was a brilliant priest who took his Ph.D. in chemistry but nearly was weaned away by the interest in botany which Greene engendered in him. Nieuwland was appointed professor of botany at the University of Notre Dame. He wrote Greene, "I am doing my level best to build up a botanical library and have lost all interest in chemistry. . . . My chemical course at the Catholic University was a great mistake and I am sorry I quit botany with you."[94] At Greene's urging, Father Nieuwland started the *American midland naturalist*. Later, he was the main influence on Greene's decision to bequeath his coveted herbarium and library to Notre Dame.

Greene was an ardent bibliophile. He recorded a love of books and learning from childhood, and apparently he began his library during his military service in the Civil War, for his mother referred to "those old books you brought from the South"[95] when she was packing to move the Greene household from Wisconsin to Colorado in 1871.

During his years in the West, Greene had accumulated a large herbarium and a unique botanical library, which he moved to Washington at his own expense. His library eventually contained about 4,000 volumes. Much of his earliest plant collections had been given to the California Academy of Sciences, and a doleful letter to him from Alice Eastwood relates the terrifying events of the destruction of the Academy in the San Francisco earthquake and fire.[96] His library was his particular pride, and he comments that it was "the pioneer among American collections of pre-Linnæan, and rare though important post-Linnæan, books of botany."[97] His involvement in the controversy over plant nomenclature was an important stimulus for his interest in botanical history and the acquisition of books.

In November 1903, after his controversy with the trustees of Catholic University, Greene submitted his resignation, effective in September 1904. He then removed his library and herbarium to the Smithsonian Institution under the terms of a contract to leave them in its National Museum for ten years. This contract also gave the Smithsonian an option to purchase the library and herbarium for $20,000 if they could secure the money from the Congress. Greene was appointed an unsalaried associate and assigned working space at the National Museum.

Although he continued his work in systematics, Greene's major interest and effort in this period was devoted to the history of botany. He rightly judged that his "sketches of the lives of the great promoters of our science who lived in other centuries will be received by many as among the most welcome and instructive of my paragraphs."[98] Greene was enabled to pursue this work by a small grant of fifty dollars a month (later raised to sixty dollars) from the Smithsonian. He completed the first part of his projected multivolume work with dispatch in 1907. He was anxious to see it in print, for he wrote to Secretary Walcott of the Smithsonian:

> I devoted in the aggregate as much as a year to my Theophrastus chapter and nothing more important on botanical history has been written. Do not forget this manuscript of mine until some German gets the glory!"[99]

Greene called this work *Landmarks of botanical history*. Part I appeared in 1909. He started on the second part, which, though nearly completed at the time of his death, is only now being printed. Greene's historical writings are a last and lasting tribute to his scholarship and his devotion to botany. A contemporary commented on Greene and the *Landmarks*:

> Easily the best classical scholar among contemporary botanists, he brought to this book a certain fluent and delightful style. The combination of such broad scholarship and the attractive presentation of his subject makes it difficult to speak with restraint of a work that has already become a classic.[100]

In a report on his work at the Smithsonian, 1904–9, Greene noted with some acerbity that he had done and published twice as much botanical work as the entire staff:

> Had I done for any University the high-grade critical work in science for five years, which I have done gratuitously for the United States Government, and in the name of the Smithsonian Institution, these last five, probably I should have been paid for that time about $20,000. Also my library and collections would still have been my own to command.[101]

In the same manuscript he wrote also that the situation was no longer tolerable and asked that the authorities of the Smithsonian "will kindly, but seriously ponder it." In spite of financial difficulties he continued his historical and systematic studies at the Smithsonian for another five years. He apparently planned a history of American botany, for which he began collecting materials, but his work made little progress and only a small fragment of manuscript is extant.

Greene was, naturally, a prominent figure in the botanical circle of Washington. He was a charter member of the Washington Botanical Club and its only president until it merged with the Botanical Seminar to form the Botanical Society of Washington. He was also a member of Washington's prestigious Cosmos Club. He was a participant in an informal group that met regularly to consider botanical matters, especially nomenclatural problems. This group included many active and prominent taxonomists in federal service, including O. F. Cook, F. V. Coville, W. R. Maxon, A. S. Hitchcock, P. L. Ricker, W. F. Wight, F. H. Knowlton, and E. S. Steele. Greene was looked on by some of this group as a little tin god who could do no wrong.[102]

FIG. 3. Greene during his last years. A photograph taken in Washington, D.C., apparently for a projected autobiography. (Print at Hunt Institute)

He is remembered by a younger associate[103] as a large man about six feet tall, with large features, white bushy hair, and small, blue eyes, who was rather egocentric, but very friendly and helpful.

Greene was a disciplined scholar. A close friend of his Washington days, Margaret Downing (1917), recorded his description of a perfect day:

> He attended an early mass, and after breakfast retired to an upper chamber to write out some notes which he had been collecting for days. This engaged him until long past noon. He prepared his meal and then, feeling a little fatigued sat at his piano and for two hours played Beethoven and Grieg until he was as refreshed as if from an icy bath. He then read an hour or so something light and entertaining, then something heavy and edifying—his well thumbed Greek testament and his favorite edition of the Psalms . . . he returned to the Church in time for Benediction. Supper and another attack on his writing followed. At ten, he found his task completed and prepared for rest. "And throughout the day," he exclaimed fervently, "not a human being came to my door, and to and from the Church, I met no one with whom I had to exchange a syllable."

Greene was not a recluse, but Mrs. Downing described him as reserved and introspective. He was apparently a congenial guest and an excellent conversationalist. One invitation, shortly before his death, said, "We should be very glad to have you come to supper if you can get your own consent."[104] The affection and regard which he inspired in many of his associates throughout his life by his personal charm and sincere friendship are no less apparent in his last years than in his earlier life.

During his last years in Washington Greene began to suffer ill health, and he mentioned a chronic liver ailment. However, he was still a rather active man, as indicated by a notebook that listed the mileage he walked on his botanical jaunts around Washington. The vitality and infectious enthusiasm for botany which had always characterized him did not leave him, for letters from young students attest to the pleasure and interest with which they regarded these trips. Greene at 70 wrote, "What man, in whom there remains the love of life and the fond hope of enjoying and doing, desires to contemplate that in years he is old? He wants to forget the fact; would rather not believe it."[105] Long life apparently intrigued him, for he compiled lists of names and addresses of persons who lived 80–100 years and collected newspaper clippings about centenarians.[106]

The disposal of his herbarium and library was much on his mind also in these later years.[107] Several institutions, including the University of California, the California Academy of Sciences, the Field Museum in Chicago, and the Smithsonian, were interested. Greene was bound by contract to leave the collections at the Smithsonian until May 28, 1914. However, as early as 1907, at the urging of Father Nieuwland, he evinced an interest in going to the University of Notre Dame. He wrote Nieuwland, "I should be very happy were it so ordered that I should join you at Notre Dame for weeks, for months, for years. . . . The Smithsonian Institution is but a school of materialistic agnosticism—which term I always think of as a more obnoxious synonym of atheism."[108]

In response to Father Nieuwland's repeated urgings, Greene in 1909 visited the University and went on collecting trips with Nieuwland and his students. This was the first of almost annual visits he made to South Bend before he moved there permanently. That he was pleased with his reception and prospects at Notre Dame is evidenced in a letter to Nieuwland. He said that he had asked the Smithsonian to release him from his contract so he could move his books and herbarium. Two places on the Atlantic Coast and one on the Pacific were waiting "to claim me, my equipment and all," but none of them interested him. He wrote further, "If I were assured the University [Notre Dame] would accept the above as a gift, publish all I write without cost to me and give me shelter and food—all this to the end of my days—I am not sure I would not prefer that to any other place."[109] Thus, some six years before his arrival at Notre Dame, Greene was considering the move to South Bend.

During this period, Greene was obviously concerned by his situation at the National Museum and was anxious to move. In January 1910, a bill was introduced in Congress to appropriate funds to buy Greene's herbarium and library. Greene and Nieuwland were much concerned about this. Continued efforts to procure the appropriation prompted Greene to take action. He wrote, "I had to leave all yesterday and go lobbying to prevent that."[110] In 1912, Greene wrote to Father Nieuwland that he would be free in a year more. "Will it be Notre Dame that will then adopt me? Will it be the Field Museum at Chicago? One of the two I hope for I am more fond of the West. For many years I have wished to write Ill.[inois] and Wis.[consin] botany which I was so much in love with before you were born."[111]

In November of 1912, he wrote to Father Cavanaugh, then president of Notre Dame, the statement of an offer of his herbarium, library, and correspondence in return for a modest annuity and living quarters for the duration of his life. He remarked in this letter that the proposed annuity was "less than half what I am sure of being able to take from one or another of four different great seats of botanical study and research. . . ."[112]

In spite of the inducements offered him by other universities and museums, Greene persisted in his resolve and arrived at Notre Dame in the spring of 1915 at the age of 71. His herbarium was established in cases of his own design, and he hoped to work on it undisturbed. He also planned to undertake a flora of Wisconsin, to continue his history of botany through several more volumes, to write a history of American botany, and to write his own autobiography as well. Failing health plagued him, however, and often he only sat in the sun.

Greene had devoted his life to his science, systematic botany, which to him expressed "simply and briefly the philosophy of the plant world as a whole."[113] To some he had appeared contentious; to others he was a warrior for scientific truth and freedom. He was conscious of his own abilities to the point of egotism, describing himself in his later years as the acknowledged master of his science. It is ironic that Greene, who asserted, "The sum total of truth is a natural positive and the essential characteristic of error is its negative quality,"[114] should be commonly regarded as maintaining negative stands. It was alleged that he was against evolution, and he protested against the acceptance of a Linnæan base for plant nomenclature. He passionately defended the right to seek and publish the truth, and he had earned the motto that appears on the bookplate presented to him at the banquet honoring his seventieth birthday: "Altiora Petivimus" ("We have pursued the higher things").

In October 1915 he returned to Washington, hoping to work there for a time on his history of botany. However, his health failed rapidly, and on 25 October he entered Providence Hospital, where he died in his sleep on 10 November 1915.[115] He was buried at Notre Dame in the cemetery of the brothers and priests of the Society of the Holy Cross, among whom he had elected to spend his last days. His epitaph describes succinctly the moving force of his life:

> A MAN WHOM NATURE IN ALL HER
> PHASES ATTRACTED AND ENGAGED
> AND FOR WHOM SHE OPENED A DOOR
> LEADING UNTO THE HOUSE OF GOD

Chronology

20 August 1843	Born at Hopkinton, Rhode Island
5 June 1855	Left Hopkinton, R.I., for Illinois
1859	Met Thure Kumlien
1860–1862	Attended Albion Academy, Wisconsin
21 August 1862–13 July 1865	Service in Union Army
June 1866	Received Bachelor of Philosophy Degree from Albion Academy
1867–1870	Taught rural schools in Illinois
8 April 1870	Arrived in Colorado
3 September 1871	Admitted to Sacred Order of Deaconry, Protestant Episcopal Church

October 1871–February 1873	Served church at Greeley, Colorado
10 August 1872	Met Asa Gray at Empire, Colorado
31 January 1873	Ordained Priest, Protestant Episcopal Church
February 1873–February 1874	Served church at Pueblo, Colorado
March 1874–February 1875	Served church at Vallejo, California
March 1875–March 1876	Served church at Georgetown, Colorado
April 1876–March 1877	Served church at Yreka, California
April 1877–August 1877	Silver City, New Mexico
August 1877–1879	Cresswell, Colorado
April 1880–May 1881	Silver City, New Mexico
17 May 1881	Arrived at Berkeley Pastor St. Marks Parish
5 February 1885	Received into Roman Catholic Church
1885–1895	Professor of Botany, University of California, Berkeley
2 June 1886	Formal deposition from the ministry
15 February 1887	First number of *Pittonia* published
1892	Member, International Committee on Botanical Nomenclature
August 1893	President, Madison Botanical Congress, Madison, Wisconsin
25 August 1894	Sailed for England to visit Kew Herbarium
3 December 1894	Landed in U.S. on return from Kew
9 April 1895	Resigned from University of California, Berkeley
12 June 1895	Received honorary LL.D. from the University of Notre Dame
August 1895–September 1904	Catholic University of America, Washington, D.C.
24 November 1903	First number of *Leaflets of botanical observation and criticism* published
1904–1914	Associate of National Museum, Smithsonian Institution
1909	Publication of Part I, *Landmarks of botanical history*
1910–1915	Consultant to United States Department of Agriculture

2 December 1913	Special program of the Botanical Society of Washington, D.C., in honor of seventieth birthday
December 1914	First number of *Cybele columbiana* published
February 1915	Greene's library and herbarium arrive at University of Notre Dame
April 1915	Moved to University of Notre Dame, Indiana
10 November 1915	Died in Providence Hospital, Washington, D.C.

Notes

I am indebted to numerous organizations and individuals for information and material concerning E. L. Greene that have been incorporated into this article. Most of these sources have been acknowledged in the footnotes. Mr. Erl Ellis of Idaho Springs, Colorado, graciously placed at my disposal his records of Greene's stay in Colorado, which were particularly useful in authenticating his movements and activities in that period of his life. Mrs. Margaret Bergseng very kindly took time to locate materials pertinent to Greene in the files of the Jepson Herbarium, at the University of California, Berkeley. Professor Joseph A. Ewan made available an unpublished manuscript concerning the association of E. L. Greene and Katharine Brandegee. Particular appreciation is accorded Father Thomas T. McAvoy, Archivist of the University of Notre Dame, whose historical expertise was useful on several occasions, and who was most helpful in making the Greene papers available to me.

1. Ricker (1914).

2. Letter, Greene to the Botanical Society of Washington, D.C., 3 November 1914, p. 2. Archives University of Notre Dame; hereafter abbreviated as Arch. UND.

3. Greene, undated manuscript notes for response at banquet. Arch. UND.

4. Rodgers (1944b), p. 307.

5. For evidence of Jones's irascibility, see his comments on contemporary botanists in successive issues of his privately published journal, *Contributions to western botany*, especially vol. 14 (1912a), pp. 48–50, from which these quotations were taken.

6. Ewan (1955), p. 30.

7. *Cf.* Brandegee (1893a); Ewan (1950), p. 65. Greene wrote an essay on Linnæus as an evolutionist (1909c), but did not use that opportunity to argue his own convictions on the subject. Although it is frequently alleged that Greene was opposed to the theory of evolution, there is little in his writings to indicate this.

8. Letter, Greene to the Botanical Society of Washington, D.C., 3 November 1914, p. 6. Arch. UND.

9. The material in this paragraph derives from a manuscript of an incomplete autobiography of 21 pages that recounts Greene's recollections of his life until about 1855. Arch. UND.

10. *Cf.* the manuscripts on Greene's early botanical correspondence (two versions of about 25 pp. each). These review his early correspondence with Asa Gray, John Torrey, and George Engelmann *ca.* 1869–70. Arch. UND.

11. Greene (1888), p. 252.

12. Letter, Thure Kumlien to Greene, 24 March 1872, p. 1. Arch. UND.

13. *Cf.* Greene, manuscripts on early correspondence. Arch. UND, and Main (1929).
14. Greene (1888), p. 252.
15. Emery (1924).
16. Anderson (1915).
17. Information on Greene's military service is derived largely from his letters to Thure Kumlien quoted in Main (1929).
18. Letter, Greene to Rasmus Anderson, 30 August 1867, p. 3. The letters to Anderson provide the best glimpse of Greene in the years immediately following the Civil War. The Rasmus Anderson papers were made available by courtesy of the State Historical Society of Wisconsin at Madison.
19. Letter, Greene to R. Anderson, 1 March 1870, p. 3. Anderson papers.
20. Greene, manuscripts on early correspondence. Arch. UND.
21. Personal communication from Edgar Anderson, who reported this from a conversation with J. M. Greenman.
22. Greene, (n.d.).
23. Letter, Greene to George Engelmann, 4 November 1880, p. 1. Greene's letters to Engelmann were made available by courtesy of the Missouri Botanical Garden, St. Louis.
24. Letter, Stephen T. Olney to Greene, 30 August 1870, p. 1. Arch. UND.
25. Greene (n.d.).
26. Curtis (1909).
27. Letter, A. Greene to E. L. Greene, 17 October 1870, p. 4. Arch. UND.
28. Greene (n.d.).
29. Letter, Asa Gray to Greene, 6 August 1871, p. 2. Arch. UND.
30. Letter, Gray to Greene, 17 October 1874, p. 3. Arch. UND.
31. Letter, Greene to Gray, 27 February 1876, p. 2. Greene's letters to Asa Gray were made available by the Gray Herbarium, Harvard University, Cambridge, Mass., through the courtesy of Dr. Reed C. Rollins, Director.
32. Bartlett (1916), p. 163.
33. Greene (n.d.).
34. Greene, notes for response at banquet. Arch. UND.
35. *Cf.* Kistler (1936).
36. Letter, Greene to Gray, 18 July 1876, p. 2 (Greene's underscoring). Gray Herbarium.
37. Letter, Gray to Greene, 3 August 1880, p. 1. Arch. UND.
38. Letter, Gray to Greene, 12 August 1880, p. 1. Arch. UND.
39. Letter, Greene to Gray, 4 January 1881, p. 1. Gray Herbarium.
40. Letter, Gray to Greene, 14 February 1881, p. 2. Arch. UND.
41. This comment was attributed to Greene by the late Norman C. Fassett in lectures at the University of Wisconsin.
42. Letter, Gray to Greene, 10 December 1881, p. 2. Arch. UND.
43. Letter from Greene to Gray, 19 December 1881, p. 2. Gray Herbarium.
44. Letter, Gray to Greene, 20 June 1881, p. 1. Arch. UND.
45. Curtis (1909), p. 242.
46. *San Francisco Call*, 30 July 1883.
47. Jepson (1918).
48. Rideout (1958).
49. Letter, C. C. Parry to Greene, Good Friday 1886, p. 3. Arch. UND.
50. Letter, Greene to Gray, 16 August 1882, p. 3. Gray Herbarium.
51. Letter, Gray to Greene, *ca.* September 1882 (the first leaf is missing). Arch. UND.

52. Letter, Greene to Gray, 3 February 1886, p. 3. Gray Herbarium.
53. Bioletti (1929).
54. Letter, Gray to Greene, 3 January 1882, p. 1. Arch. UND.
55. Personal communication from Dr. Edgar Anderson, who records this comment from a conversation with Dr. J. M. Greenman. See also E. Anderson & Whitaker (1934).
56. Personal communication from P. G. Russell, who records this from a conversation with Greene at the Smithsonian Institution.
57. Letter, Greene to Gray, 20 July 1885, p. 3. Gray Herbarium.
58. Letter, Greene to Gray, 20 December 1886, p. 2. Gray Herbarium.
59. Letter, Gray to Greene, 8 March 1887, p. 3. Arch. UND.
60. Letter, Katharine Brandegee to W. L. Jepson, 7 October 1894, p. 1; letter from M. E. Jones to W. L. Jepson, 11 October 1895, p. 1. These letters are from the archives of the Jepson Herbarium, University of California, Berkeley, by courtesy of Dr. Rimo Bacigalupi, Curator.
61. Rodgers (1944b), p. 259.
62. Letter, M. S. Bebb to Greene, 24 March 1885, p. 1. Arch. UND.
63. Letter, C. C. Parry to Greene, 11 November 1885, p. 4. Arch. UND.
64. Letter, T. C. Porter to Greene, 27 March 1886, p. 1. Arch. UND.
65. Letter, Greene to N. L. Britton , 28 November 1888, p. 5. Arch. UND.
66. Letter, N. L. Britton to Greene, 17 May 1889, p. 2. Arch. UND.
67. Personal communication from F. W. Stearns, recording a statement of M. L. Fernald in his lectures at Harvard University.
68. Letter, Greene to Gray, 26 November 1886, p. 2. Gray Herbarium.
69. Rodgers (1944a), p. 154, quoting a letter dated 22 January 1882.
70. Letter, C. E. Bessey to Greene, 2 March 1910, p. 1. Arch. UND.
71. Letter, L. M. Underwood to Greene, 2 October 1893. Arch. UND.
72. Letter, Greene to N. L. Britton, 21 May 1889, p. 3. Greene's letters to Britton, now in the Britton Papers, were made available by the New York Botanical Garden.
73. Letter, N. L. Britton to President and Board of Trustees, University of California, 28 May 1889, p. 2. Britton Papers.
74. Letter, Greene to W. L. Jepson, 24 August 1894, p. 1. Arch. Jepson Herbarium.
75. Letter, Greene to Jepson, 2 December 1894, p. 2. Arch. Jepson Herbarium.
76. Letter, Greene to Jepson, 21 September 1894, p. 3. Arch. Jepson Herbarium.
77. Letter, Greene to Jepson, 30 September 1894, p. 2. Arch. Jepson Herbarium.
78. Letter, Greene to Jepson, 2 December 1894, p. 2. Arch. Jepson Herbarium.
79. Letter, Greene to the Botanical Society of Washington, D.C., 3 November 1914. Arch. UND.
80. Letter, Greene to Bishop Keane, 21 January 1895, p. 2. Courtesy of Department of Archives and Manuscripts, Catholic University of America.
81. Greene (1887).
82. Greene, manuscript notes for a lecture on the anniversary of the University of California. 38 pp. Arch. UND.
83. Letter, Greene to Sereno Watson, 28 February 1887, p. 3. Gray Herbarium.
84. Letter, Greene to N. L. Britton, 30 March 1897, p. 2. Britton Papers.
85. Letter, Greene to Rev. J. Nieuwland, 3 January 1911, p. 2. Arch. UND.
86. Letter, Greene to Rev. J. Nieuwland, undated, p. 1. Arch. UND.
87. Greene (1893), p. 122.
88. Letter, Greene to N. L. Britton, 12 October 1889, p. 2. Britton Papers.
89. Letter, Greene to Asa Gray, 5 April 1884, p. 4. Gray Herbarium.
90. Bishop of California, notice of deposition from the Ministry, 9 June 1886. Arch. UND.

91. Letter from Greene to the Rector, Catholic University of America, undated (*ca.* December 1901), p. 2. Arch. Catholic University of America.

92. Barry (1950), p. 169.

93. Letter, Greene to the President of Catholic University, Feast of St. Thomas Aquinas 1902, p. 3. Arch. Catholic University of America.

94. Letter, Rev. J. Nieuwland to Greene, 1 October 1907, p. 1. Arch. UND.

95. Letter from A. Greene to E. L. Greene, 16 September 1871, p. 3. Arch. UND.

96. Letter from A. Eastwood to Greene, 30 April 1906, 4 pp. Arch. UND.

97. Letter, Greene to the Botanical Society of Washington, D.C., 3 November 1914. Arch. UND.

98. Greene, *Landmarks*, Part I, Preface.

99. Letter, Greene to C. Walcott, 16 February 1909, p. 1. Arch. Smithsonian Institution, Washington, D.C.

100. Obituary notice, *Torreya* (1915).

101. Greene, On work done at the United States Museum between the years 1904–1909. Manuscript dated 6 February 1909, p. 3. Arch. UND.

102. Personal communication from P. L. Ricker.

103. Personal communication from Paul G. Russell.

104. Note, undated, unsigned. Arch. UND.

105. Greene, undated manuscript notes for response at banquet. Arch. UND.

106. Greene, notebooks. Arch. UND.

107. Most of the information concerning Greene's disposal of his herbarium and library is found in his correspondence with Rev. J. Nieuwland. Arch. UND.

108. Letter, Greene to Nieuwland, 29 December 1907, p. 1. Arch. UND.

109. Letter, Greene to Nieuwland, 19 December 1909, p. 3. Arch. UND.

110. Letter, Greene to Nieuwland, 9 April 1910, p. 2. Arch. UND.

111. Letter, Greene to Nieuwland, Whitsunday 1912, p. 3. Arch. UND.

112. Letter, Greene to J. Cavanaugh, President of the University of Notre Dame, 26 November 1912, p. 2. Arch. UND.

113. Report to members of the Faculty of Philosophy, Catholic University of America, about its Department of Biology, 12 March 1902, p. 2. Arch. UND.

114. Greene, manuscript notes for a lecture on the anniversary of the University of California. 38 pp. Arch. UND.

115. *Washington Star*, 10 November 1915.

References Cited

Allison, Edith Mary. 1908. Bibliography and history of Colorado botany. *Univ. Colorado Stud.* 6: 51–76.

Anderson, Edgar, and Thomas Wallace Whitaker. 1934. Speciation in *Uvularia. J. Arnold Arbor.* 15: 18–42.

Anderson, Rasmus Bjorn. 1915. *Life story of Rasmus B. Anderson with A. O. Barton.* Madison, Wis.: privately published. 678 pp.

Anonymous. 1882. The botanist in Arizona. *Bot. Gaz.* 7: 8–9.

——— [probably Katharine Brandegee]. 1893. Notes and news. *Zoe* 4: 310–13.

———. 1894. [Review of] Greene's Manual of the botany of the region of San Francisco Bay. *Bot. Gaz.* 19: 162–63.

———. 1915a. Dr. Greene and the botanical course. *The Notre Dame Scholastic* 49: 13. (Portrait.)

———. 1915b. [Obituary notice.] *Proc. Calif. Acad. Sci.* ser. 4, 5: 199.

———. 1915c. [Obituary notice.] *San Francisco Chronicle*, 15 November.
———. 1915d. [Obituary notice.] *Science* n.s. 42: 711.
——— [Norman Taylor?]. 1915e. [Obituary notice.] *Torreya* 15: 249–50.
——— [probably Julius Nieuwland]. 1916. In memoriam. *Amer. Midl. Naturalist* 4: 228. (Portrait.)
Barber, J. 1945. Edward Lee Greene. *Catholic World* 160: 444–49.
Barry, Colman James. 1950. *The Catholic University of America, 1903–1909*. Washington, D.C.: Catholic University Press. 298 pp.
Bartlett, Harley Harris. 1916. The botanical work of Edward Lee Greene. *Torreya* 16: 151–75. (Portrait.)
Bessey, Charles Edwin. 1910. [Review of] Greene's Landmarks of botanical history. *Science* n.s. 31: 69–71.
Bioletti, Frederic Theodore. 1929. Reminiscences of an amateur botanist. *Sci. Monthly* 29: 333–39.
Brandegee, Katharine. 1891. Contributions to the knowledge of West American plants. I. *Zoe* 2: 75–83.
———. 1893a. The botanical writings of Edward Lee Greene. *Zoe* 4: 63–103.
———. 1893b. Botanical nomenclature. *Zoe* 4: 182–84.
———. 1893c. E. L. Greene versus Asa Gray. *Zoe* 4: 182–84.
———. 1893d. International botanical congress at Madison. *Zoe* 4: 294–96.
———. 1894. [Review of] Greene's Manual of the Bay Region botany. *Zoe* 4: 417–20.
Catholic Citizen (newspaper). 22 November 1912, p. 1 (Portrait). Milwaukee, Wis.
Colorado Miner (newspaper). 22 August 1872, account of Gray's Peak expedition. Georgetown, Colo.
Curtis, Georgina Pell. 1909. *Some roads to Rome in America*, pp. 187–245. St. Louis, Mo.: Herder.
———. 1911. *The American Catholic who's who*, p. 256. St. Louis, Mo.: Herder.
Downing, Margaret Brent. 1917. Edward Lee Greene. *Catholic World* 106: 13–24.
Dupree, A. Hunter. 1959. *Asa Gray, 1810–1888*. Cambridge, Mass.: Harvard University Press. 505 pp.
Eastwood, Alice, Douglas Houghton Campbell, and Eugene Woldemar Hilgard. 1916. Resolution of respect. *Amer. Midl. Naturalist* 4: 335–36.
Eisenhart, W. G. 1916. In remembrance. *Amer. Midl. Naturalist* 4: 336–38.
Emery, J. Q. 1924. Albion Academy. *Wisconsin Mag. Hist.* 7: 301–21.
Ewan, Joseph Andorfer. 1942. A bibliographical guide to the Brandegee botanical collections. *Amer. Midl. Naturalist* 27: 772–89.
———. 1950. *Rocky Mountain naturalists*. Denver, Colo.: University of Denver Press. 358 pp. (Portrait.)
———. 1955. San Francisco as a Mecca for nineteenth-century naturalists. In Edward L. Kessel, ed., *A century of progress in the natural sciences, 1853–1953*, pp. 1–63. San Francisco: California Academy of Sciences.
Gray, Asa. 1887. Botanical notices [Greene's *Pittonia*]. *Amer. J. Sci.* ser. 3, 34: 493–94.
Greene, Edward Lee. 1880. Notes on certain silkweeds. *Bot. Gaz.* 5: 64.
———. 1887. Wherefore *Pittonia*? *Pittonia* 1: 51–52.
———. 1888. Sketch of the life of Thure Kumlien A.M. *Pittonia* 1: 250–60.
———. 1893. [Review of] The Metaspermae of the Minnesota Valley. *Erythea* 1: 117–23.
———. 1909a. *Landmarks of botanical history*. Part I. Smithsonian Miscellaneous Coll. vol. 54. Washington, D.C.: Smithsonian Institution.

———. 1909b. *Extracts from personal letters on the Landmarks of botanical history.* Washington, D.C.: Smithsonian Institution. Reprinted 1912.

———. 1909c. Linnæus as an evolutionist. Proc. Washington Acad. Sci. 11: 17–26.

———. n.d. Early botanizing in Colorado. Unpublished manuscript. 95 pp.

Herre, Albert William Christain Theodore. 1960. *Katharine Brandegee. A reply to a fantasy by J. Ewan.* [Seattle Wash.:] privately published. 5 pp., mimeographed.

Hogan, Peter Edward. 1949. *The Catholic University of America, 1896–1903.* Washington, D.C.: Catholic University Press. 212 pp.

Holm, Herman Theodor. 1893. Recent literature. *Zoe* 4: 186–95.

Hopkins, Milton. 1942. *Cercis* in North America. *Rhodora* 44: 193–211.

Howell, John Thomas. 1943. Sertulum Greeneanum. *Amer. Midl. Naturalist* 30: 6–39.

Humphrey, Harry Baker. 1961. *Makers of North American botany.* New York: Ronald Press. 265 pp.

Jepson, Willis Linn. 1894. Professor Greene's mission in Europe. *Erythea* 2: 153–55.

———. 1918. Edward Lee Greene, the man and the botanist. *Newman Hall Review* 1: 24–29.

———. 1928. Botanical explorers of California. Daniel Cleveland. *Madroño* 1: 267–68.

———. 1931. Edward Lee Greene. In *Dictionary of American biography*, vol. 7, pp. 564–65. New York: Scribner's.

———. 1943. Edward Lee Greene, the individual. *Amer. Midl. Naturalist* 30: 3–5.

Jones, Marcus Eugene. 1912a. Greene and his species. *Contr. W. Bot.* 14: 49–50.

———. 1912b. Sextuple botany. *Contr. W. Bot.* 14: 48.

———. 1929a. The Brandegees. *Contr. W. Bot.* 15: 15–18.

———. 1929b. Greene. *Contr. W. Bot.* 15: 25–27.

———. 1933a. Mrs. T. S. Brandegee. *Contr. W. Bot.* 18: 12–18.

———. 1933b. North American botany. *Contr. W. Bot.* 18: 131–35.

Kearney, Thomas Henry. 1958. Botanists I have known. *Leafl. W. Bot.* 8: 275–80.

Kistler, Ellen Downey. 1936. Bibliography of the botanical writings of Edward Lee Greene. *Madroño* 3: 328–48.

Lange, E. F. 1955. Edward L. Greene and Howell's "Flora of Northwest America." *Madroño* 13: 1–4.

Lincoln, Elmira Hart. 1842. *Familiar lectures on botany, practical, elementary and physiological.* New York: Huntington. Parts I–IV, 246 pp.; Part V–Appendix, 186 pp.

Main, Angie Kumlien. 1929. Life and letters of Edward Lee Greene. *Trans. Wisconsin Acad. Sci.* 24: 147–85.

———. 1943–44. Thure Kumlien. Koshkonong naturalist: I–III. *Wisconsin Mag. Hist.* 17: 17–39, 194–220, 321–43.

McIntosh, Robert Patrick. 1960. Edward Lee Greene correspondence. *Madroño* 15: 222–24.

———. 1961. The Greene-Nieuwland herbarium at the University of Notre Dame. *Proc. Indiana Acad. Sci.* 71: 235–38.

Moran, Reid Venable. 1962. Rufus Davis Alderson. *Madroño* 16: 224–28.

Morning Times, The (newspaper). 22 September 1895, p. 11. Washington, D.C. (Drawing.)

National cyclopedia of American biography. 1926. Vol. 19, pp. 332–33. New York: White.

Reifschneider, Olga. 1965. *Biographies of Nevada botanists, 1844–1963*. Reno: University of Nevada Press. 165 pp. (Portrait.)

Ricker, Percy Leroy. 1914. The Botanical Society of Washington. *Science* n.s. 39: 109–12.

Rideout, L. U. 1958. Priests, pistols and polemics. *Calif. Hist. Soc. Quart.* 37: 51–61.

Rodgers, Andrew Denny, III. 1944a. *American botany, 1873–1892: Decades of transition*. Princeton, N.J.: Princeton University Press. 340 pp.

———. 1944b. *John Merle Coulter: Missionary in science*. Princeton, N.J.: Princeton University Press. 321 pp.

Rose, Joseph Nelson. 1916. Edward Lee Greene. *Bot. Gaz.* 61: 70–72. (Portrait.)

San Francisco Call (newspaper). 4 July 1883, p. 3; 28 July 1883, p. 1; 29 July 1883, p. 6; 30 July 1883, p. 1; 4 August 1883, p. 3; 25 August 1895, p. 8; 21 July 1895, p. 11 (inked portrait from photograph).

Sargent, Charles Sprague. 1895. *The silva of North America*, vol. 8, p. 84. Boston: Houghton Mifflin.

Scannell-O'Neil, D. J. 1907. *Distinguished converts to Rome in America*. St. Louis, Mo.: Herder. 179 pp.

Setchell, William Albert. 1926. Townshend Stith Brandegee and Mary Katharine (Layne) (Curran) Brandegee. *Univ. Calif. Publ. Bot.* 13: 155–78.

Stevenson, John Albert. 1962. The Botanical Society of Washington: A brief history. *J. Wash. Acad. Sci.* 52: 219–20.

Thomas, John Hunter. 1961. The history of botanical collecting in the Santa Cruz Mountains of central California. *Contr. Dudley Herb.* 5: 147–68.

Edward Lee Greene: An Appraisal of His Contribution to Botany

ROGERS McVAUGH

Edward Lee Greene has been dead 65 years, and his more significant writings were published long before his death, but perhaps even now we are too close to him to judge him objectively. Few now living remember Dr. Greene as a person, but many have been influenced by critics who knew him well. Harley Bartlett, who as a young man in Washington had been closely associated with Greene, never changed the opinions he had expressed at the time of Greene's death:

> He had given his life to the cause of science. Some there are who doubted the value of his work and questioned the motives which actuated his caustic criticisms. Of enemies he had many; of friends but a few. During a large part of his career the forces arrayed against him would have daunted a less courageous man . . . his enemies were for the most part those who did not know him personally; those who had met and talked with him bore no resentment for anything he may have said. And whatever may be said of his words, his deeds were always kind.

This is a picture of a courageous, selfless man; a strong man, one who has enemies "arrayed against him"; a man out of the ordinary mold. So it seems to me Harley Bartlett always thought of Greene. In his obituary account Bartlett (1916, p. 175) continued: "Nor will it be worth while for any of us, for many years to come, to attempt an exact appraisal of his contributions to science. Certain it is, however, that few, if any, exerted a greater influence than he in the development of systematic botany in America."

Now let us look at the other side of the coin. Those who knew Marcus Jones cannot but be influenced by his pronouncements, characteristically overstated as they are. During Greene's lifetime, Jones (1912) said of his work:

> E. L. Greene complains of the way the botanical public has received his proposed new species, intimating personal prejudice. . . . Because a generation ago half of the flora was almost unknown is no evidence that the present increase of species is due to changed views among real botanists as to specific limitations. Greene is always sidestepping the main issue, which is this, Do his species represent stable plant forms, or are they individual variations? We who study our plants as they grow in the field and

by thousands know that 95% of his species have no foundation in fact, and no amount of botanical juggling or dust throwing can obscure that fact. The indictment western botanists bring against him is that he is grossly ignorant of primary facts in plant ecology or deliberately dishonest. His work on Eschscholtzia and Apocynum is the acme of botanical drivel, as every one but himself knows, and differs in no ess[e]ntial particular from a typhoid fever patient in the last stages picking at the bedding.

Jones was no more merciful after Greene's death, although by the time of publication of his next *Contribution* (1929) he had had something more than a decade to cool off: After branding Greene a pest, a botanical crook, and a cur, he concludes by saying: "Recently I have been going over Greene's Leaflets and notice his treatment of Rhus trilobata, which makes one feel like committing murder, but fortunately Greene has passed beyond human retaliation. His case makes one half inclined to believe in Hell, for no other place would be suitable for him."

Nor was Jones the only one to be so outspoken. Early in Greene's botanical career Katharine Brandegee (1893, p. 64) took him to task in a withering diatribe, condemning his lack of insight, accuracy and judgment:

Mr. Greene has described "as new" about 700 species, and resurrected something like the same number of groundless synonyms, nearly all relating to the Californian flora, and thus adding to our already inflated list at least 1000 names. It is safe to say that not more than one in ten of these species is tenable, and probably one in fifteen or twenty would be nearer the mark.

In the same vein Mrs. Brandegee continued:

The specific descriptions of Mr. Greene are a disgrace to botany. Even in the few instances where he has named valid species—and in such a multitude it sometimes happens—he uniformly fails to grasp the salient points and mistakes most of the rest. Some of these errors are so gross as to be, for a man holding the position of the author, almost inconceivable, and leave the reader to choose only between deliberate misstatement and an ignorance of methods of scientific study unparalleled in a Professor of botany of a modern university. It is, indeed, to be suspected from his descriptions that, though he can write learnedly of embryological observations made by others, his only method of getting at even the cotyledons of any seed smaller than a bean, is to sprout it.

It is not easy to reconcile Bartlett's view of Greene with those of Jones and Mrs. Brandegee, nor can one easily assess the total contribution that Greene made to botany in the midst of the cloud of controversy with which he surrounded himself. It is possible, however, to describe the several disciplines in which he interested himself, to indicate briefly the extent of his interest in each, and then to make some judgments on the ways his accomplishments have affected botanical thought and botanical decisions in the last 75 years.

Various biographers have called attention to Greene's early interest in botany; his long period of maturation after the Civil War when he combined his work as a pastor with his avocation for botany; his early identification of himself with the American West and its botany; his altercations with Gray and Watson; his conversion to Rome; and the possible connection between his botanical work and his evident bent toward mysticism. It has been pointed out that he was a constant and passionate defender of certain "causes" in nomenclature, and a long-time exponent of field study as opposed to the practices of what he called "closet botanists." He was immoderate in argument; his published statements were sometimes insulting if not actually libelous. Through much of his writing runs a thread of what seems to be defensiveness stemming from personal insecurity; always he seems to have felt the need to justify his own course in the eyes of others; to show up the mistakes of others with small barbs concealed in his apparently objective criticism; to make the botanists at Harvard (and so the rest of the world) concede the importance of the American West and its botanical autonomy. Often his need for self-justification made his contemporaries less than fully appreciative of his undoubtedly considerable contributions to botanical science.

Greene's botanical life may be conveniently divided into two periods. After the Civil War and his move to the West, he dabbled in botany for a decade before beginning active publication in 1881. What may be called his Californian period began at that time and continued until 1895, when he left the University of California to accept the professorship of botany in the Catholic University of America. His second, or Washington, period began before 1900 and continued until the end of his active life. Of these the former represented his more productive and influential years. Ewan (1950, p. 65) has well noted that with his moving to Washington, D.C., and his increasing withdrawal from society, there is a perceptible decline in the quality of his work with an increasing publication of spurious species based on his rigorously entrenched views on the fixity of species. Actually such a decline may be observed not only through his publication of superfluous or spurious species, but in other ways.

Greene was a prolific writer. His botanical writings, according to the list published by Ellen Kistler (1936), included 565 separate items over a span of some 45 years. His botanical interests, as reflected by his writings, were mostly in local floras (particularly the flora of California); in history and biography and so perhaps inevitably in nomenclature; in taxonomy *per se*, especially of families and genera in California and adjacent parts of the United States; and in describing what he took to be new species whenever they came to his attention. In the last quarter-century he has been remembered chiefly for two things, his *Landmarks of botanical history* and his predilection for describing new species, both of which reflect his activities in later life.

During his Californian period Greene was an active traveler and plant collector; he was only 52 years old when he left California for good. As a professor he conducted field classes with an active group of students. Some of these, like Bioletti and Jepson, later made their own places in the California botanical scene. Much impressed with the wealth and diversity of the then little-known Californian flora, Greene set out to study it in the field and to make it his own, quite independent of any studies that others might be carrying on elsewhere. Because such a large proportion of the flora was at that time still unknown to science and undescribed, Greene's perceptiveness and his flair for noticing the unusual stood him in good stead; even the law of averages worked for him, as Mrs. Brandegee unkindly pointed out, in a state where so much was new. Actually it is quite clear that during this period of his life, his taxonomic work was sound and his production of new species not overly enthusiastic; more of this anon.

During all the years he spent in California, Greene maintained an active interest in plant nomenclature, a subject that was neither popular nor well studied in his day. Perhaps partly as a result of his preoccupation with the work of long-dead authors and their contribution to nomenclature, he became increasingly interested in botanical history and biography, and increasingly a bibliophile. His personal library was greatly augmented during this time, and eventually included some 4,000 volumes. During his first years in the West his critics had not been slow to point out that he might have avoided some mistakes by consultation of the older literature, but by the time he left for Washington most of the important early books were available to him in his own collection.

His Californian period was thus a time of growth, a time of field study and teaching, and a time of devotion to causes. Whether because of his youth, or for some other reason, his perspective seems to have been better then than in later years, and his work more acceptable to his colleagues.

When Greene moved to Washington he began a long period of what Ewan called increasing withdrawal. Although at first he had students, including some like Tidestrom and Nieuwland who continued in botanical work, he gave up teaching in 1904. More and more he devoted himself to herbarium specimens and to the study and interpretation of the past. During the last twenty years of his life he wrote little or nothing on nomenclature and little on the more philosophical aspects of taxonomy, but approximately doubled his output of descriptions of new species. Much of his time in the decade before the publication of the *Landmarks* must have been spent in reading and studying the Herbalists and the authors of antiquity; the quotations, translations, and interpretations on every page of the *Landmarks* attest this. As noted below, his more ambitious generic revisions published during this period lacked the insight he had demonstrated as a younger man in his studies of the California flora. With undiminished perseverance he worked as much as his strength would permit, but with his most productive years far behind him.

Greene as a Taxonomist

Greene's biographers and his critics alike have probably overstressed the connection between his belief in the immutability of species and his ability to see new species where none existed. There is no necessary connection between these. It would not be hard to name a dozen botanists who, in old age and increasing preoccupation with individual herbarium specimens in place of whole populations of living plants, have become known for their ability to see new species. I do not find in Greene's own work any direct statement to support the contention that he was much impressed by the immutability of species. It is not rare, on the contrary, to find in his writings a statement like the following, published (1906, p. 49) in his later years when his species-describing was at its height:

> But in this first decade of the twentieth century everyone will recognize that such a range as has been accorded to Bentham's middle Mexican *P. angustifolia* is impossible—at least, to any mind having an understanding of all those extremes of diversity as to soil, climate, altitude, and *other potently influential conditions* which exist between southern Mexico and such regions as Texas and Oklahoma on the one hand and Arizona and northern California on the other [italics added].

Mrs. Brandegee (1893) early alluded publicly to Greene's beliefs:

> The underlying reasons of Mr. Greene's devotion to "new species" are not far to seek. The most important one is his attitude concerning their origin. He openly condemns, as inconsistent with the Mosaic record, the theory of evolution held in greater or less degree by almost all biologists, and proclaims his belief in the special creation and fixity of species, taking occasional opportunity* to sneer at the misguided mortals who differ from him. How this belief affects his botanical teachings is evident at once. Rejecting the clue which would lead him through the tangled labyrinth of overlapping forms which so especially abound in the extreme variation of environment found on the western coast of North America, nothing is consistently left to him but to make a new species of every variation, no matter how trivial. That he has not made five times as many is due to his really, in spite of frequent claims to the contrary, slight knowledge of the forms belonging to our flora. . . .
>
> * "And if so is it another of that class of facts which our friends the evolutionists press into service, as indicating that species, and even genera, are created by soil, climate, or in one oft-repeated word, environment?" [K. B.'s quotation in footnote, p. 64.]

Greene's later biographers, or those copying from them, have sometimes focused unduly on this aspect of his work. Jepson (1931) wrote: "Disregarding almost entirely the effects of climatic and edaphic influences, he thought of species as immutable and so was able to discover some three thousand new ones, but only a small proportion of them have been accepted by other botanists." In Dupree's competent biography of Asa Gray (1959, p. 399) the sense of this is made even stronger in paraphrase: "Few of the three thousand or so new species . . . have won the approval of botanists." The currently accepted status of Greene's "new species," however, especially of those published during his Californian period (about 1885–1895 or even 1900), indicates a somewhat different situation.

The total number of new names for plants proposed by Greene seems to have been about 4,500, including approximately 3,000 names of newly described species, and about 1,500 new combinations based on earlier names proposed by other authors or by Greene himself. He published a few names for varieties, but he did not make any extensive use of this category. Beginning in 1881, when he published his first new species, he produced on the average about 90 new specific names a year for 15 years. About half of these were new combinations. Beginning in 1895, about when he left the University of California, and continuing until the end of his productive life in about 1912, his annual output of new names was about twice what it had been before. For the most part these later names represented new species; only now and then, as in the first volume of the *Leaflets* (1903–6), did the number of new combinations become large in proportion.

Partly because he had a discerning eye for novelties, and certainly in part because of his early arrival in western North America, Greene described many species that were unknown to earlier authors, and which are still passing under the names he gave them. In Jepson's *Manual of the flowering plants of California* (1923–25), for example, about 4,000 species of plants are treated; of these 362, or about 9 percent, bear names proposed by Greene. In the *California flora* of Munz and Keck (1959), published more than 30 years after Jepson's *Manual* and based on greatly increased knowledge of the flora, about 5,635 species are treated; of these 494, or 8.8 percent, bear names proposed by Greene.

During his Californian period, Greene's proposals of new species were not only relatively few, as noted above, but relatively conservative. If, for example, we consider the new names proposed by him in the *Bulletin of the Torrey Botanical Club* from 1881 to 1886 (volumes 8–13), and in the first two volumes of the *Bulletin of the California Academy of Sciences* (1885–86), we find a total of about 300 new names, of which about 242 were proposed for new species. About two-thirds of the names of new species applied to Californian plants; the others were from Baja California, New Mexico, Arizona, or elsewhere in the western United States. A summary is given in the accompanying table, where the word "accepted" means recognized as valid in current floras or, in some instances, recent taxonomic revisions. It is clear from the table that at least two-thirds of Greene's propositions were valid ones by the test of time. Some of his new species are still recognized under the names he gave them, some are passing under other generic names, or are accepted in infraspecific categories, so a creditably high percentage of his new names not merely referred to individual specimens, but evidently represented real taxa, that is to say natural populations. Hardly a botanist of his time, or of ours, could have done better in the California of 1885.

Source	Currently accepted unchanged	Accepted in another genus	Accepted as variety or subspecies	Doubtful	Synonyms	Total	Percent of valid taxa
California	60	24	26	5	44	159	69±
Baja California	14	6	6	6	7	39	70±
New Mexico	12	—	—	—	6	18	66.7
Arizona	7	1	1	—	4	13	70±
Various	6	—	3	—	4	13	70±
New combs.	14	33	4	4	6	61	—

During this same early period Greene's taxonomic work (as distinguished from the mere naming of new species) seems to have been not only conservative but always discerning and usually highly commendable. The evidence for this may be found in a number of monographs, revisions, and floras published in recent years. For example, as pointed out by John Thomas Howell (1943, p. 18) in a little paper called "Sertulum Greeneanum," Greene described, or proposed by name only, 36 new species and varieties in the genus *Phacelia.* Of these, 11, or almost a third, pertained to the group recently revised by L. R. Heckard (1960) under the name of the "*Phacelia magellanica* polyploid complex." It is interesting to note that only one of Greene's names, all proposed between 1887 and 1904, is reduced outright to synonymy; all the others are maintained by Heckard as species under the names proposed (but not always published) by Greene, or as subspecies of previously described species. In other words Greene, working with the Californian flora that he knew well in the field, was for the most part describing existing populations that he had seen and studied as populations, although in his day it was not fashionable to call them that.

Instances like the above may be cited from other groups of plants in which Greene had an early interest. The boraginaceous genus *Cryptantha* includes many species in western North America; according to the revision of I. M. Johnston (1925) there are 57 species. Greene proposed some 29 new species in this genus; of these only eight are relegated by Johnston to synonymy; 15 are recognized as valid species and three as varieties; one is transferred to another genus. Of all the names proposed by Greene, about two-thirds thus were needed new names pertaining to valid taxa, even though Greene's circumscriptions of the taxa may not have agreed with modern interpretations.

Turning to another plant family, the Campanulaceae, we find that in the primarily Californian genera *Nemacladus* and *Downingia* a total of 21 species are recognized in recent revisions (McVaugh, 1943). Of these, ten species were first recognized and described by Greene, between 1885 and 1895. In *Nemacladus* Greene proposed one additional name now regarded as a synonym; in *Downingia* he proposed two additional taxa, one of which is now regarded as of infraspecific status only, and the other a synonym of a Chilean species that was unknown to him.

His percentage of success, as measured by the extent to which botanists have accepted his proposals, is high in this instance. Mrs. Brandegee's contemporary dismissal of "*Downingia concolor, humilis, insignis, montana, ornatissima* and *tricolor* Greene [as] forms of a single polymorphous species" (1893, p. 93) is so far from the truth as to cast doubt upon the validity of the rest of the taxonomic judgments expressed in her wholesale reduction of Greene's species of other families and genera.

In Steyermark's revision of the North American species of *Grindelia* (1934, p. 518) the author says, in naming *Grindelia greenei*: "It is a pleasure to name this new species in honor of Dr. E. L. Greene, who perhaps had a keener insight and more understanding of the genus than any other botanist having worked with *Grindelia*." Greene proposed 14 species of *Grindelia*; of these but one is relegated to synonymy by Steyermark, who accepts nine of Greene's new species as valid, treats four as varieties, and one as a *forma*.

In his Washington period, when he was describing sometimes as many as 200 new species in a year, Greene was basing his studies and concepts primarily on herbarium material. His days of fieldwork in the West were far behind him, and like many others before him he began to see species in the herbarium that his contemporaries could not see in nature. Much of his later reputation as a "botanical splitter" can be traced directly to his work from about 1904 or 1905 until the end of his career, for during this last decade his publication of new species was inordinately rapid, and many of his proposals have been ill received. Some examples from the second volume of the *Leaflets of botanical observation and criticism* (1909–12) will make this clear.

The first article in this volume was entitled "New species of the genus Mimulus." Described as new were 17 species and three additional varieties. In the revision of *Mimulus* by A. L. Grant (1924), 13 of these species and three varieties are relegated to synonymy. In succeeding articles Greene described new Californian Asteraceae, new western species of *Arabis*, new species of *Apocynum*, and new species of *Erigeron*. Out of 19 new "Asteraceae," six are currently recognized as varieties of other species, and 13 are reduced outright to synonymy in current floras of California. Of Greene's 27 new species of *Arabis*, 22 are reduced outright by Rollins (1941), who at the same time recognizes two Greenean species as valid and two others in the category of variety. Of 44 "Accessions to *Apocynum*," 41 are reduced outright by Woodson (1930), and three form the nomenclatural bases for varietal names. Of 33 new species of *Erigeron*, 31 are reduced to synonymy by Cronquist (1947), one is taken up as valid, and one treated as a subspecies.

In the papers listed above, Greene was simply listing "new" species that came to his attention in the various groups that he happened to be studying. These studies were not monographic or revisional except in a very general way, but were thought of as preliminary to more definitive works. In some of his papers, however, especially after about 1905, Greene attempted ambitious revisions of entire genera, *e.g.*, *Ptelea* and *Eschscholzia*.[1] These formal papers have fared little better in the judgment of botanists than have his lists of new species. In his revision of *Ptelea*, for example, Greene (1906) recognized 59 species, all but a few described by himself as new. He stressed the taxonomic importance of features of the bark, twigs, and fruits, and indicated that the groups of species as delimited by bark color were also well restricted geographically. V. L. Bailey (1962), although acknowledging the validity of Greene's observations, differed markedly in her interpretation of the situation. She recognized but three species, including in one of these a number of infraspecific taxa recognizable at least in part by the features first noted by Greene.

Greene's revision of *Eschscholzia* ("*Eschscholtzia*") (1905) seems likewise to have failed the test of time. Although he recognized well over a hundred species of this widespread and well-known Californian genus, botanists generally have not been able to recognize more than about ten. In this as in other papers, he was alert to individual differences between specimens, and supposed that such differences often served as indicators of specific identities. He seems to have been overly impressed in his early days by the attitude of Gray and others that little remained to be found in the West. (For example, Bartlett quotes Gray as writing to Greene: "I hope you may find some new things; but you will be sharp if you do.") Greene may have been overly ready to publish novelties in order to prove his case to the "closet botanists" in what he often referred to disparagingly as "Cambridge." On the other hand Greene was uniquely situated for many years in an area where new species abounded, and even by random sampling he must have discovered a large number of things new to science. Working as he did with the combination of a rich and virtually unexplored flora, the desire to prove himself, and a ready discernment, it is not surprising that even in the early days Greene described a great many of what he took to be new species.

In Greene's taxonomic work, as distinguished from his activities as a describer of novelties, he concerned himself primarily with revisions of genera, or of groups of species within genera. He was also particularly interested in the Boraginaceae, the Polemoniaceae, and the Compositae, and over the years published many articles dealing with intrafamilial arrangements in these groups. His interest in the theoretical aspects of taxonomy seems to have been expressed chiefly at the level of the genus. His judgments as to generic relationships were often good ones, although equally often obscured by his vague descriptions and rambling accounts. He was not backward in proposing new generic status when such action seemed appropriate to him, and he is responsible for such well-known names as *Athysanus*, *Bebbia*, *Blepharizonia*, *Gymnosteris*, *Forsellesia*, *Holozonia*, *Iliamna*, *Langloisia*, *Leucelene*, *Lilaeopsis*, *Malacothamnus*, *Microsteris*, *Osmaronia*, *Petradoria*, *Sibara*, and others, as well as for a large number of generic names that have never been accepted.

Sometimes Greene stated clearly his reasons for supporting generic segregations, but often in a chatty manner that assumed on the part of the reader a considerable acquaintance with the groups he was discussing. In taking up *Polycodium* Raf. and *Batodendron* Nutt. as valid genera distinct from *Vaccinium*, he (1898) argued as follows:

> We have in the Eastern and Southern United States two groups of vacciniaceous shrubs either of which is at variance with all genuine *Vaccinium* in two important points of floral structure. The corollas in both groups are campanulate, while in both *Vaccinium* and *Gaylussacia* they are urceolate . . . *Vaccinium* and *Gaylussacia* are now everywhere admitted as distinct, yet, . . . there is not the slightest difference of floral structure between the two. But these other shrubs depart widely from the characters of both *Vaccinium* and *Gaylussacia* not only in their open-campanulate corollas, but in respect to their stamens, which organs are doubly marked by extremely long and slender anther-tubes, and two prominent horn-like projections on the back; so that nothing approaching these characters is found in any other genera allied to *Vaccinium*.
>
> . . . one new and most significant peculiarity of the *V. stamineum* group [*Polycodium*] . . . I alone seem to have observed. It is this, that in this group the corollas are *open in the bud*! For from ten days to two weeks before the actual flowering, and even from the time that the buds are green and scarcely larger than a pin-head, the corolla is open and campanulate. This is another character otherwise unknown in the family of plants to which these belong. Certainly in *Vaccinium* and *Gaylussacia* the buds are tightly closed, in an imbricate aestivation, until the corollas are full-grown and the anthers mature.

Because Greene's proposals, both of genera and of species, were often couched in such informal terms, his contemporaries were not inclined to accept them. Mrs. Brandegee's sweeping denunciation of all his proposals has already been noted. Even genera that have since been generally accepted were not at first favorably received. Rollins, for example (1941, p. 292), noted that in proposing *Sibara*, Greene failed to point out reliable differences between *Sibara* and *Arabis*, although rather vaguely contrasting his new genus with *Arabis*, *Cardamine*, *Sisymbrium*, and *Nasturtium*. It is not hard to see why the genus was long considered obscure; the original description of *Sibara* (*Pittonia* 3: 10. 1896) was of seven lines of text only, preceded by the following introduction:

> Of the family above named [Cruciferae] there is a considerable group of neat and elegant annuals, belonging to the Mexican and south Californian coasts and islands, the members of which, all of them discovered and brought to the knowledge of botanists within the last decade, have been temporarily referred to almost as many genera, severally, as there are species in the group. The species are six, and in publication

they have been distributed among the four genera, *Arabis*, *Cardamine*, *Sisymbrium* and *Nasturtium*; and it is a wonder to me that two of them were not consigned to *Thelypodium*, rather than to any of the others. To the annual species of *Cardamine* having pinnatisect foliage these plants would bear a closer likeness than they do, if they were not all conspicuously glaucous; but they show not the faintest indication of that sudden and elastic dehiscence of the pods which is so essentially characteristic of that genus. Their subterete pods and wingless seeds exclude them totally from *Arabis*, to no one section of which, however, do they bear any general resemblance. There is only one of the species which any one has thought of as possibly referable to *Nasturtium*. There is no rational way of treating such a group but that of conceding to it the rank of a genus, and for this one I propose the anagrammatic name *SIBARA*.

Greene was a strong proponent of the use of habital characters in the recognition and delimitation of genera; the following, for example, was sandwiched in among general observations on the Compositae (*Erythea* 2: 55. 1894):

A genus is first recognized, and afterwards defined; and that its acceptability, as a genus, depends more upon the intuitions of the experienced botanist, than upon the characters of flowers and fruit which by examination and comparison he may be led to assign it; in a word, that a genus bears some general features—presents some scarcely definable physiognomy—which, as a whole, signifies more to the specialist than does any technical peculiarity.

Even earlier, in a critique of a new species published by B. L. Robinson, Greene had said (*Erythea* 1: 237. 1893):

That by which *Nemacladus* differs so widely from all other genera of this order [Lobeliaceae] that no botanist at first glance would refer them to it, is its very unique habit. The plant when full grown and in flower or fruit presents nothing that could be called herbage. . . . There is no flowering plant of any alliance to which this genus makes any near approach in point of habit and general aspect. We were therefore a little disturbed in mind when we observed that one of Mr. Pringle's Mexican plants of this family, a plant totally incongruous with *Nemacladus* in habit, had been described as a new species under this genus, by Dr. Robinson of Cambridge.

Dr. Robinson's defense of his proposition is that the plant has the distinct and stellately radiating anthers of this genus, though he admits that in habit the herb is utterly at variance with *Nemacladus*. In our judgment the technical character by which it agrees with *Nemacladus* can not outweigh the fact that in habit it is so very different. On the contrary, habit is always the test to which characters are put, to try their value. Radiating anthers might run through a dozen genera, . . . but the remarkable habit of *Nemacladus* would, we suppose, in the judgment of most botanists, exclude from it any plant with erect, simple, racemose stems and an abundant herbage of opposite leaves.

Aside from the habit of the plant, Greene looked for other characters wherever he might find them. He was a pretty faithful follower of Linnæus's dictum that the character does not make the genus. Again and again he pointed out that, in this or that family, certain features may be valuable in generic or specific recognition, whereas in other families the same features may be quite worthless. Sometimes his reasoning was hard to follow. In an early paper (1887g, pp. 95–99), he opposed the separation of *Ivesia* and *Horkelia* from *Potentilla* on the grounds that the characters used to distinguish these genera, namely habit, depth of the calyx, reduction in the number of stamens, and dilation of the filaments, had proven to be of little value in distinguishing similar splinter groups from *Saxifraga*! Reasoning in a similar fashion (1887c, p. 24), he justified his handling of the taxonomy of the onagraceous genus *Zauschneria* on the grounds that it was closely related to a group of *Epilobium* that could be handled in the same manner:

It is altogether unscientific to assume that where flowers are large they must, for that reason, furnish characters, and that only where they are minute one may safely leave them and betake himself to pubescence and foliage for the marks of species. In a word, *Zauschneria* is very intimately related to that principal part of *Epilobium* in which, in generally admitted species by the dozen if not by the score, the corollas present no characters whatever, are never mentioned in describing the plants specifically, and all is rested upon pubescence, taken along with the insertion, venation and toothing of the leaves.

In a pioneer paper on Polemoniaceae (1887h, p. 121), Greene stated his firm views on the distinctions between genera in that family:

The one general conclusion reached by me, after eighteen years of field experience with these plants is, that characters of form of corolla and length, insertion, direction, etc. of stamens may be set aside as wholly incompetent to furnish means of defining genera; and that, by the calyx alone, especially as it appears not in flower, but in its after developments, in and of itself and in its relation to the fruit, we may limit and define good acceptable genera, made up of plants agreeing in habit, and in some other minor points.

A few years later (1892a, p. 251) Greene retreated from this position and in fact seemed almost to reverse his stand. In writing of the polemoniaceous genus *Linanthus*, he said:

The only constant character by which they differ from *Phlox* is that of the equal insertion of their stamens. The corolla in very many species is just that of *Phlox*, while in others it is much more open, coming out, in some, to the open-campanulate or almost rotate. But, as I have remarked in the first of these articles, characters of the form of the corolla and insertion of stamens have so depreciated, in the family, in recent years, as more and more species have been discovered, that only upon habit, strengthened in four or five instances by some characters of the calyx, can any genera of Polemoniaceae find confirmation.

As to the delimitation of genera, once they are recognized, Greene expressed himself clearly. In writing on the genera of Cruciferae, he said:

. . . for the satisfactory arranging of species in generic groups, a certain very marked type shall be chosen, and the genus be allowed at once to illustrate and define itself; the species showing close affinity for the chosen type, if they exist, being admitted and all others excluded from the genus; this to be done as decided when it happens to leave a genus monotypic as upon any other occasion.

Because Greene often changed his mind on matters of generic limits; because of his insistence upon the importance of habital characters in the recognition of genera and even more his dependence upon a sort of intuition that certain generic characters necessarily *had* to be important; and because of the tendency toward splitting that pervaded all his work, his judgments on genera varied greatly from one instance to another, but I think it may fairly be said that a large percentage of his generic proposals were based on reasonable assumptions, even if his interpretation of the facts differed from the more modern ones. He had a keen eye for characters that do run through groups of species and indicate some evolutionary discontinuities, but his ideas of the importance of individual characters or groups of characters were his own.

It is in the delimitation of species that Greene's concepts seem to have been farthest from reality. His inordinate multiplication of species in some groups has already been touched upon. Although some of this activity on his part was certainly influenced by his philosophical concept of the species as an entity, it seems likely that some of his reputation as a splitter may have been due to his failure to consider and revise his work before publication. Often his contrasts between supposedly related species are vague and not clearly distinctive; the emphasis in one species, for example, may be placed on the foliage, but in the other on the flowers. His keys (the few that he prepared) are not in fact keys but merely enumerations of parallel and partially contrasting characters of species after species, the whole suggesting that he himself never accomplished the rigorous separation of taxa that is necessary before a workable dichotomous key can be constructed.

Greene's Floristic Studies

The flora and vegetation of California were but little known when Greene began his work in that state about 1880, and his are among the significant early records of a plant life now extensively modified. In 1885 appeared his "A catalogue of the flowering plants and ferns of Guadalupe Island" (*Bull. Calif. Acad.* 1: 220–28. 1885). Similar floristic lists followed for Santa Cruz Island (*Bull. Calif. Acad.* 2: 388–416. 1887); San Miguel Island (*Pittonia* 1: 85–94. 1887); Cedros Island (*Pittonia* 1: 200–208. 1888); the San Benito Islands (*Pittonia* 1: 261–62, 263–66. 1889). The vegetation of the summit of Mount Hamilton was treated in 1893 (*Erythea* 1: 77–97), and that of Mount Diablo a little later (*Erythea* 1: 166–79. 1893). In 1891 appeared the first parts of *Flora franciscana*, subtitled "An attempt to classify and describe the vascular plants of middle California"; this was a pretentious flora, of which only the first four parts were ever published (Parts 1 & 2, pp. 1–280. 1891; Part 3, pp. 281–352. 1892; Part 4, pp. 353–482. 1897). In the meantime there appeared his *Manual of the botany of the region of San Francisco Bay* (1894), a work of some 350 pages.

The floristic works, like Greene's more strictly taxonomic contributions, suffer from having been over-hastily prepared and from their author's failure to make full use of the published works of others. Also notable is Greene's apparent avoidance of several major plant families almost through his entire career. For example, any flora of California must now list, as representing valid taxa or as synonyms, hundreds of names proposed by Greene; but these include no ferns or gymnosperms (*Cupressus arizonica* is a Greenean species, but obviously extra-Californian), no grasses (except *Distichlis spicata* (L.) Greene), no sedges, no cacti. For some reason known to himself he quite eschewed the study of these groups and some others. His San Francisco Bay *Manual* was intentionally restricted to 90 selected families, with the statement:

> These ninety embrace, however, all in our district which the novice in plant determination would be likely to take up, and a number of genera and species much greater than the beginner may master in one season's study, or in three. Thus no complaint will arise that the sedges and grasses, the pondweeds and the ferns have been omitted from this brief and more or less tentative treatise.

Even with these deficiencies, Greene's *Manual* was the first attempt at a major floristic treatment in California since the publication of Brewer and Watson's *Botany* (1876–80), and was in many respects an original work. It was not, however, favorably received. This was partly because of the author's attribution of generic names to pre-Linnæan authors and, paradoxically enough, because of his acceptance of the now universal convention of including in parentheses the original author of a name subsequently transferred to another genus or status. Its incompleteness, and Greene's other idiosyncracies, were held up to scorn in a contemporary review (Brandegee, 1894, quoted by J. H. Thomas in *Contr. Dudley Herb.* 5: 155. 1961):

> The title should have been *A Phanerogamic Flora of—counties in the State of California, omitting Typhaceae, Lemnaceae, Naiadaceae, Alismaceae, Juncaceae, Cyperaceae, Gramineae, Coniferae and numerous species in other orders; with thirty "new species" none of which are new, and nearly all vaguely characterized both as to character and station; and with every change of name which the author's present knowledge admits.*

Regardless of what Greene's colleagues may have thought of his more ambitious floristic papers dealing with the plants of California and the nearby islands, these contributions would have constituted an important part of the literature of botany in western North America even if their author had been less discerning than he was. His floras of the Bay Region, incomplete as they may have been, served a useful purpose even up to the publication of Jepson's *Manual* in 1923–25.

Greene on Nomenclature

Perhaps more than in any other field of botany, Greene's influence was strongly felt during his lifetime in the field of nomenclature. In a sense this was an interesting paradox, as Greene was temperamentally unsuited to the dispassionate judgments required of the nomenclator.

His writings on nomenclature are highly colored by his own emotions and by his subjective judgments. Here, however, he was not much out of line with his own time. The first "international" code of nomenclature, Alphonse de Candolle's *Lois de la nomenclature botanique* of 1867, was younger in years when Greene began his studies in California than World War II is today. There was in 1885 no real international agreement on rules of nomenclature, nor indeed was there any real international understanding of the need for an impersonal, inflexible and at the same time comprehensive code. The existing laws were accepted as a guide to good practice, much as we accept the rules of grammar. This was before the publication of Kuntze's *Revisio generum plantarum* (1891–98), which as C. A. Weatherby once said, "lifted the rule of priority from the status of a handy device for facilitating the choice of names to that of a sort of natural law." Today, when the *International code of botanical nomenclature* is a real force, when most botanists accept without much emotion (even if they do not all approve) the provisions of priority with fixed starting points, the type concept, and the conservation of generic names, it is not easy to realize that in 1885 none of these concepts had any real acceptance. Even priority was set aside for the sake of convenience or sentiment. Greene thus came into the arena of nomenclature at a time when much was unsettled, when the foundations of a rigorous system had not yet been laid, and last but not least when the contestants in the arena were less restricted in their choice of weapons than they are today. Most of the "enemies" that Greene made were antagonized in the first place by his patronizing or even downright insulting opinions of their intelligence or their probity, conveyed through the pages of one periodical or another. In view of Greene's many disagreements with his colleagues, it is remarkable that he managed to exert such a considerable influence in the development of the so-called "American School" of nomenclature. Perhaps even more impressive than his carping was his often violent partisanship of what he took to be "correct" in nomenclature; his support for certain causes never flagged, even if they sometimes seemed to be causes of the emotion rather than the intellect.

As early as 1885, for example, he was writing the following:

> Myosurus, Dillenius. . . . A sense of common justice to, and perhaps uncommon veneration for, great botanists who lived and labored before Linnæus, or contemporaneously with him, forbids my following the modern rule of crediting to that illustrious name those genera which he did not found but only adopted. I am willing to be singular in this matter, if so be that in these iconoclastic days no others will unite in the protest that no genus indicated and named by a Tournefort, for example, or by a Dillenius, ought to be wrested from its legitimate and very worthy paternity.

In the next two years Greene was stimulated to a number of utterances on the subject of nomenclature. He achieved some international notoriety, and probably strengthened his position at home, by a note contributed to the British *Journal of botany* (1887d). Earlier the same year B. Daydon Jackson had published an account of the scope and state of completion of the then-unnamed *Index kewensis*. Although "strenuously advocating strict priority as the only sure foundation for nomenclature," Jackson continued with the statement:

> . . . our practice is to take the name under which any given plant is placed in its true genus as the name to be kept up, even though the author of it may have ignored the proper rule of retaining the specific name, when transferring it from its old genus to the new. . . . To wantonly set aside the joint name thus given, and to publish a new name by joining the oldest specific name to the true generic, is a mischievous practice which should never be condoned; it is adding to the already vast mass of useless synonyms, and is more likely to be the offspring of vanity than of a sincere desire to promote science.

Jackson's position was, of course, merely an expression of the so-called "Kew Rule," which had been followed by most British botanists, and in America by Asa Gray, for exactly the reasons set down by Jackson. Greene vigorously opposed this point of view, asking: "What consistency is there in protesting that men ought to retain the old names, and yet everywhere seconding and approving their rejection of them?" He continued:

> It is easy to foresee that this evasion of the responsibility of insisting on the restoration of old specific names where they have been rejected, will only prolong this troublemaking era of mutability. The manifest evil tendency of the method in question is seen in this, that under it, people who like to change names in transferring species are without effectual reprimand or check of any kind. To quietly say to such authors that they ought not to do so will avail nothing; but let others who respect the rule which makes for fixity follow them up with the rejection of their newly-coined names, and the restoration of the old ones, and a few years will put an end to the abuse. . . .
>
> One is not well able to see how the kind of practice here contended for is "more likely to be the offspring of vanity than of sincere desire to promote science." Is not vanity more likely to find more congenial occupation in making new and supposedly improved names than in adopting the ready-made ones? There will surely be plenty of cases in which the keeping of an old one would give him who transfers, and by transferring adopts it as his own, an almost heroic exercise in humiliation, if he were not released from responsibility by a sense of obligation to use the badly-chosen or ill-constructed name.

Finally Greene provided what seems an unanswerable argument; it is of interest today to note that his point of view has everywhere prevailed, and the concept of absolute priority is written into the Code:

> [But the question is] whether old specific names ought or need not to be preserved. If really they ought to be kept, then all authors would seem bound, in so far as it lies in their power, to practise the retaining of them. If, on the other hand, the rule for keeping them is of doubtful expediency, there is no need of calling it a rule; and every botanical writer who will may return to the irresponsible and easy method of eminent authors of three or four score years ago, who in serene complacency rejected

old names and made new ones, each according to his fancy. Between their practice and that of those who, in deed as well as word, content for fixity by priority, there appears to me no middle way which at all commends itself to reason or augurs future good to science.

The venerable Asa Gray, stung by Greene's aspersions of "bad logic," "loose ethics," retorted (1887) that:

[the changing of a] name after it has got into general circulation because it has been discovered that some other botanist had previously published it in a wrong genus, is only adding a synonym without any advantage whatever, and is not even restoring an old name, *for the specific adjective is not of itself the name of a plant* [italics added]. . . .

To keep up the name under which any plant is first placed in its true genus is simple, thoroughly practicable, and in my opinion most conformable to accepted rules, as well as most conducive to the fixity of names. It is reasonable enough, under the stringent rule of priority, to resuscitate neglected older specific names pertaining to their proper genus; but surely it is unreasonable and inconsiderate to conclude any such right to specific names out of the genus to which they are subordinate.

Greene, not at all overawed, continued not only to condemn what he called "incorrect nomenclature" in general, but also to call attention to specific instances of it. Some of his writings of this period illustrate not only his insistence on priority but also his patronizing and infuriating style, with its implications of feeble-mindedness on the part of the Cambridge botanists (*Pittonia* 1: 143. 1887):

The limitation of genera is largely a matter of individual opinion. . . . But there is one kind of liberty which is granted to none of us, and that is, to take for our genera whatever names suit our fancy.

Separate or combine plants generically as we will, or as we think we must, we are bound to take for our genus the first name that happens to have been given to it; it may not be the fittest or the most euphonious of the several which may have been employed, it may not have been framed and applied by the most eminent of scholars; but if it is the first in history, that is the name our genus must bear. The principle is a fundamental one in botanical bibliography.

I may at some future day be brought to share the views of Dr. Sereno Watson, regarding the proper generical status of those Pacific American cucurbits about which he and I, and some older and better botanists than either of us, have written somewhat. Were he and I today of one mind upon the subject, it would not be left to him to select one name for these plants, and to me to select another. Were this freedom conceded to us two, it would have to be granted to every other living botanist, and so our genus might receive, next year, as many names as there might be botanists ready to add a species to it, or write a monograph of it. What is demanded of each and all of us is, that we search the printed records of our science, and find the earliest unpreoccupied name which the genus received from a botanist. . . .

Two reasons only, not based in fancy, seem to be given us by our author [Watson] for standing by the name *Megarrhiza*. One is that he himself once before in his life took it up and named species under it. Referring to that earlier day he says: "I had no hesitation in adopting Dr. Torrey's name for the genus." The known priority of *Mara* did not then stand in his way; and now he has his own former action for a precedent.

But a man so true to principles of scientific justice as Dr. Torrey, had hesitated, and not only hesitated, he had ignored the name, although of his own making, and steadily refused to publish a genus or a species under it, from the day that he became aware that, as a name it was precluded by an earlier. He would not attempt to place himself superior to the first law of generical nomenclature; and it seems hardly to argue much veneration for Dr. Torrey's memory, to force him now, into an attitude which, living, he would not assume.

Dr. Watson's second reason, to give it in his own words is this: "The genus *Megarrhiza* was recognized by Dr. Gray in 1859." The name and description of the genus *Mara* had then been some five years before the botanical public. Is it, then, in Cambridge, Massachusetts, Dr. Gray's recognition, and not historic fact, which determines the names of genera? That is what we had long been compelled to think; but we had hesitated to speak our thought. Dr. Watson's has now spoken it for us.

Sometimes in his criticism of contemporary work, Greene was even more direct (1887a, p. 3):

M. Cogniaux calmly dispossesses Dr. Kellogg's *E. muricata* of its name and gives it a new one, merely to save the name *muricata* in its connection with a Brazilian plant which he will bring into the genus from without: a kind of practice which, it seems to us, can hardly be condemned too strongly, having nothing to recommend it, and tending only to the further complication of our troubles with synonymy.

The "kind of practice" that Greene condemned was, of course, quite contrary to rule then as now, but was openly condoned for the sake of convenience. Probably Greene's continued insistence on the letter of the law hastened the crystallization of American opinion on this point, even when as in the following example (1887e, pp. 217–18) he accompanied his exhortation with an implied personal criticism:

Nuttall . . . discovered . . . the genus *Nemacaulis*, deeming the species two, and naming them respectively *N. denudata* and *N. foliosa*. When, a few years afterwards, it fell in Mr. Bentham's way to take up the genus, he seemed to find that the species was but one. Then, instead of following the ordinary practice of authors in such cases, and leaving the descriptive and quite unobjectionable *N. denudata*, Nutt., in its place, writing *N. foliosa* as a synonym, he displaced both the names of Nuttall and made himself the author of the species by writing *N. Nuttallii*; and this, which a fearless critic might call a bit of scientific iniquity, has been adopted and made his own by each American author who has, since Bentham's time, handled the Eriogoneae, as all our books and catalogues will show.

This little innuendo did not pass unnoticed. In the very next number of the Torrey *Bulletin* Asa Gray replied:

Professor Edward L. Greene . . . is certainly "a fearless critic," and in that capacity he attributes to (of all men) the late Mr. Bentham "a bit of scientific iniquity" which has been adopted and made his own by each American author who has, since Bentham's time, handled the "*Eriogoneae*"—that is the venerated Dr. Torrey, myself and Dr. Watson. As those of us who still survive are not likely, after this, to take notice of such statements, it may be well to consider for a moment what this iniquitous proceeding is. When Nuttall established the genus *Nemacaulis*, he indicated two species, *N. denudata* and *N. foliosa*. Bentham regarded the two species as one, in which he was probably right, the more so as Nuttall says they grow together and perhaps are not distinct; and to the combined species he gave the name of *N. Nuttallii*, obviously because neither of Nuttall's names was properly applicable to the species as he regarded it. This mode of proceeding has not been rare in similar cases, and in the present instance seems to be quite justified, if not actually demanded, by the rule that false names are inadmissible.

Sereno Watson was a favorite target for Greene's barbs. Watson, for example, described in 1875 a species which he called *Amarantus leucocarpus*. Twelve years later, deeming it unsuitable, he proposed to replace *leucocarpus* by *leucospermus*. Greene wrote (1887f, pp. 226–27):

. . . the author of these names says: "Opportunity is here taken to correct the name which was at first carelessly given to the species." But what is thus called a correction plainly amounts to the creation of a new name by which to displace an old one. . . . *Amarantus leucocarpus* is an entirely valid name, and as such rejoices in a dozen years of priority over the more nicely compounded new one. It has been the common property of the botanical world during these years, and the botanical public may not feel disposed to give it up; rather, may think it ought not to be given up. There have been in times past a few advocates of the principle that authorship of a botanical name is naturally invested with certain peculiar privileges of altering, recalling and replacing—a principle of which the disastrous results upon the literature of science, by inviting to an excessive multiplication of synonyms, are easily foreseen; and there is probably no author now living who would contend for any species rights of this kind. A name, if it be in one view so much the author's property that others must respect and use it, is also public property from the date of its publication, and he who attempts to recall a name should bear in mind that he has a public consent and approval to gain or lose; that whether he lose or gain his point with the public, he has made a synonym, and that, under the continually strengthening and prevailing conviction of the need of paying all respect to priority, it is most likely, indeed altogether inevitable, that the new and more accurate name will be the one to take a useless place among the synonyms.

Unfortunately Greene's position was not always so clear and logical, but sometimes dependent upon his inner convictions, which often derived from faith or emotion rather than logic. While condemning Watson for changing a name of his own, Greene about the same time scored another of Watson's proposals and offered his own substitute (1887i, p. 141). In proposing *Cuphea mesochloa* to replace *C. viridostoma* S. Wats., Greene wrote:

Now and then a name gets made which, it may be assumed, not even priority can save; and *viridostoma* must be relegated to that class, I think. It is incapable of any grammatical correction which will not alter its form beyond recognition as the same name. Or will the eminent critic [A. Gray, in *Proc. Amer. Acad. Arts* 22: 284. 1887] who last year allowed *Calyptridium quadripetalum* to pass, tell us that *viridis* has even "precarious lodgment" in the Greek lexicon, or *stoma* in the Latin?

Greene's arbitrary judgments about propriety of form, orthography, or derivation were often invoked by him in cases involving priority. For example, in a paper on the genus *Mara*, he included an explanatory footnote (1887a, p. 1):

Erroneously written Marah, by Dr. Kellogg who, as if writing an English name, follows the anglicised spelling of it given in the King James version of the Sacred Scriptures; and his mistake is copied by Watson and by Cogniaux. But the final aspirate in such Hebrew words is always omitted in Latin writing; and this not only because its presence is an obstacle to the declension of a name; it does not, even in Hebrew, represent any sound, and could be safely omitted in English as well as Latin.

At about the same time Greene, in proposing the name *Allocarya Echinoglochin* Greene (1887b, p. 15) to replace *Echinospermum (Echinoglochin) greenei* A. Gray, wrote: "In transferring this to what I am sure is its proper genus, I have felt at liberty to choose between the subgeneric and specific names imposed by Dr. Gray, selecting, for obvious reasons, the former."

Just why Greene supposed it proper to flout the rule of priority, merely because the plant happened to have been named for him, I am not sure, but this is by no means an isolated example of this kind of thinking on his part. Consider the following comment (1893b) on Sargent's (or Britton's) proposal to replace the name *Halesia* by a new name, *Mohria*: "In this choice he does not discover that he had made another homonym; but this is certainly the case; for as a spoken name—and the language of plant nomenclature is unquestionably a spoken language—*Mohria* is identical with *Morea*. . . ."

One of Greene's most notorious squabbles over fine points of nomenclature resulted from his insistence on a point that now seems trivial to many nomenclaturists, namely the grammatical composition of the specific epithet. When E. P. Sheldon published a number of new species of *Astragalus*, using epithets that Greene considered improperly formed, the latter (1895a, p. 76) sounded off scathingly:

This aspiring man has not only assumed that a mere beginner in systematic botany may unblushingly announce himself an authority upon so large and difficult a genus as *Astragalus*; he has shown such innocence of the principles of nomenclature as conceives that any pronounceable jumble of letters and syllables may pass for a specific name.

The first four which I shall here cite are each compounded of words taken from two languages. Syllables thus run together do not form words; hence they cannot be names, and, although for convenience I must cite them as if they were synonyms, they are not such; for what was never a name can not become a synonym.

Greene, logically enough according to his own understanding, proposed new names to replace those of Sheldon. The latter (1895), who was indeed at the time a young man, as implied in Greene's criticism, replied to the "ill-advised personal attack," saying that "Professor Greene's failure to keep pace with modern Latin scholarship, pardonable indeed in his case, has betrayed him into an uncalled for display of language." Greene had objected to such names as *lanocarpus* on the ground that they were half Greek and half Latin, but according to Sheldon *carpus* had indeed been taken over into Latin: ". . . while earlier and insufficient study had excluded the word from the Latin vocabulary, later research has shown its place to be as unquestionable as *brevis*, for example, which is also from a Greek root. Yet Mr. Greene does not object to *brevifolia* or *breviflora*."

This slur on his competence must have galled Greene, who replied (1895b, p. 122) with the quibble that:

Before printing this, our editorial friend [that is, the editor of the *Botanical gazette*] should have instructed himself on the important point, that dictionaries of Latin and of latinity are books of widely different scope and character. Latinity is Latin plus many other things, including whatever rubbish of barbarian and mongrel verbal jugglery and clap-trap with Latin endings of *us*, *a*, *um*, etc., those innocent of grammar have fabricated and left on written record in ancient, or mediæval, or modern times.

Returning to Sheldon, Greene continued:

QUERY. Will not old friends of the *Gazette* hide in deep humiliation confused faces, as they read on a page of this valued journal such brazen statements as that *brevifolia* and *breviflora* are mongrel terms, half Greek and half Latin, and that *cuspidocarpus* and *lanocarpus* are Latin legitimate and pure?

QUERY. When our two friends of the northwest, the Editor and the Astragalologer, combine their grammatic and literary skill in efforts to make another man appear given to hasty and unseemly speech in criticism, do not their sentences recall with delightful freshness, the homely half-forgotten fable of the pot that called the kettle black?

Another controversy in nomenclature, in which Greene became involved, was that centering around the use of the generic name *Jacksonia* Raf. (1808) in place of *Polanisia* Raf. (1819). Greene (1891, p. 174) had taken up the name *Jacksonia*, with the following remarkable statement: "There can be no doubt that the American plants were what Rafinesque regarded as the type of *Jacksonia*, notwithstanding that he named *Cleome dodecandra* as that type; for he supposed that name to have been given by Linnæus to the common American species; in which supposition he appears to have been in error."

As *Jacksonia* Raf. antedates *Jacksonia* R. Br., Greene (1893a) shortly transferred some 35 species of the Australian genus (*Jacksonia* R. Br.) to *Piptomeris* Turcz. About the same time N. L. Britton (1893) reiterated the fact that Rafinesque had written quite clearly "Jacksonia (trifoliata)= *Cleome dodecandra*, L.," and arguing that "in matters of nomenclature we must be exact," suggested that *Polanisia* be restored.

Greene (1894a) answered with an elaborate explanation of Rafinesque's real intention. His reasoning was as follows: Rafinesque, according to his own statement, proposed to publish about 30 new genera of plants discovered by himself, already "mentioned by authors," but not agreeing with the old genera in which they had been placed. *Jacksonia* was no. 7 of these 30 genera. It was presumably the plant (as already noted by Britton) that had mistakenly been referred by Michaux to *Cleome dodecandra*, and Greene argued that Rafinesque intended to write "Jacksonia (trifoliata) = *Cleome dodecandra* Michx." instead of "*dodecandra* L." Greene said we may be sure:

> . . . the author of *Jacksonia* had no foreign type in mind; that what he knew, by his own observation, to be a thing distinct from *Cleome*, and called *Jacksonia*, was the North American plant; and that it never at any time entered into his thought to give any Indian species of *Cleome* the rank of a new generic type. His only error—a small one, and of a kind which botanists are often betrayed into—was that of writing an L. when he should have written a Michx. after a certain plant name.

Such excursions into the psychic side of nomenclature must have given Greene something of a reputation for eccentricity. It is hard to realize today just what might have been regarded as an eccentric or unreasonable stand in nomenclature in Greene's time. Responsible botanists debated with heat—often more heat than light—over questions like that of the proper form of citation of names after an author has made a combination in a new genus; shall the name be cited with or without an author's name in parentheses? And is the second author's name to be cited to give him due "credit," or merely as a convenience in bibliography? Should a "starting point" for botanical priority be established, and if so should it be 1735, 1753, or 1762? The furor over this last, and very important, point, continued for many years.

Greene's own stand was for absolute priority first, last, and always. His insistence upon proper credit for authorship of genera has already been noted. He opposed any arbitrary "starting point" on the ground that such action denied due credit to the "real" authors of many taxonomic groups, but especially of genera. In his *Manual of the botany of the region of San Francisco Bay* (1894) he credited genera, when these had been known and named in pre-Linnaean times, to such authors as Dioscorides, Pliny, and Theophrastus, and to such lesser-known botanical lights as Catullus, Vergil, Nicander, Varro, and Galen.

In regard to the nomenclature of species Greene took a similar stand. In commenting on John Ray's 1660 catalogue of Cambridge plants, he showed that some 48 species had names of two words each, these having been "applied by Ray and his antecedents to the same species to which Linnaeus applied them. The Swedish nomenclator seems, indeed, studiously to have adopted the old binary names, in the place of framing new ones, wherever he could well do so" (1888, pp. 177–78).

In this matter of absolute priority the tide set against Greene. James Britten (1888, p. 262) showed the unreasonableness of any attempt to adopt these "binomials," saying:

> No one knows better than Prof. Greene that these names being composed of two words only is a mere accident. Many cited by Ray are of one word only, others are of three, four, or as many [as] was considered necessary for distinguishing the species from its neighbours. The binominal method, the reduction of nomenclature to a system, is one of the greatest of the reforms introduced by Linnæus, and the attempt to deprive him of it is not likely to be sanctioned by botanists.

During this same period, while the *Index kewensis* and Otto Kuntze's *Revisio* were in course of preparation, botanists began to realize for the first time that where hundreds of thousands or even millions of binomials were involved, sentiment for the past had to take second place. It was imperative to establish some practical starting place in time, beyond which modern botanical nomenclature might not go. A minority group in America, including the vocal Greene, felt that a stable nomenclature would be reached only through adherence to absolute priority. L. H. Bailey, always a believer in law and order but never a strong supporter of artificial systems, wrote in 1894:

> I do not believe that we are one whit nearer a stable, or what Professor Greene calls a "correct" nomenclature now than we were ten years ago. One generation cannot legislate for the next, particularly in matters of the usage of language; and I expect that the time is not very far distant when the old herbalists will have their day and the Linnean standpoint will be upset. But I hope that, as long as we are in the muddle, the new rules may be adopted on all sides as a soporific for the present generation—if haply they wear that long.

Greene's biographers have pointed out that in 1892 he was one of the three Americans, along with N. L. Britton and J. M. Coulter, named by the International Congress at Genoa as members of a standing committee on nomenclature. The Americans attending the Congress—Greene was not among them—presented a platform advocating a strong rule of priority and the adoption of 1753 as the starting point for botanical nomenclature. As Greene consistently opposed arbitrary starting points of all kinds, saying "history has none," his appointment to the committee seems to have stemmed from his long-continued interest in nomenclature, his always vigorous support of the strong rule of priority, and the mere volume of his writings on nomenclature, which in his time was not the independent discipline that it has since become.

The following year, 1893, Greene was named president of the Botanical Congress held at Madison, Wisconsin, but apparently his participation in the activities of the Congress, even those of the debates devoted to nomenclature, was small, and his interest in the formal rules of nomenclature even less. R. P. McIntosh (1961), quoting Greene's own words, implies that the latter took no part at all in the deliberations on nomenclature: "As far as the rules of nomenclature of the last botanical congress I have never read them; although I am a member of the International Commission. If you knew how these congresses are composed and how the business of them is done you would have no great deference to pay their rules."

The "Rules for Citation" adopted by the Madison Congress were reprinted in *Erythea* (3: 85–87. 1895), with editorial comment by Greene (p. 88). These rules recommended forms for abbreviations, punctuation, and other details for citation of botanical literature; they have since been adopted in all essentials by many national and international organizations, but Greene was not pleased:

> We notice that in case of joint authors the "and" is given instead of the usual short &. Why should any one conform to this sort of innovation in citation? Again, "Proc. Phil. Acad." is bad for the "*Proceedings of the Philadelphia Academy*," because the "Phil." ordinarily stands for Philosophy or Philosophical in bibliographical citation. It is quite a new thing to force this "Phil." to stand for Philadelphia. "Philad." is by long custom established for Philadelphia; and the custom grew out of a necessity, such as any accomplished bibliographer would readily perceive. . . .

This comment drew sparks from the editor of the *Botanical gazette* (20: 272. 1895):

> Professor Greene indulges in some characteristic hypercriticism in connection with the rules for citation adopted by the Madison Botanical Congress and Section G, A.A.A.S. Without finding fault with the rules themselves (for which relief much thanks), he takes a page to pick flaws in the abbreviations used in the examples which are supplied for the convenience of readers in interpreting the rules. Had Professor Greene even taken occasion to read the rules and the reference therein given, he would have seen that the abbreviations of authors' names, journals, etc., to be used in citation have not yet been reported upon by the committee. Criticism of this sort is an expression of querulousness, and degenerates into mere faultfinding, "as any accomplished bibliographer would readily perceive." We may, without offence we hope, remind Mr. Greene that kindliness of word and charitable judgment are vastly more effective than the petulance of this criticism or the rancor of his ungentlemanly attack on Mr. Sheldon in the same number of *Erythea*.

It is apparent from a survey of Greene's writings on nomenclature that they abound in inconsistencies and emotionally derived judgments; that they include many instances of the petulance, querulousness, faultfinding, and insistence upon trivia that his contemporaries found so annoying. It is nevertheless probably true that Greene's forceful and vocal insistence upon the importance of a "correct" nomenclature; upon the almost sacred necessity of giving due credit where credit was due; and in short upon the importance of an ethical and at the same time universal system of nomenclature, helped to focus the interest of American botanists upon nomenclature and to impress them with its importance. The forcefulness of his writing can nowhere be seen better than in his comments on Kuntze's *Revisio* and on "Dr. Kuntze's reviewers."

Criticizing a review by B. Daydon Jackson, Greene (1892b, p. 273) took a stand that seems vigorous and logical, although today taken more or less for granted:

> Dr. Kuntze has assumed too many of the prerogatives of a bold reformer, no doubt; but in all his volumes one may not find an example of a certain kind of pure and transparently vacuous assumptions such as of which Mr. Jackson's half-dozen pages are prolific. Here is one: "Genera which have been set aside because of their obscurity, can not be revivified by any later study." In the abstract this is simply saying that a piece of identification which one man has undertaken and failed to make, can not be made by any one: or equally, that what an earlier generation has failed to make out, no subsequent one need undertake. But, according to Mr. Jackson, even if it were otherwise, and one futile effort in such a direction did not preclude a later, "no sensible person would wantonly inflict a wrong on the botanical commonwealth by ascertaining the genera of old authors; as it can only be of antiquarian interest." Here, in the first clause, it is made a wrong even to acquire a certain kind of knowledge, if its acquisition be possible. Is then the present state of botany so precarious, that some kinds of knowledge must be absolutely avoided, lest wrong be inflicted on the science? This sounds like an echo of that proscription of research which men have said belonged to "dark ages." Nor is the second clause of the quotation more fortunate: knowledge of old authors in botany "can only be of antiquarian interest." I had always supposed that the antiquities of astronomy, for example, were of astronomical interest, and that the antiquities of another very old science, botany, were really of botanical interest. I can but wonder, after what Mr. Jackson here sets forth, if conforming to his proscriptive views, the great botanical libraries of London have lately made over to the antiquaries their hundreds of fine folios of the "old authors" in botany. . . .

Greene never acceded to what he considered to be the immoral proposals of those who would have conservation of generic names. When the German botanists, as an answer to Otto Kuntze, proposed conservation among other nomenclatural reforms, he entered a vigorous protest that expressed at the same time his tolerance (1892c, p. 285):

> While from one point of view it is somewhat alarming to hear that a considerable number of able and very prominent German botanists have thus openly professed a virtual rejection of every ethical aspect of the situation, the frankness with which the position is taken tends to make it half-respectable. Their attitude is certainly an honest and honorable one when contrasted with that of many British and some noted American botanists, who, while professing to be guided in general by the principle of priority, scrupled not on any occasion to suppress the genus of an opponent in favor of the later one of some friend, nor hesitated to denounce other people's restorations of "forgotten" genera, while often practicing the same virtue themselves. The man who, adopting wrong principles, is open and above-board about it, has some title to respectful consideration.

In his later years Greene wrote little on the subject of nomenclature, perhaps having become disillusioned with his American colleagues. When L. H. Bailey (1894) commented somewhat disparagingly on the current rules but advocated their general adoption "as a soporific for the present generation—if haply they wear that long," Greene (1894b, pp. 12–13) summed up the attitude that seems to have remained his for many years:

I can not but feel somewhat hopeful for the cause of correct nomenclature in America; not the cause of the "Rochester Code," necessarily, nor of any other formulated set of rules more or less complicated and inter-contradictory, but the cause of simple historic priority, exceptions being made only when a purely binominal nomenclature demands the suppression of an intractable generic name or a too long specific one; and these would not be found so very numerous. While I feel the warmest sympathy with, and allegiance to, every regulation voted in any Congress or Club of botanists, that makes for historic truthfulness, and is not at war with binary names, sound reason and good grammar, I still do not share Professor Bailey's expectations that the rules passed of late in America, may bring peace even for a time. That many of these have been too hastily formulated, and too inconsiderately passed by vote, is evinced by the fact that the very framers of them are soon found to ignore them, upon occasion. . . .

Greene as a Historian

As yet I have hardly touched upon Greene's interest in the history of his subject, yet it was this interest, and his scholarly exploitation of it, that brought him lively approbation in his lifetime and continues to enhance his reputation today. His principal historical work, the *Landmarks of botanical history*, for more than half a century has had no rival as a source of information on the development of botanical thought and botanical practice in Europe. His peculiar talents found their best outlet in the study, interpretation, comparison, and evaluation of past centuries. Few botanists have equaled Greene in his facility with languages, his wide knowledge of plants, his unstinting admiration of the past, and his ability to paint a moving word-picture, to set the scene that enables the modern reader to see through the eyes of Fuchs and Tragus and Valerius Cordus and the rest.

In his preface to the *Landmarks*, Greene set forth his philosophy of botany and his ideas on the writing of history. He distinguished sharply between botany for its own sake—"whatever has [a] bearing on the philosophy of plant life and form"—and the industrial or economic aspects of the study of plants. In quest of a starting point—a first landmark—in the progress of botany, he passed by the early writers on farming, gardening, and medicine

not because no traces of genuine botany occur in them; it is because we are in search of him with whom the leading idea is that of a philosophy of plant life and form. The first botanist is the first man who undertakes research upon plants *as plants* rather than as things useful or deleterious to man and beast; and the first landmark in the history of botany is the earliest book in which plants and plant organs are discussed each in relation to others.

The preface makes clear that *Landmarks* is not a history of botany. As Greene says, that would be the work of two lifetimes, one to master the science and the other to write the history: "For the history of no science can be made out, and presented in its perspective, but by him who first of all has mastered that science itself, in its completeness." Furthermore, knowledge of the past is very incomplete, and many data have been lost, so the decision to undertake the *Landmarks* was made as "permitting [the author] to evade the responsibilities of the consecutive historian, and leaving him free to bring into clearer light—and especially for study on the part of American botanists—the lives and teachings of those and those only among botanists of the past whose names are more familiar.

In a way this personal approach to history is a difficult one, Greene continues:

> for in this case quite as in the other, one must everywhere investigate individual merit, which is less apt to exist in proportion to a man's great contemporaneous popularity than in the inverse ratio of it. . . . But the annalist who leaves all these things as he finds them, reiterating popular laudation of the parasitic propagandist, and burying inventive genius yet more deeply in oblivion, deplorably falsifies history. Quite as little does he deserve the name of historian if his mistakes in this regard be those of ignorance; if they come of his having failed to discover merit because of its having lain under the pall of forgetfulness for a century or two.

Following this plan, then, Greene found his first landmark in the life of Theophrastus of Eresus, a Greek, a student of Aristotle, who lived in the fourth century before the Christian era. Like Greene himself, Theophrastus was a voluminous author, said to have written more than 200 treatises, of which but a dozen or so survive. In the nine books of the *Historia plantarum* are to be found the elements of Theophrastian botany, and from these same books Greene drew his account of the state and progress of botany in the time of Theophrastus. It is all here; the influence of Aristotle upon his greatest pupil; the state of civilization at this period; the elements of Greek botany; the morphological and taxonomical conclusions to be drawn from Theophrastus's methods and arrangements; the new discoveries in anthology and plant geography; the "list of facts botanical which Theophrastus saw, and in the main discovered, [which] embraces well-nigh all the first rudiments of what even today is universal scientific botany."

In succeeding chapters Greene followed the dim trail of botanical science through the Middle Ages, then devoted the rest of the book to the sixteenth-century "German Fathers." Beginning with the publication of Otto Brunfels's folio titled *Herbarium vivae icones* in 1530, when "the science of botany was born again," there is no dearth of landmarks in the sixteenth century; in fact the book deals mainly with the first half of that century. It is Greene's merit that he deals with each author individually, yet manages to convey, as far as knowledge will permit, the sense of continuity and the historical setting of each botanical work and its author.

In his biographies Greene is at some pains to trace each botanist's contribution to science through his innovations in phytography, the describing of plants; in anthology, the understanding of floral structures; in vegetative organography; in the morphology of the inflorescence; in taxonomy and nomenclature; and even in anatomy and physiology. He takes three pages to discuss the question of whether Brunfels regarded the dead nettles as congeneric with the true nettles, or merely lumped them all together as a concession to the half-taught root and herb dealers. Fifteen pages are devoted to the concepts of taxonomy as developed by Tragus, and ten to the discoveries of Valerius Cordus in flower structure alone. The reader follows the mental processes of the young Cordus as he begins to "perceive in the flowers of the whole long line of beans, peas and lupines, and of broom and cytisus, and laburnum tree, only more or less marked deviations from—mere modifications of—that pea-blossom type which Gesner had compared to the insect called a butterfly," and so goes on to coin the adjective *papilionaceous.* There is a breathless quality in Greene's prose, and if perhaps the people about whom he writes assume a certain heroic stature in his eyes one can forgive him, for it makes good reading and it rings true.

Another landmark, although not so labeled, is Greene's *Carolus Linnæus*, a small volume of 91 pages published in 1912, and based chiefly on memorial lectures delivered in 1907 at the celebration of the Linnæan bicentenary. When preparing a lecture to be given at a large gathering, Greene deliberately chose to emphasize not the work of the great master, but the man himself, "with citation of his struggles with adverse circumstance, and of the almost incredible patience, industry, zeal and resolution with which he conquered and rose to high renown." In the book the boy Linnæus begins his schooling and his botanical career; undertakes his university career and his trip to Lapland that was to prove so important to him; meets Boerhaave and Burmann and Clifford; and ultimately achieves his destiny. Less scholarly than the original *Landmarks*, this book is more readable. Still of interest to botanists is its final chapter on "Linnæus as an evolutionist." No trace of Greene's alleged prejudices against the doctrine of evolution appears in this chapter; instead he shows the liveliest and most objective interest in establishing the development of Linnæan thought. Consider for example his own statement after chancing on a remark in the *Species plantarum.* Of *Thalictrum lucidum* Linnæus had written, "It seems to be the product of its environment."[2] Greene was electrified by this: "this one discovery upon which I had now stumbled, seemed so much like a new revelation of the mind of Linnæus that within a very few days I had read every one of the 1682 pages of the edition of the year 1764 in search of other kindred expressions regarding the possibility of the descent of some species from others."

In the interpretation of the classical Greek of Theophrastus or the late Latin of Linnæus and his predecessors, Greene seemed to find the best expression of his genius. Here he was under no pressure to translate these texts that had endured for centuries or even for millennia; he had no equal in this interpretation of the past, and he knew it. His approach to history was through vivid biography; as he said, historians have been remiss in this respect; "mere epitaphic statements . . . seem only to bury more deeply out of sight the once living and active personality, and to relegate his very name to a still remoter place in the region of myth and shadow." The subjects of Greene's biographies are living people who go to school and church and visit one another, and love plants. No slightest implication of the old text escapes Greene's attention; perhaps he assumes too much, and puts too many thoughts into the minds of his subjects, yet almost always do these thoughts seem to follow from known facts. Brunfels was intensely a lover of nature and of plants, says Greene. This is clear, he says, because Brunfels at great expense to himself published new and lifelike engravings of plants, "as a prop to fallen botany." Here was devotion to an ideal, "a love of plants that was bent upon procuring faithful representations of them in books."

The *Landmarks*, and the little book on Linnæus, might have been written by quite a different person from the author of *Pittonia* and *Erythea*. In his historical writings there is little evidence of the faults that marred Greene's other botanical works: the pettiness, the defensiveness, the decisions hastily arrived at and too often subjectively justified; the presentation as gospel of that which was insufficiently documented. In writing about the work of others Greene took leisurely opportunity to explain, to present full justification and documentation, to present debate and rebuttal and to write fully and clearly.

Conclusion

The man that emerges from the pages of his own writings is a complex and paradoxical figure: scholarly and learned but prone to emotionally dictated decisions; meticulous in detail but in haste overlooking detail; troubled and some say tortured in mind, but kind in deed; perceptive to the highest degree but often unable to interpret his findings and curiously unaware in himself of the very qualities he criticized in others.

Perhaps it is true, as Bartlett wrote in 1916, that "few, if any, exerted a greater influence than he in the development of systematic botany in America." If this is true, then it is because Greene stimulated, goaded, and irritated his contemporaries. He was not a Gray, who by force of character and intellect dominated American botany in the nineteenth century; nor was he a Britton who could build a new institution into international importance in a lifetime. He was instead a pioneer who insisted that independent scientific work could be done as well in California as in Boston or in Europe; that plant nomenclature deserved some thought and study; that modern systematic concepts stemmed directly from the concepts of the past, and that accordingly it behooved all botanists to study the past. If Greene's own botanical work did not always measure up to the standards he set for himself and for others, it was nevertheless a catalytic agent that affected thought and aroused American awareness. In a very real sense he was the first American botanist to break away from European influence, and at the same time his influence helped bring about in America an increased desire to participate in botany on an international scale.

Notes

1. This is the correct (original) spelling; Greene always spelled the genus *Eschscholtzia*.
2. The Latin is *videtur temporis filia*. Linnæus, *Species plantarum* (1753), p. 547.

References Cited

Bailey, Liberty Hyde. 1894. Another side of the nomenclature question. *Erythea* 2: 10–12.
Bailey, Virginia Long. 1962. Revision of the genus *Ptelea* (Rutaceae). *Brittonia* 14: 1–45.
Bartlett, Harley Harris. 1916. The botanical work of Edward Lee Greene. *Torreya* 16: 151–75.
Brandegee, Katharine. 1893. The botanical writings of Edward L. Greene. *Zoe* 4: 63–103.
———. 1894. [Review of] Greene's Manual of the Bay Region botany. *Zoe* 4: 417–20.
Brewer, William Henry, Sereno Watson, and Asa Gray. 1876–80. California. Geological survey. Botany. 2 vols. Cambridge, Mass.: Welch, Bigelow, & Co., Univ. Press.
Britten, James. 1888. Recent tendencies in American botanical nomenclature. *J. Bot.* 26: 257–62.
Britton, Nathaniel Lord. 1893. In New or noteworthy North American phanerogams—VII. *Bull. Torrey Bot. Club* 20: 277.
Cronquist, Arthur. 1947. Revision of the North American species of *Erigeron*, north of Mexico. *Brittonia* 6: 121–302.
Dupree, A. Hunter. 1959. *Asa Gray, 1810–1888*. Cambridge, Mass.: Harvard University Press. 505 pp.
Ewan, Joseph Andorfer. 1950. *Rocky Mountain naturalists*. Denver: University of Denver Press. 358 pp. [For Greene, see pp. 45–66, portrait facing p. 45.]
Grant, Adele Lewis. 1924. A monograph of the genus *Mimulus*. *Ann. Missouri Bot. Gard.* 11: 99–388.
Gray, Asa. 1887. Botanical nomenclature. *J. Bot.* 25: 353–55.
Greene, Edward Lee. 1885. A revision of *Myosurus*. *Bull. Calif. Acad. Sci.* 1: 276–79.
———. 1887a. *Echinocystis Megarrhiza*. *Pittonia* 1: 1–4.
———. 1887b. Some West American Asperifoliae. *Pittonia* 1: 8–23.
———. 1887c. The species of *Zauschneria*. *Pittonia* 1: 23–28.
———. 1887d. The permanency of specific names. *J. Bot.* 25: 301–3.
———. 1887e. Bibliographical notes on well known plants. IV. *Bull. Torrey Bot. Club* 14: 215–18.
———. 1887f. Bibliographical notes on well known plants. V. *Bull. Torrey Bot. Club* 14: 225–27.
———. 1887g. West American phases of the genus *Potentilla*. *Pittonia* 1: 95–106.

———. 1887h. Some American Polemoniaceae. *Pittonia* 1: 120–39.
———. 1887i. New or noteworthy species. *Pittonia* 1: 139–43.
———. 1888. Botanical literature, old and new. II. *Pittonia* 1: 176–94.
———. 1891. Some neglected priorities in generic nomenclature. *Pittonia* 2: 173–84.
———. 1892a. Some American Polemoniaceae. II. *Pittonia* 2: 251–60.
———. 1892b. Dr. Kuntze and his reviewers. *Pittonia* 2: 263–81.
———. 1892c. The Berlin protest. *Pittonia* 2: 283–87.
———. 1893a. Corrections in nomenclature. I. *Erythea* 1: 114–15.
———. 1893b. [*Mohria* a homonym.] *Erythea* 1: 236.
———. 1894a. Dr. Britton and Mr. Britten and *Jacksonia*. *Erythea* 2: 6–9.
———. 1894b. Correct nomenclature, etc. *Erythea* 2: 12–13.
———. 1895a. Corrections in nomenclature. VII. *Erythea* 3: 75–76.
———. 1895b. Sundry propositions, commended to the consideration of the most northwesterly editor of the Botanical Gazette. *Erythea* 3: 122.
———. 1898. The genera *Polycodium* and *Batodendron*. *Pittonia* 3: 323–26.
———. 1903–6. *Leaflets of botanical observation and criticism*. Vol. 1. Washington, D.C.: n.p. 253 pp.
———. 1905. Revision of *Eschscholtzia*. *Pittonia* 5: 205–93.
———. 1906. The genus *Ptelea* in the western and southwestern United States and Mexico. *Contr. U.S. Natl. Herb.* 10: 49–78.
Heckard, Lawrence Ray. 1960. Taxonomic studies in the *Phacelia magellanica* polyploid complex. *Univ. Calif. Publ. Bot.* 32: 1–126.
Howell, John Thomas. 1943. Sertulum Greeneanum. *Amer. Midl. Naturalist* 30: 6–39.
Jepson, Willis Linn. 1923–25. *A manual of the flowering plants of California*. Berkeley, Calif.: Associated Students Store. 1,238 pp.
———. 1931. Edward Lee Greene. In *Dictionary of American biography*, vol. 7, p. 565. New York: Scribner.
Johnston, Ivan Murray. 1925. The North American species of *Cryptantha*. *Contr. Gray Herb.* 74: 3–114.
Jones, Marcus Eugene. 1912. Greene and his species. *Contr. W. Bot.* 14: 49–50.
———. 1929. Greene. *Contr. W. Bot.* 15: 25–27.
Kistler, Ellen Downey. 1936. Bibliography of the botanical writings of Edward Lee Greene. *Madroño* 3: 328–48.
McIntosh, Robert Patrick. 1961. The Greene-Nieuwland herbarium at the University of Notre Dame. *Proc. Indiana Acad. Sci.* 71: 235–38.
McVaugh, Rogers. 1943. Campanulaceae (Lobelioideae). *N. Amer. Flora* 32A: 4–12, 15–25.
Munz, Philip Alexander, and David Daniels Keck. 1959. *A California flora*. Berkeley: University of California Press. 1,681 pp.
Rollins, Reed Clark. 1941. Monographic study of *Arabis* in western North America. *Rhodora* 43: 289–325, 348–411, 425–81.
Sheldon, Edmund Perry. 1895. *Astragalus lanocarpus* and *A. bajaensis*. *Bot. Gaz.* 20: 277.
Steyermark, Julian Alfred. 1934. Studies in *Grindelia*.—II. A monograph of the North American species of the genus *Grindelia*. *Ann. Missouri Bot. Gard.* 21: 433–608.
Woodson, Robert Everard, Jr. 1930. Studies in the Apocynaceae.—I. A critical study of the Apocynoideae (with special reference to the genus *Apocynum*). *Ann. Missouri Bot. Gard.* 17: 1–212.

Landmarks of Botanical History

PART I

Preface

Any discussion, or any indication even, of landmarks in the history of botany must needs be preceded by a somewhat careful enquiry into the nature and purposes of the science as such. Wherein does botany, as a science, essentially consist? With this question unanswered it were impracticable either to indicate the origin or trace the progress of it.

In the most extended use of the term, all information about the plant world or any part of it is botany. According to this view, all treatises upon agriculture, horticulture, floriculture, forestry, and pharmacy, insofar as they deal with plants and their products, are botanical.

What many will consider a better use of the term is more restricted. In this use of it there will be excluded from the category of the properly botanical whatever has no bearing on the philosophy of plant life and form. For example, that wheat, rice, and maize agree together as to that anatomical structure which is called endogenous would be a fact of purely botanical interest. Quite as clearly such would be the proposition that all three belong to the natural family of the grasses; or this, that each represents a genus; or that the roots in all these plants are fibrous, and of only annual duration. But if it be said that wheat, rice, and maize as food products are of supreme importance to mankind, the affirmation is as completely unbotanical as the several foregoing examples are perfectly botanical. It is strictly an economical consideration.

If such a distinction between botany and plant industry as I have sought thus to illustrate be received as legitimate, the province of botany is easily circumscribed and its scope clearly definable. In any event, for the purposes of the present work our definition of this science shall be that it occupies itself with the contemplation of plant as related to plant, and with the whole vegetable kingdom as viewed philosophically—not economically or commercially—in its relation to the mineral on the one hand, and to the animal on the other.

Such a distinguishing between the philosophical study of plants and the industrial does not dispute, but rather establishes, the existence of a wide border domain between science and industrial art where botanist and industrialist work side by side upon plant subjects; it may be sympathizingly and intercommunicatively, or it may be ignoring each other's presence; a domain within which nevertheless each should be in touch with the other, because each may be, and ought to be, helpful to the other, as supplying some data useless for his own purposes but of significance in relation to the other's aims. The recognition of this border-land domain illustrates also, if it does not again directly argue, the distinctness of the two realms of botany and plant industry. Here one may observe that the distinction itself would seem less marked if he who is to set himself to the work of an economic or industrial botanist would first of all equip himself with a knowledge of the principles, and cultivate an interest in the aims, of philosophic and scientific botany; so that the industrial botanist as author might always have two reports to make upon any piece of research, one that should be of economic interest, the other one of interest botanical. It may be that this idea will be found to presuppose the conjunction of the philosophic bent of mind with the industrial, a combination of two qualities of mind as rare in the world as genius itself, and no less desirable.

In quest, therefore, of a starting point—a first landmark—in the progress of botany, in my understanding of the science, one may pass those authors by who professedly treat of plants from the utilitarian point of view, whether they write of agriculture, horticulture, or the materia medica.[1] Passing these by, I say, though by no means as not meriting the botanist's attention; for all matters relating to the qualities of plants naturally interest him, unless he be of that school in power a century ago, but now declining in influence, according to whose teachings nothing but dry morphology was of any import. Moreover, to him who, like the farmer, the woodsman, and the primitive pharmacist, has much to do with plants industrially, philosophic ideas may occur about the vegetable kingdom as a whole or in part; and every such idea, though crude, perhaps even erroneous, is a concept essentially botanical. Quite as perfectly so is the distinguishing of different kinds of plants, and the practice of grouping like kinds together under one common (generic) name, which is not only universal, but even a necessity, with those who, like the farmer and gardener, have much to do with a considerable number of plants of different sorts. People following these occupations have actually a systematic botany, with a nomenclature, families, genera, and species, all their own. So then if, in the search for a possibly early botanical landmark, the writers upon farming, gardening, and medicine are to be passed by without serious consideration, it is not because no traces of genuine botany occur in them; it is because we are in search of him with whom the leading idea is that of a philosophy of plant life and form. The first botanist is the first man who undertakes research upon plants *as plants* rather than as things useful or deleterious to man and beast; and the first landmark in the history of botany is the earliest book in which plants and plant organs are discussed each in relation to others. If there is any attempt to distinguish and define plant organs, or any suggestions about the probable functions of any of them, any indications of how plants may be distinguished from minerals on the one hand, and from animals on the other, any attempts to correlate plants as like and unlike, and that upon some recognized principles—in any and all such endeavors, we recognize the activities of a philosophic mind in its attempts to solve problems not economic but scientific. In the author of any such treatise upon plants, however imperfect or even crude his notions may seem to us, we have nevertheless the author to whom belongs the name of botanist, as in the vocabulary of the sciences that name ought to be defined.

What is here undertaken is not a history of botany. There is no purpose of presenting in chronological succession the long line of the contributors to the upbuilding of this science, with an account of the best contributions each has made. That would be the work of a lifetime—indeed, of two lifetimes—for the history of no science can be made out, and presented in its perspective, but by him who first of all has mastered that science itself, in its completeness; and the domain of botany however philosophically restricted remains vast, insomuch that one lifetime seems requisite to the mastery of it in its several departments. A second lifetime should, then, be given to him who should be required to write its history.[2] And still the presentation of a complete and accurate history of botany would remain impossible. Important data are wanting, and hopelessly so. For one example, more than two millenniums ago a highly philosophic and very extensive treatise upon plants was indited which alone among books of its kind has survived the passing of all the centuries. The author of it cites other authors on the same topic whose books, then extant, are long since lost. This writer had also in early life a very illustrious teacher who instructed him orally in botany among other subjects, and who also wrote two volumes of botany both of which passed into oblivion more than two thousand years since. How much, then, of the Theophrastian botany is that author's own? What of its principles are his only as having been imparted to him by his great friend and tutor Aristotle? What passages of the work are but compiled from writings of a more remote antiquity, with which Theophrastus may have been familiar, of which even the authors' names have perished? Questions like these serve but to admonish one of this, that the earliest beginnings of the science do not admit of discovery. The same is in a measure true of comparatively recent periods. The annals of our science, as gathered in hitherto, reveal no more thrilling epoch than that of the sixteenth century. Some of the best known authors of that period, Brunfels, Tragus, Fuchsius, studied, besides the not so very many printed books about plants that were then extant, numbers of old mediæval manuscripts from which they brought forth and quoted many a botanical idea, several of them well advanced beyond the ideas of the ancients as we know them. No annalist of a later age seems to have had time or disposition to ascertain how much of the assumed new and original botany of those German fathers—so they style them—was taken out of old mediæval manuscripts which, although they may still be extant, later historians have neither consulted nor troubled themselves to enquire after.

Contemporarily with those German herbalists there flourished in Italy a learned professor, first at the University of Padua, then at Bologna, afterwards at Pisa, whom people regarded as the one peerless botanist of the time. His university lectures were received as oracular, and students came to him from almost everywhere in Europe; yet Professor Luca Ghini published nothing. His supremacy as botanist of the first half of the sixteenth century is attested by tradition only. In the very next generation after him, several of the chief luminaries of the science were men whom he had trained, and to one of them, Cesalpino, there is now everywhere accorded the praise of having created the epoch of modern botany. To what extent is Cesalpino's great work, *De Plantis*, a product of the mind of Ghini?[3] The question is one that forces itself upon us and is perhaps the more interesting because hopeless of ever being answered.

Such are a few examples of what the annalist who would be just and truthful will often find himself in need of knowing, yet can never ascertain; and they intimate but too pointedly the impossibility of any such thing as a complete and faithful history of any period when once that period is past. It was as realizing this, and also as wishing to avoid presumptuousness, that the present writer declined to undertake a history of botany and chose the title of "Landmarks" as permitting him to evade the responsibilities of the consecutive historian, and leaving him free to bring into clearer light—and especially for study on the part of American botanists—the lives and teachings of those and those only among botanists of the past whose names are more familiar. This plan bears on its face the appearance of an easier task, and such it really is; though that it is a less responsible undertaking may be doubted, for in this case quite as in the other, one must everywhere investigate individual merit, which is less apt to exist in proportion to a man's great contemporaneous popularity than in the inverse ratio of it. It will indeed be found to have happened now and then that the genius who has discovered principles has also elucidated them, applied them to the construction of a system, and gained for himself and his principles the credit and the honor that were due; but perhaps rather more commonly the genius discovering principles has but quietly made the simple announcement of them, has died scarcely honored, and has been almost forgotten, when some other, just far enough above mediocrity to see the value of the principles, and possessing industry and ambition to bring them forth and build on them the system which the principles themselves suggest, gets the credit of the whole, is thought to have created the epoch, and enjoys the fame. But the annalist who leaves all these things as he finds them, reiterating popular laudation of the parasitic propagandist, and burying inventive genius more deeply in oblivion, deplorably falsifies history. Quite as little does he deserve the name of historian if his mistakes in this regard be those of ignorance; if they come of his having failed to discover merit because of its having lain under the pall of forgetfulness for a century or two.

The historian who is both conscientious and discreet will give small heed to popular opinion about any particular man or epoch. Neither the adulation of the multitude is of any profound import, nor its voiceless indifference. Its outspoken opposition and denunciation may even be the highest praise. Such being any writer's estimate of popular opinion regarding botanical eras past, his readers will be surprised neither by chapters that are iconoclastic, nor by such paragraphs as reveal immortal honors due to men whose names had almost faded from the roll of fame.

It has seemed to me desirable that, in the tracing of these outlines of botanical history, some prominence ought to be given to biography. The reader or the student of a book can never take the deepest possible interest in it so long as its author is unknown to him, or, as it might be said, known by name only; though that is but an empty phrase, for to know a person by name only is not to know him at all. A fair knowledge of the whole career, early and late, of the author of a literary or scientific masterpiece not only intensifies, as I said, one's interest in the work, but is most helpful to the understanding of it, if not indispensable to the full comprehension of it. To this natural and reasonable demand on the part of those who would like to learn something of the history of botany, the historians have not well responded. In most cases they give in a single paragraph, or even in a short footnote, the year of a man's birth, that of his demise, perhaps the name of the institution whence he had his degree, and of those in which he occupied a professor's chair, and so ends the biography of a man who may have been a genius and the creator of an epoch in science; mere epitaphic statements, which seem only to bury more deeply out of sight the once living and active personality, and to relegate his very name to a still remoter place in the region of myth and shadow. There are probably few botanists of this twentieth century who have the most vague conception of what a single one of the earlier master builders of our science was like in his personality and character. To most of us they are too nearly mythical, and mayhap less livingly pictured in our minds than are some of those really mythological personages that men believed in four thousand years ago. It will be seen that in these studies of the landmarks, I give some prominence to the biographic aspect of botanical history. This has been done at great expenditure of time and thought; but I have felt that the end was an extremely desirable one; and I have little or no doubt that these sketches of the lives of great promoters of our science who lived in other centuries will be received by many as among the most welcome and instructive of my paragraphs.

E.L.G.

United States National Museum
Washington, D.C.
2 July 1907

Introduction to the Philosophy of Botanical History

Any history, in order that it shall merit well the name and answer the requirements, must have its definite philosophy. History presupposes some end awaiting attainment, and in itself it would seem to be a well-connected record of the thoughts, the words, and the deeds that have either furthered or hindered the attainment of that end. It does not, however, assume that the actual makers of history recognize the ultimate end. That is something which not even the wisest can foresee otherwise than dimly and with much uncertainty. The aim of the science of botany, for example, being the fullest and most perfectly systematized knowledge of the plant world philosophically considered, it still is true that not one in a hundred among the rank and file of real contributors toward this ultimate purpose has had it definitely in view. The great bulk of botanical work hitherto accomplished has been done in detached pieces, and by such as had only proximate ends before their mental vision. And for the very best of research work no more is needed. He who carefully investigates and puts on record the whole life history of a dandelion or of a violet; who gives the whole anatomy of a few mosses, reeds, or sedges, or indicates the morphologic distinctions between the pollen grains of hollyhock and those of thistles, or traces the development of either one; who brings out the philosophy of the twining of a morning-glory stem, or indicates the organogeny or the functions of the stipules of vetch and pea; or who after years of critical field study catalogues, with original notes and observations, the flowering plants—or the flowerless ones—of a single county, or of the watershed of any lake or stream, every such laborer contributes to the stock of botanical knowledge, and this without reference to personal comprehensiveness of botanical view, or a looking to far-off ultimate ends.

Upon the historian of botany, however, it seems to devolve that he shall have some forecast of what botany in its perfection as a science shall be like; for in practice he sits in judgment on each epoch and decides whether as an epoch its tendency was more to the advancement of the science than to its retardation; from which kind of procedure it becomes certain that some ideal of perfection is in his mind. Every writer on botanical history must have his philosophy of that history, unless he content himself and hope to satisfy his readers with disconnected historic fragments.

It may be useful to survey in this connection, though with the utmost brevity, the methods of several representative historians of botany.

Tournefort (1700), eminent among even the greatest promoters of botany, was also its historian. The first fifty pages of his *Institutiones*[1] are occupied with an abridged history of the science during two thousand years preceding his own date. The history is prefaced by a definition. There are two parts to botany: the knowledge of plants, and the knowledge of the uses (virtus) of them. It is a distinguishing between systematic botany and economic. He says the distinction must be carefully noted. He denies to the properties or uses of plants any part in, or influence upon, the systematizing of them. A systematized presentation of the known facts constitutes the first beginning of every science. There can be absolutely no botany at all without systematic botany. These are Tournefort's ground principles. From them we shall gather his philosophy of the advancement of botany. The plant world can never come to be well known until sounder principles of classification shall have been established, and the whole aggregate of known plants shall have been grouped over again upon those better principles. The long line of the most noted authors before him had classified plants in all kinds of ways, some according to characters of the roots, some by differences of stems and leaves, one by fruits alone, another by the qualities and uses of the plants; another grouping them according to their places of growth, or ecologically, as we now say. Seldom were the systems of any two authors at even approximate agreement. Often that of an individual author was a compound of inconsistencies, utterly inharmonious within itself. As to that very first necessity of botany, rational system, confusion seemed to reign. The flower was an organ hitherto little studied, and scarcely yet appealed to in the art or science of plant grouping. Two or three botanists of a century earlier than Tournefort had suggested that, after all, not in roots or stems or leaves, but in the flower there might perchance be found the key to a more satisfactory method of plant classifying. He undertook now a new systematization of the world of plants, everywhere appealing to anthology insofar as by the presence of flowers and fruits the appeal was possible. Ceasing to take as criteria the qualities of plants, or even the characters of their vegetative organs, and by giving special and close study to both flowers and fruits instead, with judicious co-ordination and use of the characters of both, will botanical system henceforward obtain best furtherance.

With neither the strong points nor the weak ones in Tournefort's system, nor with its success or failure, are we here concerned. All that will engage us now is his conception of botany as a science in process of further development and improvement; in other words, what he would have taken to be the leading philosophic threads of botanical history. They would probably be two, at least as chiefly conspicuous; for during his career his mind had been much occupied with (1) the thought that better and more firmly established generic groups had been the most crying need of botany from the earliest times, and (2) that such more acceptable and more securely established genera would result from the defining of them according to morphology of flower and fruit, the consideration of vegetative organs being omitted as far as possible. So then, from his own outlook over the past of botany and from his best forecasting of its future, they have helped it forward most who have most contributed to a better anthology and carpology, and such obtain with him foremost places in his epitome of botanical history. The fullest credit is given to all botanical travellers to distant shores who have contributed to the enrichment of botanical gardens, and to the making of illustrated folios representing flowers and fruits of plants alien and rare. Meanwhile how small consideration Tournefort accorded to plant anatomy and physiology is evinced by this, that in his history he has not a line to spare for the names of Grew and Malpighi, great promoters though they were of the cause of plant organography in general, and well entitled to rank among creative botanists.[2] We may chance to find historians of less comprehensiveness of view than Tournefort, and some with greater.

Two generations later a countryman of his, Michel Adanson, sketched less succinctly than Tournefort had done the history of botany. It forms the more important part of a voluminous preface to Adanson's *Familles des Plantes*.[3] Eighty-five years after its first publication this History was reprinted, with many augmentations which the author had left in manuscript at the time of his death.[4]

A man whom all nature in her every phase attracted and engaged, but still first and last and always a botanist, Adanson's horizon was a broad one. He was also a botanist with a specialty, that of discovering how genera naturally stand together in larger groups that may be called families. On the whole, and if such distinction be allowed as legitimate, he was a systematic botanist; most pronouncedly such. But the sketch that he gives of the history of botany is neither partial nor one-sided. He reviews the science as having progressed along many lines, not one of them unimportant. But since it is families of plants that he is now to treat of at length, the foremost thought in his mind in the writing of a history of botany as a preface to the book is, that he may demonstrate the early rise and tardy progress of this very idea of plant families. It is not, however, the history of that one aspect of botany merely that he writes. Something a little too near the one-idea history was what Tournefort had presented; even as one may today say of the latest of all the historians of our science, that he came rather too near to excluding from very thoughtful consideration almost everything except the history of plant anatomy and physiology, and of the taxonomy of the cryptogams.[5] Adanson appears to have realized that no one part of botany is alienable from any other part; that the history of a part of it cannot be written as disconnected from that of the other parts; and therefore, connectedly with the presentation of whatever had been done before his time towards a natural correlating or grouping of genera, he brings into view not only that line, but others along which botany has made progress, paying due respect to every kind of effort that makes for a fuller knowledge of the plant world.

With the main purpose, then, of finding early traces of the recognition of something like natural families, Adanson analyzes briefly and in chronological succession more than sixty leading authors, beginning with Theophrastus and ending with some who have been contemporary with himself in the middle of the eighteenth century. Assuming that these analyses are correct, one may read connectedly, with small sacrifice of time and as it were step by step the progress which, up to Adanson's time, had been made in the grouping of genera into families—or whatever else one may choose to call such groups; and, while it will be regarded an important one among several threads that the philosophical and impartial historian is bound to follow, I know not who besides Adanson has ever attempted to trace this one except for a very short distance.[6]

And the next thread of botanical story which Adanson picks up and follows is one that lies close alongside the aforementioned. The earlier endeavors to indicate groups of genera were made anteriorly to the time when structure of flower and fruit had come to be accepted as the guide. By what marks did those pioneers of classification guide themselves in their attempted groupings? By way of answer I give a short selection from Adanson's own more detailed report of the matter.[7] Lobel (1570) he says was guided by general resemblances, size, qualities, and uses;[8] Porta (1588), by ecology, forms of roots, of leaves, and vegetative organs generally;[9] J. Bauhin (1650) has 40 classes, by appeal to all organs, as well as to properties of plants and their ecology;[10] Rivinus (1690), inflorescence, calyx, and corolla;[11] Boerhaave (1710), general aspects, ecology, leaves, fruits;[12] Haller (1742), cotyledons, calyx, corolla, stamens, seeds;[13] Gleditsch (1749), flowers, insertion of stamens;[14] and so on through a list of some sixty writers, each a celebrity in his day as the author of some new attempt at system in botany. Of a situation like this, and one so necessary to be brought forward in any history of the science, Sachs knew nothing, neither even Sprengel.[15]

There is another outlook upon the progress of botany that is almost peculiarly Adanson's. At the beginning of the analysis of each author's treatise he notifies us how many different kinds of plants each man knew, or had under discussion in his book—Theophrastus 500, Hermann 5,600, Tournefort 10,146, Ray 18,655, as examples—thus recognizing at every step the important consideration that, other things being equal, the greater the number of plant forms a man knows, the safer his conclusions as to the interrelations of all, or of the members of any group of them.

Of course the specialist in plant anatomy, little interested in the whole chain of plant relationships—he to whom 500 species were enough for his own purposes—may chance not to be in sympathy with these searchings of all corners of the earth for new plants.[16]

But to what comprehension of the whole of botany has such a mind attained? It would have something like its parallel in the astronomer, if such astronomer there had been, who had deprecated the labor involved in the discovery of the planets Uranus and Neptune upon the plea that there was already enough to do with the rings of Saturn and the canals of Mars. At least somewhat like that is the attitude of the historian who makes light of the work of plant discovery and plant description. To ascertain, as Adanson was at the pains of doing, what number of species a given systematist had known, was the only possible way of informing himself of the comprehensiveness of the man's view of things. And as to the ideal and ultimate perfection of knowledge of the vegetable kingdom, that is manifestly impossible of attainment, so long as a single type, either living or fossil, remains undiscovered and undescribed. It is a principle which not only justifies, but, in the interests of the science as viewed without partiality or prejudice and comprehensively, imperatively demands the most thorough exploration of every field, the equipment of the best possible botanic gardens and herbaria, and also the highest possible perfection of the art of phytography, that is, plant diagnosis or description.

Of incalculable usefulness to the student of systematization is phytography. Its purpose is that of enabling the botanist to measurably complete his knowledge of this and that group of plants, only some proportion of the species of which he has been able to see, inspect, and study in the living state. All that a man may learn about plants in twenty years of field work, supplemented by all that gardens and herbaria have to show, will not amount to the knowledge of any more than a fractional part of the specific membership of as much as one of the many families or considerable genera of higher plants. For the rounding out of his knowledge—general, even superficial knowledge—of whatsoever plant alliance, one is always dependent on descriptions. It is one of the most important conditions of all general botany; one that was fully recognized at the beginning; also one that will forever remain. It has always been and it will always be, that a good plant description, placed before one mentally equipped for the exact interpretation of it, is decidedly more satisfactory than the usual herbarium fragment of a plant.

Yet one word as to correct and incorrect phytography. One who has a new plant in hand, and who knows it thoroughly from root to seed, may use the whole of an octavo page and the half of another in what will be supposed to have been an attempt to picture this type in words. This same plant may be much more distinctly pictured to the mind of the trained and habituated phytographer in one-fourth that space or even less, by using the set terminology of descriptive botany. This was invented for the two-fold purpose of saving space and increasing perspicuity in plant definition. In its most nearly perfected state it is quite modern; and the history of this terminology is a very significant part of botanical history. The discovery of each term was, in its day, a distinctly botanical discovery and an important one; yet the Sprengels and Sachses have given rarely a hint of the evolution of terminology.[17] To have made out lucidly its history would have been a heavy tax on precious time. Adanson almost alone, it may be said, has not neglected it. It was seen by him that in a well-devised scheme of botanical history an account of the development of descriptive terminology and the art of describing should find place. Accordingly in these mere outlines for such history he charges certain authors with having described plants poorly; others he remarks upon as having described them fairly, while to here and there he gives the praise of having described them well.

One must not pursue further the subject of Adanson's topical divisions. Those presented may suffice for what I wished to illustrate, namely his appreciation of what ought to enter into the making of a history of botany. Synoptically placed, those few of his topics of which I make mention are:

1. History of grouping of genera as classes or families
2. History of accepted criteria of affinity
3. Progress in discovery of new types
4. Development of phytography and its terminology

This mere beginning of Adanson's scheme of history will enable me to indicate the contrast that subsists between his and that of Sprengel, whose not unpretentious work in two volumes was given to the public one year after Adanson's death.[18] Out of the four Adansonian topics named above, only one, the third, obtains good treatment at the hands of Sprengel. The first and second are blank with him; while under the fourth one may gather little beyond some salient points in the history of anthology. Anatomy and physiology are so discussed as if not inseparably connected with botany proper. Indeed in his partitioning of the science into the two compartments of the Systematic and Structural he expresses his mind to the effect that while Botany proper is a part of Natural History, the consideration of the inner structure and physiology of plants belongs rather to Physiology.[19] His treatment of these, as developed in the course of the seventeenth century, is nevertheless full and explicit. But it is progress in the discovery of new types, history of botanical exploration at home and abroad, and the enrichment of botanical gardens, which more particularly engage Sprengel; and, as Adanson had been more interested in the development of the idea of plant families,

Sprengel, as a devoted Linnæan, gives himself to the investigation of the history of genera and species. All the long way from Theophrastus to Linnæus Sprengel lists new types generic and specific as discovered and published by prominent authors; so that a fair chronological history of at least the European Flora is furnished; and these lists of each man's discoveries form so large a part of the body of his work that its principal index is an index of genera and species.

There is no need of pursuing beyond this brief initiative our examination into the somewhat diverse philosophies of botanical history that have hitherto found expression. Every one may be permitted to have his own. In the present treatise exception will be taken to one assumption made by all earlier historians, that for the earliest intimations of anything looking in the direction of the science of botany we must have recourse to those oldest pieces of literature in which plants are more or less freely mentioned. Adanson, for example, does not begin botanical history without naming Orpheus, Musa, Solomon, Hesiod, Homer, Metrodorus, and Hippocrates, who as poets or as physicians wrote of plants. Sprengel has among his initial chapters one bearing the title "Flora Biblica" another "Flora Homerica," another, "Flora Hippocratica"; and these chapters of Sprengel are botany, even very interesting botany;[20] but this is not saying that there is botany in the Sacred Scriptures, or in the poems of Homer, or in the medical writings of Hippocrates. They are, however, classic texts upon which a man of Sprengel's rare accomplishments may write botany. And yet I seem to apprehend certain rudiments of a science of botany in those ancient pieces of literature, the real substance of which those authors of botanical commentary on the Bible, on Homer, on Virgil, and the classics generally, have completely overlooked. Let me repeat it, that in several pieces of very old literature there are legible traces of a science of botany, traces of which even learned and botanically instructed commentators seem to have failed to take due note.

Here, let any reader who has supposed that certain sciences had their beginnings in the minds of men who wrote books, banish, if possible, that idea. No opinion ever held by a multitude of people was more groundless. If, according to the definitions given by authorities, science is classified knowledge and classification is the process of distinguishing and separating between things like and unlike, then there are certain of our sciences the earliest rudiments of which are almost among the very necessities of human speech. It will not be easy to imagine a tribe of wandering savages on any continent or in any age unused to the distinctions of plain, hill, mountain, or spring, brook, river, lake, and ocean. Their very languages will show that their mind had wrought out these perfectly solid and immovable foundations of the science of Geography. Long subsequently the man of enlightenment, he who knows how to commit thought to writing, takes this old and hitherto unwritten classification of the diversities of the earth's surface, gives it logical statement, dignifies it with the Greek name Geography, and then proceeds to build as on very old yet firm foundations his nobler edifice. He may or may not recognize it that those indispensable group names, plain and mountain, lake and river, are but a heritage to his scientific geography from a very far off antiquity, from an antiquity the history of which neither has been written, nor ever will be. It were well, however, that the geographer should perceive it that the real first beginnings of his science are not with the author of any book, but that they antedate all writing.

Botany, as certainly as geography, had its initiative in primal man's distinguishings and separatings between objects appertaining to the world of plants. The fact that in the rudest and simplest dialects of primitive peoples there exist group names for botanical entities establishes this. It is improbable that there ever was a primitive language, other than that of some arctic tribe, in which there did not occur words equivalent to tree, bush, grass, or to trunk, branch, leaf, fruit, root; and every one of these is the name, not of an individual object, but of a group of like objects. Each is a general—a generic—name, and each testifies most clearly to observation, comparison, reflection, generalization, and also either the invention of a new word, or else the more extended application of an old one, which, in as far as science is concerned amounts to the same thing. It is possible to trace to a time that lies well within the period of modern botany the first detection and first naming of that kind of organ which we call a stipule;[21] but no one will have the hardihood to propose that we may trace to its first employment the term leaf. Yet this term, which one may never hope to trace to its origin, is as strictly botanical as the later term stipule, and more important. Furthermore, there was a time when the very term leaf—or at least its equivalent in some lost language of a primal race—first came into use. And still further, the mental processes by which a Malpighi arrives at the distinguishing between the stipule and the other parts of the leaf, and those by which the unknown primal investigator came to distinguish between leaf and the stem or branch that bears it, are the same. Neither was more nor less scientific than the other. Each equally with the other had done a piece of strictly botanical research. This is not affirming equality of intelligence for the two, or questioning that he of the later time was capable of solving many problems of plant life impossible of solution by him of the earlier era. Also the motives leading to examination and distinguishing may have been quite different: he of the more recent period was actuated it may have been by that scientific curiosity, that mere zeal for knowledge, which often fires the cultivated mind; he of the primeval time was impelled perhaps by sheer necessity. He is much dependent on the plant world for life's comforts, even for its necessities. One part of a tree is of great use to him for one purpose, another part for a very different purpose, a third being of no use. Therefore from his utilitarian point of view it becomes manifestly needful that the different parts of plants be distinguished and each different part named. Language demands the introduction of such terms. But the mental processes, I repeat it, are the same in either case, and without respect to the actuating motive. It is all work of examining, comparing, distinguishing, segregating, and naming the segregates. Every step in the procedure of either is scientific. If one is tributary to a science of botany, so is the other. And if these reflections seem to indicate that scientific botany may be, as to its first elements, older than all literature, what of it? There is but one point of view from which it will be disputed, namely that which regards man as having made his first appearance on earth in a condition of advanced intelligence, with a well-developed language, and also bearing a divine commission to assign names to all manner of natural objects at first sight.[22] When one notes the perfect silence of the historians as to the possible origin of the most common and universal botanic terms, one seems forced to conclude that they accepted this doctrine of the sudden and inspirational derivation of them; and that then, as if unwilling to say so, they evaded the subject by going about the completely different and really quite irrelevant task of cataloguing the trees, shrubs, and herbs mentioned in the Bible, giving them the appellations due them according to the nomenclature of Linnæus's *Species Plantarum.* This was all a mere matter of giving the Linnæan Latin names of certain plants in place of their more ancient Hebrew names. It was not approaching by so much as one step the origin of botany, but rather, as I have said, evading the search.

Assuming that the simplest and most universally employed botanic terms entered into human speech not all at once by sudden and supernatural illumination of one particular mind, but one after another as a part of the natural and gradual evolution of language, it will be conceded that they had been formed and in use during long ages of human existence that preceded the invention of writing. And the chief botanical interest attaching to very early writings will be, not in that they furnish a few score Hebrew or Greek names of plants which the well-skilled botanist of a recent period may translate into the terms of modern nomenclature; it will be in this, first of all, that they incidentally record names of some plant organs. Such words as fruit, seed, branch, leaf, and root occur, and these seem to reveal it that plants in numbers have been looked into and studied organologically, and with such success that these names of different parts of trees and herbs are already an indispensable and a firmly settled part of every language. Moreover, the terms tree and herb, grass and grain tell as plainly another story, that of a prehistoric distributing of plants in groups according to resemblances. These two kinds, or at least two phases, of botany are in the writings of Moses and of Homer, and perhaps more valuable because there only incidentally, that is, without botanical thought or purpose in the minds of the writers themselves. They only happen to give us, as through a window accidentally left open, a view in which we see individual plants consisting of named parts or organs, and also assemblages of individual plants, some spoken of as grass, some as herbs, some as thorns, others as thistles, some as bushes, others as trees. Though it be no more than a passing glimpse that one has gained, it is enough to excite curiosity, and to suggest a number of queries legitimately botanical as to just what, in so primitive a time, may have been the full meaning and acceptation of this or that morphologic or taxonomic term as thus early in general use. Such questionings may not necessarily be idle or useless. There being no room for doubt that as far back as the time of Homer, and even of Moses, there was at least here and there a person somewhat specially skilled in the knowledge of plants, how would such a one have applied, for example, the term root? How many things, in his mind, would have been included under that name? What, in a word, might have been his definition of a root? Possibly we shall never know. Neither is it wholly impossible that we may some day ascertain it, at least approximately; for not so very many centuries after Homer specialists in plant knowledge began to write books upon the subject. Some of those books are still extant, and in print; though they have been made too little use of thus far by our historians, some of whom appear to have been disposed to divide the honors of elementary plant organography between Adam and Linnæus, which was an easy way of evading an important though most difficult part of botanical history. In the writings, I say, of the earliest of professedly botanical authors there would be reasonable expectation of finding a clue to that primitive conception of the root which was theirs who introduced the word into speech; for always the first work of him who is ready to reform and rebuild a science is that of showing wherein the prevailing opinions are at fault. To him nothing is more necessary than this. Our appeal in this instance must be made to Theophrastus of Eresus, whose writings on the philosophy of plant life and form are the oldest that are extant. As a controversialist this philosopher is of the mildest type, more apt to suggest, urbanely, that an old opinion may be wrong than bluntly to pronounce it false. His whole treatment of the subject of the roots of plants reads as if he had gone to work stealthily to undermine an old and everywhere received opinion that roots are simply the underground parts of plants. He names two or three familiar species which, as he reminds his readers, produce roots that are aerial, or at least not subterranean. Then he cites, and very well describes, certain subterranean parts—bulbs and corms, we call them now—which he thinks hardly ought to be considered roots. That Theophrastus openly discredits the doctrine that a root is a root because of its being subterraneously located is proof enough that it was the doctrine commonly received in this time. We are also perfectly warranted in believing that the exceptions he takes to it are his own and new; because no writer in ancient times was more careful than he to attribute to their proper authors any new or remarkable opinions which met with his own approval. But, that the primeval understanding of the root was that which I have supposed is again attested by its universal prevalence in our own time among people who have not been initiated into, or influenced by, the botany taught in our schools. Such peoples, dwelling in all parts of the world, if engaged in farming or gardening hold to a certain classification of farm products, and are wont to speak of grain crops, root crops, etc., using the last named expression without ever a suspicion that a potato is a kind of branch, and an onion a kind of bud. Beyond doubt a very great majority of the inhabitants of the earth today, if questioned upon the matter, would answer promptly, and fearless of contradiction, that whatever part of a plant grows beneath the soil is its root; and if any remotely domiciled rustic between Nova Scotia and Patagonia should remark that a white potato is a tuber and that onions are not roots but bulbs, we should know without parley that his abandonment of the principles of primeval plant organography had been brought to pass under the influence of modern book or school.

The survival of these primitive notions about the subterranean organs is more interesting than the origin of those notions. The tardiness of their displacement by a more rational organology is to my mind one of the curiosities of botanical history. That most complicated and difficult of organs, the flower, began to be well understood as early as the dawn of the eighteenth century; but at a time when, by the aid of better microscopes, the important function of stamens had been brought to light, and the doctrine of the flower thereby revolutionized and nearly perfected, it still remained that the rhizomes of iris, the bulbs of lilies and tulips, and the corms of crocuses were called roots by all the botanists; this also some two thousand years after Theophrastus of Eresus had suggested that it might not be very good organology.[23] And as for our historians, I have not found with one of them any intimation of who it was who at last solved for us the hard problem of an acceptable definition of a stem, the definition which at once compelled the recognition of subterranean stems as being stems, not roots. In my view this has always appeared to be one of the most signal triumphs of organographic research. Using the term understandingly and comprehensively, organography is more than half of botany. It is the whole foundation and framework of the science, and a good deal more than that. The progress of botany all along is largely identifiable with the advancement that has been made in the knowledge of plant organs; yet it is just this which one is able to learn least about from the historians. This statement must be qualified by the admission that, as regards that sudden leap forward which anthology made early in the eighteenth century, Sprengel is quite explicit, though he gives little indeed of its earlier history. It is also acknowledged that the story of the rise of microscopic organology, and its progress down to the middle of the nineteenth century, was given by Sachs, and with such fullness as to make it occupy more than half his entire volume of the *History of Botany*. Still these are but separate and disconnected chapters in the real history of organology.

If I here indicate this incompleteness of the history of botany as hitherto presented, it is not because I dare hope in these landmark chapters to make good the deficiency, though earnest and laborious effort is made to show how I think it may be done.

That prehistorically and primevally there existed not only an organology of plants but also a classification of the familiar kinds has already been suggested; and the proposition may here receive clearer statement. Moreover, certain somewhat stilted and pedantic views rather widely prevalent respecting systematic botany as of recent origin, no less than the interests of a truer philosophy of botanical history, seem to call for a vindication of this thesis.

Owing to the profusion of plant individuals on the face of the earth everywhere, the bewildering diversity of their forms, and the usefulness of them to man, it was never possible for men, at whatever stage of mental development, to intercommunicate concerning plants without having group names for them. Words that should have application to particular assemblages or kinds of plants were among the earlier necessities of language; and to speak of plants under group names is nothing less than to speak of them as already classified. The classification has necessarily preceded the invention, and the adoption into language, of the collective name. By way of illustration I select, out of a hundred or two of plant names which in our English speech are as old as the language itself, the word "clover."[24] It tells its own story. It was applied to certain plants which were seen to have this common characteristic, that each leaf was made up of—cloven into—three separate, equal, and in every way consimilar leaves. I say leaves in order to avoid being anachronistic, because leaflet is a term of really very modern invention, one unknown in English, and without its equivalent in any other language, at least of Europe, until the middle of the seventeenth century. Now clover was from the first the name of an assemblage of individuals; if of individual plants of several somewhat different kinds all exhibiting the common leaf character then it was what it now long has been—a generic name. It ought to seem superfluous to say that clover is just as much a generic name as *Trifolium*, and that white clover, red clover, and alsike clover are as perfectly binary specific names as *Trifolium repens*, *T. pratense*, and *T. hybridum*; but, as I have intimated already, the curious notion is here and there prevalent that a genus is not a genus, nor a species a species, until it obtain a Latin name. I have thought desirable to indicate thus plainly the incontestable fact that to the most primitive and untaught of herdsmen and cultivators, in their close dependence upon many members of the plant world, generic names and specific are as much a necessity, and as certainly in every-day use, as they are with us their school-taught posterity who call ourselves botanists. The true philosophy of botanical history seems to call for special insistence on this fact, as also that the viewing of a number of related genera, and the speaking of them under a family name, is likewise of a very remote antiquity. The English collective plant name "pulse" is as old as the language itself,[25] as covering under a monosyllable all the sorts of peas, beans, vetches, and lentils. It is nothing less than a family name, invented as a means of briefly designating the whole natural group of those cultivated plants of various genera which, in recent botany, are called Papilionaceæ.[26] Ancient Latin writers, to whom many genera of umbelliferous plants were known familiarly, saw plainly their interrelationship and called the whole assemblage of them the Ferulaceæ, naming it after the well-known genus *Ferula* which, as a genus, is represented by several species in the Mediterranean flora. And all this the Latins had only borrowed from a still more ancient Greek botany, for the Greeks had known as well the genus *Ferula* under the name *Narthex*, and were used to speak of the whole line of related genera as the Narthecodes.

From these two or three lucid examples of the naturalness of plant classifying taken from the records of antiquity, let us proceed to make some enquiry into like usages as they obtain among the most untaught in our own time. It is improbable that there may not be found in every country of the Old World peasant peoples who, entirely uninfluenced by books or schools, have nevertheless each some rudimentary system of botany, some terms expressive of their own classifying of plants, at least such kinds of them as they have much to do with, whether as herdsmen, as cultivators of the soil, or as woodsmen. An American student, however untravelled except in his own broad country, may have gathered even here illustrations enough of the principle now under consideration.

To the colonists and early settlers of a new country no native products of the soil are more important than the trees. Timber, lumber, wood for all kinds of building and fencing purposes, for the construction of bridges, vehicles, and household furniture, not to speak of fuel, bark for tanning purposes, and in autumn mast of nuts and acorns for the fattening of swine for the slaughter—these are among the reasons why early settlers always located their first domiciles along the edges of great forests. And now, if we remind ourselves of certain conditions of the first colonists who came to these shores from western Europe three centuries ago, we shall realize that, while they found themselves in the midst of a land bountifully supplied with timber, the particular kinds were new and strange to them. Nothing was quite the same as anything that they had known in the Old World; and no kind of information would have been more welcome to these colonists than that relating to the enduring and wearing qualities of the woods of these different kinds of strange trees. Every kind was untested, and there was no one to teach them. All had to be learned by trial and experience. Yet not quite all; for, to a band of colonists of three centuries ago, coming to these shores from England, there must be credited such knowledge of English trees and timber as was usual with Englishmen of that period, a knowledge that would be of some service to them as American colonists notwithstanding that American trees were of a much greater number of species, and none quite identical with any European kinds. They had brought with them across the sea a knowledge of oak, walnut, chestnut, beech, elm, linden, and some other trees. As for the chestnut, the beech, and the linden, they found but one kind of each here, and these not very notably unlike their congeneric European species. The settlers would naturally expect to find the American trees of these sorts available for the same economic purposes as their European allies. Neither as to the aspect of the trees nor the qualities of their wood was there so much difference; but with those very important timber trees, the oak and the walnut, the case was different. In place of the one European kind of walnut, the Virginian forests presented them with at least a half-dozen, each strikingly unlike the Old World type, both as to characteristics of foliage and fruit, and as to color and qualities of the wood. If one type of these peculiarly American walnuts bears today the name of White Walnut,[27] it is undoubtedly because the first settlers of Virginia, taking it for a probable equivalent of the English Walnut for lumbering purposes, found its wood to be by comparison much lighter in color, and named the tree, after the usages of lumbermen, by the color of its wood. The Black Walnut[28] in like manner obtained its name from the almost blackish hue of its wood compared with that of the tree of Europe.[29] And both these names bear distinctly the marks of an early colonial origin; for no native American woodsman of however early a period would have known the wood of the European Walnut so as to have made the comparisons.

From this representation of colonists as practical woodsmen—beyond all cavil an essentially faithful representation—it appears that men without the least training in school botany, exploring the woodland resources of a new continent with none other than utilitarian ends in view, find systematic botany an indispensable necessity, create for themselves a serviceable system of woodland taxonomy, make themselves the pioneers of taxonomic research in the new field; this not, however, as using the terms taxonomy and classification, not even as necessarily knowing so much as the name of the science which they are practicing. Let us distinguish mental processes. Nothing more is here needful. He who is occupied with testing wood or timber as to its economic usefulness is doing the part of the industrialist. He who compares one sort of living tree with another, noting by what marks of habit, of bark, of foliage or of fruit the two may be distinguished, and who determines to call one of them by one name and the other by some name that is different, is doing exactly the work of the botanical systematist. This man may never have learned a syllable of the terminology employed in schools of botany. He may not have heard the Latin name for oak, for maple, for poplar, or any other genus of trees, or even the word genus; but he is a botanical systematist nonetheless; and since his business obliges him to be this he proves the utility of botanical system. It is not possible for the occupations of the farmer, the herdsman, or the lumberman to be carried on without botanical classification and a fixed nomenclature of both genera and species. Therefore those thus engaged have never at any time in history waited for the schools and the school-trained botanists. They have made their own botany, have established both system and nomenclature; and these, insofar as they had proceeded, the professionals when they came upon the scene adopted. The two, that of rustic, of mountaineer, of herdsman, and of woodsman, and that of the schools, are as essentially one botany, as certainly one in kind, as wild pears, wild apples, and wild grapes are respectively one in kind with their cultivated and improved offspring of the orchards and vineyards. If this be true, then the annals of botanical science have another beginning than that which our annalists have assigned it.

When once it is seen that group names for plants are as old as language, and that these very names establish it that men always in all ages classified the many plants with which they had to do, there is another matter which immediately calls for careful investigation, that is, the parts of the plants to which rude primeval botanists looked for the marks by which to range their plants in convenient groups. We have already seen that Adanson alone among historians perceived that attempts had been made down through all the centuries to group plants by other data than those of flower and fruit. In bringing this fact into view, and by citing a long line of early authors in attestation of it, he was fearlessly contradicting, and at the same time successfully controverting, what his contemporary, Linnæus, had said when in the warmth of his zeal for the great Cesalpino he had pronounced him first in the order of time among real systematists.[30] The truth about Cesalpino was simply this, that he had been the first to attempt an orderly arrangement of the plant world by universal appeal to the fruit and seed; and that alone would still have been the superlative of praise, doubtless well merited. But that the Cesalpinian system seemed incomparably superior to every one that had preceded it could never become a warrant for saying that those systems antedating it might be left out of view altogether, as never having been systems at all. I can conceive of nothing which science more inflexibly exacts of every scientific man than truthfulness. She cannot permit an enthusiastic fancy to take the place of fact. But there have been successive generations of botanists since Linnæus who, as if swearing by his authority as if he had been infallible, have seemed to have no idea that any plant classifying ever was attempted upon any other than that anthocarpological basis which now for some three centuries has been steadily in use. As a mere prejudgment it is deeply seated in the botanical mind of today, that nothing can be done, or ever could have been done, in the direction of an orderly arranging of the world of plants but by appeal to characters of flower and fruit. And along with this prejudice there dwells another as deeply ingrained, namely, that the flower was the same thing to botanists of four hundred years ago, if not to those of three thousand years since, which it is to us; whereas not yet two centuries have passed since the flower began to be known. Our classifying by flower and fruit was fairly established while as yet the corolla was regarded as the principal part of the flower, and was in fact the synonym for flower, without even its name corolla.

Something may be done towards undermining these prejudices by giving a few examples of primitive systematizing as undertaken while as yet the flower, as to its essentials and its functions, remained an undiscovered organ.

For a good illustration of classifying by leaf characters alone, those of flower and fruit being quite ignored, we need go no farther back than the later years of the sixteenth century and the first quarter of the seventeenth. Let us limit ourselves to the forty years intervening between 1583 and 1623; also inspecting certain pages of two of the widely known and thoroughly approved professional botanists of the time, Rembert Dodoens[31] and Caspar Bauhin.[32] The genus Clover, in ancient Latin *Trifolium*, in Greek *Triphyllon*, already referred to in this chapter, is an ample one with the authors named. Bauhin's book contains names and descriptions of some sixty species; and since both he and Dodoens are almost as strict adherents of binary nomenclature as was Linnæus himself who came into this field of nomenclature a century later, it will be easy to show what they received into the genus *Trifolium* by presenting here two opposite columns of binary names. In as far as they admitted to their Clover genus genuine clovers as we now understand them, the reproduction of their names need not be made.

Dodoens (1583) and Bauhin (1623)	Recent Names
Trifolium bituminosum	*Psoralea bituminosa*
T. odoratum	*Melilotus officinalis*
T. corniculatum	*Lotus corniculatus*
T. cochleatum	*Medicago orbicularis*
T. palustre	*Menyanthes trifoliata*
T. acetosum	*Oxalis acetosella*
T. hepaticum	*Hepatica triloba*[33]

It ought here to be noted that for the combining of *Melilotus* with *Trifolium*, Dodoens is responsible but not Bauhin, who at this point saw fit to abandon the trifoliolate leaves as essentially and without exception conclusive of membership in *Trifolium*. He does not, however, as others had done before him, accept the melilot species as constituting a genus of their own, but places them all as members of the genus *Lotus*, where also some species are trifoliolate, others not so.

I observe also that if only the first four of the species of the above list had gained admission to *Trifolium* along with the clovers proper, one might have thought it probable that some dependence, after all, had been placed upon the floral structure; for in that case the authors would have had a *Trifolium* composed of papilionaceous plants exclusively. But neither in the defining of the genus nor in the description of a single one of the about sixty species of Bauhin's *Trifolium* is any mention made of the morphology of the flower. And by the admission of gentianaceous, oxalidaceous, and ranunculaceous types into that genus it is placed beyond question that in his mind the genus was limited by nothing else but the herbaceous nature of the plants, ternate foliage, and dry fruits. I say dry fruits, because in Bauhin's book the strawberries, as typically trifoliolate as the most genuine of clovers and as herbaceous, stand in closest juxtaposition to them, and it is manifest that their juicy, berry-like receptacles, with seeds all on the outside, saved *Fragaria* from being merged in the *Trifolium* of that author. And in this giving so much attention to the fruit where flowers were wholly ignored, we see the influence of Cesalpino's great treatise; for Bauhin and Cesalpino were contemporaries, in a manner, the former younger by thirty years.

All through such books as have here been cited one reads the resoluteness of purpose and the hard perseverance with which men labored to improve botanical system by studying and comparing texture and duration of stems, and above all else the morphology of leaves; a very crude system at its best, but system of some sort there had to be; the flower was still virtually unknown; the fruit was barely beginning to be appreciated in its usefulness to taxonomy; therefore the vegetative organs, chiefly the leaves, were most commonly allowed to be decisive.

The appeal to leaves was not, however, first thought of in either the seventeenth century or the sixteenth. Even then it had been more or less in vogue for three or four thousand years that we know of.

As a good example of an antique genus of trees based on leaf characters alone, with no dependence on those of flower or fruit, one recalls the Daphne of the Greeks. The original Daphne with them was the sweet bay, *Laurus nobilis* L., the type species of the genus *Laurus* as now received. Its foliage, being lance-shaped, of a somewhat leathery texture, with no marginal indentation, represents a rather common leaf type, and the tree is evergreen. Now there was a period of Greek history within which there was not only some travel to foreign lands, but even some written reports about the natural products of other countries; and there is the most convincing evidence that every new tree or shrub that came to light that was both evergreen and clothed with oblong or lanceolate entire leathery leaves was at once relegated to the genus Daphne, was named as another kind of laurel. Here is a list, possibly not a complete one, of trees with *Laurus* foliage which ancient Greeks, on that account alone, had referred to that genus:

Laurus nobilis, *Nerium Oleander*, *Nerium odorum*, *Avicennia officinalis*, *Rhizophora mucronata*, *Ruscus Hypophyllum*;[34] six species of so-called *Daphne* or laurel, belonging to five different genera, and representing as many different families, all anciently accepted as of one genus, because the foliage in all was much the same, and for no other reason whatsoever. And this again reminds us that in eastern North America, where there is no laurel, we have a number of kinds of native shrubs that have always been called by that name, just as if they had been members of the genus *Laurus* to which they are in no wise related. If we ask ourselves how this false naming came to pass, the answer is, that at the time of the early colonization of this new continent the old Greek idea was still dominant, that whatever bush or tree had a laurel foliage was a good enough laurel. The colonists brought that idea hither, and naturally enough applied the name to our kalmias and rhododendrons one and all.

This classifying by foliage was never received as anything like a general principle everywhere to be applied. So far from that, the rude primitive groupings were accomplished here and there under the sole guidance of the fruit, though in the main only as to none but its most superficial characteristics. Among fruit-bearing trees the apple tree was perhaps the oldest and most familiar type, unless the pear be excepted; and as new kinds of fruit trees in the course of history became introduced into Europe from other lands, every kind, the fruit of which was of considerable size and of something like the sphericity of the apple, was called a kind of apple tree, a member of the genus *Malus*, as it was called in Latin speech and writing. Here is a partial list of the kinds of *Malus*, or apple tree, that find record and description with several early authors, excluding however, the true apples, which were of many varieties all with binary names:

Malus Armeniaca	*Armeniaca vulgaris*[35]
M. Persica	*Amygdalus Persica*[36]
M. arantia	*Citrus aurantium*
M. limonia	*Citrus limonium*[37]
M. medica	*Citrus medica*
M. cotonea *M. cydonia* *M. aurea*	*Cydonia vulgaris*[38]
M. Punica *M. granata*	*Punica granatum*
M. Indica	*Zizyphus jujuba*

In outline, the history of the development of such a genus *Malus* as the above is this. In primeval southern Europe they had the common apple tree, *Malus communis*, and, from the beginning of the historic period at least, they had it in many cultivated varieties. The fruit was *malum* with the Latins, the tree *malus*. Then, as other kinds of trees were introduced from the East having spherical or ovoid fruits not too small for apples, their fruits were also designated as kinds of apples, and the trees as species of malus. To us who, with also several generations of our botanical ancestry, have become accustomed to a greatly improved classification, such a piece of systematizing as the above list of apple-tree species exemplifies cannot but seem absurd; but the presentation of something of that kind was necessary, partly in order that we might realize from what small and simple beginnings our later and better systems of carpological classifying have been evolved, also partly as demonstrating the groundlessness of the Linnæan hypothesis that classification by fruit characters took its rise with Cesalpino, and as late as the end of the sixteenth century.

I am unwilling to dismiss the subject of early and practically pre-Cesalpinian classifying by fruit without having given one more illustration of it. For this purpose I shall again advert to the taxonomic procedures of the early Virginian colonists. I have cited the case of their having found there, in place of the English Walnut, two allies of that tree, and that they named these new kinds Black Walnut and White Walnut respectively; also that this naming of the kinds was made not in reference to any morphological characters of the trees, but to that of the colors of the wood, this having been done quite after the manner of self-taught woodsmen, whereas school-taught botanists would have assigned names suggested by organographic marks. But this all relates to nothing else but the making of specific distinctions and the assigning of specific names. When we ask ourselves by what marks they were able to refer these new trees to the genus of the walnuts, we obtain but the answer that it was by those of their fruits, these in such degree resembling that of the one kind of walnut before known to them as to warrant the conclusion that the trees were of the walnut kind, as they would have expressed it, rather than of the oak or chestnut kind.

But our colonists' experiences with the native American oaks, if they had been more fully recorded than they were, would have been still more interesting. As English woodsmen only one kind of oak can have been well known to them. In Virginia they can not have failed to meet at once with about a half-dozen sorts, most of them in aspect exceedingly unlike the English Oak, so much so that they can not reasonably be supposed to have identified them with that genus of trees at all until after close inspection. One of the sorts displayed to them the foliage of the chestnut tree, another that of the laurel, still another the leaves of a willow. The chestnut-leaved kind had not at all the bark nor the wood of chestnut trees, but of oaks, rather; therefore these first observers of the tree would hardly have needed to appeal to the fruits in order to satisfy themselves that this new tree was but an oak, merely imitating the chestnut as to its foliage. But among the other kinds, such as had neither foliage nor bark nor wood in any way answering their idea of an oak tree, they can not have determined them to be oaks by any other note in each but that of its fruit.

That which I have thus far hypothecated concerning early Virginian colonists in relation to native Virginian oaks is demonstrable as something more than even the most rational of hypotheses. There is documentary evidence of the historic truthfulness of all, and more than all, that I have here but intimated as probable. That these men, forced by circumstance to make trial of the timber of trees new to them, did early recognize as oaks certain kinds most unlike what they had known as oaks, in all except their fruits, is attested by a colonial list of names of new American oaks which was published when the colony was but two generations old. I refer to Banister's *Catalogue*.[39] This contains a list of binary names of Virginian oaks, such as *Quercus alba*, *Q. rubra*, *Q. Hispanica*, *Q. castaneæfolia*, *Q. salicifolia*. Now while a casual reader of the catalogue cited would, without a second thought about the matter, attribute those five names to Banister, it is extremely unlikely that any one of them was invented by him. It is next to certain that the whole five are his mere translations into Latin of the oak names that he found in use among the colonists. Perhaps the plainest proof of this is, in that by turning those five binary names back into English you get precisely the names by which five common oaks are known to dwellers in that same region now, two hundred and twenty-seven years after Banister's having written his list. It is really evidence that is incontestable. To dispute it would be to affirm that the names were made by Banister himself, in Latin, then turned into English for the use of the woodsmen settlers; that these had been waiting sixty years or more for the professional botanist to come and tell them by what names to call their several kinds of oak; each part of which proposition, like the whole of it, is absurd. Under pressure of necessity, and from the outset, they must have begun to learn the different qualities of the wood or timber of those strange new kinds of oak. One or two of them were found comparable with the familiar oak of the mother country as being hard, durable, subserving the purposes of the builder, the wheelwright, and the cabinet-maker; another, not subject to decay when set into the ground, useful for posts; still another durable only when used for bars, rails, and like purposes; and there may have been a fourth and fifth kind excellent for winter fuel, but nearly worthless otherwise. No man will pretend to believe that colonial woodsmen and handicraftsmen, learning by degrees the different qualities and uses of our various American oaks, did not immediately assign a particular name to each particular kind. The important industries of house-building, boat-building, cabinet-making, the constructing of vehicles, the building of fences, and the providing of the winter's fuel, all demanded quite imperatively that there be a well-ordered and generally accepted system of woodsmen's nomenclature of oaks as well as of other genera of timber trees. So it came to pass that all important trees everywhere, in America quite as elsewhere, had their established names before the arrival of the writers of floras and silvas; and there is many a kind of tree the Latin name of which bears the mark of its having been originally the woodsmen's vernacular name for it.

Hitherto this fact of the universal existence of a crude primitive system of plant classification—one that antedates all botanical writing, a system that is in vogue today all over the world in out-of-the-way places, in complete isolation from the influence of colleges and universities—appears to have remained unnoticed by botanical writers. At least, I have met with no allusion to the fact. I therefore doubt that it has entered into the minds of botanical thinkers in recent times that such untutored yet effective and useful plant taxonomy exists, and must have existed primevally.

There will be readers enough to whom this thought will be new and somewhat startling. The fond conceit has long prevailed, that there was never anything in the world that could be called science until some three centuries ago, or four, at the farthest. Among several ideas about the botany of the past—ideas very widely, almost universally, entertained, though without the least warrant from history—I shall here mention but the following: that plant genera did not obtain fair recognition until Tournefort, nor species as distinguished from varieties until Linnæus, nor families before Adanson. Now if, according to the present thesis, the beginning of the receiving and naming of common plants in groups is ancient beyond all possibility of discovery, then no author can be credited with, or any date be assigned for, the beginning of the recognition and naming of either genera or species. What great men like those just named accomplished for the improvement of system in botany was, the better delimitation of several anciently accepted genera, and the laying down of certain rules and principles by which they thought all plants, known and unknown, might be arranged in groups more nearly according to their affinities. Assuming that the rules and principles were philosophic, all this was immensely to the advantage of classification; but when for the two-fold purpose of emphasizing the principles and making the new system easy to learn, they caused each genus name to be printed in large type, in the middle of the page, occupying a line by itself, then close under that the formal statement of its characters as a genus, and after that and only less conspicuously the species names, each occupying a separate paragraph, they were by this rigid formalism inaugurating, though they knew it not, an era of didacticism which now after two centuries has degenerated into an almost gross pedantry which rules systematic botany at present well-nigh universally. To illustrate the supremacy of this pedantry let me suppose that some accomplished botanist of Italy or France or Germany, having a new genus to propose, ignores all the usual post-Tournefortian formalities in his naming and defining of it. This will mean only that, declining to follow established typographical usages, he does not place the new generic name in large letters in a conspicuous place above all that he has to say of it, but begins his paragraph with a statement that the type is new, thence proceeding without a break to name the marks by which the genus is distinguishable from all its allies, then directly adding, in the same type and without the formality of an initial capital, even in the middle of a line, if it so happen, the name by which he proposes that the genus shall be known; all this followed, still without breaking the paragraph, by whatever else he may have to say about the plant or plants of this new genus. Between such simple uncapitalized compact taxonomic paragraph as I have supposed, and the familiar stereotyped form of naming and defining a genus, there is at first glance the appearance of great dissimilarity. As to the meaning of the two, and the information that is conveyed, there is no shade of difference between them. The plant type, supposing it to be the same, is as fully described and as certainly named in the more simple paragraph as in the one that is ostentatious. All that the botanical scholar can learn from the one he may learn just as perfectly and just as promptly from the other. Really the differences between the two are hardly more than typographical; yet notwithstanding this, it is probable that forty-nine out of every fifty botanists of today, if not even a much larger proportion of them, would in part fail utterly to perceive that the simple unostentatious paragraph which I have supposed, with generic name in small type set in the midst, had been intended as the publication of a new genus; and it is as probable that those of the forty-nine who did perceive the author's intention would deliberately ignore the paragraph, under the plea that the name and characters of a genus printed in a style so very unconventional must not be admitted to answer the requirements of publication. The genus must be treated as unpublished. This, be it noted, will be the same as to order that a new scientific fact be, in as far as possible, suppressed for the reason that certain familiar usages as to type and paragraphing were not followed in the publication of that fact. It will be regarding form of expression as superior to the facts expressed; will be allowing individual whim or fancy to ignore important matter; will make for the establishment of shallow pedantry in place of solid information and the use of plain good sense. I have stated this case hypothetically; but there are provinces where the situation is today actual.

But the greater number of my hypothetic forty-nine, as I have said, will even fail to discover that a plant group has been named and defined on any page where the modern typographical conventionalities are not in evidence. I think that among such as are quite proficient in systematic botany there are many whose impression of the printed pages of Ruel,[40] of Dorsten,[41] and of several other classics of sixteenth-century botany would be that they give no account of families, genera, or species; and this only because great authors had not then learned to make a separate paragraph for every group, and to print the names of genera and of species in type different from and more conspicuous than that used for the descriptive passages. There is no question of the superior convenience of our modern style of printing taxonomic matter; still, for the mistaking of mere incidentals for essentials by people professedly scientific, it is not easy to frame excuse. But the psychologic fact is well established that men do in this wise err, and that there are multitudes of biologic taxonomists whom familiar usage has completely deceived into thinking that no name is generic unless printed in large letters; multitudes of botanists who will have been startled by the proposition incontestable that clover, parsley, hazel, and birch, all as here printed, are names as perfectly generic as TRIFOLIUM, APIUM, CORYLUS, and BETULA. Moreover, there have been learned historians of botany in post-Tournefortian times whose minds appear to have been under the same delusion, and who thereby missed one of the fundamentals of the philosophy of botanical history.

It is impossible that men, even the most primeval and unlettered, manage their affairs with various denizens of the plant world without classifying them. Names of plants, generic and specific, and also other names more comprehensive, are a part of the vernacular of every tribe of the uncivilized, as well as of that of every rural province within the bounds of civilization today. The very names attest the fact of classification, for no name is that of an individual plant. It is that of a group of plants, always, a group specific, generic, or more comprehensive than either.

It may occur to some that the named groups recognized by the untaught do not in their delimitation correspond to those that obtain with the professional plant taxonomist, as if that, if it were true, would in the least alter the situation or affect the argument. It will be difficult to understand how the vernacular genera of the ruralists can possibly correspond to the Latin-named genera of the scientific botanists when the latter themselves are seldom at agreement among themselves as to the exact limits of any considerable genus, or of any polymorphous species. When we ourselves may have learned to agree as to where one of our Latin-named groups is to end and the next is to begin, may we with some propriety criticize the same kind of doing as accomplished by the unindoctrinated.

And now let me demonstrate it, that in the history of classification the unlettered vulgus now and again has been first to arrive at the satisfactory delimitation of a natural group, the learned doctors having arrived at this same judgment later by one or more generations, and so as to seem to have adopted it from the untutored laity. This point may perhaps be most easily made plain by returning to the contemplation of North American colonial botany and the colonial dendrologists. It was shown above, that all the several American trees of the walnut alliance with which they became acquainted, although all, in certain particulars, different enough from that one Old World walnut which they had known, they called walnuts; precisely the same as if they had denominated them species of JUGLANS, which would have been the case assuredly, had they but known and used the Latin terminology in place of the English. We, of three centuries later, dispose of these American trees differently, referring nearly all of them to another genus; but what is remarkably to the credit of that colonial and primitive taxonomy is, that so exalted an authority as Linnæus found no fault with it, but simply adopted it. With him all the different kinds figure as good enough species of JUGLANS, and bear with him even the same specific names which the colonists had assigned, but of course Latinized.

When, in a preceding paragraph, I gave early American colonists the credit of having recognized and named as oaks a considerable list of native acorn-bearing trees—even as having determined them to be oaks by their acorns alone—I felt that there might be demurrers to the opinion that these had not learned this mark of the genus *Quercus* from the schools in some more or less indirect way. I may well, therefore, here place it beyond dispute that in this case also the unlettered men of field and forest did arrive at the proper delimitation of a genus of trees quite in advance of the professional taxonomists, and these last virtually adopted the genus, as we now have it, from the ruralists. In the first decade of the seventeenth century, when the Virginian colonists were beginning to learn the industrial uses, and at the same time the characteristic superficial—as we should now say morphological—marks, of many new trees, there was yet no book of botany extant, in as far as I can learn, in which it was taught that any and every tree that has an acorn for a fruit is hereby known to be a *Quercus*. Neither Pliny nor Tournefort, nor any author of the fifteen centuries that intervene between those two, has fewer than three distinct genera of acorn-bearing trees. With each and all of them a tree bearing acorns, in order that it may be of the genus *Quercus*, must be deciduous, and its foliage sinuately lobed. In other words, none but deciduous white oaks are properly called oaks by these old authorities. Trees bearing acorns, but evergreen as to foliage, the leaf margins prickly, are of a separate genus, *Ilex*; while those oaks of southern Europe with peculiar foliage, along with a thick spongy bark—cork oaks, we call them—are of a third genus, *Suber*. At the time, therefore, when Banister was turning into Latin those English binary names which colonists had given to Virginian trees bearing acorns, there was not yet a book of botany extant in which it was taught that its yielding acorns was a sufficient warrant for naming a tree an oak. It was only the country people, the woodsmen, who held that view as to the extent of the genus Oak. The learned John Ray, the same who received from Banister the manuscript catalogue of Virginian plants and caused it to be published, himself never swerved from the doctrine then time-honored and classic, that we have here three genera, *Quercus*, *Ilex*, and *Suber*, yet expressly states that "the common people so extend the name *Quercus* as to include under it all kinds of trees that bear acorns."[42] In as far as I have been able to trace the history of oak classifying on the part of professional botanists, Linnæus appears to have been the very first to repudiate what had been the opinion of all his predecessors, and to adopt as more true to nature the more comprehensive genus *Quercus* which the vulgus had invented. And so, if we of the present, following Linnæus as to the limits of *Quercus*, are in the right, then let us concede freely the fact, from which there is no escape, that during long centuries the unlettered vulgus was taxonomically correct, while all the learned botanists were wrong.[43]

By means of the popular nomenclature of common ornamental plants, one is able to see how those uninstructed in botany readily classify things according to floral structure. Everywhere lovers of flowers have a group of plants which they call by the collective name of lily. This happens to be many times more comprehensive than the lily genus of the modern professionals, though not so widely different from that of the books of botany of some centuries ago. It embraces *Lilium*, *Fritillaria* (Checkered Lily), *Hemerocallis* (Day Lily), *Amaryllis* (Belladonna Lily), *Vallota* (Scarlet Lily), and many another genus of liliaceous and amaryllidaceous plants, besides *Convallaria* and the members of several iridaceous genera. All of these have showy flowers of the same morphological type as that of the true lily. And to this floral type even the white funnel-form spathes of certain araceous plants have been associated, as the name Calla Lily plainly betrays; though it is not to be doubted that the entire, narrow, veinless foliage of all these plants has helped in the making of this popular generic synthesis. And then, on the part of the botanists, the analyzing, assorting, and systematic arranging of these diverse elements of the primitive genus lily—the genus as even now, I say, accepted by a great multitude of mere flower lovers—have occupied a great number of taxonomic specialists during later centuries. The carefully gathered records of the gradual evolution of our present taxonomy of the lily-flowered plants would fill a thick volume; would most perfectly establish the fact that those botanically untaught sometimes classify by the flowers; would illustrate how different generations of professed taxonomists have made their various appeals to different organs, some to the flowers chiefly, others giving much weight to considerations of roots, bulbs, and corms, while others were more influenced by the considerations of the pericarp and seed. And such a book, in its completeness, would form an instructive epitome of the whole history of botany.

It would be as easy to produce instances of a primitive classifying by characters of the root, or, at least, of those subterranean parts of plants which, until within a very recent period, were universally confused as roots. But it may be unnecessary to multiply proofs of the existence of an almost more than fragmentary, and really rather extensive, system of what one may paradoxically denominate pre-botanical botany. Enough may have been said already for the accentuating of the opinion that there are beginnings of our science which the historians should not have overlooked. It has been out of those crude beginnings that learning and philosophy have developed what we now call the real systematic botany. They are even the prothallium from which at length there has arisen the frond of whatever strength and symmetry and grace there may be in the now-accepted taxonomy of plants.

This condition of things being once seen and admitted, we shall for the first begin to be able to understand, to appreciate, and to interpret the earliest botanical authors, not those only of ancient Greece and Italy, but the fathers of botany in England, in France, in Germany, and Switzerland; a worthy company whose true position seems to me never to have been half understood, and whose works have therefore more or less completely baffled the attempts of profound scholars and eminent botanists who, in the capacity of historians, have sought to elucidate the texts of these forefathers and to show what they severally accomplished. Botany did not begin with the first books of botany, nor with the men who indited them, though every historian of the science whom I have read has assumed that it did. The most remote and primitive of botanical writers, of whatever country or language, found a more or less extensive vocabulary of elementary botany in the colloquial speech of all. The chief organs of plants—stem, trunk, branch, leaf, flower, fruit, pod, seed, root, tendril, thorn, and a multitude of others—had been discriminated and named; the organs even known by all who had acquaintance with plants and trees, and the names were everywhere in use. Even the functions of several of the organs had been correctly ascertained before ever a line of botany had been written, most probably even before letters had been invented. The improvement of wild things by cultivation, the propagating of the newly acquired sorts of cuttings, by division of perennial roots, and, in the case of trees, by grafting, are likewise arts that seem to antedate history; as do also the designating of different varieties or species that are evidently nearly akin, by two-fold names, one generic, the other specific or varietal.

All these conditions being recognized, a new and peculiar difficulty will confront the critical student of a protobotanical author. It will in exceptional cases seem doubtful as to whether a given fact or generalization is the fruit of that author's own investigations, and therefore new with him, or whether it be something already long understood and accepted, which he is but placing upon written record. It is, however, a kind of difficulty that gives zest to the study of classic texts; and many such doubts may give place to certainty, or something near it, after persevering examination, and comparison with other passages that are not of doubtful import.

I am unwilling to conclude this introduction without repeating it, that the essence and substance of botany proper are organography and the logical deductions that we draw from organography. They may not be said to be the whole of the science; yet duly and comprehensively considered they will be found to come near it. The line of the development of organography—organography as necessarily including terminology—is that along which a truly coherent and philosophic history of botany must needs be written. The condition is one that will entail the expenditure of incalculable time and unremitting toil; but the cost of time and energy must not be counted if anything beyond disconnected and fragmentary paragraphs of history are to be the outcome.

I. The Rhizotomi

It is characteristic of all very early phytography that the root, that least obvious and most hardly accessible of plant organs, is as carefully described as are the stems, the leaves, and the fruits. This fact that the first of all describers of plants should have taken the root into account, and that so uniformly and so particularly, must seem strange enough to every thoughtful botanist of later centuries; it is in such marked contrast to the descriptive usages with which we of the present are better acquainted.

In the voluminous and carefully technical phytographic works of the eighteenth century and the nineteenth, one may chance upon successive pages filled with descriptions of scores of species, about the roots of which not a word is said. Nothing like this occurs in any book or chapter of Theophrastus, Dioscorides, or any other classic botanical writer. In the case of every species of herbaceous plant, and of many that are woody, they do not conclude a description without telling us what the subterranean parts are like, whether fibrous or fleshy or tuberous or bulbous, usually informing us as to the colors of these organs, as well as the properties of them when they are known to have any. And so carefully did the fathers who wrought a revival of botany in the sixteenth century follow those classic models that, in their illustrated folios, never a plant is figured the root of which is not as faithfully delineated as the foliage or the flower. Even in the letterpress accompanying the plates of Brunfels, Fuchs, Tragus, and others, the root is as well described as the foliage, and much better than the flower. All this for the simple reason that the great masters of remote antiquity had set them the example. But how did it come to pass that the ancient Greek botanists were so almost singularly familiar with the underground parts of plants, and that they so accentuated the importance of them to phytography? One would not have expected this, and it seems almost anomalous. No one who ever went forth to make philosophic conquest of the vegetable kingdom was confronted at the outset by the roots of plants. These parts are recondite. He who for any purpose would inspect a root must undergo the labor and inconvenience of digging about the plant, wresting that root from its hiding-place, and then cleansing it, so that its characteristics may become visible and tangible. When all is done, there is a great sameness about roots. They do not promise much aid to him who would find marks by which to distinguish between like and unlike. Stems present a much greater diversity, and either leaves or flowers or fruits a hundred times more differences by which to distinguish plants, than do these underground parts. It being granted that the function of the root as a vital organ was thoroughly understood, as it appears to have been from the earliest historic period,[1] still, as regards early descriptive botany, there would have been less reason to be surprised if the early fathers had commonly ignored it; or at best had made as little account of it in their descriptions of species as most of the modern systematists have done. And there must be a philosophy of this very ancient and once universal appreciation of the root as a subject of phytographic notice equally with stem and leaf. The cause must, if possible, be ascertained.

If the ancestors, even somewhat remote, of the first botanical philosophers had been savages such as, in dearth of animal food, had found the subterranean parts of many a wild plant available in its stead, then would there have been some show of reason for that singular prestige which roots had obtained so almost primevally. In the transition from savagery to civilization such root-food plants would have come into cultivation, where they would have held their place and been well known to enlightened posterity. But at the time when writing about plants began, at least with the Greeks and Latins, roots and bulbs constituted but an inconsiderable part of their table fare. The bulk of their farm and garden products were the cereals, orchard fruits, pot herbs, and salads. The ancestry of the philosophers of centuries, possibly for millenniums, had been highly civilized, perhaps to the degree of having lost the traditions of nomadic life and the feeding upon wild products of the plains and woodlands.[2]

In this civilization, however, the art of medicine held an important place; and in this circumstance we have a clew to that predilection for describing so faithfully the roots of everything which is so almost peculiar to the phytography of the ancients and their sixteenth-century imitators.

Throughout the whole period of Greek antiquity there was a class of men who followed as a regular business the gathering, preparing, and selling of roots and herbs that were of repute in medicine. It was of course naturally inferred and easily ascertained that whatever properties perennial herbs possess are concentrated in their subterranean parts during the season of the plants' rest in autumn and winter; so that what were called the roots of plants formed the bulk of the materia medica. All this the very name of rhizotomi, root gatherers, sufficiently declares.

The botanists of antiquity, that is to say, they who investigated the plant world as philosophers rather than as economists, inform us that the rhi-

FIG. 4. Discovery presents a mandrake to the physician Dioscorides. The mandrake is still tethered to the hound whose life was sacrificed to obtain the root. Singer, Greek biology (1921), p. 62, restored from the manuscript for Dioscorides, *Codex Aniciæ Iulianæ* (1906), fol. 4ᵛ, about A.D. 512.

zotomi were mostly unlettered men, usually more superstitious than scientific, observing an extensive ritual in the digging as well as in the after preparation of their simples; evidently mixing medicine and magic after a manner almost universal in the early history of the healing art; as often attributing to their preparations magic virtues as medicinal. Concerning some of the ceremonies of the rhizotomi we have information.[3] There were various prayers and incantations to be said or sung. Some kinds of roots were to be dug in the daytime, some others by night only; the powerful plant hellebore, only after the observance of various precautions. The point of a sharp sword must be drawn three times around certain roots to make them more efficacious. The gatherer of some sorts must be careful to face the east while digging. In the case of others he must stand on the windward side of the plant. Some require to be collected by one newly anointed with oil. As preparatory to the culling of other kinds, the rhizotomos must eat garlic and drink wine. As modern and as learned an author as Kurt Sprengel relegates all those observances without discrimination to the category of foolish superstitions;[4] this, as it seems to me, inconsiderately. There are plants enough the exhalations of which are so deleterious that persons of delicate organism may be uncomfortably affected by the mere passing close to them on the leeward side, of a breezy day. Any discreet American botanist not immune against *Toxicodendron vulgare*,[5] if tempted to gather specimens of it, would use among other precautions that of holding himself to the windward of the plant if there happened to be a breath of air stirring. His act would be adjudged sane and reasonable. Another such precaution might be that of using gloves while handling any parts of the plant; whereas an old-time oriental, with whom the anointing of the body with scented oil was a common practice, might successfully use oil instead of gloves while gathering particularly acrid herbs and roots. Also when one is informed that the ancient drug gatherer never proceeded to dig certain roots but with breath laden with an odor as intolerable as that engendered by garlic and alcoholic drink, there is still no reason to charge that to superstition; any more than the surveyor or handicraftsman, whose work for the day is in the midst of a steaming and unwholesome marsh or fen, makes free use of tobacco smoke as, by the physician's counsel, tending in some degree toward immunity from malarial influence. With us who believe so much in the efficacy of malodorous disinfectants as bringing immunity from infectious and malarial disease it should seem natural to attribute similar precautions to Greeks of 4,000 years ago, especially when assured, as we may assure ourselves, that even at that remote period one of the rhizotomi propounded the theory, now in our day revived, that myriads of germs, minute, invisible, permeate every atmosphere.

Such partial apology for some of the so-called ritual observances of the rhizotomi is no digression. The historians have usually referred to them as in large part a body of superstitious fakers. Such, to a degree, many of them may have become in the long course of centuries during which their profession flourished. Superstitious observance is often enough the end of that which in the beginning was a reasonable and sensible measure of precaution; and it is not a legitimate office of history to exaggerate the differences subsisting between an earlier and a later age or race. The age of superstition even as regards medicine and pharmacy, though passing it may be, is not yet quite past. If the scholarly Sprengel cites the ceremonies of the rhizotomi with impatience, it is because he is influenced—as many another passage in his work makes it evident—by an almost morbid abhorrence of everything that to him has the appearance of a superstition. If anything appeared to be an empty ceremony, he could not tolerate the thought of it long enough to examine into the possibility of its having had an origin that was scientific and utilitarian.[6]

If the rhizotomi were mostly illiterate men and quacks, still there were exceptions. Here and there among them there seems to have been a man of letters; and a few investigated plants more or less scientifically, and wrote books. The names of several such have been handed down through history, together with some of the more original and remarkable of their sayings. THRASYAS MANTINENSIS is mentioned by Theophrastus as one of the worthiest among them, because as if ignoring the common beliefs about magical effects he gave himself to the investigation of the properties of plants. He seems to have been the original proponent of the doctrine that the good or bad effects of a medicine may depend upon the temperament of the individual patient, a proposition which has met with some acceptance, at least outside the profession, if one may judge by its having been long since crystallized into a proverb, that what is medicine to one, may be poison to another.[7] The idea is revolutionary, though without yet having brought about much of a revolution. To this same Thrasyas is ascribed the compounding of that vegetable poison, so frequently in use with the ancients, which never failed to bring a speedy and absolutely painless death.[8] Theophrastus devotes two chapters to an account of the pharmacological researches of this Thrasyas, and those of his eminently successful disciple Alexius, and of those of a third of the same school of intelligent and really scientific rhizotomi, Eudemus of Chios.[9] One and another of these men, living at periods so remote as barely to fall short of being prehistoric, tested in their own persons the adaptability of the human system to the harmless use of drastic and poisonous vegetable substances. Beginning with small doses and increasing them gradually, it was ascertained that one might after a time consume without bad results such a quantity of hellebore, for example, as under ordinary conditions might have proven fatal. Using at first earthen pots and pans in the probing of questions about possible or probable antidotes to certain poisons, they would proceed, under the light gained by such experiments, to the using of their own stomachs as the crucibles.[10] And the reports of these instructive and daring experiments, together with the names of the men who made them, were either written and subscribed to at the time or else handed down by tradition to the time of Theophrastus, who gave them permanent record.

Among the earlier rhizotomi there was a famous one named CLEIDEMUS, who wrote upon the subject of electrical storms so as to have been quoted by Aristotle in his *Meteorology*.[11] He also investigated diseases of plants, especially of the fig-tree, olive-tree, and vine. Cleidemus is therefore the earliest of vegetable pathologists. And what may be more interestingly significant is this, that Theophrastus credits him with having maintained that there is the closest analogy between the organs of plants and those of animals.[12] Information like this can not fail to awaken regret that the writings of Cleidemus have not survived; it would now be so very interesting to know whether this genius of a forgotten time went so far beyond those of later periods as to have apprehended the existence of breathing organs and those of sex in plants.

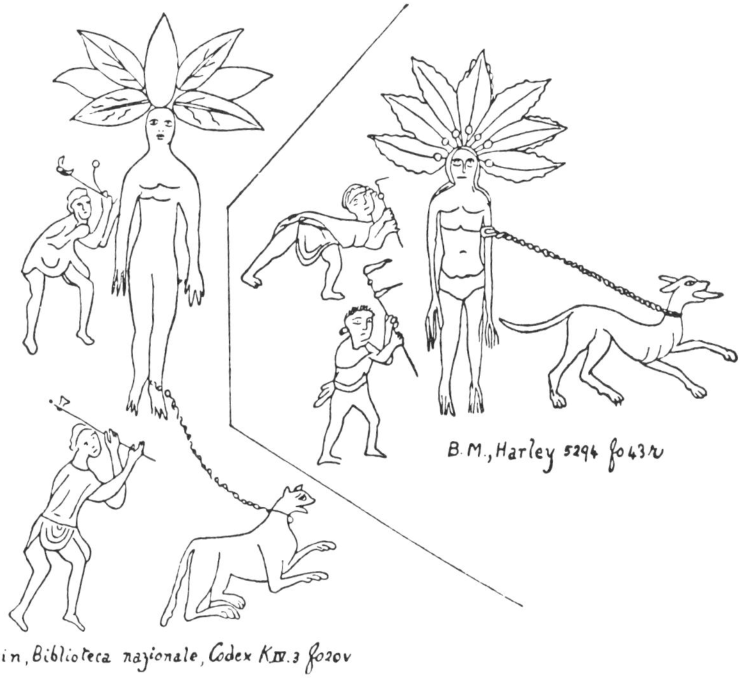

FIG. 5. Outlines of the mandrake and its gatherers. Tracings from a twelfth-century English herbarium in the British Museum and a contemporary Italian herbarium to illustrate the similarity. Singer, Greek biology (1921), p. 71.

HIPPON was among the rhizotomi who philosophized about plants in general, and wrote books. His writings are quoted by both Aristotle[13] and Theophrastus,[14] and he appears to have been the earliest among students of plant life and form to venture the opinion that all cultivated trees, shrubs, and herbs have been derived from wild ones, and are susceptible of reversion to their pristine condition. It is the earliest hint—and a very early one, apparently unknown to the annalists of evolution—of what cultivation may accomplish in the way of transformation. But the doctrine must have had the sound of a heresy verging toward atheism in the ears of a populace that had never questioned the proposition that every cultivated plant and tree had been coeval with the human race, and had been so created at the first.

But it is not that small, better-educated, more reflective, and philosophizing contingent of the rhizotomi, or the possible influence of these few upon early botanical theory, that we are just now chiefly concerned with. It is rather that in this whole body of those who, for so many pre-Theophrastian centuries, followed the root gatherers' calling, we have the men who securely established that precedent, from which the earliest philosophic students of

FIG. 6. The rhizotomist in this fifteenth-century picture was part of a tradition that probably had changed little during a thousand years. The other man is climbing the oak tree for mistletoe. This drawing is traced from pl. 22 of the atlas to Giacosa, *Magistri salernitani* (1901), originally in a manuscript since destroyed. Singer, Greek biology (1921), p. 59.

and writers about plants did not break away, of taking full cognizance of those among plant organs which nature had most deeply concealed, as if they were perhaps the last and the least to be considered. It was the example of the rhizotomists, in their books of plant description extant in the times of Aristotle and Theophrastus, that impelled Theophrastus and others after him to give the form, texture, color, odor, flavor, as well as the active properties, when these were known, of the roots or underground parts of almost every plant. And when, as already noted, it is seen that from Dioscorides and Pliny down through the middle ages, and out to near the end of the seventeenth century, authors in general described and figured the roots of every weed and grass and bush and tree, it will be conceded to have been the lot of the half-illiterate rhizotomi to make their peculiar impress upon the character of descriptive botany, an impress that should last for well nigh two thousand years.[15]

The occupation of the root gatherers is by no means peculiar to Greek antiquity. In every part of the world it may be as old as, or older than, the beginnings of civilization. Nor is it probable that in Europe there was any interruption or cessation of the occupation during the two thousand years intervening between the time of Homer and Hesiod and that of the Renaissance. The botanical writers of the sixteenth century, particularly those of middle Europe, refer to the practices, and even to the opinions, of such as ransack the woods to gather roots and herbs, sell them to the druggists and to the peasantry whom they serve in the capacity of physicians, and from whom the educated and philosophic students of plants themselves sometimes gain valuable information. Nor would it be safe to say that the rhizotomi are even now everywhere obsolete.[16] Their traces are very plainly legible in the popular nomenclature of North American plants. Every common name into which root enters as a component is one that had its origin with the "herb doctor," or "root doctor," as he was called; perhaps not a few of the names were borrowed, along with some information about the plant's virtues, from the aborigines.[17]

2. Theophrastus of Eresus 370-286[262?] B.C.

Linnæus, in the practice of his favorite art of systematizing, classified not only plants but the writers about them. The writers he distinguishes primarily as Botanists, and Plant Lovers, recognizing as Botanists only such as treat of plants from some philosophic or scientific point of view. Choosing his illustrations from the annals of remote antiquity, he names among the earliest of the Greeks who wrote of plants Hippocrates;[1] but because he wrote of plants only in the interests of medicine Linnæus styles him Father of Medicine, a title that had been conceded to that worthy ages before Linnæus, and will be accorded him until the end of time, no doubt.[2] Similarly Aristotle, who is also known to have written upon plants, but whose volumes on that subject have been lost, is down in the Linnæan list of ancient celebrities as Prince of Philosophers. To Theophrastus, however, he accords the title Father of Botany. From this opinion, far from having been newly promulgated in Linnæus's time, there has been no dissenting voice. On the contrary, Albert Haller, one of the most learned men in Europe in his day, and a botanist of such renown that Linnæus held him in reverence, and also in some fear, denominates Theophrastus "the first of real botanists in point of time."[3] Kurt Sprengel in the nineteenth century, having rehearsed the names of a long line of ancient authors who had written more or less concerning plants, says: "But the most illustrious of them all, and the true father of botany, was Theophrastus Eresius."[4] If the author of the latest of nineteenth-century volumes of botanical history, Julius von Sachs, makes but passing mention of Theophrastus, along with the names of Galen, Dioscorides, and Pliny—as if he had not been otherwise a botanist than they—he may be more or less excusable upon the ground that he was not engaged upon a general history of botany, but only dealing with the short period intervening between the years 1530 and 1860 of our era. However, that Sachs had no acquaintance with Theophrastus, or even of Dioscorides, is proven by this, that he credits his sixteenth-century German compatriots with having gone straight to nature and described plants originally,[5] whereas the truth is, that nearly all the plant descriptions occurring in Brunfels and Fuchs are almost word for word translations of the ancient paragraphs of Theophrastus, Dioscorides, and others, sometimes with a few words of their own added, as often with none. And as regards that exact and intimate knowledge of plants which comes of the careful study of them alive and growing, it is safe to say that all which Sachs's sixteenth-century German fathers knew combined would have amounted to but a fractional part of Theophrastus's knowledge, and that much of their own observing had been suggested to them in their reading of his books. An abstract of Theophrastus's work should enable the unbiased and impartial to judge for themselves whether scientific botany had its beginning with those good German herbalists of the sixteenth century, or with an immortal Athenian or two who had lived, studied plants long and carefully, lectured to thousands of students,[6] and written down the substance of their botanical lectures seventeen centuries earlier than they.

Life

Mitylene, a large and rich island in the Ægean Sea close by the coast of Asia Minor, was famous millenniums ago as having given birth to many an illustrious personage. Arion and Terpander, ancient masters of the art of music, Alcæus and Sappho, unrivalled among lyric poets, as learned critics gather from the fragments of their masterpieces that remain—these names but head the list of celebrities that had been born on that island in the earlier half of the thousand years next preceding the beginning of the Christian era. How very famous this island was for the excellent quality of its products, material, artistic, and intellectual, is shown in the fact that Greeks and Romans of a somewhat later period, wishing to bestow the highest praise on anything, whether it were a piece of music, a verse of poetry, or a cask of wine, were accustomed to pronounce it Lesbian—that is, fit to have come from Lesbos, the name by which the modern Mitylene was known anciently. Such are a few of the available hints of the environment in which the protobotanist was born in the year B.C. 370. His birthplace was Eresos, the most important town of the island, whence he has been styled Eresius—the Eresian, perhaps to distinguish him from others of his time who bore the not uncommon name Theophrastus. His father's name, Melanthus, and that of his boyhood's teacher at Eresos, one Leucippus, have found a place in history only as associated with that of this child and youth whom they called Tyrtamus. We shall be warranted in inferring that the child was of unusual gifts and marked by nature for the intellectual life; also that Melanthus, the fuller, was in comfortable circumstances financially, for the educational advantages that were given the boy were then somewhat rare and costly. How well the child had improved his opportunities is attested by this, that while as yet but a youth, he was away beyond the farther shore of the

Ægean Sea, at Athens, and there numbered among the disciples of Plato. All historians of the period credit him with having been under that philosopher's instruction before coming to Aristotle; and as Tyrtamus was only twenty-two years old when Plato died, it is plain that the enrollment among Plato's pupils must have been made when the subject of our sketch was but a youth—possibly a precocious, eager, ambitious boy only.

The histories all read as if Aristotle's marked friendship and efficient patronage had had very much to do with establishing the fame and directing the luminous career of Theophrastus. There must be a large measure of truthfulness in this representation, though it is more than possible that it is somewhat exaggerated in Aristotle's favor; and history should take cognizance of the universal and even necessary fact, that in great friendships the influences are mutual, just as when, in the heavens, two planets move to their conjunction each influences irresistibly the orbit of the other, draws it somewhat aside from what should have been its path. The story, as always rather too briefly told, leaves an impression, not intended to be made, of great disparity between the two both as to years and some other controlling influences; seeming to represent Tyrtamus as the brilliant young favorite, and Aristotle the elderly admiring teacher and foster-fatherly patron. That the youth, as if irresistibly obedient to an old and revered master's mandate, should have renounced the name Tyrtamus that he brought with him from the paternal home in Eresos, so that henceforward he should be Theophrastus, is something to create almost a conviction that the one was old and masterful, the other young and submissive, and not to be thought of as an influence upon the thought and action of the elder. Such impressions are wrong, and must vanish by a comparison of certain well authenticated dates, which show that Aristotle was Theophrastus's senior by only fifteen years; also that the two were fellow students under Plato. Aristotle from the age of seventeen to thirty-seven had been Plato's pupil. Tyrtamus, it may reasonably be assumed, entered Plato's discipleship at seventeen if not earlier. If so, he may have been Aristotle's student companion there for from five to seven years before the day when Aristotle left Plato and opened a Lyceum of his own. Shortly after this Plato died; and then, unless he had done so even earlier, Theophrastus became Aristotle's student. And as for this new name, it is not necessary to suppose that it was bestowed merely as the flattering compliment paid a highly promising young student by an old preceptor. The two were in truth much upon an equality. They were companions and much attached friends, with no signal disparity between them as to years; so that the change of name may well be thought to have been brought about by mutual agreement.

I can not but wonder at the boldness with which Meyer pronounces this change of name from Tyrtamus to Theophrastus to be a fable.[7] No one else has questioned the authenticity of this part of the biography; and he has not been able to adduce so much as one valid reason for his pronouncement against its truthfulness. One of his supposed reasons is, that Aristotle was no flatterer. To have rendered this an argument, Meyer should first have disputed the sincerity of Aristotle's friendship for Theophrastus; for between genuine and devoted friends flattery is impossible. But he says the name Theophrastus was not uncommon among Greeks of the period, which is equivalent to saying that, if the philosopher had been going to give his disciple a new name he would have selected some uncommon name, or else that the Eresian had always been Theophrastus and never Tyrtamus. That "such a changing of names was unknown" is quite as inane as the rest of this historian's argument upon the subject. Not one in ten thousand of the ancient Greeks has been known to us by any name at all. Even of that comparatively very small number whose names we have heard, who shall say that none besides Theophrastus ever underwent a change of name because the event if it happened was not recorded? There is not the least reason for thinking that with Greeks, in passing from one age or condition of life to another, the taking of a new name was uncommon. There were distinguished examples of it among some of their neighbors, the Hebrews, for example. Meyer concedes that this "fable" about Theophrastus was universally received as a fact by all the ancients, and we add that it is as perfectly authenticated as any other fact or incident in Theophrastus's life. He who can make fiction of this part of it may upon similar grounds invalidate, piece by piece, the whole biography. No reason having been given for doubting about this incident, every one to whom the appellation is not meaningless will see that Tyrtamus was a grotesque name, if not a ridiculous one, to be borne by a scholar and orator of commanding presence and predestined great renown. The young man himself, as well as his master and friend, must have realized this; and it is hardly to be doubted that the displacement and rejection of the unsuitable name was the first object which the change had in view; and that what the new name should be was but a secondary consideration—a matter of less importance. Indeed, the biographer Laërtius relates that at first the Eresian began to be called Euphrastus, and then later Theophrastus. Posterity should be grateful that the change was made; and also grateful for that devoted attachment between the two philosophers by which it came to pass that the elder of them, dying in middle age, had his own work taken up, and carried forward with success during almost another half-century.[8] It was such a friendship as led Aristotle to give to Theophrastus his own library, said to have been the richest one then in existence, and to have included the manuscripts of his own works, a treasure which by means of Theophrastus's jealous care was almost singularly preserved, and handed down to posterity well-nigh complete. Also the botanic garden which Aristotle had established at Athens was made a gift to Theophrastus, by whom also it was newly equipped, variously improved and adapted to greater usefulness; this, too, on a scale so extensive, that a wealthy friend of Theophrastus and benefactor of science is named in history as having borne the expense of those improvements.[9] The fact of the existence of this Athenian botanic garden will explain how Theophrastus, occupied as he was with the management of, and also engaged in teaching in, a school of two thousand students, with no time or opportunity for travel, gained so intimate a knowledge of the life histories of many plants as he surprises us with in certain chapters of his books. He had studied in that garden at morning, noon, and evening for

FIG. 7. Floral design over the doorway of the Erechtheum at Athens, showing stylized parts of a flower. Mauch, *Greek and Roman orders* (1910), pl. 92.

perhaps sixty years or more when, almost a centenarian, he wrote such clauses as the following in his will. They should help us to a realization of the scientific zeal and activity of a forgotten time.

I bequeath to my friends, who are specially named in this my Will,[10] and to those that will spend their time with them in learning and philosophy, my garden, walk, and houses adjoining; upon condition however that none of them shall claim any particular property therein, or alienate them from their proper use; but that they shall be enjoyed in common by them all, as a sacred place where they may familiarly visit one another and discourse together like good friends.[11] And further: I desire to be buried in any part of the garden that they shall think most suitable; charging them not to be at any superfluous expense either upon my funeral or upon my tomb. Which being done, my will is that Pomphylus who lives in the house take care of everything, as he did before.[12]

That Pomphylus was an overseer, directing the labors of bondsmen gardeners owned by Theophrastus, comes out incidentally in another clause.

As for my boys, it is my will that Molo, Cymo, and Parmeno be forthwith set at liberty. As for Manes and Callias, I will not have them given their freedom until they shall have labored four years longer in the garden, so that there be no fault found with their labor and diligence; but after that, let them have their freedom.

Besides these five, two others are mentioned. "I give Cano to Demotimus, and Donax to Neleus."[13]

By means of this testamentary document, the transcript of which has been fortunately preserved, one is able to realize something of the extent, and even of the perpetuation during perhaps three generations, of this pristine garden for biologic research. And this realization will be exceedingly helpful toward a comprehension of the magnitude of Theophrastus's work along these lines. The philosopher was never, like many of his class and in his time, a traveller. He did not devote any more than a fraction of his time to botany. His writings on this subject amount to perhaps not more than a twentieth part of all that he did in the line of written authorship. And there are chapters in the *Historia Plantarum* that are so crowded with facts about seeds, seeds in process of germination, young seedling plants and older ones, observations upon this plant and that shrub as they appear in spring, summer, autumn, and winter, that, all being considered, we should have wondered greatly how this most untravelled and sedentary of the great philosophers had gained all this minuteness of knowledge about the little things of plant life had we not been informed concerning this great garden in the midst of which he dwelt,[14] taking his daily recreation along its paths, and among its seed beds, and within the bounds of which, obedient to his last request, they buried him.

Theophrastus was a voluminous author, having written upon a great diversity of topics. The biographer Laërtius gives the titles of 227 treatises.[15] Not many of these have reached us; but among those preserved are the *Historia Plantarum*, in nine books, with the fragment of a tenth, and the *De Causis Plantarum*, originally in eight books according to ancient records, of which the last two have long been lost. The following paragraphs are the result of a prolonged and laborious study of the principal work, the *Historia*, the edition quoted being that of van Stapel,[16] published at Amsterdam in 1644.[17]

As we have already seen,[18] there existed in the old Greek literature that was before Theophrastus, many a trace of properly botanical observation and reflection, so that he is not in such wise the father of botanical science as that no one before him had recorded a philosophic thought or suggestion about the plant world separately considered. Yet he is, in most cases which he cites, the sole perpetuator of the name and fame of such as Menestor, Hippon, and Leophanes whose passages he quotes and in quoting has saved from oblivion. There is, then, no reason to suppose that in his philosophizings about plant life he had been helped by any predecessors beyond that for which he has given them full credit.[19] It has been observed by historians and critics that a few passages in Theophrastus are also in Aristotle, unaltered, and uncredited to their real author. This hardly merits notice. It is undoubted that the enlistment of Theophrastus's great talents in the service of botany was secured by Aristotle; and it is as certain that the alliance between these two celebrities of antiquity was that of the most devoted friendship; that at Aristotle's demise all his manuscripts, published and unpublished, complete and fragmentary, were gladly bestowed on Theophrastus. They became his property. He was free to dispose of them, to publish any parts of them, according to his own judgment. Modern science and erudition have no example to show us of like community of even intellectual property between two illustrious friends.[20] Moreover, from the fact of Theophrastus's having lived and studied and written during half a century after Aristotle's demise, we are warranted in thinking of him as of one who had acquired a great store of knowledge about plants beyond all that to which Aristotle in his briefer day had attained.

As having been the author of the oldest distinctively botanical treatise that is extant, the place of Theophrastus is unique, and invites to special and careful consideration. He writes from the midst of an advanced civilization; a state of society in which there is much farming, extensive cultivation of the vine and olive, fruit growing, market gardening, and cultivating of medicinal, aromatic, and ornamentally flowering herbs, shrubs, and trees; a time when many improved varieties of all sorts of things have been derived through cultivation, and when it is already perfectly well known that such improved varieties can not be depended on to come true to seed, but may be preserved, and the stock of each increased by division of roots, by cuttings, and by grafting. It is also a time when the very masterpieces of literature—some of them even in Theophrastus's time ancient and classic—abound in facts and fancies and myths and fables about flowers and fruits, shrubs and trees.[21] Of course all obvious and familiar parts of plants—their organs—have their names. These are a part of the common vocabulary of things. Also group names for growths that are alike are in as universal requisition. If a genus evidently consists of several different kinds, be they what the botanist of the present would denominate species, or be they notable varieties only, each such kind is designated in speech or writing by a cognomen; so that a binary nomenclature, precisely that which all farmers, gardeners, and foresters find needful and have always created and employed, is perfectly established.

Such, in brief outline, is what Greek civilization had attained to in the way of experimental knowledge of plants, independently of all philosophy, and without help from the philosophers. And if Theophrastus had been less than a botanical philosopher, and if as a mere annalist he had but recorded the untaught industrial and experimental botany of his period, together with that very considerable vocabulary of botanical terms which then formed a part of the Greek language, he would still have done us an inestimable service; though masters of the science in the seventeenth, eighteenth, and nineteenth centuries of our era would not have styled him Father of Botany.

In our study of this maker of the first landmark in the History of Botany the main object must be that of discovering in what ways, under what limitations, and yet how well, he accomplished the placing of knowledge of plant life and form upon the list of the sciences.

Method

That a treatise be recognizable as scientific it must be methodical.[22] It is even a necessary characteristic of it. There must be a principle, or a set of principles according to which the facts or propositions find an orderly arrangement. This does not, however, imply that the method or system be of some particular kind, as, for example, that in botany one should be required to arrange the matter of one's treatise according to what are conceived to be the natural affinities between plant and plant. The author who writes botany from the industrial point of view, if he so elect, may discuss his plants in the order of their relationship in families, genera, and species; or, ignoring taxonomy of that sort, he may arrange them according to the nature of their serviceability in household economy and what are called the useful arts; may discuss in successive chapters food plants, drug plants, textile plants, vegetable dye-stuffs, trees, as supplying timber, fuel, oils, gums, sugars, resins, nuts, etc. That botanical matter so arranged may be scientific can not successfully be controverted.

Theophrastus of Eresus might have adopted such a method as this last. He was abundantly capable of discussing the plant world from the economic and utilitarian standpoint;[23] indeed, had he so arranged the substance of his work he would but have been following established precedent. Every treatise on plants which was extant in his day was of the nature of agricultural, horticultural, or medical botany. There was not yet any other method of arrangement for botanical writing but the economical. Furthermore, his own chapters everywhere abound in references to the qualities of plants, and their uses in the economy of human life; though such references are commonly supplementary to the statement of other and different considerations. The very title of his work, *History of Plants*—in more idiomatic English, *The Story of the Plants*—seems to look toward an investigation of this realm of nature for its own sake, the vegetable kingdom thought of philosophically rather than industrially. No work was yet in existence, unless one by Aristotle which has been lost,[24] in which plant organ was discussed in relation to plant organ, and the kinds of plants in relation to other kinds. The prevailing attitude of mind respecting the plant world was not such as would tend to the encouragement of other than utilitarian views of it. That these objects were brought into existence only with reference to man, and for his use and benefit, was even a part of the religious belief. And so deeply seated and generally prevailing was this sentiment that it is said the students of Aristotle and of Theophrastus became objects of ridicule with some of the literaries, poets, satirists of the time, because of their going about the country picking up and curiously peering into the least little things of nature, such as were of no possible use.[25] And one may not attribute to antiquity alone these prejudices against philosophic nature study; for they rule the mind of untold millions even now. Antiquity, in this phase of it, is with us still, in the ideas of the uncivilized races, and also in rural districts of the lands of the enlightened. Not many a botanical traveller and explorer along the frontiers and in the remoter country sides has failed to be accosted with friendly queries respecting plants of which he has been seen to be making specimens: What use has this plant? What is that kind good for? And what betrays in these good-naturedly inquisitive rustics their complete subjection to the pre-Theophrastian utilitarian botany is, that when the man of science answers frankly that he knows no use whatever for the plants in question, he is not believed, but is silently credited with wishing, for his own pecuniary advantage, to keep their use a secret. The adherents of this archaic philosophy of the vegetable kingdom are, I say, doubtless numbered by thousands on all the continents; people who have not heard of any other; and we have no proof that another had been any more than merely suggested before Theophrastus.[26]

In the Second Chapter of the philosopher's First Book there is presented the following list of the external and obvious organs of a highly organized plant, *i.e.*, a tree: root, stem, branch, bud, leaf, flower, fruit.[27] I recall that upon first reading these initial chapters of Theophrastian botany I was quite startled to find here this complete and faultless statement of the external organs of a tree or shrub; to be confronted with it even here, and be made to realize that it is so very, very old; that our own masters and tutors of a few years ago did not invent it, neither their own immediate botanical ancestry, but that all so-called modern botany has this fundamental of plant-morphology from old Theophrastus, and all unawares. But the surprise passes. The sentence is so simple, so natural, so logical, the connection between term and term so perfect, that one doubts that it could have been done otherwise; and may be disposed to say to himself that any botanist of whatever epoch might have gone out on a morning walk, looked at a number of different kinds of growths, come in and written down that methodical proposition offhand. It is extremely improbable that it was so done. It is next to certain that it cost its author a great deal of careful observation and prolonged study and reflection. The sentence is studiedly methodical, and no less a dictum of science than a work of art. It is easy to forget that, as in art, the production from which all evidence of the artist's anxious care and hard work has been eliminated is the masterpiece, so in science and philosophy the axiom or the aphorism which when finished reads as easily, smoothly, and convincingly as if every one always must have seen its transparent truthfulness, and as if almost any novice in that same science might have written it down in just those words—that this is the little sentence which may have cost its author the expenditure of time and mental energy with which he might have written a whole volume upon some topic that was not difficult.

If the origin of the simplest elements of universal botany is to be shown, this Theophrastian list of plant organs will have to be looked into rather particularly, will need to be studied with great care and caution. Those six or seven important terms, as our philosopher links them together, constitute the most classic piece of elementary botany in existence. The sentence has also much to reveal about the author's botanical method in general.

First of all, the terms of the sentence, root, stem, leaf, bud, etc., have not been created or invented by Theophrastus. As the names of those things they are part of the common vocabulary; botanical terms, assuredly, yet in their framing and use doubtless far antedating all written botany. That our first philosopher of plant life, he who first brought the terms together and placed them in line, altered the meaning of certain of them by giving them a more comprehensive or else a more restricted application, is easily possible; though that does not here concern us.[28] We have but to note that this fine equipment of most fundamental botanic terms, the first botanist—as we must denominate him despite the suggestion of the paradoxical—found ready made. They and other such terms had been the legacy to human speech made by a primeval race of nature-ponderers, almost infinitely remote and prehistoric even in Theophrastus's time.

Observe now, that while root, stem, bud, leaves, flowers, etc., are familiar and obvious parts of a tree, they are far enough from being a complete list of such obvious parts. Viewed at close range the bark is very conspicuous, more so than either the buds or flowers in the majority of arboreal and arborescent growths; and since he has here left it out, one is obliged to think that this point was well inquired into by him, that the omission was deliberate, and the result of sound reasoning. We shall find proof by and by that he investigated the inner structure of trees; and there, among the anatomical parts of the tree as they disclose themselves in the cross section of its trunk or stem, we shall find him cataloguing the bark. To separate organography into the two divisions, morphological and anatomical, is, then, also classic. It is another part of universally approved botanical method which originated with and was established by Theophrastus.[29]

With that Theophrastian list of organs under consideration, modern botany at the very outset divides it into two parts, labelling one division of them the vegetative organs, the other the reproductive. The Greek has incidentally given us to know that he, too, pondered very seriously indeed the question of a natural division of his series, and that he effected one. It is as far as possible from corresponding to our modern classification of the same organs, and must needs have been so; because the only reproductive organs of plants known to Theophrastus were seeds and buds. Of the sexual organism of the flower he had no information. He was without a microscope. His dividing line between the two classes of organs is drawn, not as with us toward the upper end of the series, but near the middle of it. Root, stem, branch, bud, form the first division; and the perfect naturalness of it may be realized by observing that precisely those organs, and no more, are what one enumerates as constituting deciduous woody plants in their winter condition. No leaf, no flower, no fruit is there; yet the organism as it stands betrays no imperfection. From its deepest rootlets to its remotest twigs and scaly buds it is alive, in health, perfectly normal in every particular. What is more, every such tree and shrub on the face of the earth passes half the period of its life in just that condition, no difference whether that life period be fifteen years or fifteen hundred.[30] It is a classifying of the external organs of a plant as permanent and transient, a very remarkable division of the series, as true to nature as any classification of organs that ever has been proposed anywhere by any man, and yet, as I suppose, entirely peculiar to Theophrastus's method. How thoroughly natural and scientific this proposition is, may be further shown; as also how studiously the learned Greek nature student reached his conclusion about it. It is not the deciduous tree only whose different organs fall into the two classes of the permanent and the transient. He has discovered that the evergreen trees, like the rest, acquire one set of new leaves each year, and as unvaryingly lose an old set;[31] so that their perpetual verdure is due to this only, that each set of leaves remains on the tree during two or more seasons. It is clear to his mind that if leaves, flowers, and fruits are to be catalogued as plant organs at all, the division of the whole series into constant organs and inconstant must be maintained. But he is even perplexed with a question of whether those inconstant and scarcely more than occasional parts are to be listed as plant organs at all,[32] a position which most twentieth-century readers will think very singular and strange. But must there not also have been with him a time of doubt as to the placement of certain other very common and external parts of plants? Such things as prickles, spines, thorns, tendrils, excrescences of several kinds which imitate fruits but are not—it was by no accidental oversight that these were omitted from his catalogue of plant organs. Since in the phytographic and taxonomic parts of his writing he evinces his perfect familiarity with them, it becomes certain that he surveyed with his wonted carefulness the ground of their possible right to enumeration among these other organs, and that he deliberately ruled them out. The ground of his doubt concerning leaf, flower, and fruit as admissible into the line he states fully. Being himself first a zoologist, then a botanist, and always interested in making comparisons between these two kingdoms of nature, he is aware that the fœtus of a gravid animal is no part or organ of that animal. The fruit of plants, being analogous to the animal fœtus, should be denied any place in the list of plant organs. To make this part of the Theophrastian argument quite clear to the reader, it will be needful to anticipate our study of his anthology in so far as to say that the conception of ovary and ovules as being parts of the flower is one that never entered into the mind of Theophrastus. With him those first small rudiments, as they appear still encircled by or concealed within their colored floral leaves, are the fruit. From this incipient stage, in which, as I say, we call them parts of the flower, the Greek held them to be the fruit, and so forward to their fuller development and final maturity, always the fruit. Whether, by the way, his doctrine of the fruit or ours is the more natural, the more logical, and the less forced and arbitrary, he may determine who may divest his mind of its every prejudgment about the case. In the light of this explanation it will be easy to see that, if the fruit be denied by Theophrastus its place in the list of plant organs, the flower, that is what we call corolla, is equally disqualified, partly on account of its intimate connection with the fruit, and partly on the score of its exceeding transiency in all cases. Furthermore, because in the thought of the Greek the flower itself was but a circle of leaves, different from the ordinary foliage only as to form and coloring, if the floral leaves must fall short of mention in the list of important organs, the green leaves must remain with them. So the catalogue would begin and end thus: root, stem, branch, bud. Indeed, his first presentation of it is in this abbreviated form. Now against his own argument for the exclusion of such parts from the list, he presents such reflections as the following. It is only when trees stand vested in their full foliage, flowers, and fruits, that they seem to have reached their fullness of beauty and perfection. That which makes for the perfection of an organism should apparently be accounted a part of that organism. And as for permanency, there are exceptional cases among animals in which certain parts are transitory. Fowls periodically shed their feathers, and stags their horns; and his last observation here is that animals and plants are in many ways so very different in their constitution that arguments from analogy must not be pressed too far.[33] And so, after much observation and astute reasoning upon the subject, he convinces himself that leaf, flower, and fruit are entitled to places in the list of plant organs, where nevertheless they by their nature form themselves into a separate division. One can not but admire this piece of Theophrastian method, wrought out originally and laboriously by himself, and so unique; in its deepest meaning fairly amounting to a division of the organs as vegetative and reproductive, and drawing the line between bud and leaf—mistakenly, of course—instead of between leaf and flower.

There is another mark of deep study in the making of this list. It lies there, rather well concealed from our first glance, in that elegant sequence according to which the names of the organs are placed: root, stem, leaf, flower, fruit. Its mere orderliness pleases one, even captivates the mind to that degree that one readily believes that it is of both natural and logical necessity, and that any single modification of that arrangement would make the whole unmethodical and altogether bad. To begin at one pole of the plant axis and proceed thence without a break to the opposite pole is artistic, and therefore satisfies our æsthetic fancies. It may also be that such procession of names of plant organs is called for by scientific principle; and since our protobotanist was keenly in quest of principles and a little indifferent to matters of phraseology, we must inquire into his scientific reason for writing the root first on the list of plant organs, rather than the stem.

We have already seen that primevally the root received more attention morphologically than any other organ of the series;[34] that early descriptive botany is rather anomalous in this particular, and that all this came to pass through the influence of root gatherers and their patrons the physicians and the pharmacists. Such a merely economic and commercial consideration as that which influenced men of the time in their descriptions of roots can not be supposed to have the least effect upon the mind of Theophrastus at this particular juncture, where he is engaged upon a study that is in nature purely scientific, biologic. No botanist has lived in any age of the world more capable of distinguishing between the economic and the biologic in nature study. Now in those parts of his work which are descriptive, written for the purpose of enabling the reader to identify the plant, his sequence of the organs is different. It is now stem, leaf, flower, fruit, root. Another not unnoteworthy item of Theophrastian method: one sequence for the treatment of organs phytographically, and another sequence for the discussion of them biologically. The former is well suited to its purpose; for, to the great majority of observers, nothing is seen of any plant or tree but its stem, foliage, flower, and fruit; and any reader would be discouraged if not repelled by a description beginning with a full account of the root, about which part he neither knows nor particularly cares to know anything. As to the other and biological sequence, it is evident that the philosopher arrived at it only after careful and prolonged investigation. "In all plants the growth of the root precedes that of the superior parts."[35] The allusion is to young plants, whether growing from seeds or from cuttings, which latter means of propagation was only less extensively resorted to by the ancient Greeks than it is by ourselves; and they knew as well as we that when a cutting set into the ground has remained for weeks and maybe months without having been increased by so much as one leaf, that only means that it is slow in forming its roots, and that until these are somewhat grown, other growths will not begin. It is improbable that this kind of fact was not of common knowledge with gardeners and orchardmen ages before Theophrastus; so that all he had to do was to verify it by experiment. But when he states that even within the seed it is the radicle which first begins to swell and grow, and that this is invariably the first part of the growing seed to appear outside the shell,[36] we feel assured that these are fruits of observation on the part of no gardener, but of a biologist.

In the year 1672 of our era Nehemiah Grew, aided by a microscope (!), had repeated those investigations of the seed as germinating, and published what was but a confirmation of the view that Theophrastus had presented some nineteen hundred years before, namely, that biologically considered the root is first among plant organs, then the stem, and after them, leaf, flower, and fruit.[37]

One must now leave the Theophrastian method in merely elementary organography, and survey briefly the outlines of his dealing with plant as compared with plant.

The philosopher has several different viewpoints, from each of which he perceives the vegetable kingdom as a whole to be divisible into two parts. The first of these divisions is according to texture and duration of stem and root, a chapter in botany which he wrote for all time—the distinction between woody plants and herbaceous. Trees and shrubs, alike as to their woodiness, are distinguished by him precisely as in the most recent botany, not neglecting the word of caution that no hard and fast line separates the two; that many, like the filbert and the pomegranate, are naturally shrubs, the stems growing in clumps, and are seen in the form and dimensions of trees only when under the cultivator's care and art. Even the pear tree, olive, and fig when left to themselves become many-stemmed and shrub-like. He also apprises the reader that certain pot herbs of the gardens have always the one-stemmed and arboreal mode of growth, even approaching trees in their dimensions, and all within the time of a few months; but that these have not the duration of trees or even of shrubs, and therefore prove that they are neither. He mentions certain mallow and cabbage kinds as of this category.[38] He denominates them *δενδρολάχανα*, which his Latin translators have rendered olerarbores, and would appear in English as tree pot herbs. All herbaceous plants he classifies as perennial, biennial, and annual; also carefully stating that the annual and biennial are hardly distinguishable, inasmuch as both die, root and branch, within the space of a year, and as soon as they have once perfected seeds.

While this important example of botanical method is in hand, it will be pertinent to take note of those two mental processes, analysis and synthesis, of which every piece of method is the outcome, this for the purpose of clearing our own mental vision for an inquiry into the question of how much of this classifying according to texture and duration was found by Theophrastus ready-made and in common use, and to what extent, if at all, he revised, augmented, and improved it.[39] Problems of this kind are most difficult, even impossible of exact solution by the mere botanist. The erudition of the specialist in philology and the history of language is here called for, without the aid of which the early history of botany never can be written. Happily, however, great linguistic learning is not requisite to a few reasonable inferences respecting Theophrastus's part in this classic piece of method. The distinguishing between woody growths and herbaceous is doubtless older than history. It is also evident that with remote enlightened antiquity tree and shrub were distinguished. The words representing these ideas are very ancient; but the half-shrub, or suffrutescent growth as we of today speak of it, appears to have been set apart as a group, and assigned a distinctive name by Theophrastus himself. There are many old and classic names for plants distinctively herbaceous. There is weed, grass, herb, vegetable, even the word "plant" itself as originally used, every one implying the herbaceous as to texture and short-life period; and so much for very primitive analysis and synthesis; but the putting of all these things together, the synthesis of them under one comprehensive term, was, if I mistake not, a Theophrastian contribution to botanical method. Moreover, and what is of even more profound interest, it appears as if the synthesis of everything that vegetates—tree, shrub, half-shrub, and herbaceous plants, including even sea-weeds and fungi—into one vast comprehensive assemblage of living entities called plants, is also to be attributed, if I mistake not, to Theophrastus. Primeval and prehistoric observations and reasonings, I say, doubtless established and named such lesser groups as tree, perhaps bush, certainly garden vegetable and field and garden weed, and grass and reed, rush and mushroom; but the synthesis of all as in some way alike and interrelated, and as spoken of comprehensively under one word—that would seem to have awaited the advent of a master mind like that of Aristotle. The written record of so significant a piece of method in nature study I find for the first time in Aristotle's greatest botanical disciple. The fuller investigation of this topic may be deferred until we come to Theophrastus as the founder of Botanical Terminology.[40]

From another viewpoint Theophrastus beheld the world of plants as divisible into the cultivated and the wild;[41] and he formally approves this line of separation, though almost as compelled by circumstances; for he admits that it is not natural, and that the differences are in the main such as result from cultivation. It presents, nevertheless, a forcible example of rude primitive plant classifying. Untaught peoples of all countries, and many all around us, hold to such a division, firmly believing, even superstitiously, that the wild parsnip or wild carrot—differing from its parent plant only as growing spontaneously by the wayside, rather than within the garden wall under cultivation—is a poisonous thing, perilous to the life of him who would dare to eat it. Our present nomenclature of plants, the vernacular as well as the Latin, presents countless clear vestiges of the former popularity of this antique parting of all the plant world into these two divisions. Such specific adjectives as *agrestis*, *silvestris*, *trivialis*, *arvensis*, *pratensis*, *horten-*

sis, *sativus*, *urbanus*,[42] and many more tell of a time past when about the first question concerning any plant was whether it was wild or cultivated. There is no need of citing examples of those hundreds of vernacular plant names the first term of which is "wild"; but all of them, as relics of pristine botanical ages, attest the once-universal prevalence of this partitioning of all things that grow out of the ground into the two groups of the cultivated and the wild.

As for Theophrastus, out of the some five hundred species and varieties of plants of which he treats, only an insignificant proportion are other than domesticated; and he says that the uncultivated things of wildwood and mountain are mostly still unknown and have no names. To have assigned space in his book for the consideration of many wild plants must have appeared like a marked innovation, one that indeed looked in the direction of a widening of the field of botanical investigation, and was therefore of promise for the future of the science; but it must have given occasion for carping critics to ask how wild plants, such as have neither names nor history, are entitled to a place in the History of Plants. But while this newly suggested distinction had within itself one element that would eventually accomplish its obliteration, Theophrastus did not perceive this. Among wild trees and other growths which it had been sought to introduce into gardens, there were some which had baffled every effort to transfer them, whether by root or seed. From this he reasoned that there existed a line of natural demarcation, at least between plants that were susceptible of domestication and such as were not. But some trees which he named as apparently impossible of domestication are now successfully cultivated.[43]

Having held the status of an acceptable part of botanical method for a millennium and a half, at the revival of botany in the first part of the sixteenth century this distinction began to decline in popular favor, and within two centuries more it became so nearly obsolete that, in books descriptive of the plants of particular regions or districts, those of field and garden were wholly omitted. Only wild plants were now taken note of; and so an extreme squarely opposite to that of Theophrastian times had been reached. And what lends deeper interest to these observations now, is the circumstance that of late years there has been awakened the keenest passion for the study of cultivated plants that history has known. I speak, of course, of that purely philosophic and scientific investigation of them which, either together with or apart from the industrial in motive, engages the attention of many botanists.[44] And here, realizing that the very father of written botany was chiefly attentive to domesticated growths, and upon these as principal subjects wrought out his scientific system, one wonders whether or not in botany the first cycle of its history is being completed.

From yet another outlook over the vegetable kingdom as a whole, all the subjects thereof range themselves under the two assemblages of the "flowering" and the "flowerless." Theophrastus records this;[45] but assuredly the invention of such a division can not be ascribed to him. It must have formed a part of the universal prehistoric botany. Never since human intelligence came into the world, and lived in converse with nature, can people have failed to remark the presence here, and the absence there, of those differently shaped, differently colored, and singularly congested foliar parts that they call the flower. It must have been observed immemorially, and accepted as a nature-taught fact, that such and such trees, weeds, and other plants never display flowers. They are flowerless. What I have been careful to state, namely, that the primitive notion of a flower is that it consists of leaves, must be insisted on. The reader must not permit himself to think of the pristine idea of flowerless plants as embracing anything less than the cryptogams plus the whole body of the apetalous phanerogams. With this fixed in our minds, we are then ready to inquire whether or not Theophrastus made any improvements of his own upon this part of pristine botanic method; whether perchance he looked into the matter far enough to have discovered that there exist such things as flowers destitute of flower leaves, and thereupon enlarged the boundaries of the flowering plants and correspondingly restricted those of the flowerless group. The answer must await our study of Theophrastian anthology.

In close connection with his separating between the flowering and the flowerless, the Greek divides the whole plant world again into the two categories of the "fruit-bearing" and the "sterile."[46] Judged by certain criteria that are of comparatively recent adoption there would be no call for this last distinction; for the flowering and the fructiferous would exactly correspond to each other, as would also the flowerless and the sterile; so that this last seeming distinction would be but a different naming of two primary groups before indicated. But this is not true; for when we have learned his doctrine of the flower we shall perceive that he had in mind fructiferous plants, and trees even, the flowers of which he had been unable to detect and which therefore of logical necessity he must classify as flowerless. Also, since he knew nothing of such sexual distinctions as have their ground in floral structure, there presented themselves to him what we of today know as the males of certain diœcious plants, which flowered freely, and yet were sterile invariably. There was, then, no correspondence at all between those two items of his method as he saw and indicated them.

As regards the scientific merits of these two groupings our estimate must be formed according to their correspondence with the facts known at the time the groups were proposed, and not by bringing, for example, Theophrastus's groups of "flowering" and "flowerless" abruptly into contrast with those groups as they stand in recent botany under those same names. Such a mistake as this the historian Meyer[47] must have made when he pronounced the Theophrastian distinguishing between "flowering" and "flowerless" to be "of little value." Its worth or worthlessness should have been determined—in the judgement of a historian, at all events—by the measure of its answerability to the state of knowledge actual at the time of its promulgation, and also with reference to its usefulness as an incentive to further inquiry into the nature of plants as flowering and flowerless. Or, even as the historic starting point in the designation of two groups, in name at least equivalent to Phanerogams and Cryptogams, this Theophrastian and very suggestive expression is of deep interest, and of no small value.

These mere outlines of his general method must be concluded with the bare mention of some other aspects of the vegetable kingdom as it presented itself to the comprehensive and deeply thoughtful mind of the protobotanist. He discusses, and in much more than mere outline, meteorology and climatology in relation to plant life, has chapter after chapter upon ecology and geographic distribution, and even touches more than lightly the topics of plant pathology and the transmutation of species.[48]

Vegetative Organography[49]

Theophrastus begins his botany at the beginning. The remark is pertinent; for with recent writers it is no uncommon practice to begin somewhere toward the middle of the subject, leaving the foundations of the science out of sight. It is an easy way, avoiding as it does the difficulty and the responsibility of laying down first principles. To recognize three separate realms of nature seems necessary; yet to indicate clearly the boundary lines between them is confessedly anything but easy. Theophrastus at the outset acknowledged the difficulty of distinguishing universally between the vegetable and animal kingdoms, but he faced it. There was no evasion. He addresses himself at once to the task of defining the plant as differing from the animal.[50] And first he presents what must have been the popularly accepted method of indicating the distinction, though only to show its insufficiency for the purposes of science. Animals of whatever description have at least a mouth and a stomach. But it can not be said that all plants have roots, stems, branches, buds, leaves, or fruits. To circumscribe the plant world by listing the common organs of plants may enforce the exclusion of many things which are neither of the animal kingdom nor of the mineral, and are therefore plants. Here it should be remarked that well toward the beginning of the seventeenth century there were people—among these at least one very eminent botanical author—who did not allow the fungi to be classed as plants at all. Theophrastus knew all these lower forms of vegetable life, fungi both terrestrial and hypogeous, lichens and algæ in abundance, and proclaimed it unhesitatingly that they are plants, plants indeed without stem or leaf or root or seed. "The phases of plant life are so exceedingly diverse in nature and constitution that, to give a general (*i.e.*, morphological) definition of a plant, and that in few words, is not possible,"[51] he says. He ventures, however, one distinctive peculiarity of plants. "They are not, like animals, endowed with ethical susceptibilities and the power of voluntary action."[52] This is metaphysical, indeed, yet no whit more so than that which Linnæus gave in the middle of the eighteenth century when, pressed for one single mark of distinction between the genus *Homo* as separate from those anthropoid mammals next in rank, he was compelled to cite the "Nosce tepsum," that is, man's consciousness of his own existence.[53] Between animal and vegetable kingdoms the man of the eighteenth century could offer no better distinctions than this: "Plants grow and live. Animals grow, live, and feel."[54] But he of twenty centuries earlier had been more circumspect and cautious. Such movements as heliotropism and nyctotropism, and others seemingly akin to nervous irritability, were not unrecognized by Theophrastus, and may well have seemed to indicate something too closely allied to feeling.[55]

In defining the manifest organs of higher plants the philosopher proceeds with like caution, preferring physiological to morphological characteristics; in this, he is like the first precursor of the modern biologist, who, if required to name one distinction between plant and animal, confines himself to a point in physiology.

The root, Theophrastus says, is that by which aliment is taken up,[56] not a satisfactory definition to us moderns who demand morphological distinctions. Yet very safe is Theophrastus in his reserve; for what he names as characteristic of roots, albeit a merely functional characteristic, is one that holds. It is also apparently a new definition, framed by himself, and intended to displace another definition which, though time-honored, was fallacious. That antique fallacious definition of a root has not yet been universally displaced. Without going quite out of sight of some of our colleges one may elicit that same old characterization of the root, and from the voices of people who have a certain familiarity with this particular plant organ. Multitudes in every land under the sun today will answer promptly that the root is the underground part of a plant. The Lesbian boy, while yet untaught by either Plato or Aristotle, it is most likely would have given this same answer. In maturer years, after much careful questioning of nature, he has found that popular definition of a root does not hold good. The shrub or tree which the Greeks call *Helix*, the common ivy, he has observed minutely and experimented with until he has established it that those threads by which it climbs rocks, walls, and tree-trunks have nothing of the nature of tendrils, but are perfect roots, exercising in some degree even the usual function of roots.[57] He knows quite as intimately the shrub *Ixos*, the mistletoe, and that its seeds refuse to so much as sprout elsewhere than on the bark of living trees, into which bark it strikes its roots.[58] He had not seen the banyan tree of the Indies; but there were Greeks, educated Greeks, who had both seen and described it, with its many lesser subsidiary trunks grouped around the

FIG. 8. *Crocus* L. Theophrastus, *Historia plantarum* (1644), p. 667.

large central and original one. In their descriptions the travellers had stated so definitely the origin of the accessory trunks as starting out from main branches and growing straight downwards, that Theophrastus without hesitating declared that such things, however trunk-like they may at last appear, are roots.[59] Thus did the first master of organology completely invalidate the ancient world's definition of the root, and at the same time indicate with clearness the two categories of roots, subterranean and aerial. He did not, however, name them. As nomenclator of even his own most brilliant discoveries he was usually delinquent. We seem to read it between his lines that there were in his mind some suggestions of root characteristics which, had he been less cautious than he was, he might have added to that very reserved and merely physiological definition. The downward growth from the branches, in case of the banyan, and also the statement that those pendents at first and while young and tender are of a light color and hairy(!)—both these reported facts influenced his decision that these were roots. We shall also see presently that he had noted other morphologic marks, but such as he did not think universal.

FIG. 9. *Allium* L. The onion and related species were important for both food and medicine from antiquity down through the seventeenth century. Theophrastus, *Historia plantarum* (1644), p. 787.

The roots of herbaceous growths he in general classifies as ligneous, fibrous, and fleshy; and these include many things recognized in recent botany as subterranean stems, that is, bulbs, corms, tubers, and the more thick and fleshy forms of rootstock or rhizome. All really fleshy underground parts he distinguishes again as vertically elongated, spherical, depressed-globose, and what he calls nut-like, this last class embracing such incongruities as the small and solitary shell-covered corms of the crocus, and those tuberiform enlargements that appear, as if strung like nuts, on the roots of the asphodel. The rhizomes of arundo, and of what he names as arundinaceous plants in general, and which he remarks are sometimes partly above ground, he denominates jointed roots, but notices that these all have fibrous roots attached to them. Others, like the bulbs of squill and onion, are composed of a multiplicity of scales or tunics which can be removed one by one, so that these differ from other fleshy roots in that they exhibit two different kinds. Their nature is so peculiar that one might be excused for doubting that they are roots at all; for if in that they are subterranean they would at first seem to be such, they are in other particulars of quite another nature; because roots properly so-called diminish in size toward their lower extremity and end there acutely, whereas these bulbs and their like are widest at base and grow smaller in the opposite direction; moreover those fibres which descend from the bases of some and from the sides of others are the real roots which take up aliment; but the extuberant part is more like a fœtus, or a fruit.[60] However, after still further discussion, he seems to rest in the conclusion that, as roots are of various kinds, and even bulbs and other fleshy roots are functionally much alike, all may well enough be continued under the category of roots. If to any botanists of the twentieth century it may seem a strange thing that the Greek, having distinguished between roots as subterranean and aerial, should have failed—and after all his study of them—to classify stems also as aerial and subterranean, let them recall to mind that philosophic conservatism which led Theophrastus to make more of the function of an organ than of its form; that he was sure that corm and rhizome and tunicated bulb attract nutriment and are by that token roots. Let them also contemplate the fact that for twenty hundred years and more all botanists accepted the decision of Theophrastus, and that even with Linnæus, those organs which we have somewhat recently been learning to regard as subterranean stems were nothing but roots; and that Linnæus in this particular was even so far back of Theophrastus that he had no doubts about their being roots.[61]

That in the most primitive phytography roots received almost singularly minute attention has been adverted to, and the probable reason assigned. Accordingly we shall find Theophrastus, when done with their morphology, anatomy, and physiology, giving a full account of their differences as to color, odor, flavor, and their qualities as wholesome or deleterious. In color, some are white, others black, not a few yellow, some tinged with red, and some quite intensely red. As to odors and flavors there is again much diversity; and some that are sweet and pleasant to the taste are deadly poisonous, while several kinds that are of disagreeable odor or bitter are harmless, and even of medicinal value.

At the correct definition of a stem as being that part of a plant which bears leaves, Theophrastus did not arrive. His imperfect conception of the leaf, for one thing, stood in the way. Those merely scale-like short leaves, upright and even appressed to the stem, such as those of asparagus and orobanche, were mere scales in his view of them, and the stems of such plants he considered leafless. Again, to his vision there was a horde of stemless plants the leaves of which arise not from any stem at all, but directly from the roots. Here I can not forbear remarking that we of today, despite our better characterization of the stem, and our recognition of it as present in all except the very lowest plants, yet contradict our own definition in our practice, and have fallen back upon that of Theophrastus whenever we speak or write, as we freely do, about acaulescent plants and radical leaves. The ancient author defines the stem primarily as that part which is the main vehicle of aliment to the other parts, adding that it rises up singly from the ground,[62] which is of course to distinguish it from the branches and leaf stalks, both of which he knew to be also channels for aliment. This definition, equally with that of the root, evinces his distrust of morphological characteristics as definitive, and his feeling that the physiological are safer. But, the stem once defined functionally, he proceeds with care and skill to indicate its many morphological diversities; and here not only classifies but also names several of the classes. He acts as if he thought the language of science would be defective if such dissimilar things as an oak log, a mullein stalk, and a rye straw all were to bear the name of stem. In practice, therefore, he called the stem of a tree its trunk, *στέλεχος*, the light hollow or pithy stem of all grassy and reedy things was a culm, *κάλαμος*; thus the ordinary word stem or stalk, *καύλος*, was mostly limited to designate those of what we know as herbaceous exogens; and so our modern botany has these three kinds of stems as designated by Theophrastus. Culm is even very manifestly a modification of the Greek *kalamos*. Embracing as it does almost all endogenous stems, it is more comprehensive than the English words straw, reed, and rush all combined; and in our botany we were obliged to borrow and make over just the Greek term which Theophrastus invented—unless he, too, borrowed it; or, what amounts to the same, extended the use and gave a new and scientific meaning to an old and familiar term. As to their forms and modes of growth he distinguishes many kinds of stem among herbaceous plants. And, as woody growths are classified as trees and shrubs, according as their trunks are one or several from each root, so the herbaceous are distinguished as one-stemmed or many-stemmed. The numerous kinds of bulbous plants both wild and cultivated he understands as being invariably one-stemmed, and therefore does not speak of this in his descriptions of such; but of other herbaceous growths his custom is to mention, in his descriptions of them, whether the root sends up a single stem or many. It is a distinction of importance to phytography; and if the anthological extremists of one and two centuries ago thought it superfluous, and neglected it, its value is now again beginning to be clearly seen and freely admitted. Again, herbaceous stems are upright, or reclining, trailing, climbing, or twining.[63] He also has observed that one-stemmed herbs are apt to be erect, the many-stemmed otherwise; or at least that the reclining or trailing are always many-stemmed. Among upright stems he perceives how different those of the umbellifers are from most others in that they are fluted, or at least striate, and, giving these and all their like a name that really points to their anatomical structure rather than to their external appearance, he denominates all such plants nervose-stemmed. Through having missed the discovery of the good morphologic mark of stems, he is constrained to offer the category of what he calls smooth-stemmed plants. Such "stems" are nothing more or less than the flower stalks of acaulescent perennials, called smooth stems because devoid of those unevennesses now designated as the nodes and internodes, and received as the most universal mark of real stems. His examples of the smooth stem are those of the onion and leek,[64] good illustrations of the scape, as named and defined in later organography.

The leaf is a thing so almost infinitely diversified that he does not attempt to characterize it morphologically, or even physiologically, for he can not with any degree of certainty name its chief function. So without vouchsafing any definition of it, he goes about the rehearsal of its many aspects. Although as an organ it heads the list of that division of them which he has distinguished as transitory, and therefore in a manner secondary, no part of a plant would seem more deeply to have interested him, or to have been more carefully observed. He is even somewhat diffuse in his writing upon it, more so than in the case of any other organ, unless the fruit and seed are to be excepted; and, since the opinion now prevails almost too widely that little was done in the direction of plant organography until within the last two centuries, the interests of truth can not at just this point be better subserved than by giving the substance of this ancient Greek's morphology of the leaf somewhat in detail.

> Leaves are commonly attached to the stem or branch or to whatever else supports them, by a stalklet; this either firm and holding the leaf steadily in a certain position, or else slender and feeble, allowing the leaf to hang downward and perhaps tremble with the passing breeze, or even to become inverted, turning the usually paler lower face upward. But there are also leaves with no stalklet, these adhering directly to the branch. Some leaves arrange themselves only in opposite pairs, with regular intervals between the pairs, while others are scattered singly and without order up and down the stem.[65]

It will be seen that these beginnings of Theophrastian teaching about the leaf are precisely what one finds in every primer of botany today. Every beginner has to be taught the importance of the distinctions between a petiolate leaf and one that is sessile, and between the opposite and the alternate in leaf arrangement. We have improved the descriptive phraseology, and are able to say the same things in fewer words; but that is about the only difference. Some general differences in the configuration of leaves are also adduced. There are those of rounded periphery, or even somewhat elongated, all without angles, and there are the angular in outline, some of them like those of the fig deeply cleft, others like those of the oak sinuated all around, still others with saw-like teeth all around; and some are sharply pointed at the apex, the slender leaves of pine and others even ending in a prickle. In certain thistles he notes that spines take the place of foliage. He is convinced that certain leaf-like organs in a number of asparagus allies are not leaves, yet he gives them no name; nor had they obtained a name—that of cladodes—even as late as the time of Linnæus, who, as if he had been of a pre-Theophrastian age, still called them leaves. The hollow and fistulous foliage of the onion and some of its kindred elicited remark from Theophrastus as being very exceptionally curious. So did that of the sedges, as being conduplicate and keeled. The essential characteristics of the leaf that is pinnately compound he also seems first to have detected; for he is at the trouble to argue the case before those who, as he seems to acknowledge that he himself also once did, regard this as a leafy branch. He has observed the autumnal falling of the foliage in the ash tree, elder, and sorbus, and reports that the whole of that which seems a branch falls away piece by piece, thus establishing it beyond dispute that the whole is one leaf. He even speaks of it, afterwards, as the pinnate leaf.[66] And, as there are kinds of tree in which leafy branches, by being somewhat lengthened and having their leaves set closely in two opposite ranks, resemble the compound leaf, Theophrastus in one place that I have noted cautions the student against being deceived. The case is that of the elm, the pinnately leafy twigs of which might be mistaken by the unwary for compound leaves. The observer is warned that the elm has but a simple leaf, φύλλον ἀσχιδὲς.[67] The leaves of the rose bush I do not find described in Theophrastus. These shrubs were so universally familiar that description of their foliage was needless, but that he recognized this as a pinnate foliage is evident by one comparison that he makes. Wishing to convey some notion of the leaves of an interesting tree of the Orient (*Tamarindus Indica* L.), he says it is "many-leaved, after the manner of the rose bush,"[68] meaning that the individual leaf of the tamarind tree is of many leaflets, like that of the rose. The philosopher devised now and then a new botanical term, but did this with reserve; and the discovery of compound leaves does not appear to have called, in his thought, for any such distinction as that of leaf and leaflet. He applies the term leaf, φύλλον, to the compound leaf as a whole, and to the individual leaflet indiscriminately. As regards the differences of compound leaves and the classification of them he did nothing; nor, indeed, does anything appear to have been attempted in this direction, after Theophrastus, until the time of Jung, who in the middle of the seventeenth century strongly advanced the morphology of the leaf.[69] But Theophrastus in his environment cannot have met with anything like that diversity of compound leaves with which Jung was familiar. He must have known the bipinnate fronds of certain ferns, but did not essay any description of them; and when a certain bipinnate-leaved tree from Egypt (*Mimosa polyacantha* Willd.)[70] was in need of a description, he evaded the difficulty of the situation by saying that its leaves were like those of a fern.[71]

We shall be furnished later with some proofs that the venerable Greek could make significant discoveries in connection with such very little things of the plant world as small apetalous flowers, and even the inner structure of small seeds. Without microscope, hand lens, or spectacles, he seems to have been almost microscopic-eyed sometimes, as well as always alert for the detection of the exceptional or unusual in the grosser morphology of things. We therefore wonder that, after his having noted so carefully that some leaves, even small ones, are compound and even doubly compound, he should not have taken stipules into account; for he makes no mention of them. During it may have been fifty winters he had seen the branchlets of the plane trees, under which lay his daily walks, encircled at intervals by their wheel-shaped stipules still persistent after the body of the leaf had fallen. During as many summers of his centenarian career he had observed the foliage of many garden Leguminosæ, in some species of which everywhere in cultivation anciently more than half the foliar area is stipule; and yet this organ is unmentioned by Theophrastus. As to its presence on the boughs of the most common Athenian wayside shade tree, and its more conspicuous showing amid the herbage of every sort of pea and vetch and lentil, this father of plant organography is as silent as if the organ had not existed. This seems anomalous. To the average reader it will seem like a curious hiatus in the Theophrastian leaf morphology. The situation seems nevertheless readily to explain itself. One has but to recall to mind the ancient botanist's strong inclination to regard function first and form last everywhere in his organography. The excessively enlarged and leaflet-like stipules of *Pisum* and some other leguminous herbs have not the least appearance of being functionally different from the other leaflets. They are larger and also located a little differently from the others, but that is all; and there is the best of evidence that a thorough training in modern organographic refinements is requisite to the determination of the enlarged basal leaflets of the pea vine as stipules. The evidence is this: that the most original and logical of great organographers and terminologists, Joachim Jung, as late as the year 1662 of our era knew nothing of any such organ as a stipule. Tournefort in the year 1700 knew them not, and Linnæus claims them as among his own organographic discoveries, though unwarrantedly, as we shall see later.[72]

There is, however, one particular kind of stipule which, unless I misunderstand Theophrastus, drew his attention and elicited his comment. It was a case in which there is about the strongest possible contrast in appearance between the leaf proper and the stipular appendage, that of certain umbellifers in which, while the decompound blade is deep green and almost capillarily dissected, the large stipular development below it is pale, membranaceous, wholly uncut, sometimes cup-shaped, and hollow. Here again it would seem to require our modern refinement in organography to perceive in such a thin bladdery stipule and green, capillarily cut blade the different parts of one and the same organ. Theophrastus, at all events, knew some umbellifers of this description, and wrote concerning one species of the ge-

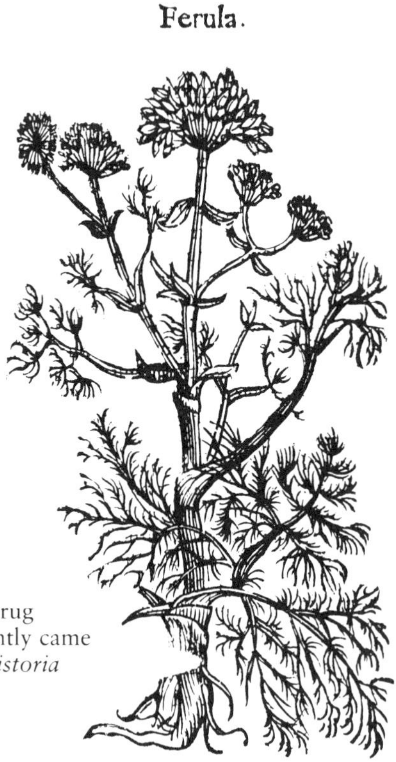

FIG. 10. *Ferula* L. Silphium, a drug widely used in antiquity, apparently came from a member of this genus. *Historia plantarum* (1644), p. 583.

nus *Ferula* that it puts forth, at the nodes of its stem, leaves and *βλαστοί*.[73] The blastoi, one from each node of the stem, have perplexed some of the botanical commentators upon the text. I think they are the stipules.

If Theophrastus does not anywhere formally define the leaf, that may have been for the reason that, not at all comprehending its function, it was not possible for him to define the organ, as he had defined root and stem, physiologically. Nevertheless he did state, in a most informal way, its very best morphological mark, that by which it is almost always readily distinguishable from a stem or any part of it. He observes that most leaves have an upper face and a lower. The upper, exposed to the sun, he notes as being of a deeper green and smoother; the lower face is paler, roughened by the greater prominence of the veins, and apt to be pubescent.[74] From the fact that the lower face of leaves when pubescent is found to be moist he says it has been inferred that it absorbs nutriment and feeds the rest of the leaf. This he confidently declares to be erroneous, affirming that all parts of a leaf are fed by way of the veins and fibres which are carried to every part, and which he knows to be connected with the bark. This was the thing which he could easily demonstrate and prove. The other proposition was not in his day demonstrable; for he could not know either the structure of a plant hair, or the existence of stomata. Theophrastus, like his class in all ages, is likely to be correct in that which he affirms, and wrong as to that which he denies.

Anthology

Without any understanding of flowers as organs of sex, and quite in the dark as to their significance in the economy of plant life, Theophrastus applied himself assiduously to the study of their morphology, and that with a measure of success compelling the admission that he is the founder of anthology; for several of his distinctions remain fundamental in the anthology of today, notwithstanding that the theory of the flower has been completely revolutionized within the modern period.[75] And his success is the more remarkable because of his having made his researches at a time when, for mere lack of optical equipments, the discovery of the functions of the essential floral organs was impossible.[76]

From ages antedating all history cultivators must have observed that in such trees, garden shrubs, and herbaceous field crops as flower conspicuously, no fruit or seed develops but as an aftergrowth from the flower, that a young tree never fruits until after having for the first time flowered, and that any mischance befalling the flowers of the tree in their season extinguishes, for that year, all hope of fruit.[77] Upon a considerable array of facts of this kind, the first philosophic investigator who came along might naturally propound such a theory as this, that wherever there is now a fruit or seed, at some time there must have been a flower, a proposition which the cultivators at once and with one voice would have disputed; for in the husbandry of antiquity no tree was more esteemed, nor any more familiarly known than the fig tree, and this, as all the world understood, produced its fruits two crops a year, without a trace of flower. And Theophrastus, after all his searching for and philosophizing about flowers, seems to have found no way of controverting the universal opinion. He thought that the fig produces its fruit without flower of any description.[78] But in his philosophic quest for flowers of some sort as the forerunners of all fruits and seeds, he appears to have discovered true flowers, though sometimes recondite, in other trees that had been supposed to be like the fig, flowerless. The flower, in prehistoric thought and speech, may most reasonably be assumed to have been a thing showy on account of its being made up of leaves colored differently from ordinary foliage, and differently arranged. It must have been essentially that which modern botany knows as a corolla. This inference as to what a flower was before either botany or history began to be written is confirmed by our experience with untaught rustics and mountaineers of today, as to their understanding of what trees and plants have flowers and what have none. They are the modern counterpart of those unlettered ruralists of remote antiquity whom Theophrastus cites as denying that oak and walnut trees, hazel bushes and chesnut trees and junipers have any flowers at all.[79] The philosopher, the man of science who is truly such, has this among other characteristics, that with him negations are apt to go for naught. Of the populace they are largely the mental stock in trade, so to speak, but himself negations do not satisfy. They say that neither oaks nor hazel bushes have flowers. They recognize it that oaks put forth clusters of loose pendulous tassels that they call oak-moss, and also globose bodies denominated galls; but oak-moss is not a flower, any more than oak galls are acorns. These are specimens of the facts, and of the reasonings upon them, which confronted the protobotanist of so long ago. Stimulated by the thought that almost always where a fruit or seed now is, there was once a flower, from the very heart of which the fruit or seed took its origin, he enters upon his researches. Now this very idea that flower and fruit are related as antecedent and consequent so that where any manner of fruit or seed is found the essentials of a flower must be sought, is the germinal idea out of which the whole of systematic botany as we have it has been engendered; and the originator of that idea would have been the father of botany even if he had accomplished nothing further.[80]

There is one thing which he who would know, even in outline only and superficially, the history of botany, must not do. When in Theophrastus he meets with the word *ἄνθος*, or in Pliny with the term "flos," he must use care not to read into that term the meaning which the word flower has in modern botany; for, if he assumes that the word stood, with those authors as with us, for a congeries of four circles of different organs, beginning with the calyx and ending with the gynœcium, he will never correctly apprehend a word they say about the flower. That view of the comprehensiveness of the flower which we now take, extremely unlike the ancient idea, was really first presented for acceptance within somewhat less than two hundred years from the date at which I write. The flos of remote antiquity, the pre-Theophrastian anthos, appears to have been simply the corolla, as we have said before; and that without a special name as such. It was but a set of leaves, shaped and colored and arranged differently from ordinary foliage, and having for its function the protecting of the future fruit and seed while in their tender and rudimentary stages.

Now oaks, walnut trees, alders, and hazels have no corollas. They had been considered flowerless because they have none, and correctly enough so long as the flower was defined as a whorl or tuft of specially altered and colored leaves; and it was so defined in the minds of the majority of people in that time, as it is in the minds of untold thousands in every land today. And the very possibility of detecting upon oaks and filbert bushes some small thing that should mark the point of origin and presage the coming of each nut and acorn involved the possibility of a revolution in the idea and definition of a flower, an extension of the term, to make it embrace anything, no matter how colorless, shapeless, and obscure, which should be found in the place where a flower ought to be.

This earliest *Historia Plantarum*, intensely interesting though it be as we have it, would have been still more so had its author given some record of his own processes of research; his successes and his failures in attempting to find flowers on trees and herbs that had the reputation of being flowerless. But the traditions of the lyceum at Athens were against that. Men were taught that knowledge is best communicated in language concise and brief; and Theophrastus's three short chapters of anthology may vie with any other three chapters of scientific matter ever written, in respect to the terseness and brevity with which important propositions follow one another in close succession.[81] His first proposition as to the general morphology of the flower is this: "Some flowers are capillary, like those of the grape, mulberry, and ivy; others are composed of leaves, like those of the almond, apple, pear, and plum trees."[82] These are trees; but he proceeds to say that quite the same is true of the flowers of herbaceous and annual plants, some of which have foliaceous flowers, others capillary. It is evident as can be that by examining the earliest germs of fruit in plants that never show flower leaves, he has found those fruit-germs at a certain early period encircled by hair-like or filamentose things quite as transient as flower leaves, and which seem in some way to take the place of them, though they have not always the usual special coloring of flower leaves. On the strength of what he has studiously observed, he has now virtually given to the term ἄνθος, flower, a new definition, a scientific one. The term must embrace whatever is intimately though transiently connected with a fruit-germ, whether laminal and colored or filamentose and greenish. This, insofar as written records show, is the earliest proposition ever laid down concerning the morphology of the flower; and it was a mighty contribution to scientific botany. It is in substance the distinction of petaliferous and apetalous flowers. It will therefore hold its place in the science of plant life and form as long as such a science shall exist.

The investigations of Theophrastus along the line of what we denominate apetalous flowers appear to have opened his eyes to the presence of the capillary organs in a large and showy petaliferous kind; for in this same chapter he states that many flowers are two-fold, showing another flower inside the main one. He cites such familiar garden flowers as the rose, violet, and white lily as examples; and, as against any suspicion of ours that his two-fold flower of rose and violet and lily might mean a double flower, as composed of multiplied petals within the main outer circle, there occurs the one word διχροα, two-colored, or differently colored. It is, of course, the stamens within the corolla of red rose, purple violet, and white lily that are colored differently from the corolla. This is the earliest recognition of the flower as other than a simple organ. It is the beginning of the classification of its parts, a small beginning, but highly significant. It is given out for the first time, that many flowers have two circles of organs, a flower within a flower, the one within readily distinguishable from the broad, leafy one outside. He does not formally name this inside flower, but he has found so many flowers that lack the leafy outside circle altogether, displaying nothing but the inner, that he names this kind capillary or woolly flowers. It must here be affirmed that Theophrastus knows nothing of the calyx as being any part of a flower. The color and texture of floral organs were what distinguished them from ordinary foliage; and by their points of agreement with the latter any green leaf-like organ or circle, however near the "flower," would fail to be included as a part of it. Also the ovary and ovules were not indicated or received as organs of the flower. They were simply the fruit or seed, in whatever stage from that of the flowering to that of full maturity; and this neither through dullness nor indifference. The colored leaves, together with the colored threads, set in the midst of them, were all there was to the flower. One

FIG. 11. Pear (*Pyrus* L.). *Historia plantarum* (1644), p. 397.

may fancy some brilliant Greek pupil asking the master if that protuberance in the middle of many flowers ought not to be regarded as a part of the flower, and called the fruit, ought not to be called by a name of its own while in the flowering stage. He who knows the keenly penetrating and severely logical mind of Theophrastus will infer without chance of a mistake, what the substance of his answer would have been. At what particular point in its development will that protuberance begin to be a fruit? I suppose that such logic might silence the ablest morphologist who has lived hitherto. Our modern term ovary is but an illogical convenience. It suitably abbreviates the following expression: "fruit at the budding or flowering stages and for an indefinite period thereafter." Our neighbors the industrial botanists even of today have no need of the term ovary and ignore it. When a hard unseasonable frost has sterilized the ovaries of their trees, whether in bud or in flower, it is the "fruit" that has been killed; and so the Theophrastian anthology still lives and is widely though unwittingly approved.

What are now known as the styles were not segregated from the other flocci, or capillamenta—that is, from the stamens—until ages after Theophrastus. He made no distinction between these two, at least when found together within the same flower; and his capillary flower might consist of stamens alone, or of styles only, or of both. What is more, there are certain arrangements of stamens under which they failed to gain recognition by him as being of the nature of floral parts, as in the aments of hazel, the walnut, the oaks, and many more. He gave them a name, *ἰούλοι*,[83] Latinized as *juli*, writing about them so minutely and describing them so well as to attest his perfect familiarity with them; but apparently the more he studied them the more enigmatic did they seem.[84] The filbert, *Corylus avellana*, not only grows wild in Greece, but it had been cultivated there doubtless for ages before Theophrastus's time.[85] It would be irrational to question that it was among the shrubs of that botanic garden in the midst of which lay his daily walks for many years. At all events, his perfect familiarity with its tassels is attested by the following account which he gives of them.

In autumn after the nuts have fallen, there appear in bunches of several certain things that look like worms, inserted on a short thick stalk. These are called juli. Each is made up of countless scales arranged somewhat after the manner of those of the nut pine (*Pinus pinea*, Linn.), the whole longer in proportion to its thickness than that, and also of equal thickness throughout. Before the end of winter it begins to grow. In early spring the scales separate and stand apart, and are then become yellow, the whole then sometimes as much as three inches long. When the leaves begin to put forth, these things shrivel and fall. Then the cups that enclose the nuts develop; one cup for each flower, and one nut in a cup.[86]

The concluding sentence places it beyond doubt that the writer knew the crimson pistils of the shrub as well as he did the yellow aments. He does not stop to describe them. They are of his class of capillary flowers, and that is enough. In a later chapter,[87] in which he brings out the habit and vegetative characters of the filberts, indicating two species (*C. avellana* L. and *C. tubulosa* Willd.) by differences of fruit, he has no occasion to mention again the flowers, but can not forego renewed allusion to those perplexing aments. "To these shrubs belong that julus of compacted scales which we have elsewhere described." The sterile aments of the oaks, slender, lax, more tufted than those of the hazel as well as of short duration, must also have been known to Theophrastus; and so were the colorless and very inconspicuous pistillate flowers; for, while he quotes the popular opinion of his time that these and other like trees are flowerless, he tersely contradicts it. "Both oaks and alders flower";[88] by which he must have meant the axillary and scattered pistils of the oak, and the cone-like clusters into which the alder pistils are congested.

Of the aments of fir and pine I observe no mention in Theophrastus. Even the young cone of pistils with their subtending scales in the conifers, from its form he denominates a julus; but it is not of the enigmatic class. Without hesitation he denominates that the flower. It is also plain that with him this must be received as a petaliferous or leafy flower, for it shows nothing that could be called a capillamentum. The rudiment of each fruit rests in the axil of an ample and highly colored leaf.[89] Nor does he here cite the opinion of any of his forerunners or contemporaries as having denied or questioned that these are flowering trees, as they had done in the case of oaks and alders and hazels. The flowers of the fig tree he could never discover. To him it was as flowerless as a fern or moss. He was loath to believe that junipers are not equally flowerless. He had investigated them; had observed that in summer their fructiferous branches bear one set of fruits full grown, and another set newly formed and not half grown; a proof that its fruits require a year and somewhat more for their growth and ripening. He can hardly have failed to see the staminate aments, small though they be, and of brief duration. They were nothing, in his view; at least nothing floral, and not worth mentioning after he had once described the like phenomenon as conspicuous and of long duration in other trees and shrubs. What he was looking for, he could never find, that is, what he would have accepted as a flower, a folium, or a capillamentum indicating the seat of the juniper berry that is to be. The pistillate or fertile juniper flower is as far away as it is possible to go from having the appearance of a flower at all. It shows even under a lens no trace of style or stigma or ovary. It is so little different from the minutest first rudiment of a merely vegetative twig, that an experienced botanist, even of these later times, may fail to recognize it, though he search with a lens. It is improbable that there is a man in the world today who, in the feeble botanical light of the Theophrastian age, and without the aid of magnifiers, would ever have found the pistillate flower of a juniper.

FIG. 12. Alder (*Alnus* L.). *Historia plantarum* (1644), p. 220.

But what is fully as interesting as Theophrastus's failure to find anything upon a juniper tree which he could demonstrate to be a flower, is his recording the opinion of those who think differently. "There are those who say that the juniper tree is of two kinds, one that flowers and bears no fruit, and another that is flowerless but fructiferous."[90] The pronouncement is interesting as being diametrically opposed to the Theophrastian doctrine that nothing not in immediate juxtaposition to a fruit rudiment is to be regarded as a flower. It is a virtual contradiction of the opinion that juli or aments cannot be flowers. These people who held that male and fruitless juniper trees have flowers, and that fertile ones have none, were people who evidently regarded those small evanescent yellow, dusty male aments of the juniper as true flowers, even the only flowers that any juniper ever has. If evidence was elsewhere wanting to prove Theophrastus a true philosopher and scientific man, devoted to the truth whatever that may be, rather than to his own theories, it is not wanting here. He publishes this adverse opinion of his neighbors for the very reason that it may possibly turn out to be the right opinion, concluding the whole passage with the recommendation that investigation of the subject be continued. "The matter should be looked into further."[91]

Quite as briefly as he had indicated the distinction between leafy flowers and capillary does Theophrastus give the suggestion that the leafy flower in certain plants is made up of but a single leaf.[92] It is practically classifying corollas as choripetalous and sympetalous. He writes of what he calls the monophyllous flower as if within the field of his observation it had been somewhat exceptional; and he warns the reader that it is not always distinguishable at a glance from the other kind. Viewed as to their periphery they will seem to be made up of separate leaves, but at the center or base they are seen to be monophyllous. In the morning-glory (*Convolvulus sepium* L.), however, the monophyllous character is readily apparent, only a certain angularity of the periphery remaining in place of the appearance of separate leaves. Even small flowers may be monophyllous. Such are those of the olive tree. Lying on the ground under the trees they are readily seen to be perforate. From his having cited the olive blossom, one is assured that he held a corolla to be monophyllous even if the leaves were united only at base. But one must guard against mentally attributing to him any general recognition of all sympetalous corollas as being such as that to which later generations have arrived; for evidently he had not proceeded far along this line of anthological research, but was only at the beginning of the subject. Having so perfectly settled the monophyllous structure of the small and quite choripetalous-looking olive blossom, we are disappointed that he has not done as much for that of the elder tree. It is a tree in which he has manifested a special interest, and with which he has grown very familiar. The thin wood of its shoots and branches, along with an extraordinary development of pith, then externally the remarkably long internodes, and the foliage, from watching the falling of which in autumn he seems to have learned the very important classification of leaves as simple and compound—all these aspects of *Sambucus* he has noted fully. Will he not perceive that its flowers, like those of the olive, are of that structure which he designates as monophyllous? They are too small, much smaller than those of the olive tree, even quite minute to one who is without a lens; and Theophrastus may not have examined them very carefully as individual flowers. Either that, or else, in condescension to popular usage, he permits the corymb or umbel of small flowers to pass for a flower. And so he describes the blossoming of the elder thus: "The flower is white, composed of many small ones all white, the whole with the appearance of a honeycomb, and attached to the summit of a shoot by a number of stalks."[93] There is another type of monophyllous flower not as small, which first and last remained to him an enigma. It was the ovate, hollow, and pitcher-shaped corolla of *Arbutus unedo*, a most common type among ericaceous plants. He says it is not a leafy flower, an expression equivalent to the modern phrase apetalous flower. He describes it as being in the form of an egg-shell with one end cut off, leaving an aperture.[94] He cannot have detected the five obscure recurved teeth at the orifice; for they would have taught him that this, like the faintly five-angled morning-glory blossom, is of five almost completely united leaves.

One chapter of the *Historia* opens with a sentence like this: "Flowers differ in respect to their origin and insertion."[95] It is one out of a number of Theophrastus's brief statements of significant fact, any one of which would have rendered famous any herbarist of the sixteenth century or the seventeenth who had been privileged to announce it as a discovery of his own. In the most recent and approved taxonomy of flowering plants, this point in anthology, first indicated by the ancient Greek, holds a most conspicuous place. Let the Greek himself explain what he means by the origin and position, or insertion, of the flower, always keeping it in mind that with him the leaves and the thread-like parts in their midst are all there is to a flower, the ovary being the fruit.

> Some produce the flower around the [base of the] fruit, as do the grape vine and the olive tree. . . . In the greater proportion of plants the fruit thus occupies the center of the flower. But there are not wanting such as support the flower on the summit of the fruit, as do the pomegranate, apple, and rose, all of which have their seeds [ovules] underneath the flower. A few bear the flower on the summit of the seed itself, such as the thistles, and all that have their flowers in that manner crowded together.

It is a clear distinguishing between the hypogynous, perigynous, and epigynous in floral structure; clear notwithstanding that the one example brought forward to illustrate the epigynous insertion, that of the flowers of the composites, was not from the modern point of view well chosen; because then what he understood to be the seed we regard as a fruit. If he had been accustomed to assign names to what have proven to be his great discoveries in anthology, he would have called this third mode of insertion the epispermatous. He learned this springing of the "flower" from the top of the "seed" to be characteristic of the whole family of the umbellifers, and of the few rubiaceous plants that he knew, as well as of the thistles and their kindred. It seems to me that what is more to be wondered at in Theophrastian anthology than his distinguishing of the hypogynous, perigynous, and epigynous modes of insertion, is the fact of his having made out so positively, that the head in the composites is not a flower, but that it is a dense cluster of separate and distinct individual flowers, each complete in itself. Less than three generations ago, eminent systematists were still writing up the scales of such involucres as "sepals," the whole involucre as a "calyx," and the circle of ray flowers as the "corolla." At this juncture the sublime old Greek will appear to have lived before his time by more than two thousand years.[96]

FIG. 13. Strawberry tree (*Arbutus unedo* L.). *Historia plantarum* (1644), p. 244.
FIG. 14. Basil (*Ocimum basilicum* L.). *Historia plantarum* (1644), p. 757.

In his study of flowers the arrangement of them was not unnoticed. He observes that in most trees they appear as scattered on all the branches, all appearing nearly simultaneously, so that the flowering period of such is but short. In many herbaceous and half-shrubby growths they are clustered together; and in describing such plants as to their flowering he makes frequent use of terms equivalent to spike, raceme, and umbel, though not with such definiteness of meaning as they convey in modern botany. In these clusterings it is observed that the flowering of the whole cluster at once seldom takes place, that usually the lowest flowers are first to expand, then those next above them, this succession continuing in some plants so long that the seeds from the basal flowers are ripe before the terminal flowers have opened. The aromatic garden herb ocymum is named as a case in point;[97] but he mentions this kind of inflorescence repeatedly. He also names one plant whose flowering begins at the top, the succession of later bloom following downwardly. Thus is Theophrastus again a botanical discoverer. He has distinguished between the centripetal and the centrifugal in inflorescences. The historian Meyer was surprised at this, remarking also that he knew of no other botanist's having noted this distinction again until the time of Link and Robert Brown.[98] It is evident that Meyer pondered the fine picture books of his compatriots of the sixteenth century, Brunfels, Fuchs, and Tragus, to the neglect of the one real botanist that there had been among them all, Valerius Cordus.[99]

Fruit and Seed

Without fully appreciating the significance of fruit and seed as furnishing the best clew to plant affinities, Theophrastus nevertheless studied them assiduously. Even flowers in their beauty and fragrance, and by their multitudinous forms, engaged him chiefly as being heralds of the fruit and seed. The perfecting of fruit he alludes to here and there as being the culmination of the plant's existence. He notes it that even such vigorous and enduring things as trees and shrubs shorten their life period by excessive fruit-bearing; that myriad annuals live but the length of one summer season because they exhaust all their vitality in the yielding of their one crop of seeds.[100] Seeds were of very special interest, in his view; and succeeding generations of botanists have been with him in that opinion.

The scientific examiner of even commonest objects finds more things in nature than there are names for in common language. The investigator of things is therefore obliged to be the inventor of new terms; and every science has therefore its vocabulary of special terms, every one of them necessary to the science and to the man of science, but to the world at large useless. In connection with his study of seeds, Theophrastus was obliged to invent new terms, and to give new meanings to old ones. Most seeds were grown and perfected under some special covering formed to shelter, contain, and nourish them until mature. For particular kinds of such coverings particular names were in common use: pod, husk, chaff, shell, and for succulent or fleshy coverings of seeds pome, berry, acine, or more comprehensively, fruits. A general term which should include all these coverings was needed, and the word pericarp was coined.[101] This done he defines a fruit scientifically. It consists of a pericarp and the seed or seeds which it encloses. Henceforward, while in agriculture, gardening, domestic economy, and the world's commerce a fruit is what it always was, in botany the term has another meaning, a meaning at once more exact and more comprehensive; and it has this new meaning universally, and from Theophrastus forward; for modern botany reiterates it from him, unaltered by a syllable; and that of the future will do the same. In practice he did not always rightly distinguish between pericarp and seed. Lecturing upon the fruit, and having a mature sprig of sage or other labiate in hand, he would have taught that the four black nutlets are the seeds, and that the green calyx is their pericarp. Or with a handful of spikes of wheat or barley before him, he would have mistaken the grains for mere seeds, and the chaff for the pericarps. Errors like these in the mere application of his terms were inevitable. They could never have been corrected without microscopically aided vision; and it was indeed a long, long time after the invention of the microscope that botanists first learned the structure of sage nutlets and wheat grains to be that of fruits and not that of seeds.

About pericarps he seems to have observed everything that lay before him within his own limited field. He notes the extreme diversity of them, but, as usual with him, and doubtless for want of time to correlate and classify he gives to the most distinctive kinds little more than an informal mention. Only a single deduction does he venture concerning pericarps in general as unlike other organs, a deduction superficial, curious, geometrical: "No pericarp shows a rectilinear or angular circumscription."[102] Yet the cursory reader of the main chapter on fruits—perusing it in the Greek original—might well wonder with what justice or propriety it can be said that the philosopher did not carefully and effectively generalize about seed in relation to pericarp, when he finds him using over and over again such exceedingly important taxonomic terms as angiosperm and gymnosperm. The fact is, he employed neither in anything like that breadth of meaning which they convey as used in later botany; but both of them very restrictedly. He used angiosperm only to designate one particular circumspection of what are known with us as capsular fruits, namely, the subglobose or urn-shaped or vase-like kinds. The example given is "that of the poppy and those like it."[103] It is correlative with λοβός, the Theophrastian name for all leguminous fruits.[104] His gymnosperms include nothing really gymnospermous in our taxonomic use of that expression. By the examples cited, "coriander, anise, fennel, cyminum," and several others among kitchen-garden plants, the gymnosperms were the umbellifers. They were naked-seeded to him, because, as already noted, he had not recognized any such organ as a calyx; one nowhere in all the plant world more recondite than in the umbellifers. The only real gymnosperms—according to our application of the term—which he knew, were the conifers; but they do not enter, with the umbellifers, into his category of that name. He expresses distinctly though modestly the idea as his own that, as to fruit and seed these stand naturally aloof from all other groups, and thinks the view may be tenable that cones are not fruits at all. "No trees bear capsular fruits, unless you can call a cone a capsule; but it is possible to regard the cone as different from a fruit." Under the head of anthology it was seen that the cone-scale at its flowering stage was a flower leaf in Theophrastus's understanding of it. Logically, therefore, he would have regarded it in its maturity as a sort of pericarp. But that he left the cones in a place apart, as unclassifiable with other seed vessels—the types of what were to be named gymnosperms twenty centuries afterwards—is yet another evidence of the profound sagacity of the protobotanist. After these things, it is no longer with any surprise that we read his accurate descriptions—descriptions of them from center to circumference—of such fruits as the drupe, the pome, the nut, fig, pomegranate, and other types, and citing as he always does familiar examples of each different kind. But that he should have been near as he was to a systematizing of the placental attachment of seeds within the pericarp is again almost startling; for when he records it that in many the seeds are as it were promiscuously crowded within the pericarp, while in others they are ranged in regular lines, or at least in separate groups—of which latter he says the squash, cucumber, and apple are examples[105]—we know that he has been carrying on research in this direction, and has been able to make this fair beginning at discriminating the different modes of placentation.

The foundations of the whole philosophy of higher plant life and form center in—are concentrated within, if one may so speak—the seed. In botany no less than in zoology is embryology indispensable to a right understanding of the interrelations of things. From the minuteness of his researches into the structure of seeds and the behavior of seedlings, it almost seems as if Theophrastus may have realized this fact. He records many observations on them all, even their most familiar aspects, not neglecting the diversity of them as to form and coloring.[106] In a few terse sentences he gives the results of what may have been years of investigation in his botanic garden, upon the subject of the different periods of time required by the seeds of different plants for their germination.

> Ocimum, blitum, eruca, and radish are most prompt of all, for they come up on about the third day after the sowing; lettuce on the fourth day; cucumber and squash on the fifth or sixth; anise on the fourth, pepper-grass and mustard on the fifth; beets sown in the spring, on the sixth, in the fall, on the tenth day, orache on the eighth, cabbage on the tenth.

Leek and shallot are such close congeners that he evidently expected they would agree as to the time required for their germination, but he finds the seed of shallot coming up at the end of from ten to twelve days while that of the leek takes nineteen or twenty. More than thirty days must be allowed, he says, for either satureia or origanum, and forty for celery. After long experience he finds it remarkable that the most favorable conditions as to the season of the year and the state of the atmosphere do not shorten the usual time required for the germination of any kind of seed, though a cold atmosphere, concurring with clouded skies, retards it.[107]

These studies in seeds and seedling plants, though by chance interesting and instructive to the gardeners of his time, are essentially those of a great botanical philosopher, with whom not the smallest fact relating to plant life is held unworthy to be placed on record. And as he proceeds, the twentieth-century botanist will be apt to read with amazement a passage like the following as occurring in Theophrastus. "Some seeds in germinating put forth their primary root and leaf from one and the same point; others, the root from one end and the leaf from the other."[108] Preserving all the terseness and brevity of the original in this very literal rendering, the statement of the fact may seem a trifle exaggerated. The first roots issuing from a grain of wheat or barley appear not quite from the base, but from near it, and the first leaf appears from a point well toward the summit of the grain. The two do not, as in dicotyledonous plants, come forth from the same point. But thus early in the history of botany, even by this first forefather, was given in these words, the first hint of a fundamental distinction between flowering plants as dicotyledonous and monocotyledonous. By way of further elucidation he continues:

> Wheat, barley, rye, and all the grains sprout from both ends; that is to say, the basal and thicker end of the grain puts forth the root, the upper and narrower end the green herbage. The two, however, are connected and continuous as one. But neither the bean nor any other seeds of leguminous plants have this way of sprouting. These put forth root and stem from the same point, namely, that at which the seed was linked to pod, as if under that point [the hilum of later terminology] lay the special seat of the growing principle. In the case of seeds of this kind the root at first appearing begins to show a downward tendency, the stem an upward. The seeds of the frumentaceous and the leguminous kinds are alike in this one particular of sprouting from the point of insertion; but of certain trees, the almond, walnut, the oaks and their allies, the seeds sprout from the opposite end.

I have omitted here one of the most important clauses, that in which he indicates his having observed in the bean and lupine allies the two cotyledons, joined to the hypocotyle.[109] Later in the same chapter he states without simile or comparison the same characteristic. "The seeds of all the latter," that is, of the particular trees he has mentioned, and of the leguminous herbs, "are in a manner two-parted." And again: "Wheat and barley make their first appearance with but a single leaf, peas and beans with several"; from which it is manifest that he counts the cotyledons as leaves, along with the one or two that appear in the plumule. Still other facts and phenomena observed and recorded by him about germinating seeds, and young seedlings, must enkindle toward Theophrastus the wondering admiration of the most accomplished modern botanist. He says that he finds it uniformly true, whatever the kind or the structure of the seed, that the "root" is first to appear, after that the leaf or leaves; also that the cereals, while as yet showing but a solitary primal leaf, exhibit quite a tuft of roots, these all simple and equal, whereas the two-parted seeds in their germinating exhibit several leaves and but a solitary "root." Furthermore he notes that the grains with their multiple roots send up culms that never branch, while the merely taprooted leguminous herbs exhibit stems that branch freely and widely.

To the beautiful work of a Malpighi[110] one gives somewhat more credit than is fairly due it, until one has read these chapters of the ancient Athenian master. Then it is clearly apprehended that the man of the seventeenth century may have received the suggestions of his own work directly from the Greek philosopher, and is almost ready to add that the beautiful drawings of sprouting grain adorning Malpighi's folio might almost have been done from the Theophrastian descriptions of the same. It must needs be conceded that the botanic garden at Athens, founded by Aristotle, and the earliest of which there is any record,[111] was wonderfully prolific of new botanical facts of profoundest import. What later one has equalled it in supplying first principles to botany as a science? Or who since Theophrastus has used an opportunity of that kind so immensely to the advantage of succeeding generations?

Anatomy

Immediately after having enumerated the principal external organs of plants, and given the first outline of a system based on these, Theophrastus takes up the subject of internal structure. Two short chapters contain the simplest elements of plant anatomy, as he is able to make them out. If these chapters commend themselves to our most careful reading, it is partly because they are the earliest in which such matter was discussed, and partly for the reason that after the writing of them some eighteen centuries elapsed before another botanist resumed the topic.

Apparently having in mind all forms of plant life except the lowest and simplest, he opens the subject with the statement that "plants are made up of bark, wood, and pith when pith is present."[112] We of the present are accustomed to this as being the structure of stems. Let none be disquieted by the fact that Theophrastus does not limit bark, wood, and pith to stems, for we are learning that he never writes a line carelessly, never indites the simplest and most fundamental proposition without rigid investigation and profound forethought. Possibly we shall find that he thinks the substances of bark, of wood, and of pith all occur in other parts besides stem and branches—in leaves, or even in fruits.

Now if the history of plant anatomy is to begin as near its true beginning as possible, those three Greek terms must be consulted which come into our language as bark, wood, and pith. The first two will not detain us. The bark and wood are, each in many different ways, too indispensably necessary to primal man, to have failed to be distinguished and named long before the advent of the most primitive philosophies. Theophrastus took the terms *φλοιός*[113] and *ξύλον* as he found them, ready to hand and well suited to his use. It was otherwise with the term which he selected by which to indicate the pith of plants. This part was not well known. Primal man, in quest of only the obvious and the useful in nature, may have been unaware of the existence of it. Woody growths, in that mature condition in which they supply the savage and the half-civilized with timber, fuel, bast, and dye-stuffs, show no pith. But that enlightened and philosophic inquiry into nature's obscure things, which Greeks had begun to pursue before Theophrastus's time, had brought to notice this part of plants which was not bark, neither wood. The philosophic must have discussed the substance, perhaps had written about it, for they had attempted to name it. This we have from Theophrastus himself, who says that some called it the heart of plants, others called it the marrow.[114] In this connection he did what with him is most unusual, almost timorously conservative man that he was; he declined to accept either "heart" or "marrow" as a suitable name for this part of a plant. This can mean nothing else but that he himself had taken the pith under special investigation and thereby had reached a new conclusion, had found that it was in no important point analogous to marrow, and farther still from corresponding to the heart in animals. We know enough about Theophrastus's temperament to warrant the assertion that he would have been the last of men to reject two names already in use for a thing, until he was able to prove that both were utterly false to nature. The new name which eventually he offered is one which can not have suggested itself from any study of the substance in that dry, whitened, imponderous, and effete condition in which it is seen in mature stems and branches, in which condition alone we of today call it pith. As philosopher, and as one whom we, if he had lived in our time, should have known as a "scientific botanist," what he wished to ascertain was the function of this substance in the economy of plant life. This he certainly did discover; but he never could have done so had he confined his investigations to the pith as no longer vitalized and operative; therefore he must have sought it in its young and growing state, when of all parts of the plant this is most tender, replete with sap, and a strong center of vital activity, really the matrix within which are generated both bark and wood. The new name which he gives it is *μήτρα*, the word mother, *μήτηρ*, somewhat altered, and in earlier Greek the womb, the mother of life.[115]

Theophrastus never records his processes of investigation; but when he selects the word metra, matrix, to be the appropriate name of the pith of plants, we see at once that it does not really apply to it in its white, imponderous, and devitalized condition, and that he must have named it in reference to its earlier state, that of living parenchyma. We may well divest ourselves of the prejudice that the beginnings of a scientific and sound plant anatomy were not made until after the invention of lenses and microscopes. This man of antiquity, he who had progressed so far in plant embryology as to have made out even the seed distinctions between exogens and endogens, could just as easily discover, in the cross section of a large and very tender shoot of elder or of maple, the first traces of the several fibro-vascular bundles standing in a circle midway between where the pith was to be and the place of the bark. Successive cross sections made at intervals would reveal to him the gradual hardening and widening of those bundles, their final apparent meeting and coalescing into the cylinder of hard wood intervening between the central pith and the now formative bark. It is undoubted that Theophrastus had thus seen, in young shoots, the early stage at which bark and pith are all one in form and substance, and that from it, even within it, both wood and bark are gradually generated;[116] and that because of this he called the juicy soft living pith the metra, the matrix. In reality, while searching to find what pith would be like in its living state, and what in that state its function might seem to be, Theophrastus had discovered living parenchyma.

Such a conclusion, when we have been driven to it in order that the man's words may not be meaningless, needs no further support; and yet there is one other circumstance which would confirm it were confirmation called for. Hitherto I have conformed to usage, very ancient as well as modern, of writing bark, wood, pith, in just that sequence. It seems natural. It is at least a geometrical succession of the terms, and therewithal not inelegant. Even the translators of Theophrastus from Greek into Latin have made him out as having named first the bark, then the wood, and lastly the pith. Unwittingly they committed an inaccuracy, and have misrepresented him; for Theophrastus wrote it thus: bark, pith, wood. This sequence is not geometric, but it is biologic. It would be illogical and awkward, were it not more exactly truthful than the other and merely geometric succession of the terms. As he placed them, the story is told again of how in early stages of development bark and pith are substantially identical. Therefore in their most widely differentiated condition they are next of kin, more nearly related each to the other than either one is to the wood.

This biologic investigation of the pith led to the detection and segregation of other elements in stem structure, discoveries that are all his own, for he expressly says of these elements that they have no names. He thinks that he can not do better than apply to each the name of some analogous part of the anatomy of animals. The whole fabric of stems he makes out as consisting of what he calls veins, nerves, and flesh. They are new uses of these terms and he defines the botanical use of each, premising first of all that vein and nerve are substantially one, differing scarcely otherwise than as to dimensions, nerves being smaller than veins, adding also that in plants these are simply elongated and do not branch. The universal mark of the fabric which is composed of them is, that it splits, and is not otherwise readily divisible. Flesh has the very different quality of being divisible with equal readiness in all directions, like a lump of earth.[117] Here, quite as we had been prepared to expect from his new naming of the pith, and his indicating its consanguinity with bark rather than with wood, we see plainly that Theophrastus discovered and distinguished well what in post-Malpighian times we have learned to know as parenchyma and prosenchyma. He is now able to add a new chapter to plant anatomy, for he can name the elemental constituents of bark, of wood, and of pith. Wood, he says, is composed of nerve and moisture; although some woods, such as of the palms and of ferula, are of flesh; such however in mature and dry condition show that the flesh has been partly converted into wood. When he cites the wood of palm and umbellifer, we realize the expression of the popular notion of wood as being that which the bark of any tree encloses, without regard to density or ponderability. His language here nevertheless evinces plainly his having taken full cognizance of the fundamental difference between the wood of palms and that of all exogenous trees. To pith he attributes flesh and sap only. Bark is composed of all three of these elementals in most cases, as in the oak, poplar, and pear tree, though in some, the grape-vine, for example, of nerve and sap only—that is to say, without flesh, or what we call parenchyma.

The structural elements of the leaves of exogens he apprehended as substantially identical with those of the bark. The stalklet, and therewith the fibrous framework of the blade, are of that which he designates as nerve, or sinew.[118] Next to this he names the epidermis, and after that, what he calls the flesh, known to us as the mesophyll. In the leaves of palms and reedy or grassy plants he finds no flesh at all, and thinks these are composed of fibre and epidermis only. The edible fruits of trees and shrubs are composed mostly of flesh, with little or no fibrous tissue; while, on the other hand, some pericarps consist of a rind or skin only.[119] He has so clearly distinguished these two or three elementals of the plant fabric that he is able to trace and point them out in every plant organ from root to fruit.

Phytography

In descriptive botany there are two different methods by which it is undertaken to convey by means of language the image, so to speak, of some tree or other growth which the describer has seen, and the reader is supposed not to have seen. That two distinct methods in phytography exist is something of which I have seen no mention either with any botanical author, or with any historian of botany; but a suggestion of them has been made very recently, with also a good account of Theophrastus's phytographic method, in an excellent treatise on some parts of Theophrastian botany by Dr. Hugo Bretzl. Let me for convenience designate the methods as the natural and the artificial. They may, however, quite as fitly be named the comparative and the positive methods of plant description. Of the two, one is very ancient, the other strictly modern; and the artificial, or positive, is doubtless the more perfectly adapted to its purpose, though only for such writers and such readers as are competent to use it, for it requires the mastery, on the part of both, of a very extensive vocabulary of special terms—in reality, the learning of a new language. What I have designated as the natural method of describing plants vindicates its right to that title by the fact of its primitive universality. I speak of it otherwise as the method of comparisons, in allusion to what it chiefly exacts of him who would use it: that is, first, familiar acquaintance with certain specific types as standards of comparison, and second, the ability to construct a mental image of the unknown by means of the describer's telling in what several particulars it differs from the given type known to both. By way of illustration, an untaught woodsman, familiar with junipers, reports to a botanist that in a new region to which he has wandered there are trees in all respects like red cedar, or juniper, except that instead of yielding berries they bear round cone-like bodies approaching the size and form of smallish walnuts. The botanist at once pictures in his own mind cypress trees, and assures his informant that his new trees are cypresses. Such is the method of comparison in phytography;[120] and it may quite as aptly be called the natural one, for it is that invariably used by the primitive and untaught, not, however, always very well and successfully. Being the primitive method, it is therefore that of Theophrastus, not, however, to the complete exclusion of a number of absolutely definitive terms such as are now in use and always have been. The Greek observed that, taking the plant world as a whole, the leaf is the most endlessly diversified of organs, and also that within the limits of a species its form is constant; from which two conditions it follows that no other organ is so readily available in making distinctions between different plants. Now as to the art of plant description by the method of comparison, it is very necessary that the types to be used in comparing be chosen considerately. There seem to be indications of his having thought upon this matter, though he has not explained, or even didactically set forth, his scheme. It is only by a certain order and fitness in the scheme itself that one infers the author's having studied it out. The most common types of leaf outline are perhaps the lanceolate, ovate, oblong, and suborbicular. He knows nothing of any such terms; but when he wishes to say that which would express what modern botany expresses by the term lanceolate, he says the leaf is that of the laurel, *i.e.*, *Laurus nobilis*. A leaf that we should describe as oblong, unless it be much too large, is with him that of the olive tree. A leaf that is of rounded contour and nearly as broad as long is compared by him to that of the pear tree. For the ovate in outline his type is the leaf of the ivy, *Hedera Helix*, in respect to which one must not fail to note that it is the leaf peculiar to the mature and fruiting ivy bush, not that of the rooting and climbing young plant; for the leaves of this are too broad to represent the ovate, and are angularly lobed.

Note now certain points of agreement among these four leaf-outline types. (1) All are leaves of trees,[121] for few herbaceous plants display such uniformity of foliage as to the individual specimen. There is apt to be one form and size of leaf near the base of such a stem, and another widely different description of leaf at the summit, with intermediate forms up and down between base and summit. With some notable exceptions—of which the Greek takes advantage here and there—leaves of herbs do not answer the purpose. (2) The trees are all selected from among such as are most universally and familiarly known, every one of them common in cultivation. Every civilized Greek of three thousand years since knew the sweet bay, the olive, the pear tree, the ivy. (3) The leaves of them all have a certain firmness of texture, either leathery, or approaching it. By this prevision any soft herbaceous plant having lanceolate entire foliage may be described as to its foliage by merely saying that its leaves are like those of the laurel, but thin. (4) The leaves of all four of the types are entire: whereby the leaf that is of lanceolate cut, coriaceous texture, and entire margin, no matter in what genus it may occur, may be described by simply saying that it is the leaf of the laurel. Supposing, however, that a tree is to be described the leaves of which are lanceolate, coriaceous, but with margin serrated, then its description as to leaf will be, that it is like the laurel leaf, but serrated. And this selecting of types that have entire leaves is manifestly better than it would have been to have selected serrate leaves for types. In such a climate as that of the Mediterranean, evergreen trees and shrubs predominate, all of them with coriaceous foliage, the kinds with entire leaves being very notably in excess of those having toothed or serrated leaf margins; therefore the choosing of as many entire-leaved types as he could was natural to a man in Theophrastus's place and environment, as well as making for economy of time and space in describing things. It was by no means accidental that this descriptive botanist selected the olive, the sweet bay, the myrtle tree, and the box as patterns of leaf outline to be referred to in his phytographic work.[122]

How well such a system of morphologic types is adapted to the purposes of description may best be shown by quoting a few examples. A wild elm tree that inhabits the mountain districts he says has "leaves like those of the pear tree, but longer in proportion, with serrated margin, and a rough rather than smooth surface."[123] This makes it plain that Theophrastus's first purpose in choosing leaf types was that of imparting ideas of general outline. To begin the account of an elm leaf by affirming it to be like that of a pear tree is awkward and even mischievous, upon any other supposition than that by such phrase he alludes to size and general circumscription only, which also the expressions immediately succeeding prove; for he who had pictured in his mind at first a pear leaf for an elm leaf must now proceed, under direction of the describer, to alter it by giving it a very distinctly saw-toothed margin, and after that a roughness of surface, of which there is no trace upon the pear leaf. What he now sees mentally as a leaf of the little known wild elm is like a pear leaf in nothing save its general contour. The elder tree,

Sambucus nigra, Theophrastus seems to have taken pleasure in describing rather minutely, although the tree was no rarity, but rather familiarly known. But it seems to have been this which taught him the existence of such a thing as a compound leaf; and he gives a particular account of the species from root to fruit. When he comes to the lanceolate individual leaflet he says it is like the leaf of the sweet bay but larger, relatively wider in the middle and at base, more pointed at summit, serrated all around, and the whole more soft and pliable in texture.[124] To one acquainted with the sweet bay and the elder, I do not know where, in even the most recent botany, he will find a more complete description of the elder leaflet. The leaves of maples, mostly wild-wood trees and less familiarly known, are compared with those of the omnipresent plane. The maple leaves are also ample, cleft somewhat after the same manner, but not to the middle as in the plane, *Platanus orientalis*, longer in proportion to their breadth, of a more delicate texture, and not rough to the touch.[125]

This system of leaf describing by comparison with types is both natural and not ill adapted to the purposes of phytography. Had it not been so it

FIG. 15. Sweet bay (*Laurus nobilis* L.). *Historia plantarum* (1644), p. 188.

would not have remained in vogue for two thousand years after Theophrastus. Greek authors after him, as well as Pliny and other Latin writers, knew no other method of leaf diagnosis. The number of the indicative types was gradually augmented, and the use of them was universal even with the fathers of sixteenth- and seventeenth-century botany; nor has any later generation wholly ceased from the usage; though it is perhaps chiefly conspicuous nowhere but in nomenclature, where such specific appellations as *salicifolia*, *alnifolia*, *betulifolia*, *delphinifolia*, and a hundred more, though ostensibly figuring but as names, may chance to be the best part of the diagnosis, at least in the estimation of any not well versed in the post-Linnæan descriptive terminology.

It is not to be inferred from anything here said, that the Greek knew nothing of any geometric terminology of leaf forms. In that chapter in which he treats of leaf forms in general he names the orbicular, the oblong, the angular, and some others;[126] but they lack definiteness of meaning, at least such definiteness as the exigencies of plant diagnosis call for; though terms that bear upon differences of apex, margin, and base as well as the superficies of the leaf are of more fixed and certain meaning.

It will be observed that even flowers are described by Theophrastus comparatively, the less known being brought into contrast with the well known; and the same rule applies, of course, in his diagnoses of fruits and seeds. He was not particularly given to describing plants. A great proportion of those which he discourses upon were well known to all who would become his readers. The common things of the gardens, of the cultivated fields, of orchard and vineyard and of academic grove, were so familiar that the mention of the name was sufficient. But when he undertakes the description of any herb or bush or tree, he is apt to give more than a rude outline of it, very often a good word picture of it; and he who does this is a master of phytography, without question about the age in which he has lived, or the method he has employed. In the case of a number of Asiatic and African trees unknown to Theophrastus except by report of travellers, he so carefully gleans all that others have said about them, and with such consummate art sums up the whole, and draws up his own description, that in reading it one finds it not easy to realize that the author of it never saw the tree. The books of botany that were composed by Greeks and by Latins within three or four centuries after Theophrastus show that the authors of them copied his descriptions whenever such were available, and in other cases made his the model of their own diagnoses. When we come to the era of strongly renewed scientific activity in the sixteenth century, we shall find botanical authors of the time employing many of Theophrastus's descriptions of plant species without alteration of any kind; some, like Brunfels, Tragus, Cesalpino, and their class, formally crediting each such diagnosis to its ancient author, others without making such acknowledgment. But they of this period who ventured upon new descriptions of plants which Theophrastus had described of old seem to have exposed themselves to public censure, as needlessly, perhaps irreverently, supplanting or amending the excellent work of the father of phytography.

Taxonomy

To teach, as it has been taught and is still taught, that Tournefort (1694)[127] first ranged the members of the plant world under genera, that Linnæus (1753)[128] first clearly distinguished species and varieties, and that Adanson (1763)[129] first proposed the grouping of genera into families—all this is to inculcate fable. It has been already suggested, and forcibly enough, that plant taxonomy was not invented in any school, or by any philosopher, that it is everywhere as old as language, that no plant name is the name of an individual plant, but is always the name of some group of individuals, and that all grouping is classifying. Botanical taxonomy began at whatever time far off and prehistoric men began to give names to plants; and it increased with the recognition and the naming of new groups—always groups, never single plants. Had some earlier Theophrastus appeared upon the scene some thousands of years earlier than this one did, in this particular at least he would have found himself in the midst of a like environment. He would have found some hundreds of kinds of cultivated plants familiarly known, spoken of always under group names. In other phrase, he would have found a certain taxonomy ready to hand, such as answered the needs of those who had to do with plants industrially.

The real Theophrastus, entering the field, not in the far-off age of Homer, whose poems are full of tree and plant lore, but many hundred years later, had much to do in the first place in acquainting himself with the vast sum of knowledge, of theory, of poetic fancy, and of superstitious fable that was then extant concerning plants. All this he accomplished, as his pages plainly show, and that with the occasional expression of something like the scientific man's impatience of the superstitious and the fabulous. As distinctively a nature student, however, exploiting the realms of nature from the philosopher's viewpoint rather than from that of the economist, it would not have been strange had he invented some new taxonomic scheme of his own, and then, thrusting aside all the commonly accepted plant groupings, had sought to install an entirely new system of taxonomy in place of the old.[130] He did nothing of the kind; and if he did not, it may have been for the excellent reason that that already in vogue, when duly examined, to a great degree commended itself to the philosophic judgment as having been deduced from nature, and that in as far as it had progressed, was often well enough done. Parts of it could not be amended, and, we may now say confidently, were destined to acceptance as sound taxonomy as long as the world of plants should endure, or a botanist remain to study it; at many points it might be amended or added to, and the whole must be extended and variously improved.[131]

We may assure ourselves by a study of Theophrastus, that something very like this was the task to which he addressed himself as regards the classification of plant organs and the systematization of plants themselves; and the careful reader of his chapters will note often his great conservatism—his manifest aversion to startling the good public by pronouncements that are new, and that will openly antagonize them as assailing their old doctrines and their deeply ingrained prejudices.

All these things being true, one ascertains with difficulty, if at all, what the historian is most in need of knowing, namely where this writer of the first book of botany is recording points of taxonomy that are of prehistoric discovery and universal traditional acceptance, and where he is introducing some amendment or improvement of his own. For example, in a very early chapter of his work Theophrastus ranged all the plants that he had ever seen, or heard or read about, under the four primary groups of tree, shrub, half-shrub, and herb—*δένδρον, θάμνος, φρύγανον, πόα.*[132] It is one of the most classic pieces of plant taxonomy, one that stood the test of all the ages and is immortal. Nothing that by any means could be elicited from out of the hazy past would be of deeper botanical interest than information as to whether this fine piece of taxonomic work had been handed down in its completeness to Theophrastus, or whether as he gives it, it represents much augmentation, or condensation, or finishing and perfecting accomplished for it by himself. We have met with like problems before;[133] to the full solution of this I am incompetent. For this, the botanist's best skill would need to be supplemented by the erudition of the specialist in Greek philology. Nevertheless the following hints may make it plain that *δένδρον, θάμνος, φρύγανον, πόα*[134] is a piece of classification that was studiously wrought out by Theophrastus himself. In the Greek of his time there were at least two words for tree, *δένδρον* and *ὕλη*, the former more particularly designating such as were cultivated, the fruit-bearing, nut-bearing, and such as were planted for shade or ornament; the latter applying more specifically to wild trees used as timber, almost the equivalent of our English terms woodland, forest, timber-tree, etc. We have elsewhere remarked upon the Theophrastian classifying of all growths as tame and wild. The idea was deeply seated in the Greek mind; and for trees in general, wild as well as domesticated, he could not well have chosen any other term than *δένδρον*, though it was more properly the appellation of the civilized contingent of arboreal growths. Similarly *θάμνος*, at least etymologically, signified a densely compacted woody growth—and not necessarily of low stature. The full-grown olive tree was sometimes called a *θάμνος*, on account of the bushy density of its head. Also as looking to the distinction between bushes of cultivation and a wildwood thicket, the latter was also *ὕλη* sometimes, if not even usu-

Lavendula.

FIG. 16. Lavender (*Lavandula angustifolia* Mill.). *Historia plantarum* (1644), p. 670.

ally. Thus again as with δένδρον the botanist selects for extended use that which signifies the cultivated and better known.

Coming now to the class of suffrutescent plants—the half-shrubby, half-herbaceous kinds—it appears to me no less than certain that Theophrastus was first to discover, indicate, and name this particular category. There seems to have been no word for this kind of thing in the older Greek; for *φρύγανον* meant nothing more than a bundle of faggots, dead and dry branchlets and twigs of trees which, either as windfalls or as left behind by the woodchopper, were laid lengthwise and tied into bundles for fuel. There was, however, the old word *θαμνίον*, which one almost wonders he did not adopt for his category of the half-shrubby. It is but a diminutive of *θάμνος* and means a little bush or small shrub. At second thought one perceives that it would not well answer the purpose. It gives no intimation of the true characteristic of the suffrutescent growths, which is this, that the lower and woody part represents a shrub, while the upper portion, that which bears more scattered foliage together with the flowers and fruits, is herbaceous—that is, dies back every autumn after the fruiting and is renewed again in the summer. In late autumn and early winter, while the dead or half-dead upright and parallel summer branches are still present, surmounting the shorter tuft of truly woody lower branches, the bush would vividly enough recall a faggot bundle. Even where abounding, as such growths do on open plain or stony mountain slope throughout all half-arid regions of the world, they must have been used as faggots always. He who knows familiarly such ancient garden plants as the lavender and sage and rue, and the wild half-shrubby artemisias and other like composites of all dry climates, will perceive readily that *φρύγανον*, the faggot bundle, lent itself to Theophrastus's scientific purpose in this instance. He might have created a new term; but the conservative prefers to make new use and application of some old and familiar term. The public never takes kindly to new names.

In distinguishing the category of the suffrutescent, the Greek had proceeded analytically. In establishing upon all herbaceous plants one comprehensive group under one name, his procedure was synthetic. It was not indicating a single new group hitherto unrecognized and naming it. It was the synthesis of a number of groups long recognized and separately named, the putting together of such, to constitute a single more comprehensive assemblage, and under one name.

A glance at the actual situation in which we English-speaking people find ourselves as to our terminology of the herbaceous will help us to apprehend clearly the Theophrastian standpoint. We have no single word by which we venture commonly to designate the aggregate of things herbaceous. If in our fundamentals of botany we still follow Theophrastus in writing or speaking of tree, shrub, and herb, that is at once the beginning and the end of our using the term "herb" thus comprehensively. Thenceforward we ignore it and write or speak about "herbaceous plants"; this for the manifest reason that "herb" used by itself has almost universally a special meaning of which it seems impossible to divest it. An herb is something, neither tree nor shrub, which is either medicinal, aromatic, or culinary. The other terms in common use for subordinate groups of herbaceous plants are vegetable, weed, grass, and worst of all the word "plant" itself; for this, as first introduced into our English speech,[135] and as almost universally employed down to our day, signifies only things herbaceous, yet not all; for neither weeds nor grasses are commonly called plants, in our tongue, except technically. Thus our category of the herbaceous includes the segregates herb, vegetable, weed, grass, plant. And the endeavors of our English scientific forefathers to make any one of those terms serve as the name for herbaceous growths in general have been unsuccessful. Theophrastus had been in the same predicament as that which became theirs. In his mother tongue there was *φυτόν*, the herbaceous plant of cultivation, at least as to its most primitive signification; there were λάχανον, the kitchen-garden vegetable, βοτάνη, the weed, and πόα, the grass, the forage herb.[136] The first of these terms, φυτόν, the herbaceous thing that men plant and transplant and cultivate, was a term that he himself appears to have rendered unavailable. He had made it to include the whole vegetable kingdom from oak and pine to seaweed and fungus. The name that to him seemed most reasonably available for designating the sum of all things green—herbaceous—was the common name of the grasses and fodder plants, πόα. When I say reasonably available, my thought is that the selecting of the term grass rather than the term herb for expressing the aggregate was most natural to any one who, like Theophrastus, had thought the matter over. The grasses form by far the greater proportion of that low-growing verdure which, outside of the forests and thickets, covers the whole earth in all temperate latitudes during half the year. It was therefore more true to nature—more scientific—to take up that term which came so near being synonymous with verdure. The term herb is comparatively unfit, as suggesting a much smaller aggregate, and that also of plants marked by odors, flavors, and other qualities which the eye cannot detect. There are even modern languages in which the word for all green herbage is the correlative of our word "grass": languages in the botanical vernacular of which, what we call herbaceous plants are known only as "grass." This was strictly true of our English of only a few centuries ago.[137] Thus again how plain it is that the forefather of all botany produced this primal outline taxonomic only after the most careful weighing and considering of every point involved. Nor must our attention be called away from this first chapter of Theophrastus's classification without our having observed the sequence of the groups. Why does he begin with the Grand Division of the trees and proceed downward to that which will include such phytologic ambiguities—for such they were in his day and for ages afterward—as lichens and fungi? There is no room for doubt that one consideration was the very rational one, that botany may not safely begin at any point where doubt may arise as to what realm of nature the subjects of study belong to. The title of trees in general to be regarded a part of the plant world is secure; that of no other growths more so; while judged by his own criteria of phytologic rank, the very highest place must be accorded to trees. In his view they were the most perfectly organized of all.

Inasmuch as the fourth and last Grand Division, πόα, the Grass Plants, *i.e.*, the Herbaceous, is the largest of all as to numbers both of individuals and of species and as regards their almost universal prevalence, it will naturally be here that we shall look for further expression of taxonomic idea, suggestions of grouping subordinate to that of πόα, the herb.

The most comprehensive of his subordinate groups is that which he denominates καλαμώδης,[138] which comes into Latin as Arundinaceæ; its type being that superb grass *Arundo Donax*. In an author so primitive one does not look for any rigidly formal diagnosis of a group. Calamodes in itself is

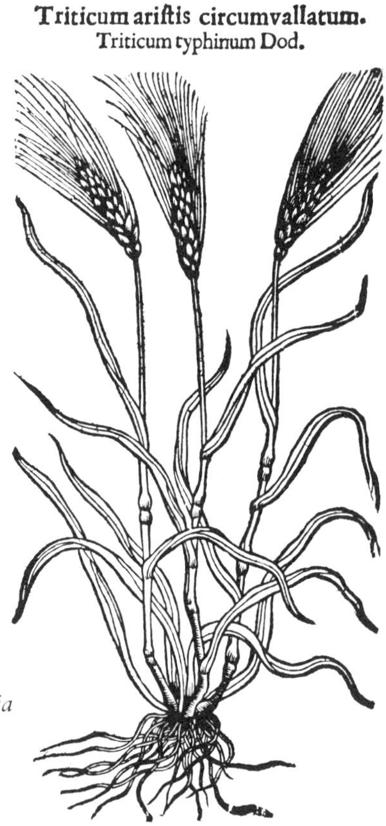

FIG. 17. Wheat (*Triticum* L.). *Historia plantarum* (1644), p. 956.

diagnostic. It will include all plants that recall calamus, that is, *Arundo Donax.* In one place he has written that the leaves of this, and other things which he cites, seem to be made up of nerves only, comparing them with those of the grape and fig, which he says have not only nerves but also flesh and epidermis in their make-up.[139] The interpretation of this is that he has become aware of the differences of anatomical structure subsisting between the leaves of endogens and exogens. His group Calamodes—better written, after the usage of his Latin translators, Arundinaceæ—embraces *Arundo* and the cereals *Triticum*, *Secale*, *Hordeum*, *Oryza*, and others; that is, he names these as types. By their leaf characters all other gramineous plants cultivated and wild, known or unknown, fall within the lines circumscribing his Arundinaceæ. And what else besides the true grasses? In one place he names as among the arundinaceous certain plants the leaves of which in so far depart from the typical as to present an angular cross section, and these seem to him as if made up of two leaves joined by their edges to something like a keel. It is the conduplicate and keeled leaves of such things as *Cyperus* and *Sparganium*, as any botanist would know from the description of them, even without Theophrastus's having named those particular genera, adding that other denizens of marshy grounds have such foliage. Even the palms—of which he knew only *Phœnix*, pinnate-fronded—he cites by name as being arundinaceous.[140] Such reeds and rushes as to him seemed quite leafless would still vindicate to themselves places within the group by virtue of their altogether pithy or else hollow stems or culms. We are sure, then, that this assemblage of the Calamodes embraced all true grasses, all sedges, besides the juncaceous, typhaceous, sparganiaceous plants, and palms. It will be observed that all are monocotyledonous; that the group embraces by far the greater proportion of such, only the showily spathaceous and the really petaliferous genera being left out. Recalling here the circumstance that in all early taxonomy roots figure very conspicuously, it becomes interestingly significant that into this great class of the Calamodes—a group which, as regards the number of species, comes near being the equivalent of our endogens—not a genus is admitted that has either bulb or corm. Every member of the vast assemblage has copious fibrous roots; these in moiety of the species supplemented, as Theophrastus might have worded it, by that which he had chosen to name the jointed root, *i.e.*, the slender rhizome; a single sedge, *Cyperus esculentus*, bearing nut-like protuberances as if at the ends of some few of the root fibres. There is not the shadow of a doubt that this pristine plant anatomist and systematist recognized the structure of leaf stalk and flower stalk of *Arum* and *Colocasia* as at full agreement with that of half his Calamodes; and the same must have been familiar to him in the case of the grassy-leaved crocus and its allies; and the "roots" of these must have excluded them from taxonomic consociation with the rest, in all probability, even had their flowers been leafless and less in contrast to those of grass and reed and sedge. The aggregate of bulbous and cormose plants, the araceous I think excepted, were known and spoken of by him as *βολβώδης*—the Bulbaceæ.[141] As a group it contained, first of all and typically, the onion and its several congeners, even the leek, a plant that though alliaceous is not bulbous; after these the bulbous Liliaceæ, Amaryllidaceæ, and the cormose iris allies. With the types and several of the species of all these he was familiarly acquainted.

There are other natural families not a few of which Theophrastus apprehended with precision, even assigning names to several of them. Such are the Umbelliferæ, for an example, to which as a family he gives the name *ναρθηκώδης*, which in the Latin versions had to be written Ferulaceæ, from *ferula*—in Greek *νάρθηξ*,[142] a stalk of fennel or of ferula—commonest umbellifers of the Mediterranean region. Plainly the thistles and their natural allies were accepted by him as constituting another such family group; for he often refers to them under the collective name of *ακανθώδης*,[143] and twice mentions that all of them are prickly leaved herbs whose flower consists of a head of florets, each floret sitting on the summit of a seed.[144] In Latin versions these are the Acanaceæ, from *Cnicus Acana* L., one of the most common thistles of Greece and Italy. Into this family Theophrastus admitted *Dipsacus*, the leaves of which are not prickly, as he concedes, but on account of its answering to the thistles in this, that its florets crown each its seed. And *Eryngium* also, and not unnaturally, with its spinescent foliage and capitate inflorescence, finds place among the thistles rather than with the umbellifers. The numerous cichoriaceous genera, with their peculiar habit, milky juice, and sameness of character as to flower and fruit, formed also a family with our Greek, who called it κιχωριώδης,[145] which the Latins wrote Cichoriaceæ. These for examples of his having given to groups of genera class names and family names. Others need not be cited; but it should be mentioned that the family relationship of small groups of genera is in many an instance clearly seen by him when no group name is used. The pines, fir, spruce, and larch are discussed in a place by themselves;[146] the various poplars, together with alder and birch, occupy successively another place,[147] and the same is true elsewhere in the dendrological chapters of the book. The intimate relationship between the poppies and the pond lilies was so clearly perceived by Theophrastus that, while in general he seems to like to group things ecologically, the aquatics in chapters apart from mesophytes, he nevertheless proceeds without a halt from the papaveraceous plants of the grain fields to their kindred of the lakes and rivers.[148] He perceived upon them all, the marks of one and the same family.

It is again very interesting to note here and there a question raised as to the extent of some family; whether such or such a genus is or is not of the same family with such another, for example: "Some affirm that cucumbers and squashes are as closely interrelated as radishes and turnips are; others deny it."[149] The passage reveals gardeners and botanists of remote antiquity in debate about the affinities of genera; and the man whose word of authority might or might not have ended the debate, diplomat that he is, as well as philosopher, expresses no opinion; though none who have studied him well can doubt that he had one, and the correct one.

This outlining of families of plants and giving them family names entails one extremely important logical sequence, which one must not fail to indicate. His Arundinaceæ, Bulbaceæ, Ferulaceæ, Acanaceæ, Cichoriaceæ, and all the rest, as established on certain organologic characters, are each and all logically and completely subversive of that distinction which he formally keeps up, between things cultivated and things wild; for each such family necessarily includes both. The few historians who have not shrunk away from the time-consuming task of studying the Theophrastian volumes have been perplexed by his seeming approval of ancient Hippon's theory about the origin of cultivated plants, which seeming approval is at once followed by a feeble argument or two against the theory. Here is what Meyer says, referring to the primary divisions of tree, shrub, half-shrub, and herb:

> Each of these four is subdivided into the groups of the Cultivated and the Wild. Hippon's pronouncement, that every plant is at first wild, and then by cultivation made tame, is thus in a general way approved, though Theophrastus immediately adds that certain wild plants are not at all amenable to cultivation, while others take to it readily, whence it will follow that such a distinction is not altogether untrue to nature.[150]

This historian's difficulty arises through his having missed two items important to the understanding of the man Theophrastus. First, that the illustrious Greek was as successfully a student of human nature[151] as he was an investigator of the plant world; and that he studied to avoid opposing with needless directness the prejudices of the multitude.[152] If he should pay no respect to those time-honored categories of the tame and the wild, but should jumble them all together, and openly, forty-nine out of fifty among his readers would adjudge him not only a bold innovator, but perhaps also a godless heretic; for, as elsewhere intimated, the staple plants of agriculture,

FIG. 18. Edible cyperus (*Cyperus esculentus* L.). *Historia plantarum* (1644), p. 451. This illustration was from Mattioli, *Commentarii . . . Dioscoridis* (1565), p. 543.

even in ancient paganism, were viewed as special creations of the gods—their immediate gifts to men. Old Hippon the rhizotomos had been an outspoken heretic of this stamp. Theophrastus quotes his bold theory. In his secret soul he believes it sound; yet for the sake of avoiding scandal to the forty and nine or the ninety and nine, he veils his belief by admitting that some wild plants refuse to be tamed, a fact which may innocently be construed as against Hippon's idea. The other fact which the historian failed to apprehend is that the Greek outlines, and gives names to, a half-dozen or more of large natural groups, every one of them embracing without discrimination plants domesticated and wild. He thus completely nullifies that distinction, yet it is all so quietly, and as regards the superficial reading so covertly, done as to escape the notice of the forty-nine out of fifty of his readers; even also of our latter-day botanical historians, learned men and able, yet with mental vision impaired by the strong light of those typographic pedantries—convenient and helpful, certainly—which the botany of the nineteenth century had as a legacy from that of the eighteenth, affected by a sort of botanico-literary dyslexia which is slow to perceive that such a name as ferulaceæ is as perfectly the name of a natural family of plants as when printed FERULACEÆ.

The recognition of genera—using the term in a modern sense—is as informal with Theophrastus as that of families. However, when we come to the word itself, γένος, genus, it is employed variously—that is, with several different degrees of comprehensiveness. Indeed every natural group is with him a "genus," whether it be of the whole assemblage of herbaceous growths, or a family group, or a genus in our sense, or a species, or a variety merely. It seems to be the exact equivalent of our English expression "a kind"; and because such use of the word "kind" is not yet obsolete, at least surviving in rural districts, it will not be difficult to make plain its meaning. If a gardener or farmer of the present day mention to a botanist that he has in cultivation a strange plant of the squash, cucumber, and gourd kind, the latter understands perfectly that this is something belonging to the family of the Cucurbitaceæ, though he can go no further. But if it now be said in rustic phrase that the plant is something in between the squash kind and the gourd kind, he has used the word in a different and much more limited sense; for now the skilled botanist at once puts out of mind seven out of the eight primary divisions—subfamilies—of the cucurbits, or, in other words, dismisses from his thought, we will say, sixty or more of the seventy genera of this family; for he clearly understands the farmer to have that in view which must lie close to the two nearly related genera *Cucurbita* and *Lagenaria*. Now for a third manner of using the word "kind," with a most distinct third meaning, the farmer shall say that he has a new kind of squash. The botanist now has not the least doubt that the genus *Cucurbita* is meant; whether a species or a variety he cannot tell; but the expression "kind of squash" at once translates itself into the school-taught expression, "species or variety of *Cucurbita*."

These three distinct old-fashioned uses of the word "kind" illustrate well the different ways in which ancient Greeks and Latins employed their word "genus." It is not a usage that makes for that perspicuity which a science calls for. For three meanings, three words are better than one. Nevertheless there is seldom room for doubt in Theophrastus's writings as to whether by "genus" he means such a group of species as we receive under that name, or a more comprehensive, or a less comprehensive group; any more than one well read in English fails to get the meaning of each of the uses I have brought forward of the equivalent word "kind." But the modern botanist who innocently should read into the term genus of an ancient author always the meaning which it has in modern botany would soon reduce his own mind to a state of utter bewilderment as to the ancient author's meaning. I have therefore been at the pains of making this attempt at an explanation. Upon this and many another important matter of terminology the historians have been silent.

Employing now the word "genus" quite as used in modern botany, the genera of Theophrastus are numerous, most of them obtaining acceptance and holding their places in the systematic botany of the present, most of them also bearing the same names under which they were written about by him. This will be best shown by a few examples, which I select from under the letters A and C of any Latin index to his work:

Abrotonum	*Anemone*	*Calamagrostis*	*Cissus*
Acanthus	*Anethum*	*Calamintha*	*Colutea*
Aconitum	*Anthericum*	*Cedrus*	*Coriandrum*
Ægilops	*Antirrhinum*	*Celastrus*	*Coronopus*
Agrostis	*Aparine*	*Cenchrus*	*Cratægus*
Aira	*Aristolochia*	*Cerasus*	*Cycas*
Alopecurus	*Arum*	*Ceratonia*	*Cydonia*
Althæa	*Asparagus*	*Cercis*	*Cyperus*
Anchusa		*Chelidonium*	

There must be more than a hundred Theophrastian genera the names of which are as familiar as these. A very considerable proportion of them were then known only as consisting of a single species, and are therefore of the kind which we speak of as monotypical; others were made up of from two to several, and the species are mentioned by name at least, when not described. If he establishes no new genus, and all of them which he enumerates or describes were of common recognition, and under those same names, even before he had penned a line on botany, this fact of itself will demonstrate anew the truthfulness of the proposition that the perception of genera and the naming of them are older than history, and that plant names, generic and specific, are a part of human language always and everywhere.

Helleborus albus fl. atrorab. Helleborus Niger.

FIG. 19. *Left*, white hellebore (Liliaceæ: *Veratrum album* L.). *Right*, black hellebore (Ranunculaceæ: *Helleborus niger* L.). *Historia plantarum* (1644), p. 1065.

As to the grouping of his genera, almost the whole story has been told, at least by implication. There were the genera of trees, the genera of shrubs, etc., in places apart; and there were ecological groupings of wild plants and particular assemblages of genera of things cultivated in field and garden; as to these last, the mere retention of antiquated popular groupings which, in deference to the cultivators, he was unwilling to ignore or displace.

There occur in Theophrastus a number of passages which seem like forecastings of a system based more particularly upon flowers and fruits, a system the development of which was of course impossible then, or even at any later period preceding the invention of the microscope. But the very impossibility of his having been able to develop such a system is something which makes his few and faint adumbrations of it interesting and remarkable. I shall cite but two or three.

Commenting on the cylindric spicate inflorescences of certain cereals and grasses, he recalls that those of the plantains are so like them and even the flowers so similar, and thinks it might not be presumptuous to regard them as being interrelated.[153] To the average botanist of today the idea of any consanguinity as subsisting between *Plantago major* and an *Alopecurus* or a

Phleum will seem crude enough; and this partly because parallel-veined leaves do not indicate to a certainty that a plant is an endogen and therefore more or less allied to grasses, and partly because we with our hand lenses and microscopes perceive between the small apetalous flowers of *Plantago* and *Alopecurus* marked differences that were impossible of discovery by Theophrastus. But the one thing noteworthy is that the Greek thus makes flower and inflorescence the criterion of natural relationship. He does not positively affirm it. It was but a pointed suggestion; and the suggestion passed unheeded during seventeen centuries.

The true hellebore and the veratrum are not more closely allied than are *Alopecurus* and *Plantago*, but as to their flowers and more particularly as to their follicular fruits, there is a strong likeness between them. It may have been this circumstance which, along with their powerful medicinal qualities, led to their being named as of the same genus, *Helleborus*.[154]

More signally indicative of his regard for fruit characters as sometimes taxonomically outweighing the vegetative, is the fact of his having associated the yellow water lily with the poppies, rather than with nelumbo. Having given account of the wild of the grain fields and stony uncultivated lands, he who is so apt to treat of plants in ecologic groups proceeds now to speak of the poppy "that is called nymphæa."[155] Evidently its milky juice, the generalities of its floral structure, and above all the external form and the inner structure of the capsule, as well as the seeds themselves, constrained him to think of this and the poppies as congeneric. Also when this same chapter ends with an account of *Aristolochia*, the capsules of which are so much like those of poppies, one can assign no other reason for it than that by their fruits he guessed that *Papaver* and *Aristolochia* were interrelated. Other like instances need not be cited; though it should not here be lost sight of that his families, the Umbelliferæ and the Carduaceæ, were in his mind characterized each by marks of flower and fruit. And so, when the anthocarpologic doctrine of affinities is traced to its beginning, one no longer may think of it as having originated with Cesalpino. The idea had been suggested to his mind, and most impressively, by Theophrastus.

Nomenclature

No such thought as that of botanical nomenclature finds expression with the Greek. When he wrote of any tree or shrub or herb he used that name by which it was known in the everyday speech of Greeks. It does not appear that it ever occurred to him that a living thing, or any group of living things, required to be named otherwise than as commonly designated in his mother tongue. When in reading his books one encounters batrachium, erigeron, leucoium, lithospermum, myriophyllum, myrrhis, narcissus, and a hundred more as familiar, it is because he knew no other names for them. Nevertheless Theophrastus has great part in what is now come to be received as the scientific nomenclature of the genera and species of plants; and if this has come to pass without forethought or purpose on his own part, it is still natural and was inevitable. During some two centuries next succeeding the writing of it, this was almost the only treatise on botany that was extant, and the names of plants therein written about obtained by that very fact great prestige. When at length the Latins began to study plants, and would write about them, they had to learn Greek in order to be able to read the works of Theophrastus, for that was the one supreme treatise on plants. All well-known plants were therefore known to Latins by their Theophrastian and Greek names, as well as by their Latin names when they had such. Pliny, the supreme Latin writer about plants, in translating Theophrastian texts by the hundred into Latin for Roman readers, made use of familiar Latin names in place of the Greek names when there were such, *e.g.*, instead of the Greek *itea* he wrote *salix*; in place of *drys*, *quercus*; Latin *ulmus*, *sambucus*, and *ranunculus* in place of Theophrastian *ptelea*, *acte*, and *batrachium*. There were still many scores of plant types which were known to Latins by no other names than those that had been assigned them in Greek; another evidence that Theophrastus by his books had been the one teacher and authority upon botany to Latins as well as Greeks. *Platanus*, *Cerasus*, *Rhamnus*, *Anemone*, *Thalictrum*, *Delphinium*, *Helleborus*, *Pæonia*, and a host of other such remained the only names of the genera, whether one spoke or wrote botany in Latin or in Greek; and so during some seventeen centuries most of the plant names in use were quoted from Theophrastus.[156] The popular fable about Linnæus as first nomenclator of botany is not yet a hundred years old, and will need to be perpetuated for sixteen centuries yet to come if the years of his nomenclatorial fame are to equal those during which Theophrastus held the prestige.

Early in the sixteenth century, when new impulses moved men everywhere to scientific research, Latin had now long been the universal language of the educated. Theophrastus had been translated into Latin for the convenience of those who had not learned Greek; but still, as to botanical nomenclature, Greek held its prestige fully. When in the course of their herborizings the botanists found plants in no wise answering to any descriptions in the ancient books, and therefore adjudged new and nameless, under the conditions then prevailing it would have been the most natural thing in the world if all new plant names of the period had been made in Latin; and this indeed often happened, *e.g.*, *Pulicaria* and *Fragaria* Brunfels (1531), *Digitalis* Fuchs (1542), *Sanicula* Tragus (1552), *Bidens* Cesalpino (1593); yet Latin names for new genera are somewhat exceptional even for that period, Greek-made names being commonly preferred. The reason was simply this. The greater proportion of plant names then in use, even in Latin botany, was Greek, and that by unbroken tradition from the Greek father of all botany; and Greek-made names for new types were more in harmony with the general tone of botanical nomenclature than Latin names. Thus has it come to pass that even down to our twentieth century the favorite etymology for new generic names is Greek. Such very modern names as *Calliandra*, *Chimonanthus*, *Chionanthus*, *Chionophila*, *Chionogenes*, *Epigœa*, and hundreds like them, all very modern, attest the perpetual influence of Theophrastus upon botanical nomenclature.

In botany as elsewhere the genus presupposes species. A genus may consist of many species, of few, or of one only. Theophrastus had very many monotypic genera, at least as they were then known. The specific representative of a monotypic genus has with him but one name, commonly a one-worded name; that is, the one species constituting such genus lacks a specific name. It really has no need of any. Where there is but one thing of a kind, there is never in ordinary speech a second and qualifying name. If neither men nor things existed but in monotypes, language would not need adjectives, and there would be none. Had there been but one race of men on the earth, the name of that race would have been man simply, and the adjectives Caucasian, Mongolian, African, etc., would not have existed. The Theophrastian nomenclature of plants is as simply natural as can be imagined. Not only are monotypic genera called by a single name; where the species are known to be several, the type species of the genus—that is, that which is most historic—is without a specific name, at least very commonly, and only the others have each its specific adjective superadded to the generic appellation. The situation may best be shown by examples.

Theophrastian	Modern
Clethra	*Alnus oblongata*
Melampyron	*Melampyrum arvense*
Dolichos	*Phaseolus vulgaris*
Ostrya	*Ostrya vulgaris*
Peuce	*Pinus picea*
Peuce Idaia	*P. maritima*
Peuce conophoros	*P. pinea*
Peuce paralios	*P. Halepensis*
Mespilos	*Mespilus Cotoneaster*
Mespilos anthedon	*Cratægus tominalis*
Oxyacantha	*Mespilus Pyracantha*
Syce Idaia	*Mespilus Amelanchier*
Aira	*Cratægus Aira*
Cydonion	*Pyrus Cydonia*
Coccymeles	*Prunus domestica*
Spodias	*P. institia*
Cerasos	*P. Cerasus*
Pados	*P. Padus*
Oié	*Sorbus domestica*[157]

The first four names above are those of genera known to Theophrastus as consisting each of a single species. It is evident he saw no occasion for any second and qualifying name in any case of that kind. To have given such second names would very certainly have exposed him to the criticism of having abandoned the attitude of the philosopher, the man of literary taste and scientific brevity, and having assumed the role of the pedant. Why do botanists of a recent time invariably append the needless second name to every monotype? I ask the question but to emphasize this point in the history of biologic nomenclature. I recall no instances of the assigning of the useless specific adjective to a generic monotype until well toward the time of Linnæus; and despite the weight of his authority in favor of it, the nineteenth century was on the dawn when there were no longer eminent botanists standing out against the practice. The assigning a species name in these instances is, of course, previsional. The monotypic genus may cease to be such; but even then, according to Theophrastian usage, the generic name alone might stand as that of the original and typical specific member; but that is too provincial. Theophrastus's work, in as far as relates to taxonomy and nomenclature, is provincial, not a universal flora, but a somewhat local one. It was not a perfect pattern for the universal. There is also a certain lack of uniformity in a system of nomenclature which fails to provide every species with a distinctively specific name; and it was nothing more than the desirability of uniformity which brought about the modern usage. But, as we shall see, this question was long in controversy, and was settled late.

Matters of nomenclature and taxonomy are almost inseparately connected. The name itself is but the expression of a taxonomic idea. Excepting those rare instances in which an individual historic tree has received a proper name, every plant name that ever was, in any language, is the name of a group. Naming is classifying. The Theophrastian names for pomaceous and drupaceous genera have been above placed in close succession, and opposite them Linnæus's disposal of the same type. By comparing those several names it is readily seen that the eleven species are distributed to five genera by Linnæus, whereas by Theophrastus they had represented nine. In what may be called the average twentieth-century classification of them—as far as the century has proceeded—the same eleven species are ranged under about eight genera, namely *Amelanchier*, *Cotoneaster*, *Cratægus*, *Cydonia*, *Sorbus*, *Cerasus*, *Padus*, *Prunus*. If we of the present are correct, the mean between Theophrastus and Linnæus is the happy one; and this in any case must be admitted, that every revulsion against the Linnæan taxonomy of these fruit trees is a step in return toward the Theophrastian. The same sort of departing and then returning finds illustration in the naming of water lilies as that was done formerly, and has been done over again of late.

Theophrastian	Linnæan	Recent
Nymphaia	*Nymphæa lutea*	*Nymphæa lutea*
Sida[158]	*N. alba*	*Castalia alba*
Lotos	*N. Lotus*	*Castalia Lotus*
Cyamos[159]	*N. Nelumbo*	*Nelumbo speciosa*[160]

The Greek, it is thus seen, received the four species as representing each a genus. With Linnæus the genus of them all was one; while recent systematists have well-nigh completely returned to the Theophrastian view, in all save the names of the genera; and the restoration of even these will follow, under the law of priority.

The shortcomings of Theophrastian nomenclature as to uniformity have not all been indicated. There is one other. While something like half his plant species have but a single one-worded name and that the generic, there are not a few of his genera that are invested with a double—that is, a two-worded—name. It is highly important that this be fully understood. This kind of name is frequent with every botanical author that I am acquainted with, between Theophrastus and Linnæus; and with this fact overlooked there is no understanding any single pre-Linnæan author's plant names either generic or specific. Nor have I found any writer of botanical history making so much as a passing reference to this. I subjoin a very few such Theophrastian genus names as samples, giving, as usual, their equivalents, with also the specific names now in use.

Theophrastian	Modern
Calamos Euosmos	*Acorus Calamus*
Dios Anthos	*Agrostemma Flos Jovis*
D. Balanos	*Castanea vesca*
Carya Persica	*Juglans regia*
Syce Idaia	*Amelanchier vulgaris*
Ampelos Idaia	*Tamus communis*[161]

Most of the names in the left-hand column have exactly the form and structure of ordinary generico-specific binaries, one term being a noun, the other a qualifying adjective. Their respective equivalents placed over against them demonstrate beyond cavil that these particular binaries are not of the usual meaning of such two-fold names, but are purely generic. To take up the first on the list: Theophrastus has a genus *Calamos*, the great reed-grass arundo its type, phragmites also being included in the genus. It is not imaginable that a botanist of Theophrastus's ripe experience and great attainments should think those large grass plants and the sweet-flag to be of the same genus. Beyond doubt, however, that name *Calamos euosmos* did originate in the notion that arundo and acorus are next of kin; for, however unlike they are as to size, foliage, and other particulars, there is a remarkably close similarity in their rootstocks, these being of almost the same size, form, and color in the two. The gatherers of roots and herbs, as we know, looked first of all to the "roots" of things, and these were their first criteria of plant relationships. To these it would be perfectly natural to place the sweet-flag alongside arundo, the true κάλαμος, by its closely imitative "root," and then on account of the aromatic properties of that root to call the plant κάλαμος εὔοσμος. It is equally incredible that Theophrastus should have adjudged the service tree which grew on Mount Ida to be a kind of fig tree. The country people who found the sweet fruits in some way suggestive of figs must have been the creators of that name συκῆ 'Ιδαῖα. But the author who desires to conciliate the public will use great reserve in the matter of suppressing, altering, or even amending established and familiar names; and Theophrastus left plant nomenclature as he found it. And what reasonable objection could have been raised against such binary generic names? It can hardly have entered his mind that it made any difference whether such name were one-worded or two-worded. Outside the domain of our Latin-worded technicalities it makes not the least difference to any of us today how many words go to the making of a generic name. Ivy is a generic name, and as certainly such are Ground Ivy and Poison Ivy. Pine, Ground Pine, and Prince's Pine are names of three genera in no wise interrelated. The same may be said of the five following: Pink, Moss Pink, Squaw Pink, Mullein Pink, and Pink Root. We have no fault to find with such generic names as Star of Bethlehem, Lily of the Valley, Grape Hyacinth, Jerusalem Artichoke, Indian Turnip, American Cowslip, and some scores of others like them. We are a living illustration of Theophrastus in this regard, except that we have two languages for our botany, whereas he had but one. We have two languages in which we use botanical names, with a separate set of rules for each. Into our Latin nomenclature we do not admit any of these two-worded generic names which we use so freely and so readily in our vernacular. In this we differ from a very long and illustrious line of our own botanical ancestry. It is less than two hundred years since what we know in English as Dogtooth Violet and in Latin as *Erythronium* was in all Latin botany the genus *Dens Canis*, *Taraxacum* was *Dens Leonis*, *Convallaria* was *Lilium Convallium*, *Glechoma* was *Hedera Terrestris*, *Helianthus Flos Solis* was *Drosera Ros Solis*, *Centaurea* was *Centaurium Majus*, and the little gentianaceous genus *Erythræa* was *Centaurium Minus*. By the same token, *Chelidonium* was the genus *Chelidonium Majus* and *Ficaria* was the genus *Chelidonium Minus*. In a word, Latin botany for more than seventeen centuries admitted two-worded generic names as freely as the simpler kind; and all after the example of Theophrastus and the prehistoric nomenclators.

This is not the place in which to give the history of the elimination of binary generic names from Latin-written botany. But it was imperative to show that such names are common with Theophrastus, Dioscorides, Pliny, and thenceforward for well-nigh two thousand years. Without knowledge of this fact the ancient names cannot be understood. Without understanding the names as applied during such period, its taxonomy is an enigma, and the setting forth of the history of taxonomy is impossible. To read into that ancient Greek name *Aster Atticus*, as has been done lately, the character of an ordinary binary plant name of the nineteenth century[162]—to fail to recognize in that a mere two-worded generic name—seems to evince a condition of bewilderment as to the whole subject of botanical nomenclature with ancient Greek and later Latin authors. That the term *Aster* is generic and *Atticus* specific cannot well be believed to have been in the mind of Dioscorides; for the genus was monotypic, and they did not then give specific names to monotypes.[163] Nevertheless, in genera of several species generico-specific binaries quite like *Aster Atticus* as to form were very frequent with Theophrastus, though he applied them to species and to varieties indiscriminately,[164] as the subjoined Latin versions of some of his binary names will show:

Salix alba	*Triticum agrigentinum*
S. helix	*T. africum*
S. nigra	*T. assyricum*
Papaver rhœas	*T. Ægyptium*
P. nigrum	*T. siculum*
P. corniculatum	*T. thracium*
Origanum album	*Olea domestica*
O. nigrum	*O. silvestris*
O. creticum	*Phlomis alba*
O. heracleoticum	*P. nigra*

Ecology

Among forecastings of method with Theophrastus, that of the natural associations of plants in particular places is very definitely presented. That which we have learned to designate as the ecologic he accepts as being a very natural kind of grouping.

> All are distinguishable as either terrestrial or aquatic, just as we also primarily distinguish animals; for there are some plants which grow nowhere but in the sea; small ones in our sea (*i.e.*, the Mediterranean), larger ones in the Red Sea. Others affect only marshes or other very wet places. Some cannot live in very wet ground, but restrict themselves to dry ground. Certain others are littoral only. A few trees thrive in either moist land or dry; such are the myrtle, alder, and willow.[165]

In another place, having wild trees and the diversity of them under consideration, he affirms that they differ according to the different nature of the localities in which they grow.

> There are lands that are overflowed by water, there are marshes, and there is dry ground; there are rocky places and smooth pasture lands and harder soils, besides other diversities. There are depressions in the landscape where all is tranquil, and there are elevated and wind-swept exposures; which varied conditions tend to the production of many different things.[166]

Hereupon follows a considerable catalogue of trees which in Macedonia he says occur nowhere but in the mountain districts: fir, wild pine, spruce, holly, linden, hornbeam, beech, box, arbutus, juniper, yew, wild fig, alaternus, phillyrea, walnut, chestnut, and holly-leaved oak. Then there is given a list of such as are common to mountains and lowlands: tamarix, elm, poplar, willow, cornel, alder, oak (*Q. robur*), wild pear, wild apple, privet, hop-

Papyrus.

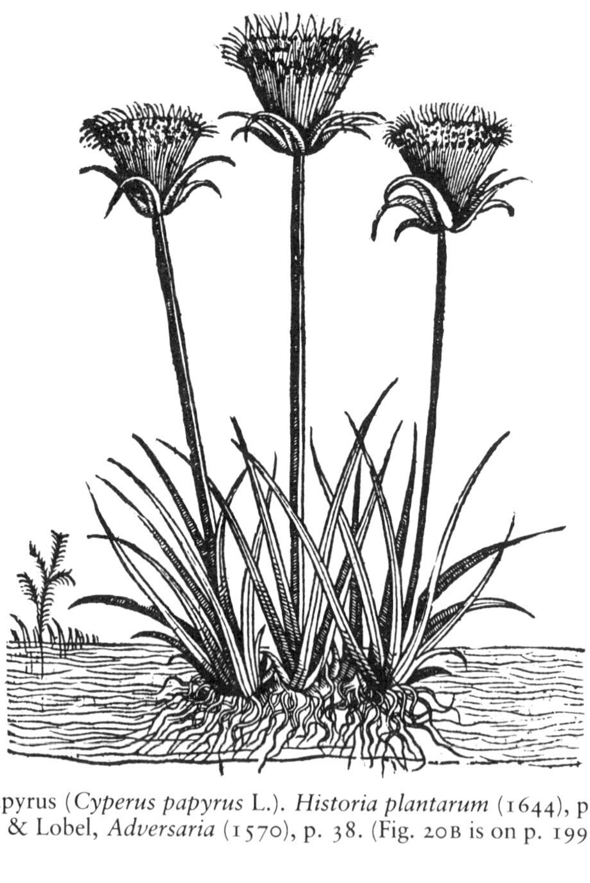

FIG. 20A. Papyrus (*Cyperus papyrus* L.). *Historia plantarum* (1644), p. 431; originally in Pena & Lobel, *Adversaria* (1570), p. 38. (Fig. 20B is on p. 199.)

hornbeam, ash, hawthorn. As to these denizens of both highland and plain he says that

> in general they are of larger dimensions and more comely form on the plains, but of better timber and better fruit in the mountains. To this rule the wild pear and wild apple are exceptions, both being of better timber and better quality of fruit on the lowlands; for in the mountains the trees are gnarled and thorny. Even as to the peculiarly montane sorts of trees, those inhabiting the lower valleys are both the largest and the most copious; and on the highest summits everything is in its most stunted condition, excepting such as by nature require the cold.[167]

Reminding ourselves of this, that Theophrastus treats mainly of field and garden plants, giving much less space to the uncultivated, it becomes particularly noteworthy that in one place eight successive chapters are given up to locating and describing aquatic and other hydrophilous growths,[168] all of them, of course, wild plants. Without any formality of naming the distinctions, yet in practice he distinguishes (1) marine aquatics, (2) marine littoral plants, (3) plants that grow in deep fresh waters, (4) plants of shallow lake shores, (5) those affecting the wet banks of streams and rivers, (6) and those of marshes. His marine herbaceous plants are mostly algæ. The submerged trees, resembling the mode of growth and branching oaks, fig trees, the palm, and the vine, and attributed to the Red Sea, are mostly corals, organisms that were still considered to be plants, denominated lithophytes, until within the last two hundred years.[169] The author, in describing the marine oak, marine fig tree, etc., is careful to inform his readers that their resemblances to the trees of the land are only those of mode branching, that they are smaller than their terrene analogues and have no leaves. In stagnant fresh waters several cubits deep thrive nelumbo, nymphæa, and trapa. Along the shores, in shallow water are reeds, rushes, the papyrus, sparganium, and typha. The banks of running streams suit the poplar, alder, and willow, the roots of which only are laved by the flowing waters. In wet sandy soil not far from streams thrives the cyperus with nut-like edible roots.

That ecology should have formed a sort of taxonomic basis for Theophrastus in his treating of wild plants was most natural. Such pronouncedly hydrophilous growths as reeds and rushes, coarse sedges and the largest grasses, phragmites and arundo, besides typha and sparganium—all are at agreement not only ecologically but in many respects also morphologically. They all have upright, smooth, and simple stems, filled with pith when young, some of them hollow when mature, but none ever woody-solidified; their foliage long, narrow, entire, never with any trace of the network of veins. Moreover, every one of the group was what would have been called flowerless, by all save Theophrastus and his students, because they had no flower leaves. Other aquatics, such as nelumbo and the waterlilies, colocasia and sagittaria, were like the rest structurally except as to foliage, the leaves being large and ample, rounded rather than narrow-elongated, with radiating veins if any, and some of them with large flowers made up of showy leaves. The Greek, I say, might have written such a diagnosis of his aquatics as a group viewed morphologically. That he saw these marks none will doubt who reads his descriptions of the species.

In the vegetation of the mountains he again lists the trees that grow on the exposed and sunward slopes, and those that flourish nowhere but upon cold northward declivities, and such as inhabit only the frigid summits.

Dendrology

Almost an undue proportion of Theophrastus's space is given to the consideration of trees.[170] He appears to have been a great lover of them; and his knowledge of them is presented with so much system that at least a considerable dendrological paragraph becomes a necessity if one is to convey any adequate notion of his botanical work as a whole. He classifies them (1) as cultivated and wild. This is one of his general divisions of all kinds of plants, one that has already been sufficiently discussed in another place. Trees are (2) deciduous or evergreen. Their diversity as to tenure of foliage is so thoroughly discussed, and withal so judiciously, that the more than two millenniums that have passed seem to have recorded but few and unimportant additions or amendments to the principles of this chapter as he left it.[171] Adhering to his classification of all things as cultivated and wild, he gives two lists of trees that are evergreen: the olive, palm, sweet bay, myrtle, the cypress, and our pine among the domesticated; for the wild, the fir and spruce, wild pine, certain kinds of oak, holly, box, and the arbutus tree. The last, he says, sheds the foliage from its lowest branches, while the head of the tree remains evergreen. This distinction of evergreen and deciduous he regards upon the whole as quite natural and valid, despite reports he has heard, and readily accepts as probably true, that in the warmest climates grape vines and fig trees shed their leaves so tardily as to seem almost evergreen. He has observed that, among perfectly familiar species, some regularly divest themselves of all foliage in earliest autumn, others later, while a few habitually retain it until winter has begun. He can therefore credit those who assert the existence, in other lands, of kinds that do not cast their old leaves until near the time of the development of the new in spring. He has even found out that the most strictly evergreen develop one new set of leaves, and as invariably lose one old set, every year. Upon deciduous trees and shrubs he seems to have kept phenologic records.[172] He relates that certain kinds come into leaf early, others late; also that such as are first in leaf in the spring are not the first to shed their foliage in autumn, and it is equally established, that those latest in leaf do not retain them longer than others. He has likewise learned that trees in a moist climate and soil retain their foliage for the longest period, while deciduous things of a dry soil and poor shed their leaves earliest of all; and finally, that a young tree keeps its

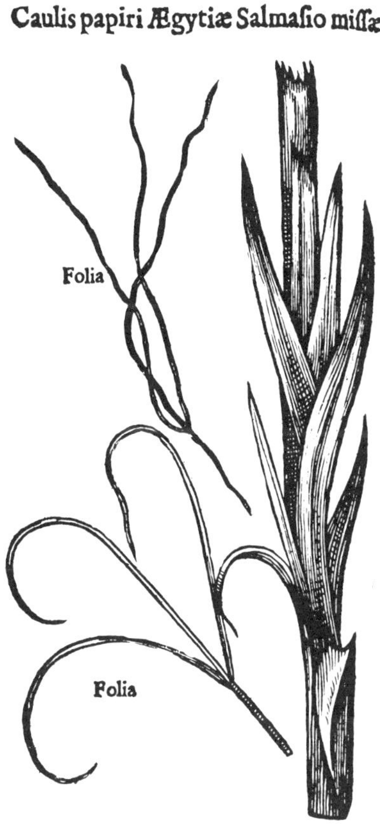

FIG. 20B. Papyrus (*Cyperus papyrus* L.). *Historia plantarum* (1644), p. 432.

foliage until a later date than does an old one. The foliage of evergreens is usually narrower than that of deciduous trees, is of a firmer texture, and in a much greater proportion of species is fragrant or aromatic.

In respect to their modes of branching, trees are (3) regular or irregular. If at any time before having studied Theophrastus I had been asked who first taught the distinction between the excurrent and deliquescent in the forms of trees, I should have attributed it by guess to some dendrologist of about the middle of the nineteenth century whom I could not name. Therefore nothing that I have come upon in this author of antiquity has more surprised me than his lucid setting forth of these two modes of tree development. The differences seem to have forced themselves upon his notice while studying the fir tree. At least, he gives fir as the best type of what we call the excrescent; and the oak is his example of the other mode as to branching.[173] The distinction (4) of flowering and flowerless in trees did not imply the recognition of such as in modern botany are called cryptogamous. It was but a matter of the author's success or failure to find what he would have allowed to pass for flowers. And the classifying them as (5) fructiferous and sterile is not at all the equivalent of the flowering and flowerless division. There is one kind of sterility that is manifestly accidental only, a consequence of something unfavorable in the environment. The palm is sterile in Greece; yet if transplanted to Babylon from Greece, it becomes as fruitful as the Babylonian.[174] Peach trees are sterile in Egypt. The wild sorbus of Greece, transplanted from its mountain habitat to the fervid low country, though flowering copiously in the new situation, never fruits there. The reason is plain to him. Its nature requires the cold climate of the mountains. But when he alludes to the black poplar of the island of Crete as sterile when introduced on the mainland, one may suspect that to have been owing to the possible circumstance of only male trees having been brought over. I do not think Theophrastus ever suspected the fact of diœcism in any plant or tree, however often we may find him speaking of them as male and female. Of this I have more to say in another place.[175] Again individual trees according to their species have (6) certain of their branches fructiferous, certain others always sterile.[176] He has observed that some, like the vine and the fig tree yield fruit on no branches but the newest, those of the season; that almond, apple, pear trees, and many more fructify upon no other branches than those that are one year old; some like the carob (*Ceratonia Siliqua* L.) produce their fruits both upon those of this season and the season before, with also a few at the same time upon old and thick branches. Some kinds are fruitful on none but their topmost branches; others are fruitless at summit, and only fruitful on lateral branches. Another distinction is that of (7) the location of fruits in respect to foliage. Some trees bear their fruits beneath the leaves; on others it is borne above them; while in some, like the sycamore fig tree (*Ficus Sycomorus* L.), it grows down on the naked trunk.[177] From our point of view this is a useless distinction; but not so with Theophrastus, who seems to have been unable to attribute to the foliage of trees any more important function than that of a protection to the young and growing fruit or seed.

Trees are extensively treated of (8) as to their ecology and geographic distribution. There are trees peculiar to mountain districts, and others confined to lowlands and plains. As of the former habitat he names the fir, wild pine, spruce, holly, box, walnut, chestnut, and many more. A still greater number of different kinds are of the plains only; among them are one of the elms, the ash, maple, alder, willow, poplar. A few kinds are common to mountain and plain.[178] Among the montane some, like the wild pine, luxuriate on slopes that look southward, and will hardly grow at all in any other places, while the fir, on the contrary, attains perfection on the cool and shady sides, and if ever seen elsewhere, has an inferior growth and is unlike itself. The tallest and largest firs known occur in a deep valley in Arcadia where they say the sun never shines. He notes it as a general rule that the kinds of tree affecting shady and cool places are tall and straight, their trunks not forking or parting into subsidiary trunk-like branches; but that arboreal growths of this latter description are those of open and sunny places. Certain trees are wont to grow nowhere but along watercourses; and certain others belong exclusively to the highest elevations of the mountains near perpetual snow.

It is evident that in those coniferous and hardwood trees belonging to cold northward slopes of southern mountains Theophrastus sees a sort of fringe, so to speak, of the great almost unknown regions of Europe northward; for what reports have been brought from that direction indicate that there, even the lower lands are clad with forests of fir, pine, oak, chestnut, and others known at the South only on the mountains; and he thinks it may be reasonable to infer that the whole North has no other silva than that thus indicated. He has not been able to learn that it has kinds of trees peculiarly its own.[179] Southward, however, across the Mediterranean, and away up the Nile, are very different kinds of trees; many that cannot be successfully transplanted from that dry and heated climate into regions where there are rains and cold weather at the winter season. There, in some localities where

FIG. 21. Sycamore fig (*Ficus sycomorus* L.). *Historia plantarum* (1644), p. 291.

it never rains, the palms attain their greatest dimensions and their best quality of fruit, not, however, indicating that they have no need of moisture. On the contrary, wherever a grove of wild palms occurs water is sure to be found at no great distance below the surface of the ground, though it is usually subsaline, a circumstance which, he says, has taught the cultivators to use a little salt with advantage in growing dates in other than their native soil.[180] In Phœnicia and in Syria there are various kinds of palm, because these like other trees differ according to differences of region and climate as well as according to the culture given them. The palms as a group interest him deeply; they are in many ways so very unlike other trees, in their best known type bearing everything—leaves, flowers, fruits—in a single terminal tuft, the caudex being without a branch. Now, with Aristotle, father of biologic investigation, and with those of his school, there was much and serious inquiry into the question of a soul, and some particular seat of life in plants. The latter was hard to locate, so many are the trees which, as susceptible of propagation by mere cuttings, thereby proclaim it that their seat of life is everywhere, so to speak. But these palms were different. Cut off the leafy and fructiferous summit of the tree and the whole is killed, just as one kills an animal by decapitation.[181] He was near thinking that in this kind of tree that one terminal part is the seat of life; but he knows of a smaller palm, native to the islands of Sicily and Crete (*Chamœrops humilis*), which, if the top be removed, or if even the whole tree be cut down to the ground, renews itself by shoots from the root.[182] He has grown from seed the few kinds of palms available at Athens; knows all about their germination and early stages of development; finds no distinction of bark, wood, and pith in the structure of their trunks;[183] recognizes in them the only plants of which he dares to say that they exist in the two sexes, a male that flowers and is fruitless, and a female that is flowerless but bears fruit;[184] but a special seat of the vegetal soul, or life, evades him even here. That he admits trees into the alliance of the palms on vegetative characters alone, when the fruits are not in the least date-like, is seen in two instances. One such is so described as to have convinced some authorities that what he had in view must have been the cocoa-nut palm of the farther Indies; but it is now no longer doubted that it is the *Hyphæne coriacea*,[185] and not *Cocos nucifera*. He also knew, and well described the type of the Cycadaceæ, *Cycas circinalis*, as a kind of palm.

Theophrastus, like an ancient Humboldt, or Grisebach,[186] takes pleasure in making comparisons between certain of those trees of arid northern Africa and certain others of southern Europe with which all his readers are well acquainted. There is the persea, as he calls it (*Cordia Myxa* L.), which in some ways suggests to him the pear tree, a large and very handsome tree, in its mode of branching, its foliage, flower, and fruit externally resembling the pear; but it is evergreen and ripens fruit at all seasons, the fruit however possessing a nut at its core like that of a prune, etc.[187] There are other Egyptian trees so unlike any known to his untravelled countrymen that he cannot contrast them with any familiar kinds; but the competent botanist of today will recognize the genera and species of some of them by his descriptions. About Memphis are trees frightfully armed with thorns in every part except the trunk. It is the arid subtropic region of several gum-bearing acacias and their allies. He attributes to all of them the leguminous fruit, *καρπὸς ἔλλοβος*, says the pods are gathered and employed as a substitute for galls in tanning leather, and also used medicinally. One kind he calls white thorn (*Acacia Senegal* Willd.).[188] To this he attributes flowers beautiful and fragrant, so that they make garlands of them. Another he denominates black thorn (*Acacia Arabica* L.). Quantities of gum are gathered from this kind. It exudes from the trunk where incisions have been made, or even spontaneously without incision of the bark. In the vicinity of Thebes there are extensive forests of these trees, and that far away from the river, where they are never irrigated. Such pen pictures of foreign dendrologic scenes are not rare in Theophrastus; and they are always so vividly drawn that the reader inevitably thinks of him as writing from the very midst of the things he is describing, whereas it is quite certain he is only compiling from travellers and historians. To this picture of Egyptian dendrology he will add yet another member of this same alliance of trees, and this the most remarkable one. Forbiddingly spinescent or thorny like the others, it has a more delicate foliage, like that of a fern. Whenever a branch of this tree is disturbed by a touch, all the leaves upon it seem very suddenly to wither away and collapse; then after a little time they revive and return to their former condition. This quaint description of what has been called sensitive foliage is probably the oldest extant. The species that was described by Theophrastus is doubtless *Mimosa polyacantha*.[189] These and as many more kinds of tree and shrub he mentions by name and short description as "peculiar to that region."

In succeeding chapters he dwells at some length upon the ligneous growths of the Arabian deserts, where it rains no oftener than once in four or five years, and trees are scarce and all of them spinescent;[190] and describes the varied and often luxuriant silva of the more distant Indies.[191] It is to be remembered here that Theophrastus was contemporary with Alexander the Great, whose expedition to the farther Orient was the first of its kind in all history to include among its officials learned men whose duty it was to write up the geography, climatology, and even the zoology and botany of the regions traversed, an enlightened thought of Alexander's, beyond doubt suggested by his boyhood's illustrious tutor, Aristotle, father of all nature study.[192] To the manuscripts brought back by this scientific staff of Alexander, Theophrastus was indebted for all that he knew of the farther Oriental plant ecology and geography; and all that remains of those reports is what the philosopher quoted from them. The originals were long since lost. In these chapters of Theophrastus we have the earliest, and very interesting and faithful, accounts of the banyan tree (*Ficus Bengalensis* L.),[193] citron (*Citrus medica* Risso),[194] the cactus-like euphorbia (*E. antiquorum* L.),[195] the oleander (*Nerium Oleander* L.),[196] the tamarind (*Tamarindus Indica* L.),[197] a tree which they reported to possess the singular faculty of folding up closely its pinnated leaflets at nightfall and going to sleep for the night, the earliest record in botany of the phenomenon now known as nyctotropism.[198]

The *Historia Plantarum* is not without its chapters on the diseases of trees, and the influences of the seasons upon them;[199] the proper seasons of the year for felling timber of different kinds;[200] qualities and special uses of wood of different trees;[201] but space must not here be given to comment on these economic aspects of the subject. A more philosophic interest attaches to questions of longevity in trees.

Under the caption of duration (9) he records a number of curious facts, and discusses briefly a question that may arise in reckoning the age of certain trees. He certifies that it is common with orchardists and vine growers of his time to renew, as it were, an old and moribund tree by cutting it down near the ground, and then training up in its place one of the new shoots that are thrown up from the base of the stump. How, he asks, is the duration of the tree upon this spot to be estimated? Do the old and new tree constitute two individuals or only one? If the main trunk be essentially the tree, then the new trunk is that of a new individual; and he adds that the very roots of the original tree perish, eventually, and that the new one now has none other than its own.[202] Yet individual grape vines, the continual growing and fruiting of which during two centuries is perfectly authenticated, have in this way been renewed by the cultivator's art, several times over within the two centuries. He finds it a prevalent opinion in the rural districts that all wildwood trees are long-lived and all the domesticated of short duration. This the philosopher does not think well grounded. It is true only in a very general way, and with many exceptions. Some kinds of forest trees live very long, others do not; and the same may be said of the domesticated, though these, upon the whole, have a shorter period. And, universally, those that fruit copiously have a shorter time of life than the unprolific; also the kinds of wild trees that affect low and wet land are shorter-lived than those occupying dry and barren ground. Even sweet-fruited and aromatic trees live longer than the sour-fruited kinds, he has observed. On the reputed great ages of certain individual trees still living in his day as well as carefully preserved and religiously venerated—such as the olive tree at Athens said to have been planted by Hercules, and the Caphian and the Delphic planes, both believed to have been planted by Agamemnon—he is somewhat incredulous. The traditions as to their origin have come down by some who also wrote fables. It might be well to investigate. He thinks it quite certain that olive trees and planes, as well as many other kinds, live a very long time. Beyond this platitude he will not go; but it is manifest that he is not in sympathy with the mind of the credulous multitude as to the extreme age of this or that individual and historic tree. He was rather skeptical on the subject, and probably would not have believed it possible that northward in Europe far beyond the Mediterranean oaks sometimes lived through ten or a dozen centuries; nor that on another and unknown side of the world there were conifers[203] of considerable dimensions then living which would be flourishing still, after the passing of twenty-two or -three hundred years.

Transmutation

In this twentieth century of our era there are farmers in the world, and not unintelligent, who believe that to some seed of wheat or barley after it has been sown in the field something may happen by which it comes to sprout and grow up into a plant of what they call chess, or cheat, a plant known to botanists as *Bromus secalinus*, this name itself now apparently destined to perpetuate forever that old opinion—older than history, no doubt—that a grain of barley, *secale*, may become the parent of a plant of chess.

The seemingly indicative facts upon which this transmutation theory appears as if it might have established itself in the minds of prehistoric grain growers were several. Neither chess nor darnel grew commonly elsewhere than in the low wet parts of grain fields. In these spots only very few, depauperate, and almost infertile were the stalks of wheat or barley, though the seed of one or the other had been sown there copiously. The explanation which a very primeval and elementary philosophy could offer was, that the grains of wheat, debilitated to the verge of decay by unusual cold and dampness, became unable to generate a better plant than the small-grained and worthless chess, or cheat, as the farmers still call it. Even Nature herself had taught them with what ease she can, and how every year she actually does effect more marvellous transformations, at least in the animal kingdom. The fish-like creatures that swim about in pond and pool in the spring months, they had seen come forth in summer, changed to such things as frog and toad. In their simple life, lived very much out-of-doors, the observing and intelligent had taken note of how underground grub and tree caterpillar by and by assume a larval state, rest in that for months together, and then suddenly are born again, the one as beetle, the other as butterfly. In his environment, the primeval nature student would not doubt the easy possibility, or the apparently strong probability, of some kind of sudden transformation in the world of plants; and the well-known frequent appearing of chess in wet ground instead of wheat where only wheat had been planted might be evidence enough of such transmutation.

That the supreme philosopher of antiquity, the father of animal biology, who knew so nearly everything about metamorphosis in lower animals, also must have investigated the case of the supposed transmutations of plants, appears most probable. For this department of botanical history it may be thought particularly unfortunate that Aristotle's botanical writings have not survived.

Theophrastus does not formally and didactically discuss this question, though he makes a number of references to this changing of one plant into another as something universally believed in his day. I shall reproduce a number of them.

Recording in one place the usages of his time as to the different seasons of the year and the several methods of sowing cereals, as well as giving a long list of leguminous plants that he names, he concludes the chapter with a remark like this: "None of the above are liable, on account of a bad condition of the seeds, to change into other plants except wheat and barley, which people say may change into darnel (lolium); more particularly wheat, and this being said to occur as the result of wet weather, and in muddy places of the fields.[204] In the same connection he records it that "Some think flax also changes into darnel."[205] Quotations like these read much as if the author had been unwilling to take the responsibility of either affirming or denying the proposition. "People say" that such metamorphoses occur. But in another paragraph, one relating to different kinds of wheat as imported into Greece from other parts, he affirms that from Pontus, from Egypt, and from the island of Sicily grain-growers of his time obtain seed wheat which matures crops free from lolium; though that from Sicily comes up infested by a different weed called melampyrum, *i.e.*, black wheat.[206] In this report about certain of the imported seed wheat as immune from changeability into lolium, one almost reads between Theophrastus's lines that he regarded the absence from or presence of darnel in wheat fields to be due to the absence or presence of darnel seed in the seed wheat at its sowing. There is another passage in which the philosopher by implication seems to question the wheat-darnel metamorphosis. With the ancient husbandmen as with the modern it was usual to sow wheat either in autumn or in spring. Theophrastus records this, and also says that lolium always germinates in the autumn. He has investigated the case, and gives some points of diagnosis by which young plants of lolium may be distinguished from young plants of wheat.[207] This autumnal germination and winter growing of the lolium almost forces upon the thoughtful reader the inference that if lolium occur in a field of wheat that was sown in spring, it was already up and growing at the time the wheat was sown. But there is one phase of this popularly credited metamorphosis doctrine of which Theophrastus is so impatient that he openly denies it. "Some say that barley changes to wheat sometimes, and wheat to barley, and that in the same field. Such statements are to be received as fables. Changes of that kind would be without a cause. It is diversity of condition that induces change."[208]

However skeptical Theophrastus may have been about all such pretended metamorphoses, he had doubtless the usual prudential reason for declining to assail them openly at every mention of them. The belief in them was universal; and the time for the elimination of such belief from even thoughtful minds was yet far distant. We find it persisting with men of intellectual attainments as late as the seventeenth century; at which time Scaliger, a most learned commentator on Theophrastus, avers that he himself has witnessed the transformation of wheat into barley and intimates that the Greek might have done better than to discredit the phenomenon.[209]

If it was the metamorphosis attending the development of the individual reptile, and the insect, which helped to elevate to the dignity of a quasi-rational belief the superstition about the changeability of wheat into lolium, it must be allowed that the reasoning was not very cogent. The cases are not parallel. One is that of the changes in an individual between youth and maturity. In the other it is the seed of one species which, between the sowing of it and the germination, mysteriously changes to that of another species.[210]

There is, however, a phase of transformation in plant life that runs parallel to the metamorphoses of lower animals. This has become generally known only recently, and by means of the compound microscope as applied to the making out of the life histories of ferns, liverworts, and other flowerless plants of lower organization. As illustrated in these plants, this kind of individual metamorphosis could never have become known to the nature students of antiquity, or even to those of the earlier modern epoch, owing to their lack of the necessary optical aids. But somewhat analogous metamorphoses take place in the individual life histories of certain higher plants, even of trees; and this fact is not so commonly known as it ought to be. In the family of the Mimosaceæ there is a considerable list of trees which only in the state of seedlings of a few years old exhibit the usual delicate fern-like doubly pinnated foliage of their family. Before the trees are old enough to flower they have divested themselves of every trace of that kind of leaf and are clothed instead with very narrow, simple, entire, firm and almost leathery organs, in cut somewhat recalling willow leaves, or perhaps better compared to those of mistletoe. Now it will not be in the least to the discredit of a circle of experienced and quite skillful botanical amateurs of the Northern Hemisphere if, placing before them two branches of such an acacia, one from the ferny-leaved young tree, the other from the mature tree with its stiff phyllodes like mistletoe leaves, and stating that these two branches represent one and the same species of Australian acacia, the whole circle of them suspect at first that I may be jesting. Some of the Australian eucalyptus species undergo as complete a metamorphosis in the individual, with this difference that the adult tree, at least in the earlier stage of maturity, exhibits both phases of branch and foliage; the lower and flowerless portion of the head of the tree seeming to represent one genus, the middle and upper branches—those that have the flowers and fruits—seeming as if they must be those of another genus, or even of another family. It is quite as if the tree at a point just below midway of its axis had become by grafting from that point upwards a tree of another genus.

Of these changeable acacias and eucalypts the ancient Greeks of course knew nothing; but they were familiar with a similar case, that of what is known in very modern botany as *Hedera Helix*, the English ivy. This, as they well understood, is most commonly seen as a trailing or climbing shrub, the stems rooting everywhere, all the foliage leathery, of angular outline and dark green. In this phase the plant may pass the whole of its existence and remain flowerless. Occasionally, when very old and having climbed by wall or tree trunk to sunlight and upper air, a new thing happens. Out of the summit of this dark-green creeping ivy mass an upright bush appears; its branches not rooting, firm, independent, bearing leaves not leathery in texture, not in the least degree angled, and even of a decidedly light green; and this bush up at the top of the ivy will in course of time bear flowers and fruits. The ancients before Theophrastus had no difficulty in explaining this phenomenon. Their firm belief in all sorts of transmutations as taking place in nature saved them any perplexity. They held these two phases of the ivy to be generically distinct, and had their fully established names for the two genera: *Helix* for the rooting and climbing plant of dark angular foliage, *Cissus* for the upright bush of the pale thin ovate leaves, that into which the *Helix* sometimes in old age transformed itself.

Again Theophrastus fails to be satisfied with the popular philosophy, and suggests one that he thinks more rational. If every plant of *Helix* under the right conditions and with fair opportunity would develop a *Cissus* bush at summit in maturity or old age, which he says some agree to as being probable, then he would be of the opinion that the distinction between the two is not a generic one, but only a matter of the age of the individual.[211] He who has taken note of this philosopher's way of advancing his most revolutionary propositions with urbane reserve will understand him as here pronouncing against the time-honored doctrine of a generic change from *Helix* to *Cissus*, and as averring that these are but the young and the old phases of one species. It was the inductive philosopher, the scientific botanist, undermining as it were by stealth an ancient botanical superstition, because he had a truly scientific proposition to put in place of it. In this instance Theophrastus was, as usual, far in advance of his own time. For centuries after him men still held to the bigeneric ivy. I should confidently expect to find this pre-Theophrastian view surviving still among the peasantry of some parts of Europe where *Hedera Helix* is common and well known.

Living in the midst of a time when a thousand superstitions prevailed everywhere concerning plants, their origin, magic powers, and their often grotesque metamorphoses, it is greatly to the credit of Theophrastus that he should have risked something of his own popularity as a teacher and author by expressing, even covertly, some of his own doubts and disbeliefs; even as to some, the truth of which seemed so probable that the belief in them is not yet in our time obsolete.

Recapitulation

Certain opinions that are completely groundless respecting Theophrastus's merits as a botanist, opinions quite opposite to any that had ever before been expressed, and such as no man who had read three chapters of that author could have entertained, have been widely disseminated during the last thirty years. One such statement of opinion is before me and reads thus:

> Greek authors built their views of the philosophy of botany on very weak foundations; scarcely a plant was known to them exactly in all its parts; they derived much of their knowledge from the accounts of others, often from dealers in herbs. From this scanty material, and from various popular superstitions had Aristotle formed his views on the nature of plants; and if Theophrastus possessed more experimental knowledge, he still saw facts in the light of his master's philosophical doctrines.[212]

Such reckless writing as that, betraying innocency not only of Theophrastus's work, but also of that high opinion of it which had been expressed by most accomplished botanists of the eighteenth century and the nineteenth, has been widely read by botanists of the present generation. In view of this, it seems more than desirable that there be presented briefly and synoptically something like an enumeration of those items, or elements, of universal botany of which Theophrastus appears to have been the discoverer and first promulgator.[213]

In this recapitulation I shall employ a few modern terms, such as petal, corolla, and andrœcium, unknown to ancient Greek botany, that I may thereby both more clearly and more briefly express the fact of the Greek's having recognized, though under other names, the things themselves.

1. He distinguished the external organs of plants, naming and discussing them in regular sequence from root to fruit; the naturalness of which sequence was afterwards pointedly denied; but in modern botany it stands everywhere approved.

2. He classified such organs as (a) permanent, and (b) transient, a division of them which may yet be shown more scientific than the modern distinguishing of them as (a) vegetative, and (b) reproductive.

3. The existence of aerial roots, as being of the nature of roots, and thus different from tendrils and other prehensile organs, was discovered by him and has never since been disputed.

4. He remarked upon the inconsistency of retaining in the category of

roots certain enlarged, solidified, jointed, and otherwise peculiar underground parts, a suggestion which lay unheeded during two thousand years of botanical history, and has only recently led to the open recognition of the category of subterranean stems.

5. He recognized, by differences of size, solidity, and other particulars of structure, three classes of stems: the trunk, stalk, and culm.

6. By never speaking of calyx and corolla as peculiar and separate organs, but always referring to their parts as leaves merely, it is evident he regarded the flower but as a metamorphosed leafy branch; to which forgotten Theophrastian philosophy of the flower[214] neither Goethe nor Linnæus had but returned, when each supposed himself the discoverer of a new anthogeny.

7. He divided the plant world into the two subkingdoms of the flowering and the flowerless.

8. The subkingdom of the flowering he again saw to be made up of plants leafy-flowered and capillary-flowered; really the distinction between the petaliferous and the apetalous; one the deep import of which was first realized and taken advantage of by the systematists of some two centuries ago.

9. He indicated the still more important differences of the hypogynous, perigynous, and epigynous insertion of corolla and andrœcium.

10. He distinguished between the centripetal and centrifugal in inflorescences.

11. He was first to use the term fruit in the technical sense, as applying to every form and phase of seed encasement, seed included; and gave to carpology the term pericarp.

12. He classified all seed plants as (a) angiospermous and (b) gymnospermous.[215]

13. Respecting the texture and duration of their parts he classified all plants as tree, shrub, half-shrub, and herb; also noted that herbs were of perennial, biennial, or annual duration.

14. He indicated with clearness several of those differences in the structure of stems, leaves, and seeds by which the botany of later times separates plants monocotyledonous and dicotyledonous.

15. He described the differences between the excrescent and deliquescent in tree development.

16. He knew how the annual rings in the stems or trunks of certain woody growths were formed.[216]

17. Theophrastus, with natural vision unaided by so much as the simplest lens, and without having seen a vegetable cell, yet distinguished clearly between parenchymatous and prosenchymatous tissues; even correctly relating the distribution of each to the fabrics of pith, bark, wood, leaves, flowers, and fruits.

This list of facts botanical which Theophrastus saw, and in the main discovered, is not complete, but it embraces well-nigh all the first rudiments of what even today is universal scientific botany. It illustrates superabundantly the fact that Theophrastus, and no man of any later time, is the father of the science as we now have and hold it. And in the light of the above partial recapitulation of his discoveries, what possible remark could be more inane than this? "If Theophrastus possessed more experimental knowledge [than Aristotle], he still saw facts in the light of his master's philosophical doctrines." When a man has firmly laid the foundation of a science, and then has added the suggestions of almost the whole superstructure, what faintest shade of pertinency can there be in asking what his philosophic doctrines were? As reasonably might one leave any scientific work, alive with new facts, quite unexamined because its author's philosophy was that of a school unpopular, or his creed unorthodox.

The most generous interpretation of the words quoted would seem to be, that their author, having no knowledge of Theophrastus, thought to absolve himself from the task of acquiring it by trusting that the Greek would never again be found worth studying.

To me it seems not improbable that historians of the future, learning to know this great founder's mind better than it is yet known, may agree in some judgment not unlike this: that all that has been added to the understanding of plant life and form—to morphology, anatomy, physiology, perhaps even to taxonomy—within the last three centuries has been due to the inventions of the opticians, and to the increased number of students and investigators, rather than to the appearing on the botanical horizon, within the modern period, of any one mind in powers of observation, penetration, and sagacity superior to Theophrastus of Eresus.

Plumier (1703) sought to commemorate Theophrastus in a newly discovered genus of West Indian shrubs, yet was so inconsiderate as to name the genus ERESIA. This Linnæus (1740) changed to THEOPHRASTA.[217]

3. Greeks and Romans After Theophrastus

Leaving Theophrastus, and going forth in search of the next landmark in the progress of our science, we seem at once to enter an almost boundless pathless waste. Or, as the outlook has been described by another:

If history be a connected succession of events, botany from Theophrastus forward to the sixteenth century has no history. Only isolated pieces of information, like bits of wreck half buried up and down stretches of seaside sand, are left. These are connected with certain names; but, beyond that, are hardly of historic import. Of written monuments of real botany in the Greek language after Theophrastus there remains not one. For the small volume of Nicolaus Damascenus, known to us only by a translation into barbarous Latin, perhaps might, and possibly might not, have been reckoned such a monument.[1]

What is here understood is, that a great multitude of scattered fragments, together with several completed pieces of writing about plants that, out of the literature of antiquity, have been preserved, do not afford material for the history of botany for so much as one only of the ten centuries that next succeeded the times of Aristotle and Theophrastus; and that this is true partly for the reason that the pieces and the fragments that have reached us are from men who were not botanists after the order of Theophrastus, but writers on medical, agricultural, and horticultural botany.

That the whole number of Greeks who wrote of plants in one way or another was very great there is evidence enough. As many as three hundred and fifty years ago the learned botanist, zoologist, and bibliographer Conrad Gesner gave out a printed list of more than one hundred names of Greeks who, in his day, were known to have written more or less botany.[2] Of ancient Latin authors of botanical works similar to those of the Greeks in kind, the same distinguished bibliographer enumerates ten or twelve, but does not seem to have quite finished that catalogue. As considerable a number of Arabian medical botanists of a less ancient period is also given.

The best of those botanical fragments, gathered in as it were from the wreck of ages, were what really inspired the first beginnings of modern botany in the sixteenth century. From the time of the establishment of universities and better schools of medicine in the middle ages, the best text-books of pharmacy were those of the ancients: Hippocrates, Nicander, Dioscorides, Pliny, Galen, and as many more less celebrated than they. The remedies in use were almost wholly vegetable, as were also the poisons and their antidotes; and the old authors' books were the topics lectured on in every school, and their plant descriptions were trusted to for the correct identification of plants alimentary, medicinal, and poisonous. And so, not even from the simplest outline of botanical history may all mention of the old Greek, Roman, and Arabian agricultural, horticultural, and medical botanists be omitted. We are not indeed able to construct out of their literary remains a botanical history of their period; but we know that they became at last, and incidentally, the inspirers of a new epoch which dawned upon botany a thousand years or so after the last of their line was dead. Sketches of the lives and works of a few of them, and only such as came after Theophrastus, will here find place.

Nicander of Colophon

This Greek grammarian and poet flourished in the second century before the Christian era,[3] was native of a small village, Claros, close by Colophon in Ionia, and was anciently known as the Colophonian Nicander by way of distinction from others of the name of Nicander. He was of great renown as a poet, and his topics were mostly such as invite to the consideration of the living things of field, forest, and wilderness. Evidently Nicander was a naturalist, also learned in pharmacy and toxicology, and chose to express himself in poetic measure.[4] That which may have been his most elaborate work has been lost, that is, the *Georgica*, a versified treatise of agriculture praised by Cicero,[5] and extensively quoted by Athenæus,[6] whose quotations are all that remain of the poem. Among these remnants, there is a long passage on flowers and other ornamental plants, an account of the Egyptian nelumbo, a dissertation on poisonous fungi—the earliest on record—and even another on the cultivation of edible mushrooms, besides many things relating to various other food plants.[7]

The only works of Nicander that have reached our time are two poems, under the somewhat lugubrious titles *Alexipharmica*, a treatise on poisons in general and their antidotes, and *Theriaca*, on poisonous animals.[8] The first of these, according to Meyer,[9] who appears to have read both of them carefully, is of 630 verses, and has under discussion 21 different poisons, of which 2 are mineral, 8 animal, and 11 vegetable products; and the remedies for them are with hardly an exception vegetable. The account of the symptoms of different poisonings is said to be both true to modern experience and vividly drawn, but the plants themselves, whether poisonous or antidotal, are hardly more than named, never described, and the book as a whole is devoid of matter properly botanical.

In the *Theriaca*, a more extensive work of 958 verses, botany, as well as zoology, fares somewhat better. After a preliminary statement of means of frightening away poisonous animals or keeping them aloof, together with certain precautions to be observed by such as sleep out-of-doors at night, there follow some descriptions of certain more common and dangerous kinds which are often drawn with remarkable exactitude and faithfulness to nature. And here again, the bites and stings of these have always their remedies in certain plants,[10] of which also in most cases only the names are given, though sometimes a few hints are given as to how the plant may be identified.[11] The three particular plants, centaurea, aristolochia, and trifolium, are together efficacious against every poisonous animal's bite or sting. The identity of Nicander's centaurion is uncertain. It may have been *Hypericum olympicum*, but that of Theophrastus, whom Nicander often quotes, is *Ferula opopanax* more probably. The aristolochia is that of modern botany, the species either *A. rotunda* or *A. longa* or both. The only trifolium known to the Greeks was our *Psoralea bituminosa*.[12] In the two poems thus adverted to Meyer counted the names of 125 different plants.[13]

Sprengel gives a list of some thirty species of Nicandrian plants which, though not in all cases identifiable with certainty, seem to have been first mentioned by this writer.[14]

Adanson in 1763, resolving to dedicate a genus of plants to Nicander, doubly distinguished himself in so doing; for NICANDRA not only commemorates a worthy name in old Greek plant lore, but the type that was to bear the name was with an exquisite sense of fitness chosen from out the family of the nightshades.

MARCUS PORCIUS CATO (B.C. 235–150)[15]

An illustrious Roman of vigorous mind and great originality, serving the public in the most exalted and responsible offices with great acceptance, affluent as to means, he led a life of great simplicity, temperance, and frugality, delighting in nothing else so much as the training of his children in virtue, and cultivating flowers and fruits.[16] Practically a philosopher indeed himself, Cato held in abhorrence the philosophies of the Greeks, was strongly averse to the introduction of Greek art and Greek customs into Rome, apprehending the destruction thereby of Roman valor and simplicity, and recalling his son from the study of Greek. Later in life Cato must have fallen captive to the charms of Greek erudition; for he himself mastered the language, and on a visit to Athens addressed a concourse of the people in their own tongue; and it is observed that his own writings have quotations from Greek authors.

The literary monument that immortalizes Cato the Censor is his *De Re Rustica*, a treatise on farming, gardening, fruit growing, etc. It is the oldest book of its kind in Latin literature,[17] and therefore is of botanical interest. We learn from its pages that almost every method of propagating choice varieties in use with twentieth-century pomologists and vineyardists was practiced by Cato long before the beginning of our era, even to the different modes of grafting;[18] and there is no intimation that any of those methods were other than ancient at that time. The number of named varieties of things which they had and were careful to perpetuate is also sufficiently interesting to merit such exemplification as I here subjoin, culled from Cato's book:

Brassica crispa	*P. cucurbitina*	*F. hiberna*
B. erratica	*P. mustea*	*F. Saguntina*
B. lenis	*P. sementiva*	*F. Telana atra*
B. lævis	*P. Tarentina*	*Vitis aminea majuscula*
Olea albiceris	*P. volena*	*V. a. minuscula*
O. Colminiana	*Myrtus alba*	*V. Apiciana*
O. conditiva	*M. nigra*	*V. gemina*
O. Liciniana	*M. conjugalis*	*V. helvola*
O. Salentina	*Ficus marisca*	*V. h. minuscula*
O. Sergiana	*F. Africana*	*V. Lucana*
Pirus Aniciana	*F. Herculana*	*V. Murgentina*[19]

All this would pass readily for good twentieth-century botanical nomenclature; but these names are easily two and twenty centuries old.

No fewer than five genera dedicated to Cato have been proposed, by as many different botanical authors, each apparently unaware of the attempts of the others. The CATONIA of Patrick Browne (1756) has priority.[20]

MARCUS TERENTIUS VARRO (B.C. 117–27)[21]

In so far as the mastery of human learning gives distinction, Rome had in Varro the most distinguished personage of all whose names adorn the pages of her ancient history. One Symmachus, who lived four centuries later, and whose letters are extant, wrote to a friend: "You know the writings of Terentius, not the comedian, but the Reatine, the father of Roman learning."[22] This Terentius was sometimes called the Reatine in allusion to his birthplace, which was the small village of Reate—now Rieti—some ten miles north of Rome. The family was plebeian, but there had been gifted scions of it before this one, and there were others after him. A century before him there had been a consul Caius Terentius Varro, chosen by the tribunes of the people for the reason that he was of the common people.[23]

Concerning the childhood, youth, and even the early manhood of Varro, and under what conditions the passion for learning was developed, nothing seems to have been recorded;[24] and we seem to obtain our first certain view of him as in the public service under Pompey in the war against Mithridates; but he is then fifty years of age. Also at seventy he is still a naval commander. Being a man of great wealth, owning extensive landed possessions in several provinces, and having acquired so costly a thing as a great library was at that time, the fact of his having devoted his energies to the military service of the Pompeys and Cæsars for so long a period has not seemed easy to account for. In these chances of war certain of his richest estates were confiscated, and his library was plundered, this event entailing the loss of not a few of his own writings. And all the while his own tastes and preferences were for the life of quiet study, with rural avocations for his pastime.

Varro was more than seventy years of age when Julius Cæsar, returning to Rome as the great victor, recognized him as the most learned man of his time, and charged him with the work of collecting and arranging a great library, a noble scheme which Cæsar's assassination a year or two later brought to naught. After that, under the triumvirate of Antony, Augustus, and Lepidus, Varro's name was placed on the list of the proscribed; but by help of his friend Calinus, who concealed him in his own villa, his life was saved until this storm was past. The remainder of his life was given undisturbedly to literary work. His industry as an author has made Varro a wonder to succeeding generations. He wrote long treatises on Antiquities, a History of Literature, another of Philosophy, another of primitive Rome, a History of Religion, a volume on Education, a Latin Grammar, a book on Navigation, and unnumbered other treatises, all, or nearly all, long since lost, though referred to by many contemporaries. His treatise on Agriculture,[25] in three books, almost alone of all his writings, has survived. He tells us in the first chapter of the work that he begins the writing of it in his eightieth year. It is replete with learning of all kinds, and is still a practical treatise; yet also evincing the author's familiarity with those Greek authors who, like Aristotle and Theophrastus, wrote on the theories of plant life and form.

As compared with Cato, the list of Varro's cultivated plants is not as long, and he does not enumerate as many varieties of *Brassica*, *Pirus*, *Myrtus*, and other genera. The choice varieties of cultivated cherries, long known in Pontus, Varro adds to the list of Roman fruits.

He is first among Roman authors to take note of certain phenomena of plant life, such as the growth and development of leaves and flowers, and also certain movements. The leaves of the olive, white poplar, and willow, whitened underneath, are apt to become inverted so as to show the lower face, and this at about midsummer, which phenomenon they take for a sign of the arrival of the solstice. The flowers of heliotrope follow the course of the sun from morning until nightfall;[26] and there are other kindred observations, with even a hint that there is a physiology of such things that it might be interesting to know something of. In a passage on cattle raising one of Varro's interlocutors is represented as saying: "The thing is so, but why it is so, that is your affair, you who read Aristotle."[27] By such tokens is Varro more a botanist than Cato.[28] True, he has derived every one of these ideas from the Greeks, not one of the observations being original with himself; but it was something to have been first to call the attention of Romans to them.

Patrick Browne (1756) sought to establish a genus VARRONIA. It seems that the name is untenable, the same genus having been named *Cordia* more than a half-century earlier.

PUBLIUS VIRGILIUS MARO (B.C. 71–19)[29]

Virgil, who has often been designated the prince of Latin poets, was born at Andes, a small village near Mantua,[30] some seventy years before the beginning of the Christian era. His early years were passed at Cremona, where his father had valuable landed possessions. These were among the lands which, after the battle of Philippi, Augustus Cæsar confiscated, distributing them to his veteran soldiery. On this occasion the future poet was near losing his life through attempting to dispute with the soldiers the possession of his fields. He escaped by swimming across a river, and then Virgil with his father repaired to Rome. It was the beginning of his greatness. His presence, manners, and accomplishments recommended the young man to the great Mæcenas, the power behind the throne of Augustus, and the latter soon restored to Virgil his lands; and the emperor's reward for this kindness was the ten pastoral poems (*Bucolica*) composed in the course of the next three years, and dedicated to the imperial benefactor.

After these followed the *Georgica*, accounted the most perfect and finished of all Latin compositions.[31]

The simple narrative of the poet's career at Rome, and elsewhere until his rather early death, is one of the most fascinating and beautiful chapters in all history, but for several reasons must not here be presented anew.

The *Georgics*, by which Virgil is even more favorably if less universally known than by his unfinished epic, the *Æneid*, treat of agriculture and gardening; but also again one must refrain from anything like a botanical analysis of the poems. It may suffice to indicate how prolific a field for botanical research the poems of Virgil long have been. The following list of works on the Virgilian botany is doubtless incomplete.

1. *Virgilii Georgicorum Libri IV*. The Georgics of Virgil. With an English translation and notes by J. Martyn. London, 1741, 4to.[32]

2. *Virgilii Bucolicorum Eclogæ X*. The Bucolics of Virgil. With an English translation and notes by J. Martyn. London, 1749, 4to.[33]

John Martyn was a physician,[34] and Professor of Botany at Cambridge, the friend of Sherard, Sloane, and Dillenius, and was the first to establish the fact of this poet's profound knowledge of plants. Both the works named were in so great demand among men of erudition as to have been several times reissued, and in octavo form. A German translation of Martyn's edition of the *Georgica* was published at Hamburg in 1759.[35]

3. *Flora Virgiliana*. Eller forsök at utreta de wäxter som utföras i P. Virgilii Maronis Eclogæ, Georgica och Æneides. Jämte Bihang om Romanes Matwäxter, by Anders Johan Retzius. Lund, 1809, 8vo.

4. *Flore de Virgile*. Composú pour la collection des Classiques Latins, by A. L. Fée. 1822; also again in 1837.

5. *Osservationi sulla Flora Virgiliana*, by M. Tenore. Napoli, 1826.[36]

Although Virgil was by profession a man of letters and a poet, he nevertheless exceeds the other agricultural writers of Roman antiquity in the number of different plants which he knows, and of which he makes mention; for Cato (B.C. 235–149) knew 125 kinds, Varro (B.C. 117–27) mentions 107, Virgil (B.C. 70–19) 164. Yet the sum total of the plants of these Romans, 245, is only about half the number that had been known by Theophrastus some 300 years earlier.[37]

The celebrated Lamarck (1793) dedicated to Virgil a new genus of African trees under the name VIRGILIA.

Lucius Junius Moderatus Columella

This very celebrated Latin writer on agriculture and horticulture flourished in the next generation after Virgil, during the reign of Tiberius Cæsar, and may have been in the midst of his years at the opening of the Christian era. He was a native of Cadiz in Spain, and was educated by his father, whom he characterizes as having been a man of erudition and also an experienced practical farmer.[38] The son declares himself to have been what one would now call an omnivorous reader, and before settling in Rome had travelled somewhat widely in Greece and Syria.

Columella is the most voluminous of all the classic Roman authors on rural topics.[39] There are thirteen books, and these aggregate some two hundred and sixty chapters, mostly long ones, and if none of the chapters are very strictly botanical, the whole work is a treasury of information about ancient husbandry, and the treatment of cultivated plants and trees.[40] From him we have the earliest account of the device called a hotbed, heated from beneath by a mass of fermenting manure, and protected from the rigors of winter weather by panes of glass. He reports that by these means Tiberius Cæsar raised cucumbers all the year around.[41]

To Columella we are again indebted for a complete account of the methods of grafting as practiced by the ancients.[42]

Sprengel has given a considerable list of plants and trees that are first brought to notice by this author.[43]

Near the end of the eighteenth century two distinguished botanists almost simultaneously bethought themselves of the propriety of consecrating a genus to Columella. The COLUMELLIA of Ruiz and Pavon (1794) appears to have the priority.

Pedanios Dioscorides (about A.D. 64)

If to have written the most practically serviceable book of botany that the world of learning knew of during sixteen centuries were the best title to botanical greatness, to Dioscorides would readily be conceded the absolute supremacy over all other botanists, not only of antiquity but of all time. Concerning the duration and the absoluteness of his supremacy Sprengel has the following:

> During more than sixteen centuries he was looked up to as the sole authority, so that everything botanical began with him. Every one who undertook the study of botany, or the identification of medicines swore by his words. Even as late as the beginning of the seventeenth century both the academic and the private study of botany may almost be said to have begun and ended with the text of Dioscorides.[44]

Almost volumes have been written in controversy as to the time when Dioscorides lived; though the extremes of opinion do not assign him an earlier date than B.C. 30, or a later than A.D. 98;[45] and the most probable seems to be that which locates him in about the middle of the first century of our era. That he lived in the time of Nero is inferred almost to a certainty from remarks of Tacitus and of Galen.[46]

Though in relation to this author biographic dates fail us, the land of his nativity does not. It is well established that he was a Cilician Greek, his native city being Anazarbos; for in order to distinguish between him and others named Dioscorides, eminent writers referred to him as Dioscorides Anazarbæus, the Anazarbean Dioscorides.[47] It is clear also from his own writings that he was a learned physician and practiced medicine; also that he had travelled widely to study plants, and obtain knowledge of other than vegetal remedial agents. In these travels he came to know many plants before unknown to Greek and Roman physicians, and was at the pains of describing many such[48]—that is, of indicating not only their qualities and remedial effects, but also something of their aspects and morphology as living plants, describing their roots, stems, foliage, and even sometimes their flowers—and the number of plants and plant products of which he gives account is about 600.[49] Such a list of merely medicinal and alimentary plants is by more than 100 greater than the sum of all plants known to Theophrastus three centuries before Dioscorides. And it was because he had described so many, and often so well, that in after ages he came to be regarded as the supreme botanist. The usefulness of his medical botany, from the phytographic point of view, was not only fully realized, but also enthusiastically somewhat overestimated. The scientific botanist among the Greeks was Theophrastus; and there is no comparison between him and Dioscorides, whose theme was medical botany;[50] but, quite as usual, the man of "applied science" was the one to meet with general appreciation and approval.

So highly esteemed was Dioscorides during the middle ages,[51] that early after the invention of printing, his work, though in Greek, obtained an editor and a publisher at Venice as early as the year 1499.[52] This edition must have obtained a ready sale, for in 1518 it was repeated. A third Greek edition appeared at Basle in 1529.[53]

Latin being the universal language of the schools, Latin versions of Dioscorides were in demand, and early became rather numerous. The very first of these, a book rare and obscure, purports to bear the date 1478,[54] thus antedating the first Greek prints. But from the year 1516, when the first excellent translation by Ruel appeared, Latin versions became numerous;[55] and for a whole century thereafter the most voluminous and most useful books of botany were in the form of commentaries on Dioscorides. Such in large part are the works of Anguillara, Mattioli, Maranta, Dodoens, Cesalpino, Fabius Columna, and the Bauhins.[56] In several of these the annotations and comments quite exceed in bulk the Dioscoridean text, and are replete with new botany; that is, they contain the names and descriptions of many plants which the commentators are convinced Dioscorides did not know, and which they therefore judge to be new. One may fairly say that the greater part of all the new botanical matter published during the whole of the sixteenth century, and a part of the seventeenth, came out in the form of annotations upon the text of Dioscorides.[57] Thus it appears that the Greek, who only meant to provide medical students with a full compend of remedies, and of the marks by which to know them, became incidentally the first master of phytography; the one every line of whose plant descriptions has been more attentively studied word by word, and that by a greater number of erudite men, than any other book about plants that has yet been written, unless one should possibly be obliged to make an exception of Bauhin's *Pinax*.[58] But even that is, first of all, a compend of Theophrastian and Dioscoridean phytography, together with such augmentations and improvements as in the year 1623 were found necessary.

Latin editions of Dioscorides are too numerous to be given a reckoning; and almost the same may be said as to early translations of him into modern tongues, for between the years 1555 and 1752 there were at least twelve Spanish editions,[59] as great a number in Italian, and there were editions in French in 1553, 1559, and 1580. There was one translation into German as early as 1546, another in 1610, and this last appears to have been issued again in 1614.[60]

Little in the way of botanical taxonomy will be looked for in a work on pharmacy that is nearly nineteen centuries old. The most comprehensive of his groups are formed according to properties; thus his Book I is devoted to the consideration of plants that are of merely aromatic rather than medicinal qualities; growths that furnish oily, gummy, or resinous products, such as enter into the composition of salves and ointments; and after these follow the trees that yield fleshy fruits of grateful though not specifically aromatic flavors. Book II, beginning with animals, and animal products that are of dietetic and medicinal use, ends with the cereals, the leguminous, malvaceous, cruciferous, and other garden herbs. Then Books III and IV deal with a vast number of plants more distinctively medicinal.[61]

It is to be observed that he does not treat first of herbs consecutively, then of shrubs, then of trees. That would have been the formal adopting of a piece of regular taxonomy even in his day, as we have seen, already long established. There are merely spicy things that are trees, others that are shrubs, and some that are herbaceous; and the same is true of things alimentary and of things medicinal. This qualitative classifying is better suited to his purpose. But that even the more strictly botanical taxonomy may now and then gain a point by ignoring herb, shrub, and tree as fundamental divisions comes out interestingly in Dioscorides' chapter on *Sambucus*.[62] He has two species, one of which is only an herb, the other woody and almost a tree. We shall see later that, after the revival of botany, at a comparatively modern period, such a hold had been gained by the old distinction between the herbaceous and woody that in deference to it systematists almost with one accord divided Dioscorides' *Sambucus* into two genera and separated them widely, *Sambucus* being located among the trees, and *Ebulus* among the herbs; and that hardly after Bauhin as late as 1623[63] had followed Dioscorides in writing them as one genus was the botanical world of that time

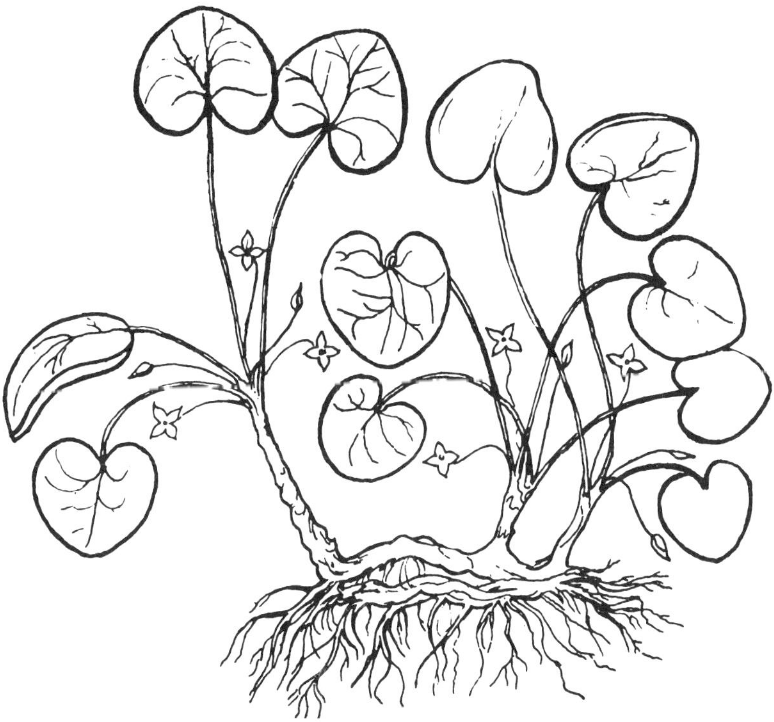

FIG. 22. Asarabacca (*Asarum europæum* L.). Dioscorides, *Codex Aniciæ Iulianæ* (1906), fol. 30v, tracing in Singer, Herbal in antiquity (1927).

FIG. 23A. *Sambucus*, identified by Gunther as elder (*S. nigra* L.). (See across.)

ready to accede to a proposition so subversive of what was deemed fundamental in taxonomy, *i.e.*, that a tree and an herb could not be congeneric.

Subordinately to the more general and qualitatively outlined divisions of the work, Dioscorides recognizes all the more familiar natural families of plants; that is to say, within each Book there is a line of labiate genera, another of the leguminous, another of the umbelliferous, and the succession of cognates is not often interrupted by the intrusion of a genus not of such natural alliance. Even among the composites *Anthemis*, *Parthenium*, and *Cotula* are in close conjunction on his pages,[64] as are *Anchusa*, *Lycopsis*, and *Echium*;[65] and the succession of the representatives of five or six genera of solanaceous plants is only broken by the intrusion of *Cardiospermum*.[66] Examples of this need not be multiplied. It is propagating fable in place of history to affirm that natural families were first recognized and indicated by any Linnæus, or Adanson, or Jussieu of the eighteenth century.

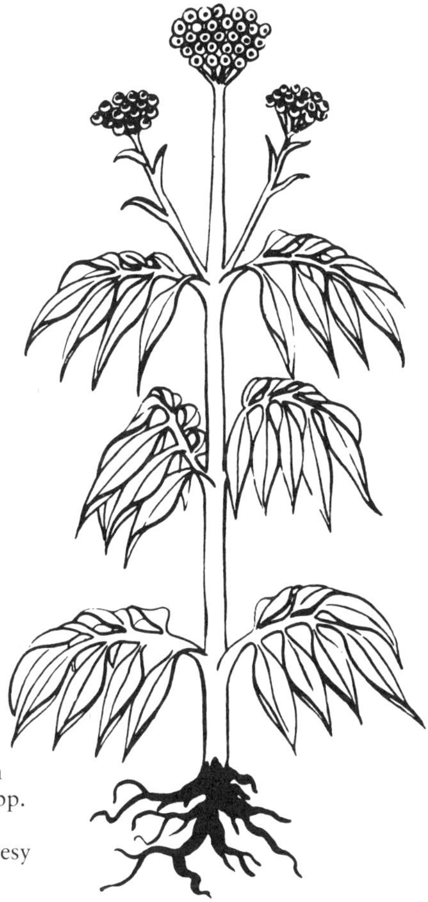

FIG. 23B. *Sambucus*, identified as danewort (*S. ebulus* L.) by Gunther in Dioscorides, *Materia medica* (1934), pp. 574–76. Both line drawings after the *Codex Aniciæ Iulianæ* (1906). (Courtesy A. E. Gunther, M.A., F.L.S.)

The whole subject of Dioscorides as a taxonomist merits a fuller development. A thorough study of his text might show that classification had progressed somewhat during those three centuries that had then elapsed since Theophrastus.

The services of this next to the last of great Greek physicians to botany are well commemorated in that fine generic type DIOSCOREA dedicated by Plumier in the year 1703.

CAIUS PLINIUS SECUNDUS (A.D. 23–79)

Two venerable cities of Italy contend for the honor of having been his birthplace, namely Como and Verona. But wherever he was born, the elder Pliny lived in Rome and called himself a Roman. He is known as Pliny the elder, to distinguish him from a nephew of his who bore the same name, and is also an author of some note. But the elder Pliny is one of the remarkable men, and among the most voluminous authors, of ancient times. His character, his manner of life, and his rather early martyrdom to knowledge were written of after his death, by his nephew and namesake in a letter to a friend:

I rejoice exceedingly that thou readest so eagerly my uncle's books, that thou wishest to obtain a complete set of them, and makest inquiry concerning all of them. I will serve you in the capacity of an Index, and shall even indicate to you the order in which they were written; for to know that is a matter of interest to the learned. *Spearcasts of the Cavalry.* One Book. He wrote this while in command of a company of cavalry, where his leadership was marked equally by courage and prudence. *The Life of Pomponius.* Two Books; a tribute to the memory of his best friend. Twenty Books *Of the German Wars*, embracing a full account of all our wars with the Germans. He undertook this work while in military service in Germany, and it was suggested to him in a dream. Three Books of *The Student*, so comprehensive as to fill six volumes, wherein the aspirant to oratory is advised and directed from the cradle forward. Eight Books on *Hesitancies in Public Address.* These were written under Nero, when anything like bold or unrestrained writing was unsafe. Thirty-one Books supplementing the (historic) *Work of Aufidius Bassus.* Thirty-seven Books of *Natural History*; a work as comprehensive, learned, and many-sided as nature itself. Thou wilt wonder how a man of so many business affairs could bring to completion so many volumes, and on such difficult subjects. You will be still more surprised to learn that for a long time he was a busy lawyer, that he died in his fifty-sixth year, that he had been much occupied with the duties of the most responsible public offices, and that his time had also been heavily taxed by the exactions of the friendship of princes. The explanation is, that he was of the keenest intelligence, and his economy of time something almost beyond belief, even to the reducing of his hours of sleep to the shortest limit. In the winter, he was accustomed to work until between one or two o'clock in the morning, or at the shortest until midnight. He could fall asleep at once, no matter at what hour; sometimes even at work he would fall asleep for a few moments and awake again. Before the break of day he would go to the Emperor Vespasian—for he too used to work at night—to receive his orders or to fulfil some commission. Returning home he would study until breakfast time. After a light breakfast, if it was summer, and he had a little leisure, he would lie down in the sun and have a book read to him, taking notes and extracts; for he read nothing without making some excerpts, being accustomed to say that no book was so bad as not to contain something useful. After sunset he would usually take a bath, then recreate and sleep a little. After that, as if another day had dawned, he studied again until dinner time. Even at this principal meal a book was read, and comments written, and this without interrupting the reading. I remember that once upon a time one of his friends present checked the reader, who had given a wrong inflection, and had him read the line over again. 'But you understood the meaning at the first reading, did you not?' my uncle interposed; to which the other nodded assent. 'Why, then, did you call for the repetition? We have lost the time it would have taken to read ten lines, by this interruption.' So avaricious was he of time. He arose from the dinner table, whether while it was yet daylight in summer, or when in winter it was after dark, always with the same promptitude, as if compelled by law. This was his manner of life amid the business and turmoil of the city. In the country the only respite he allowed himself was that of the daily bath; and when I say that I mean the actual time of the bath; for while the drying and dressing was going on he was either listening or dictating. On his journeyings, as if putting out of mind all business cares, he did nothing else but that; keeping always close beside him a rapid penman, a book, and a writing tablet. . . . For the same purpose even in Rome he had himself carried from place to place in a sedan. I remember well how once in meeting me when I was walking he said: 'You ought not to lose these hours'; for he reckoned all time lost that was not given to study. It was by such exertions as these that he brought all those volumes to completion.[67]

We have from another letter by this same kinsman annalist—a letter to the historian Tacitus—an account of the circumstances of Pliny's untimely death.[68] The facts must here be presented in brief.

In the summer of A.D. 79 Pliny had command of the Roman fleet, then in the Tyrrhene Sea, at anchor behind what is now known as the Punta di Miseno. On the 22nd of August of that year, at about noonday a terrific earthquake and rain of ashes accompanied that frightful eruption of Vesuvius which buried the cities of Herculaneum and Pompeii. A cloud in the shape of a huge Italian Pine, with straight trunk and horizontal branches, seemed to arise from the crater of the volcano. Immediately, on a small and light

FIG. 24. Feverfew (*Chrysanthemum parthenium* (L.) Bernh.). Dioscorides, *Materia medica* (1934), p. 380. Line drawing after the *Codex Aniciæ Iulianæ* (1906). (Courtesy A. E. Gunther, M.A., F.L.S.)

sailing vessel Pliny set forth in the direction of the mountain to gain a nearer view of the phenomenon. Soon discovering that by the dread progress of the eruption the lives of the whole population along the mountain's base were imperilled, and that they had no way of escape but by sea, he ordered the larger ships to sail to the rescue, and went with them. At one time, when a shower of stones and ashes was falling on his ship, and the sailors were trembling in mortal terror, the commander was seen to be quietly dictating his own observations to the amanuensis. They made land at a point where some friend of Pliny had a country seat. The villagers were found already provided with the means of escape by sea, only a strong contrary wind prevented embarkation; meanwhile the dangers of remaining on land increased with every hour. Pliny, when nightfall came, as if to inspire courage in others, took his bath as usual, then dined, and composed himself as for a night's rest; but the inhabitants of the place remained awake. Towards morning the fall of stones and ashes so increased, and the earthquake shocks became so much more frequent, that the people awakened Pliny, fearing the closing up of the porch on which he had made his bed. The demolition of buildings appeared imminent, while out of doors the stones fell still more thickly. People bound cushions and pillows about their heads and rushed down to the beach; but winds and waves are still high and there is no boarding the ships. It is now daytime, but the blackness of night is still over land and sea, interrupted only by flashes of lightning or flames bursting forth now and then from the fissures of the earth. The people rush to and fro in the frenzy of despair. The fumes of sulphur threaten to suffocate them. Pliny, helped by two slaves, arises from his couch, and falls dead.

The *Historia Naturalis* is Pliny's monument; no other work of his having survived. The following outline of the contents of the work may be useful here. Book II, Cosmology and Meteorology; Books III to VI, Geography; Book VII, Anthropology; Books VIII to XI, Zoology; Books XII to XXVII, Botany, in the main agricultural, horticultural, and medical; XXVIII to XXXII, Zoology again, but in relation to medicine; as also in connection with aquatic animals some paragraphs on the remedial efficacy of certain waters; Books XXXIII to XXXVII, Mineralogy, more especially as applied to medicine, painting, plastic art. Book I of the work is prefatory and introductory to the work as a whole.[69]

At first thought it may seem extraordinary that so vast a work, filling so many volumes, should have escaped the fate of many hundreds of less voluminous things, and should have reached modern times in its completeness. The probable explanation has been suggested by the historian Meyer, who thinks that its very voluminousness was its safeguard. The copying of the complete *Natural History* by hand was a large enterprise, and the manuscript when done was very costly. Men take care of that which is worth much money.[70] In the middle of the nineteenth century, an excellent authority reported the existence of more than sixty manuscripts of Pliny of greater or less antiquity.[71] Since the invention of printing more than eighty different editions have been printed of which number as many as eighteen appeared in the fifteenth century,[72] and more than forty in the sixteenth.

The high prestige held by Pliny throughout mediæval times was due to the fact of his having written in Latin. All the other authors of greatest importance as to natural history had written in Greek; and Latin was the language of the middle ages. The work was also extensively, if not mainly, a compilation, and was made up in large part of translations into Latin from the greater Greeks, Aristotle, Theophrastus,[73] Nicander, and Dioscorides,[74] and so had much the character of a compend of all natural history.

As a Roman among Romans, Pliny was of the utilitarian bent of mind; even a subscriber to the dogma, then antiquated, that all things that are upon earth are here for the sake of man;[75] not that he so holds it as piously to praise nature for universal benignity or generosity; for in one of his earlier volumes he says: "If nature appears to have produced everything for the sake of man, still man is often obliged to pay rather dearly for her gifts; so that it is not so easy to decide whether she has more the character of a benign mother than of an unkind stepmother."[76]

Whatever nature may or may not be to man, it is impossible to establish any natural order and sequence of things on those utilitarian principles; and he who looks into Pliny's botanical volumes for anything like what we understand by taxonomy will be disappointed. Here is what a distinguished botanical systematist of the eighteenth century said of the Roman's plant classifying: "Pliny, the indefatigable compiler, published in fifteen books all that Theophrastus, Dioscorides, and their predecessors had said about plants. But he treats this matter in a manner so strictly historical, although in such flowery language, that one may well say of the whole that it is in beautiful disorder."[77] True it is that Pliny begins his history of plants with the discussion of trees.[78] The philosophic Theophrastus had done so; but it was for the reason that trees seemed to him to claim the first place as being the most highly organized type of plants. Pliny begins with trees because he judges them to be, on the whole, more useful to man than herbaceous plants. Similarly everywhere in his writings the thread of the economic rather than the philosophic is that by which one is to trace whatever of system there is in his treatment of plants and plant lore.

It has been claimed by some, and disputed by others, that Pliny was more than a compiler, and that some of the facts which he records were from his own observation. A renewed and thorough study of all his botanical books, with such a question foremost, is still called for; but rather unusual accomplishments are also demanded on the part of the investigator; those of the ripe classical scholar and of the master botanist combined.[79] Adanson, who if I mistake not had for his thesis inaugural a study of Pliny,[80] accredits him as discoverer of the distinction between growth buds and fruit buds in trees, and says that he named the former kind germen, the latter gemma.[81] It was a botanical discovery of high rank, whoever first announced it; but I should suspect it of having been made anterior to Pliny.[82]

As if to crown Caius Plinius Secundus with a wreath of myrtle, Plumier (1703) dedicated to him a genus PLINIA belonging to the family of the myrtles.[83]

CLAUDIUS GALENUS (A.D. 130–201)[84]

In respect to natural endowments, wealth of information acquired by study at home and travels abroad, and fertility of able and learned authorship, Galen was one of the great celebrities of antiquity; as a physician ranking above all others figuring in the early history of medicine except Hippocrates.

Galen was a Greek, born at Pergamos in the year 130 of our era. Ancient analysts take note of the fact that Pergamos was the seat of the most celebrated of temples to Æsculapius, therefore of good augury as the birthplace of predestined physicians. Nico, the father of Galen, was skilled in philosophy, geometry, and astronomy, and was an architect; also evidently interested in plant life from the philosophic point of view, for the son reports it that Nico, as if doubting about the transmutability of one species into another, made some experiments with grain sowing, the result being the removal of all doubt concerning the changeability of certain grains into chess, or darnel.[85]

The career of the son seems to indicate that the architect Nico was a man of wealth, for the child received most careful education under the best masters at Pergamos. At the age of seventeen years, having chosen the profession of medicine, he was sent upon his travels, and continued them until the age of twenty-eight.[86] He spent several years in Ægypt, with headquarters at Alexandria,[87] passing thence into Bithynia, Palestine,[88] Thrace, Macedonia, Italy, and the islands of Crete, Cyprus, and Lemnos, these all noted in that early time for the wealth of drugs, and the good quality of them, that were imported from them into all the cities of the then known world. The high distinction to which he afterwards attained attests the improvement made of those rare opportunities that wealth had afforded him. One object of these prolonged journeyings had been that of profiting intellectually by converse with learned men, and the most noted physicians of every land. He was reputed also to have mastered all the dialects of Greek, as well as the Latin, Persian, and Ethiopian tongues. Also everywhere he sought the most perfect knowledge of every plant anywhere in use remedially. It was not enough that a given remedy might be purchased from any druggist. The physician ought to know all about the plant, even as living and in its native soil, and thus become qualified to distinguish pharmacologically between the false and the genuine, and to detect adulterants.

FIG. 25. Roman mosaic from the second century A.D., showing the aesthetic interest in a wide variety of flowers. 67.5 cm × 104 cm. (Permission of the Vatican Museums)

Galen's years of travel were concluded by a considerable sojourn in Ægypt, at Alexandria, then the world's greatest center of art and erudition; and thence he returned to his native Pergamos, where for some years the surgical and medical care of the gladiators was committed to him. In the year 164, when he was about thirty-three years of age, a revolutionary disturbance in his native city was the occasion of his betaking himself to Rome. Here he now practiced medicine with distinguished success, and gave lectures on anatomy; all this, however, to the arousing of a storm of jealousy on the part of the native Roman physicians,[89] to whose unrelenting sallies, or else to a malignant outbreak of the plague, or to both forces combined, he yielded at the end of three years and once again returned to Pergamos.[90] The stay there was this time short, for the Roman Emperor Marcus Aurelius, with his imperial brother Lucius Verus, going to war with the Quadi, Parthians, and Marcomanni, desired his services as their physician on that expedition.[91] The plague broke out in the country of the attempted conquest, the expedition became a failure, and the imperial majesties with their Greek physician began a retreat to Rome. Lucius Verus, stricken with apoplexy, died on the way, and the imperial philosopher and Galen reached Rome in safety. Here the latter now engaged actively and industriously in medical authorship during a number of years,[92] and when in the year 172 Marcus Aurelius set forth on a second attempt to go and conquer the Marcomanni, desiring Galen to accompany him as his physician, the latter interposed, piously, that the god of his native city, the revered Æsculapius, to whom he was under solemn vows, had decreed otherwise. The excuse availed, and gave no offence to the pious Aurelius; so that Galen continued in Rome, as physician to Commodus, son of Marcus Aurelius and heir to the empire, then very young. Later, and at a date unknown, Galen returned to his native city, where he ended his long, laborious, and most distinguished career, at the age of seventy years, or, as some authorities say, at ninety.

Some idea of Galen's industry as an author may be conveyed by a note or two on editions of his works issued since the invention of printing. One published at Basle in the year 1538, and containing the Greek text only, fills five folio volumes.[93] An edition given forth at Leipzig between 1821 and 1833, embracing both the Greek text and a Latin translation, is in twenty octavo volumes.[94] And such editions do not include certain of his works the Greek originals of which have been lost, and only the Latin versions of them have been handed down from early times; much less others which have been lost altogether. Even in Galen's lifetime certain books of his, kept in the Temple of Peace at Rome, were destroyed in a conflagration of that building and were never reproduced.

A man of as profound erudition, and of such high attainments in botany as the celebrated Albert Haller, conceding the supremacy of Galen as a man of genius, and a great master of everything relating to the healing art, seems to deny that he was anything of a botanist, affirming that almost all his botany is borrowed from his predecessors, and chiefly from Dioscorides.[95] This criticism, in as far as it is adverse, touches the writings of the man as a possible contributor to botanical knowledge. It does not—or if it does it ought not to—stand as an impeachment of his knowledge of plants. It is incautious to pass judgment against any man's attainments in a subject until he has at least in some way expressed himself on that subject. To know many plants familiarly and well is to be something very like a botanist, whether one ever write a paragraph of botany or not. A familiar knowledge of many plants Galen not only urged upon the whole medical profession; he claimed that he himself possessed such knowledge.

> In as far as possible the physician ought to know all plants, and if not all, the greater proportion, and those most useful. . . . He who knows the different kinds in all their states from young and small to fully grown, and can so distinguish between them, will in many places find certain useful plants, as I have done in various parts of Italy, where he who knows them only in the dead and dry, would never recognize them whether in the young state or the mature. There is no quacksalver who does not readily identify the herbs that are imported from Crete by their fruits; but that some of these selfsame things might be gathered on the outskirts of Rome they do not know, because the season of their herborizings does not correspond to that of the

maturity of these plants. But that time is well known to me, and I go in quest of Chamæpitys, Chamædrys, Centaurium, Hypericum, Polium, and others of that kind, at just the right time, and gather them in their perfect maturity, neither waiting until they are past that, and are sunburnt, nor going too early, that is, before the fruit is well formed.[96]

It has well been observed by one of the historians of botany, that "The man who wrote thus must have been either a consummate charlatan, or else a man of deep and thorough knowledge, and a charlatan Galen was not."[97]

When one reads in Galen over and again such commendations of an intimate knowledge of many living plants, and when it is remembered that he made long journeys by sea and land in the endeavor to familiarize himself with the medicinal plants of different climes, and when one has marked the keenness of his powers of observation everywhere, it is impossible to agree with the learned Haller that in botany Galen was inexpert; nor can it reasonably be questioned that had he betaken himself to phytography, he would have laid all botanical posterity under deep obligations to himself. Now, that he did not describe plants, but was accustomed to give their names only, or but little more, one might have been disposed to charge to the fact of his having flourished in the very next century after Dioscorides whose six hundred species, embracing the whole vegetable materia medica, may have been for the most part well identified at Galen's period, so that the mention of a name only would sufficiently recall a species. But such apology for Galen would be superfluous. The truth seems to be that he had next to no faith in phytography at all. He takes openly the ground that: "The identification of plants is better accomplished by the actual observation of them under the help and guidance of a teacher, than by the method which may be likened to the attempting to learn to navigate the seas by studying books on navigation."[98] This, then, is the main reason why Galen almost abjured plant description.

The passage is luminous with historic information about the study of botany in the Rome of eighteen centuries ago. We know already that at this period the occupation of a well-trained physician is lucrative. There are many of them; therefore the candidates for the profession are not few. The remedies in use are almost all botanical, and they all study botany; quite otherwise, by the way, than botany is studied in twentieth-century schools of medicine, and less perfunctorily. Unless in their practice of medicine they are to be at the mercy of the unscrupulous among herb gatherers and drug vendors, they must know the marks of the genuine thing. Therefore important among their regular exercises is that of identifying plants, the book open before them, the specimen it may be a withered and shrunken root or rootstock, not improbably supplemented by a fresh one newly brought in from its native soil, or from some drug garden. The standard botanical work, descriptive and pharmaceutical, is Dioscorides—its author hardly a

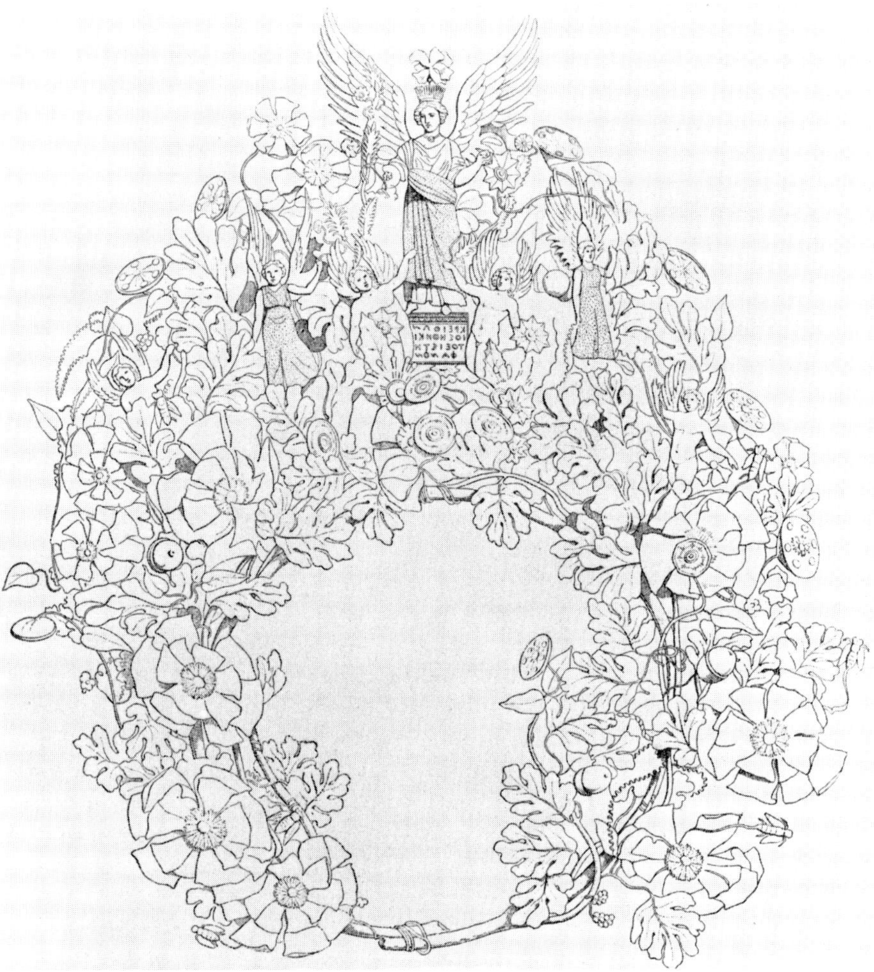

FIG. 26. Golden burial wreath from the Hellenistic period. Baumeister, *Denkmäler des klassischen Altertums* (1885–88), vol. 2, p. 794.

century dead—and there are others. The descriptions are mostly brief and often inadequate, so that mere guesses at the identity of things frequently pass instead of certainty, and about the identity of some that are of remedial importance the whole fraternity—Galen himself perhaps excepted—is wrong. At all events none but he is so discouragingly aware of the multitude of errors that have originated through placing dependence on descriptive botany. He thought there was a better way; but this proposal of his seems to imply on his part an overweening confidence in the perpetuity of things. He did not foresee a time when the race of capable phytognosts would fail, and when in default of such teachers for the identifying of plants there would be no other dependence at all but the old and often imperfect descriptions.

Men follow great leaders when the leaders are in the wrong, about as faithfully as when in the right; and if, during several centuries after Galen, lesser lights continued to mention plants hardly more than by their names and remedial qualities, it was after the example of his authority as supreme. In such manner may the most expert man of science chance to antagonize the best interests of that science, and heavily impede its progress along one line while advancing it in a different direction.

His indifference to phytography notwithstanding, Galen has been credited with having made some few additions to the list of known plants by new name, and by some sort of description.

Michel Adanson attributed the discovery and the naming of two new genera to Galen. They are *Lycopersicon*[99] and *Arctostaphylos*.[100] Both names are now in use for genera, but it is impossible to identify either one with the type which Galen had in hand; but from such description as the Greek gave out, his *Arctostaphylos* would be *Vaccinium Arctostaphylos* rather than *Arctostaphylos uva ursi*.

In the year 1737 Linnæus dedicated a genus GALENIA to the memory of Galen.

4. Introduction to the Sixteenth-Century German Fathers

That long course of ages intervening between the last decline of the Roman empire and the revival of learning in the fifteenth century is chiefly distinguished botanically by what we do not know about it. Even the historians of botany, with hardly more than a single exception,[1] instead of making intelligent and unimpassioned use of the scattered fragments of botanical record for the period, have done what they could to perpetuate their own hereditary prejudices against the whole period.[2]

However, he who is in quest of landmarks chiefly will be absolved from the task, interesting though that would be, of following the vicissitudes of botany through the middle ages. The period has no apparent landmarks of botanical history.

The tenor of the German writing of its history is, that the science of botany was born again, as it were, in the year 1530 and in Germany, by the publication of Otto Brunfels's folio entitled *Herbarum Vivæ Icones*—Living Pictures of Herbs. The Germans have always been and are the chief historians of botany. I pay full tribute of acknowledgment to their supremacy in this field of high endeavor when for the heading of this chapter I adopt what is become their own favorite caption. All of them use it: Kurt Sprengel,[3] Ernst Meyer,[4] Emil Winckler,[5] Julius von Sachs.[6] All of them name Brunfels, Fuchs, and Tragus (Bock) as the fathers of the new botany of modern times.

It has been indicated in a preceding chapter of these *Landmarks* that the real father of botany as a science was Theophrastus of Eresus. If he is the father of the science he is the father of even modern botany, though not of those developments of it that have been the peculiar achievement of modern botanists. Science is truth. The foundations of a science are its fundamental truths, and so the foundations of a science once laid are laid forever. These things are self-evident.

We shall not be able to realize in how far the "German Fathers" contributed to the superstructure of modern botany until we have examined with great care and diligence their best works; and this is something which, I shall make bold to say, not even the German historians have been at the

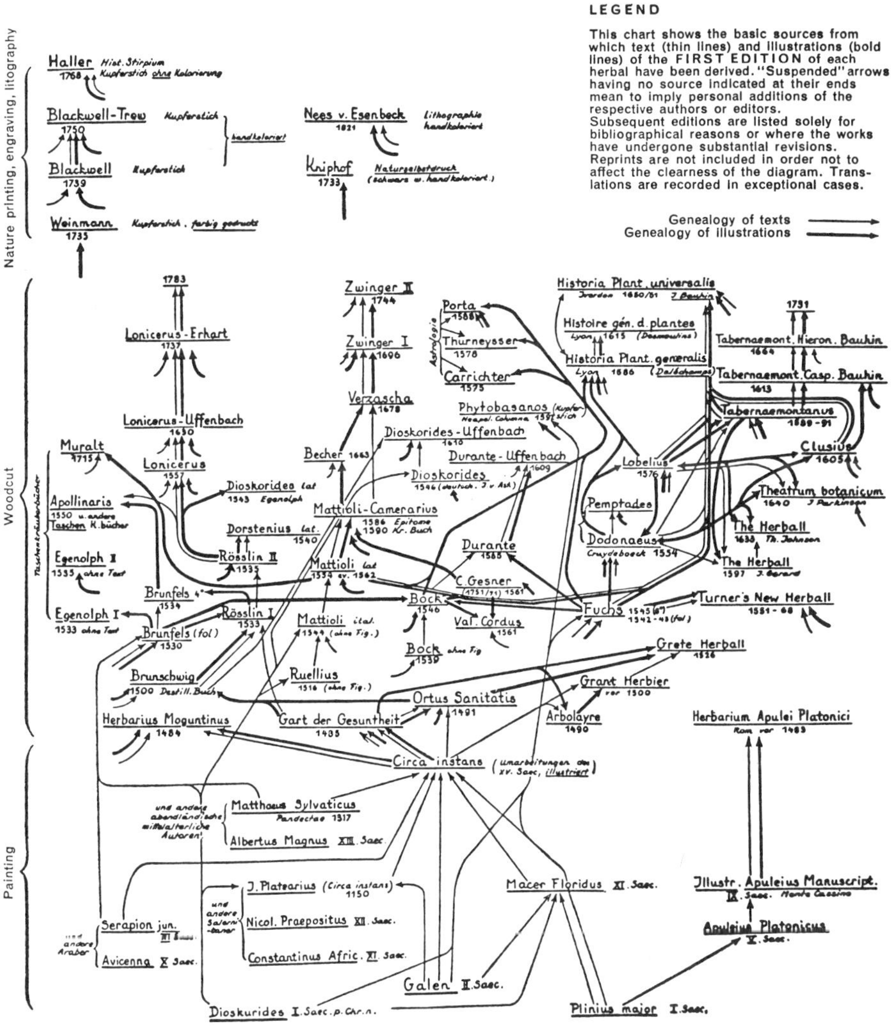

FIG. 27. Genealogy of herbals, by A. Schmid, *Über alte Kräuterbücher* (1939). This copy, with English legend, is from Nissen, *Herbals of five centuries* (1958), p. 49. (Courtesy L'Art Ancien S. A., Zürich; Robert Woelfle, Munich; Weiss-Hesse, Olten)

pains of doing; though Sprengel, first of their lineage, did much and well in this direction, while also leaving very much for others to accomplish. Julius von Sachs, the latest in the line, copied Sprengel's caption "The German Fathers," etc., but knew next to nothing of their works, even rating as unimportant Valerius Cordus,[7] who was immeasurably the greatest of them all.

The four now named represent two rather distinct kinds or grades of botanical work. Brunfels and Fuchs busied themselves almost wholly with medical botany. It is a rare thing with either of them to mention a plant of unknown or even uncertain medicinal or alimentary qualities; and their plant descriptions are almost as uniformly either compiled or literally copied from authors of centuries and even almost thousands of years before them. The books of Tragus and of Cordus abound in new and original descriptions. These demonstrate that these two men examined plants with their own eyes, and for the love of them as plants, and that they saw many things about the structure and the behavior of them to which the other two men, and even all botanists before them, had been blind.

There is another contrast. Brunfels and Fuchs, realizing the defects of many of the ancient descriptions, sought to render the identification of remedies more easy and certain by supplying engravings of the plants. This idea was very far from being new; indeed, it was almost as old as botany. Pliny knew as many as three Greek authors who, before the Christian era, had illustrated their manuscripts by paintings of the plants.[8] The traditions of still others have been brought to light. In the middle ages early and late rare manuscripts of old botanical authors illustrated by drawings or paintings of plants were known and referred to. The most noted of such ancient manuscripts, now some thirteen centuries old, has been reproduced photographically, and in this way actually published since the beginning of the twentieth century.[9] Even forty or fifty years before these fathers of plant iconography, there were printed copies of the *Hortus Sanitatis*,[10] and of its German version, *Gart der Gesundheit*, illustrated by some five hundred wood engravings of plants. Doubtless the wretched character of those first printed plant pictures,[11] along with the fact of the great popularity of the books containing them, were what moved Brunfels to undertake the production of the *Herbarum Vivæ Icones*; and the success of his enterprise stimulated Fuchs to inaugurate a larger one. These two might worthily have been styled Fathers of Plant Iconography,[12] but to name them the German Fathers of Botany is superlative; for it will have to be admitted that the mere publishing of plates of plants, with names of said plants and their uses, is not in itself the setting forth of any scientific principles beyond the few taxonomic ideas which the mere grouping of the plates may chance to indicate. What are plant picture-books for? In the case of the authors of them, they may be the refuge of those who cannot describe, or, with such as can describe, they are a condescension to such as cannot read; also to others who are incapable of mentally imaging a thing from the verbal description of it. By the large picture-books of Brunfels and of Fuchs all sorts and conditions of men, lettered and illiterate, could identify some hundreds of useful plants, a thing which never had happened in the world before that day. For this they deserve only praise. Nevertheless, had no books of botany been issued in the sixteenth century essentially different from those of the two authors named, it is difficult to see how botany could have progressed a single stage within that century.

In the works of Tragus and of Valerius Cordus we have books in character essentially different from those of the two aforenamed. Both these were deeply interested in plants of all kinds, were given to examining their organs minutely and marking the behavior of certain growths at different stages, and all this before ever having thought of writing books thereon. Also when they betook themselves to writing, it was without any purpose of calling upon artists to make pictures remedying the defects of their descriptions. They were under the inspiration of a new idea in botany, namely, that plants might be so described as to be identifiable by description. Galen's dream about a kind of apostolic succession of living teachers, one generation of whom should forever teach the next to know the medicinal plants by their right names[13]—all that had proven a very idle dream. Thirteen changeful, turbulent centuries had now passed since Galen. The succession had been

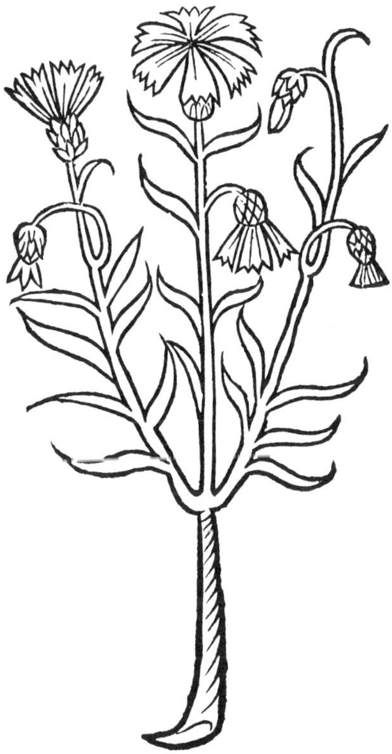

FIG. 28. Madonna lily (*Lilium candidum* L.). *Gart der Gesundheit* (1485), fol. clxxxxi[v].

FIG. 29. Elitropium. [*H*]*ortus sanitatis* (1491), ch. 171, fol. 82[r]. This illustration is apparently of blue chicory (*Cichorium intybus* L.); the modern *Heliotropium* L. is in the Boraginaceæ.

De Fragraria. Strawberyes. Ca. C.lxxv.

Fragraria is an herbe called stra bery. It groweth in woodes and grenes / and shadowy places / it is pryncypally good agaynst all euyles of the mylt. The iuce therof dronken with hony profyteth merueylously.

For the breth.

For them that take brethe with payne as it were syghynge. The iuce therof take in drinke wo white peper heleth it. Straw beryes eaten helpeth colerike persones / cō forteth the stomake / and quencheth thyrst

FIG. 30. Strawberry (*Fragaria vesca* L.). *Grete herball* (1526), fol. lxvii^{v}.

obsolete a thousand years, and the world botanical was far at sea as to the true identity of many important plants. There must be descriptions; and they must be better than those handed down from ancient times. I should not venture to credit the erratic and garrulous Tragus with having known the history of botany so well, or having planned the opening of a new era in descriptive botany. We shall probably see, by the perusal of his book, that what he achieved here, and it was not a little, was but the spontaneous outcome of his admiring curiosity about plant structures. On Cordus's part, it is unmistakable, there is the deliberate plan of creating a new phytography. Therefore, and by a study of the men and their books, I think we shall perceive that in the Germany of the first half of the sixteenth century, there were two fathers of plant iconography and two fathers of descriptive botany.

FIG. 31. Albrecht Dürer, Das grosse Rasenstück [The large piece of turf], 1503, watercolor, 41 cm × 31.5 cm. (Courtesy Graphische Sammlung Albertina, Vienna)

5. Otto Brunfels
1464[1489]-1534

First in point of time among the German botanical reformers of the sixteenth century, Brunfels is also easily first in rank respecting those educational and literary qualifications which go to the making of what one calls a scholarly book. In this particular his one botanical treatise, the *Herbarum Vivæ Icones*,[1] is peerless among the several books of botany that appeared in middle Europe within the first half of the sixteenth century. Others produced more and better botany; but there are marks of a dignified and conservative erudition that are characteristically Brunfels's own.

Life

His career was a long one, at least for a consumptive,[2] and was singularly varied. One need not here analyze the motive of that countryman of Brunfels who pretended that the man's professional life might be summed up in one sentence like the following: "At first a schoolmaster at Strasbourg, then a physician at Berne."[3] This would be good language in which to epitomize the professional life of one who had been at the early outset a schoolmaster, after that a university graduate in medicine, and then a practitioner. Such would be the natural interpretation of a sentence like that quoted; and the trouble with this pretended epitome is, that it leaves completely out of view Brunfels's occupation during the first fifty years of his life, revealing only the last twenty; for certain it seems to be, that when in default of other means of a livelihood he opened at Strasbourg a school for boys, he was well past fifty years of age; also that when at the University of Basle he won the degree of Doctor of Medicine, he was sixty-five.[4]

In the history of botany Brunfels will hold in the future, as he has done in the past, a somewhat distinguished place among the notabilities belonging to his century; and we must review, as well as at the distance of four hundred years we may, the incidents that had to do with the moulding of the youth, and helped to establish the character of the man.

The birthplace of Otto Brunfels was Mayence, or Mainz.[5] The family had

Otto Brunsfeld / Medicus.

FIG. 32. Otto Brunfels (1489–1534). (From Van Kaathoven Collection, National Library of Medicine, Bethesda, Maryland. Print at Hunt Institute)

taken its name from Castle Brunsfels[6] not far from Mainz where the earlier ancestry of the botanist had lived. At Mainz, John Brunsfels, the father, was in the business of a cooper; appears to have been in comfortable if not affluent circumstances; was well known and much respected as a man of character and high integrity; also, as we learn by his opposition to young Otto's plans, a man with a will and purposes of his own, qualities inherited by the son, as we shall see. Otto was the only son, and entertained the thought of devoting himself to the service of the Church. At that time Martin Luther was yet unborn and all Germany was Catholic.[7] A Catholic father of that period, if rich or well to do, would have been a marvel of pious unworldliness, if he had been willing that his only son should become a clergyman, for that would mean the immediate extinction of his own branch of an ancestral line. This father of young Otto Brunfels was resolute and persistent in his opposition to the son's wish, and naturally so; and this must have continued until the son was of legal age; for at last, hopeless of otherwise attaining to the priesthood, he left home[8] and became a novice in the Carthusian monastery that was in his native town. This he would not have been permitted to do had he been a mere youth, unless the father had given consent.

Meyer's inference that Brunfels remained but three or four years an inmate of the monastery[9] proceeded from several misunderstandings, one of which was that the man had not been born until a little before the year 1500. But there is now good authority for our accepting 1464 as the year of the botanist's nativity;[10] so that in 1500 he was already thirty-six years old. Then, since to assume a part in the new Lutheran movement was the object of his secret flight from the house of the Carthusians, and that movement was hardly well under way before 1517, it becomes highly probable that the man was fifty-three years of age when, renouncing monasticism and giving his learning and talents to the support of Luther's cause, he took up the sojourn at Strasbourg.[11] He betook himself to school teaching only after his voice had failed him, so that he could no longer preach. Here again the historian Meyer draws an inference. It is this, that his school must have been a financial success, because at the end of nine years at teaching he had saved money enough to pay the expenses of his degree at the University of Basle. Without doubting the financial success of Brunfels's school it is next to certain that he realized a much more considerable income from the sale of his rather voluminous Protestant theological writings; for these included, besides learned commentaries on certain books of Scripture, pamphlets for popular reading, and a catechism for children. There is a long list of them in Conrad Gesner's *Bibliotheca Universalis.*[12] Altogether his two vocations of teacher and theological author must have yielded him a very fair income during these first nine years at Strasbourg, for he was able to give employment to the best engraver of Strasbourg, Hans Weydiz (Latinized Guiditius), who did the engraving of the *Icones*, and is a man of distinction in the history of wood engraving.[13]

It must have been after having taken his degree in medicine, and within two or three years from the time of his death, that Brunfels made a journey from Strasbourg to Hornbach for the purpose of personally urging Jerome Bock (Tragus) to write a book of botany for German readers. For the record of this visit history is indebted to Tragus himself. In the thirteenth chapter of his preface to the *Stirpium Historia* he says:

> When information about the labors and the journeyings which I had undergone in behalf of plants had in some way been conveyed to the most learned Otto Brunfels of pious memory, he himself came journeying all the way from Strassburg to Hornbach, that he might see my gardens and collections. These things pleased him so much that from that day forward he ceased not to exhort, as did also others by letter, that I would reduce all this matter to order, and give it to the German public.

Not one of even the compatriot German historians of botany, in so far as I am aware, has set before us this evidence that it was to Brunfels's personal influence over Tragus that the writing and publishing of Tragus's work was due. How much botany owes to Tragus's unusual powers of observation and description we shall learn later; for the half of that story has never yet been told.

It is well worth noting that this visit to Tragus, with its fruitful conse-

Reformation der
Apotecken/ welche inhaltet vil gůter
stück/ die eynem yeglichen fast nützlich sein/ so seiner
gesundtheyt gern acht haben will/ als nemlich von
kreütteren/ wurtzlen/ safft/ samen/ blůmen/ öle/
feystigkeyten/ gebranten wassern/ Juleph/
vnd anderm/ wie man solche ding
bekommen/ behalten vnd
brauchen soll.

Von edlen steynen/ wie die zukeñen/ vñ wa zů sie nütz sein.

Alles beschriben durch den hochberümpten Otto Brun
felßen/ der Artznei doctor.

Wie man Syrupen/ Latwergen/ vnd Confect machen
soll/ verteütscht auß dem latein durch D. Hansen Eles.

Zů Straß. durch Wend. Riel. M. D. XXXVI.

FIG. 33. Brunfels, *Reformation der Apotecken* (1536), title page, illustrating in science the same reformist perspective that characterized the religion of that age.

quences, was the last service which Brunfels rendered to botany. The visit must have been made as late as the year 1532,[14] for not until that year was Tragus settled at Hornbach; and in the next year Brunfels, now newly appointed physician to the city of Berne, removed thither, where also after only a year and a half of service he died in 1534.

After having ceased from theological authorship, and subsequently to his having taken a degree in medicine, Brunfels published several medical works;[15] but both theology and medicine appear to have forgotten his name. In the history of botany only is he immortal; and this because he was intensely a lover of nature and of plants. His book gives proof of this, although the figures are the best part of it. It was because his love of plants could not tolerate the absurd pictures then common, that he resolved to produce something in that line true to nature, despite the cost; for the employing of the best artist of his time cannot have been less than very expensive to him, and there may have been no clear prospect of any return, even of that which the plates cost him. Indeed no one can assert that there ever was any. But here was devotion to an ideal, a love of plants that was bent upon procuring faithful representations of them in books. And so a well-marked epoch in the study of the plant world dates from Brunfels and the year 1530.

To the botanical memory of this ex-Carthusian, the Franciscan monk Charles Plumier dedicated the genus BRUNFELSIA in the year 1703.[16]

PHYTOGRAPHY

If by a man's phytography is meant his manner of describing plants, that is, his word-picturing of them, it cannot be said of Brunfels that he has any; and Julius von Sachs was never farther from writing history than when he set this man forth as among those who "went straight to nature, and described the wild plants growing around them."[17] Brunfels publicly disclaims all purpose of writing verbal descriptions of any plants whatever, and in the following terms:

> In this whole work I have no other end in view than that of giving a prop to fallen botany; to bring back to life a science almost extinct. And because this has seemed to me to be in no other way possible than by thrusting aside all the old herbals, and publishing new and really life-like engravings, and along with them accurate descriptions extracted from ancient and trustworthy authors, I have attempted both; using the greatest care and pains that both should be faithfully done.[18]

His meaning as to phytography is plain. He will describe nothing anew. He will ignore the contents of wretched mediæval herbals like the *Hortus Sanitatis*. He will reproduce the standard descriptions of classic Greek and Roman authors. For an example, take his presentation of the two water lilies, the white-flowered kind and the yellow. He figures them admirably, and, not having a word of his own to add to that knowledge of them which has been the common property of botanists for a thousand years and more, he supplements the two plates by three folio pages of quotations about them, taken from a list of eleven of the most approved botanical authors, ancient, mediæval, and contemporary. Here is the list:

Theophrastus
Dioscorides
Plinius
Apuleius
Georgius Valla
Avicenna
Serapion
Simon Januensis
Rases
Joannes Vigonius
Hieronymus Herbarius[19]

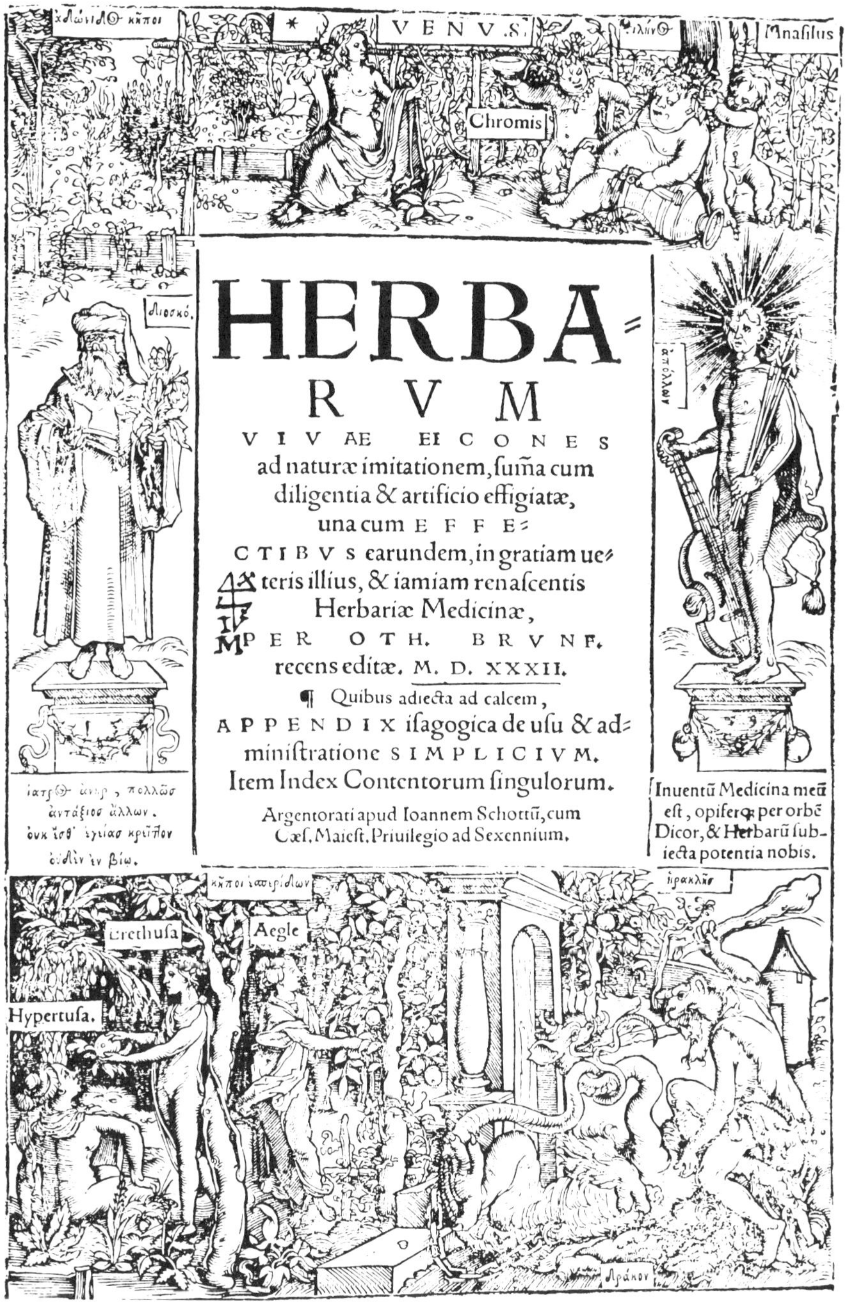

FIG. 34. Brunfels, *Herbarum* (1532–39), title page for vol. 1; this second printing displays drawings by Hans Weiditz.

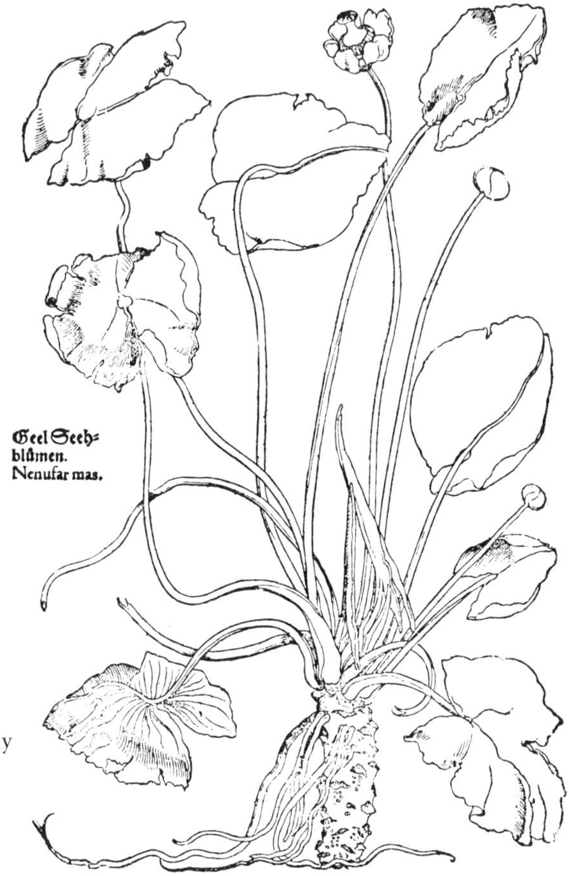

FIG. 35. Yellow water lily (*Nuphar lutea* (L.) Sm.). *Herbarum* (1532–39), vol. 1, p. 36.

And what is true as to his presentation of the water lilies holds good in the case of almost every other genus that he takes up. Rarely does he append to such a succession of quoted paragraphs a few remarks of his own; and these always indicated as his by the special caption, "Sententia nostra," or "Sententia Othonis"; nor are such original paragraphs really of the nature of descriptions. They usually express some opinion as to the identity of the plant in question, have reference to the correct application of a classic plant name. As to phytography, therefore, the Brunfelsian volumes are a treasury of select quotations from a long line of books many of which are now seldom seen. But there are no new descriptions in his volumes; and it may be doubted whether upon the whole he directly advanced the art of plant description by a syllable.

It is no impeachment of his erudition to question that he had the ability to describe plants well. There is evidence that he had not the faculty of mentally imaging an unknown plant from its description; and the ability to describe and that of making effective use of a description are twin accomplishments, if indeed they be not almost one and the same, so that he who has the one has also the other. Certain it is that Brunfels read and studied here and there a classic plant description to little purpose. Bringing together in one chapter the classic descriptions of *Aristolochia*,[20] the figures by which he illustrates the genus are *Corydalis bulbosa* and *C. Halleri*.[21] So gross an error explains itself in this way. The aristolochias were of southern Europe and not found in Germany. Here, however, the roots of the fumariaceous perennials, not so unlike those of the principal aristolochia, had usurped in medicine both their place and their name. Brunfels, though professedly reforming German pharmacy by the correcting of just such blunders, did not detect this one. Yet the very descriptions of aristolochia which he reprints from Dioscorides and Pliny must have shown, had he really read them, that

FIG. 36. White water lily (*Nymphæa alba* L.). *Herbarum* (1532–39), vol. 1, p. 37.

these things could not be aristolochias. Other such errors also remained undiscovered by him, and as inexcusably; so that when his countryman and contemporary Fuchs remarked that in Brunfels the descriptions and the plates accompanying them are not in all cases at agreement,[22] he was passing but a gentle criticism on his neighbor's phytographical shortcomings.

Anthology

I have met with no evidence that during the fifteen centuries intervening between Dioscorides and Brunfels there had been any progress made in the knowledge and understanding of floral structures.[23] There were several of Brunfels's younger contemporaries who, after the year 1530, added somewhat to anthology; but the time was yet more than two generations distant when the science of the flower was to become so far developed as to begin happily to revolutionize plant classification. There is no sign in Brunfels that such a day is near its dawning. In his attempts to range plants in groups he is

FIG. 37. Corydalis (*Corydalis cava* (L.) Schweigg. & Koerte). *Herbarum* (1532–39), vol. 1, p. 48.

Heyter Nesszeln.

Taub Nesszelen weissz.

FIG. 38. Nettle (*Urtica dioica* L.). *Herbarum* (1532–39), vol. 1, p. 151.
FIG. 39. White dead nettle (*Lamium album* L.). *Herbarum* (1532–39), vol. 1, p. 152.

no more influenced by considerations of floral structure than were the medical botanists of remote antiquity; even less so than Dioscorides, who, as we shall see, could not abide the placement of the bilabiate-flowered dead nettles in the same genus with real nettles, but segregated them, on account of their two-lipped corollas, and assigned them a new generic name of their own, and framed to express the peculiarity of their flowers. It is possible to rate the Brunfelsian anthology as more antiquated and imperfect than that of Dioscorides; for he of the sixteenth century less openly recognizes as generically distinct the *Galeopsis* and *Lamium* "nettles" and the proper *Urtica*;[24] and if he figures the thistles, the anthemideous composites, the principal borragineous plants, the bulk of the labiates, and some other such, each as a group by itself, it is done without any particular reference to floral structure, at least on Brunfels's part; for in all these instances he is but continuing groupings which the ancients themselves had indicated as being natural, and had well established.

TAXONOMY

Brunfels adopts without hesitancy the ancient primary classification of growths as herbaceous and woody. When, however, it comes to that apparently quite as ancient division of the vegetal kingdom into things cultivated and things wild, he deliberately ignores it. His first three plates represent three most common and homely wayside weeds, members of the genus *Plantago*; and thenceforward throughout his volumes he deals much more extensively with wild plants than with the domesticated.

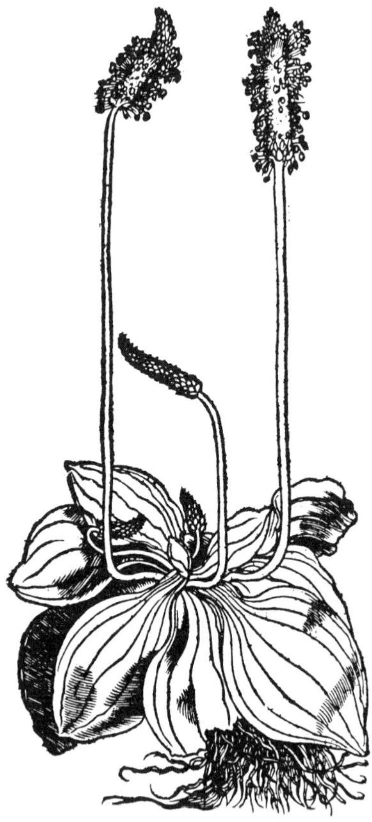

FIG. 40. Spotted dead nettle (*Lamium maculatum* L.). *Herbarum* (1532–39), vol. 1, p. 153.

FIG. 41. Hoary plantain (*Plantago media* L.). *Herbarum* (1532–39), vol. 1, p. 22.

Now this eliminating of the distinction referred to is not to be attributed to any following of the suggestions of Hippon, who some two thousand years before had declared plants wild and domesticated to be all of one lineage.[25] There is no intimation that Brunfels had made tests, and proven out of the book of nature that this old-time grouping must be abandoned. The thought had come to him solely as a deduction from theological premises. The polytheistic ancients had held that the different alliances of cultivated plants and trees were each the creation of some beneficent particular divinity,[26] and that the less useful or the altogether useless had hardly been created at all. The theology which Brunfels accepted, and, as a profession, taught, was monotheistic. One Divinity had made all the plants that are—the wayside weeds, the homely remedial herbs, as well as the beautiful things of the field, the garden, and the orchard. Such doctrine of the equality of all plants as to one divine origin finds expression in the last one of Brunfels's several prefaces, which contains a prayer, after which one reads his apology for giving to those common, lowly, and weedy things, the plantains, the foremost place in his system of botany.

> They are the very commonest of plants, he says, and are known to everybody; and being both lowly and also singularly useful, they are most apt to recall to mind the thought of God, whose way it is to work wonders through means that are usually accounted insignificant, passing by such as make more display, and which men therefore hold in more esteem.[27]

This is even showing a preference for wild growths before those that have undergone domestication, a kind of preference that has been felt by the great majority of philosophic botanists from Brunfels's time to ours; and by virtue of his being the first propagandist of this new idea he sets up another landmark in the history of botany.[28]

This idea of the equal genetic dignity of all plants seems to have come to Brunfels as a deduction from a theologic principle, rather than inductively from the study of nature; but whence he derived it signifies nothing to the disparagement of the idea itself, especially now, after all the world has come to concede its truthfulness. But it was not at once approved in Brunfels's time. There were learned men among his contemporaries who were at first startled by, and then made light of, his having brought forward some of the most plebeian and beggarly roadside pests, and introduced them as upon an equality into the company of the nobler growths of the fields and meadows, and of the vegetable and drug gardens. Among the more serious faults that his contemporary Fuchs found with Brunfels's work, one was "That he sometimes takes for subjects the most common weeds."[29]

By at least one other item of his method, over and above this of ignoring the old distinction between things as domesticated and wild, does Brunfels commend himself as a believer in some kind of a natural classification. He declines to adopt anything like an alphabetic sequence of genera, a kind of

HEPATICA.

Edel Leberkraut. Hyrßklee.

FIG. 42. Hepatica (*Hepatica nobilis* Mill.). *Herbarum* (1532–39), vol. 1, p. 190.

arrangement which was adhered to by several of his noted botanical contemporaries, as we shall see. He prefers freedom to express, if but tacitly, some ideas of a more rational grouping, such as the alphabetic succession of names almost wholly precludes; and, with the medical botanist, that arrangement may be most convenient, if not even in a sense natural, in which plants, whether alike or unlike as to morphology, are held in juxtaposition by agreement as to what are taken to be their medicinal virtues.

For an example of this kind of classifying carried to an extreme, take his two genera of liverworts, *Hepatica* and *Jecoraria*. The former is that anemoneous herb that has retained in later times the name *Hepatica*; the other is the common *Marchantia polymorpha*, a cryptogam. The two are figured and described on opposite pages, and their medicinal uses are said to be the same.[30] It may be noted that each bears alike, even in our time, the common name of liverwort. Before Brunfels *Hepatica* usually meant the plant *Mar-*

chantia, which was also called *Jecoraria*, and the restriction of the name *Hepatica* to the genus of anemone allies, and of *Jecoraria* to the lichenoid hepatic, seems to date from Brunfels, and was a distinctly taxonomic movement on his part; as if his judgment had been that types so very unlike morphologically ought not to be treated of under one and the same generic name.[31]

Because of their having been employed interchangeably in medicine, under the common designation of *Verbena*, our medical botanist figures and discusses, one next after the other, *Verbena officinalis* and *Senecio vulgaris*.[32] The botanist of a later time will see no likeness or other sign of true affinity between these; and it is very probable that Brunfels himself realized how very distinct they are when considered from the morphological rather than the remedial point of view; for when he formally designated one of them *Verbena mas* and the other *Verbena fœmina* it is beyond question that he was purposely indicating the morphological distinctions between them, quite as he had done in assigning to those two morphologically dissimilar liverworts each a generic name of its own. And according to the usage of Brunfels's time, as well as for two centuries later, binary generic names like *Verbena mascula* and *Verbena fœmina* were thought as suitable, and were as freely made and admitted, as those of one word only.

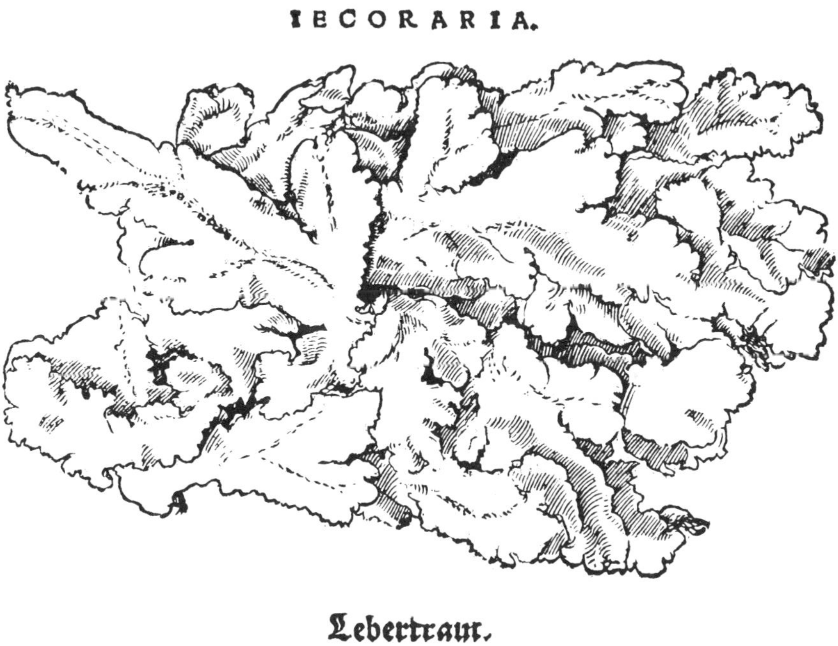

FIG. 43. Common liverwort (*Marchantia polymorpha* L.). *Herbarum* (1532–39), vol. 1, p. 191.

Of such attempted improvements in classification by appeal to considerations of morphology, one may come to a fuller appreciation by looking into Brunfels's way of presenting those many herbs which, in his time, had long been reputed to be good vulneraries, and had therefore passed under the medico-generic name of *Consolida*, with which *Symphytum*, *Sanicula*, *Vulneraria*, and *Solidago* were synonymous, each such name indicative of the property which these plants all had, or were believed to have, of promoting the closing-up and healing of cuts and wounds. Here is a partial list of these plants under their mediæval names, with their equivalents in modern nomenclature:

Mediæval	Modern
Consolida major	*Symphytum officinale*
C. media	*Ajuga reptans*
C. minor	*Sanicula Europæa*
C. petræa	*Coris Monspeliensis*
C. regalis	*Delphinium consolida*
C. rubea	*Tormentilla erecta*[33]

One thus gains an idea of how great a diversity of plants passed with mediæval pharmacists and physicians under the generic name *Consolida*. And the list must now be given again, that the Brunfelsian taxonomic betterment of it may as readily be seen:

Mediæval	Brunfelsian
Consolida major	*Consolida major*
C. media	*C. media*
C. minor	*Diapensia*
C. petræa	*Symphyton petræum*
C. regalis	*Consolida regalis*
C. rubea	*Tormentilla*[34]

One observes that out of the six *Consolida* names, three have been eliminated, and others brought forward to take their places. I say brought forward, for neither *Diapensia* nor *Symphyton petræum* nor *Tormentilla* is coined and proposed as new by Brunfels. He picked them up every one out of the ancient and mediæval synonymy of the vulnerary herbs, from which also it will appear that other men who lived and wrote botany in times long forgotten—but the history of which times must nonetheless some day be written[35]—thought as Brunfels did, that plants totally unlike in appearance, *i.e.*, morphologically very different, ought to be invested with names more than partially different, even when as to qualities and uses they were very similar. And these group names established upon the merely remedial virtues of things visibly most dissimilar must have been misleading and confusing in the extreme. It seems as if Brunfels realized this, and intended to sug-

FIG. 44. Vervain (*Verbena officinalis* L.). *Herbarum* (1532–39), vol. 1, p. 119.

gest improvement when he set aside three out of the six *Consolida* genus names and wrote others in place of them.[36] It is as if he had thought it out, that since the different kinds of plants can only be well distinguished and scientifically grouped through attending to their morphology, it is not well that they should bear names that point to their qualities rather than to their forms. Therefore, in the interests of a more sure identification of important plants, as well as at the same time encouraging the appeal to morphologic marks in classifying, it would be a good thing to at least place a check upon this multitudinous repetition of pharmaco-generic names, the first half of which is the same for a half-dozen very dissimilar genera.

If it be asked why he did not, while he was about it, proceed to the suppression of as many as five out of the six *Consolida* genus names—leaving perhaps one of the genera to bear the simple name *Consolida*—the right answer will seem to be that Brunfels was not of the temperament of the taxonomic revolutionist but only a reformer, and disposed to be somewhat conservative even as a reformer, between which character and that of the bold iconoclastic revolutionist there are differences.

Entirely consistent with his aversion for genera made up of plants qualitatively alike but morphologically unlike, is Brunfels's approval and adoption of some in which the species are qualitatively unlike, and at agreement morphologically. Such a genus as this is that which he fully illustrates under the classic name *Urtica*,[37] which in the botany of today comprises only the true nettles. The specific constituents of his *Urtica* seem to be these:

Urtica dioica
Lamium maculatum[38]
Galeopsis tetrahit
Urtica urens

And this, too, is the order in which they succeed one another in the book: first a real nettle, then two so-called dead nettles, the line closing with a second true nettle; a genus composed of two nettles, and two or three members of the very different family of the Labiatæ. And this, as intimated above, is a genus not qualitatively but morphologically constituted, a fact easily demonstrable when it is remembered that Brunfels had no anthology, that the flowers of plants not only were not at all understood by him, but were the least and the last parts of them to receive any consideration. So long as two or three herbs were alike as to roots, stems, and leaves, they might easily be designated by the same common, *i.e.*, generic, name. Forgetting, then, all anthologic differences between nettle and dead nettle, note how remarkably they are at agreement. The roots in all are small, fibrous, and not deep-seated. The stems of all are upright, almost or quite without a branch, conspicuously quadrangular, and the leaves they bear are opposite. The leaves in all are short-stalked, their blades of the same ovate or oval outline, serrate as to their margins, and are of much the same texture as well as form. The seeds in all—for though anciently flowers were neglected, seeds never were—the seeds were black, and were always clustered together in the axils of the leaves all up and down the stem. All these quite marked characteristics of all their vegetative organs *Urtica* and *Lamium* and *Galeopsis* have in common.

Since the thought is one far from being familiar to the botanical mind of the present, it must here again be insisted on, that the grouping together of several plants upon vegetative characters only, but under a generic name, is as exactly of the nature of a generic concept as that group which is rested on characters of flower and fruit only. By either method a genus may be circumscribed which shall be unnatural; and the idea is equally the idea of a genus in either case.

I do not see what chapters of any history of botanical science should be more profoundly significant, or of a more general interest, than those touching upon the development of men's ideas of a plant genus; for the idea of the genus seems to be first and last the type-idea, if one may intelligibly so speak, of taxonomy. It was because this seemed to be true, that in our "Introduction to the Philosophy of Botanical History,"[39] the earliest available expressions of such idea by even the primitive and the unlettered were dwelt upon somewhat at length; and for the same reason, one desires to examine with the utmost care leading expressions of the idea of a generic group as they occur in this, almost the earliest propagandist of what has slowly developed into the thing known as modern botany. Brunfels was of thoroughly well-educated mind, even a profound scholar, also naturally endowed with a keen insight into the beauties and the harmonies of plant life and form. On all these accounts it would be exceedingly interesting, if it were possible, to know just what his own opinion really was as to the philosophic tenability of such a genus as this which we have been inspecting, a genus *Urtica*, by name, but made up of species some of them urtical, but as many others labiate. If he has any taxonomic opinion different from that which, in as far as we have proceeded, he seems to have expressed, we shall be likely to find the evidence of it, if there be any, by reading as it were between the lines; for even a botanical genius, if writing as Brunfels professes to write, in the interests of medical botany only, inditing a work the readers and students of which are to be the physicians and the pharmacists, must not yield to every impulse he may feel to improve taxonomy; for such improvement commonly involves changes in nomenclature, and there is nothing of which the druggist, or other plant industrialist, is more intolerant than changes in names of his commodities.

The opinion, if Brunfels held it, that nettles proper and labiate-flowered nettles are generically distinct was not original with him. We observe that Dioscorides as long ago as the first century of our era segregated the dead nettle as a genus, and under a name which pointed to the character of its flower, the name *Galeopsis*; and this proposition had evidently been acceded to by some of the mediæval Latin botanists, who, instead of the Greek γαλίοψις, had employed such Latin equivalents as *Urtica mortua*, *Urtica iners*, and *Urtica labeo*, the last a most significant appellation, "nettle with a lip," evidently taking cognizance of the floral character, while the other two refer merely to the lack of stinging hairs. Now this mediæval synonymy of the plants is perfectly familiar to Brunfels. He formally quotes every item of it; and his approval of *Galeopsis*[40] as a proper genus comes out plainly enough, at least to the careful reader, and on this wise. In reproducing what Dioscorides had said about the labiate nettles he does it under the following caption: "De Galiopsi simili Urticis herba, Dioscorides."[41] Of such a caption the English version is, "Concerning Galiopsis of Dioscorides, an herb resembling the nettles."

Of course, that which resembles another thing *is not* that other thing which it resembles; and nothing that was ever printed in a book is plainer than that this author did not regard *Galeopsis* as congeneric with *Urtica*. When in his *Icones* he sandwiches the dead nettle in between two real nettles, and when as a heading to his Chapter XXIII., in which both kinds are discussed, he places that simple "De Urticis," he is purposely adapting himself to the understanding of the half-taught root and herb dealers, and the untaught old women, who call them all nettles indiscriminately. In a word, Brunfels is a man of some learning and insight in matters botanical, and also a man of discreet conservatism, holding it unwise to lay too openly before the general public every advanced taxonomic view that is his own.[42]

In his indubitable though dissembled accepting of *Galeopsis* as distinct from the nettles he cannot but have been impressed by the fact that Dioscorides in making the segregation had done so in deference to its flowers, which he described as being "slender and purple"; and it may or may not have been in deference to similarity in floral structure that closely appended to the Urtica-Galeopsis series comes an unbroken line of three other galeate-flowered labiate types. If, however, this be an example of guidance by anthology to the recognition of affinity, still it is a guide which Brunfels is as far as possible from following steadily. The flowers of orchidaceous plants have as much agreement in character as have those of labiates; but when he comes to the grouping of what are known to us as the orchid genera we find that all those which have two or three large tuberiform roots are gathered into one place by themselves,[43] while their merely fibrous-rooted kindred form a group quite apart from these,[44] various wholly unrelated types intervening between the two orchid groups. This is all quite after the method of antiquity, the method of those who, heedless of flowers, to the knowledge of which they had not advanced, concluded things to be allied because they were alike as to roots, and, it may be, as to stem and foliage also.

Let us give a moment's attention here to another set of plants which he places in juxtaposition for the reason that they all exhibit a kind of tuberiform organ as developed among their roots. The group is

Brunfels	Modern
Scrophularia major	*Scrophularia nodosa*
S. media	*Sedum Telephium*
Ficaria	*Ficaria ranunculoides*[45]

The point that is of special interest here is, that for the third member of the group, Brunfels rejects that which was one of its common mediæval names, that is, *Scrophularia minor*; though more anciently, even with Dioscorides, it was called *Chelidonium minus*. What he did with this third plant of the list seems to attest that there was in him, botanically, as there was ecclesiastically, something of the spirit of the revolutionist, or reformer. If there had not been, he would have been almost sure to have called this

FICARIA.

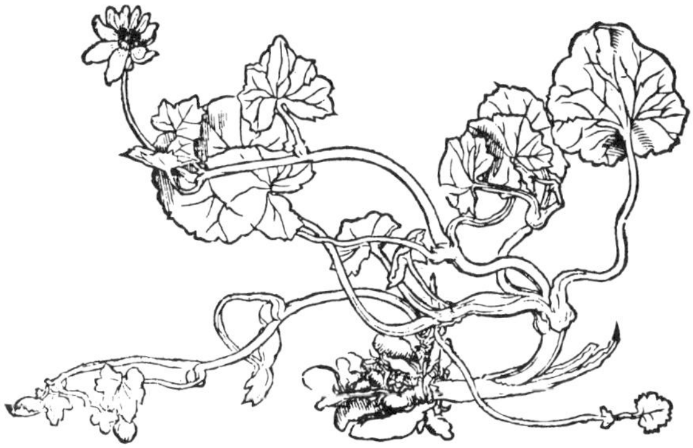

Fygwartzkraut.

DE SCROPHVLARIA,

Rhapſodia XXXVII.

¶ NOMENCLATVRAE Maioris,

Latinæ, Scrophularia maior. Caſtrangula. Vrtica mortua.
Germanicæ, Sauwurtz. Fygwartzenkraut.

MINORIS.

Latinæ, Ficaria. Millemorbia.
Germanicæ, Pfaffenhödlin. Fygwurtz.

MEDIAE.

Latinæ, Scrophularia media.
Germanicæ, Knabenkraut. S. Johanns kraut. Wundtkraut. Fotzwangk. Gauchgeyl. Fotzwein.

FIG. 45. Lesser celandine (*Ranunculus ficaria* L.). *Herbarum* (1532–39), vol. 1, p. 215. The list below the illustration shows Brunfels's concern for proper identification of the plants. Almost all of his accounts of plants had such lists of synonyms.

ranunculeous herb by one or other of its ancient and mediæval names rather than startle the herbalists and pharmacists of his time by that new name, *Ficaria*, for a type so long known under very different appellations. We shall also, I think, miss a part of what was in his mind, if we do not read here the expression of an objection on his part against the old way of naming and grouping of plants conformably to their medical qualities rather than according to their morphology. All three of the plants had been called kinds of *Scrophularia*, because they were believed to be efficacious against scrofula; and there is with me no doubt that Brunfels in displacing one of the old *Scrophularia* names by the new generic name *Ficaria* is to be understood as mildly protesting against qualitative criteria of plant affinities, and affirming the need of appealing to the morphologic.

We were observing above how Brunfels might be said to have limited his group of the orchids to such genera of them as have a certain kind of underground organ, that he excluded from the group such as have only fibrous roots, himself all the while oblivious—as all the world before him always had been—of the flowers by the structure of which all stand at agreement. Let us now observe him locating as far away from each other two groups of genera known to us as borragineous plants. In this instance he does not separate on ground of differences as to roots, or form of leaves, but of pubescence only, that is, over and above certain qualities common to all. Upon such principles are *Echium*, *Cynoglossum*, and *Borrago* made to form a group of genera.[46] We of today, after four centuries of taxonomic progress, concede that Brunfels was correct in apprehending a very intimate consanguinity between the three. But we hold them in juxtaposition on quite other grounds than those which had weight in the early sixteenth century. We judge them near allies because the plan of their flowers and the common characteristics of their fruits are the same. With Brunfels the flower was so almost wholly unknown that no such thing as the plan of a flower had been thought of. And, viewed superficially—the only view that had yet been taken of flowers at all—they were very notably dissimilar. The corollas of the genera are of remarkably distinct types, that of *Echium* being narrowly tubiform below, with an irregular almost bilabiate limb; that of *Cynoglossum* is short-salverform, perfectly regular; that of *Borrago* broadly and flatly star-shaped. Few families of plants present three genera so unlike as to the cut of their respective corollas as these three. We therefore seem to infer to a certainty that in collocating these three generic types, he had had the utmost regard to their likeness as to roots, stems, foliage, and especially to that armature of harsh, somewhat stinging bristles wherewith all three alike defend themselves; and that in the process of his reasoning the corolla, *i.e.*, the "flower," was not at all considered. And, as if to place this beyond dispute, two other borragineous types are relegated to another part of the book. One is a *Cynoglossum*, the other a *Myosotis*.[47] Both differ from the other group in that they show no trace of the stinging-bristly or any other rough indument. They are almost silkily soft-hairy. Had he not held such differences to be most significant, taxonomically, it is impossible to see why he separated so widely these two groups of what we of today understand to be near allies.

If one is to follow the progress of plant taxonomy from the year 1530 forward, it will be needful to bear in mind such things as Brunfels's failure to apprehend the consanguinity of all the borragineous genera that he knew, as well as to note, if perchance one may discover the reason, why he failed. Then afterwards it must be observed how those who came after him, one after another and little by little, brought the other genera of such a family into a continuous sequence, also all the while attending to—even carefully noting—the development of new principles, whatever they may have been, in accordance with which the better taxonomy of more recent centuries has

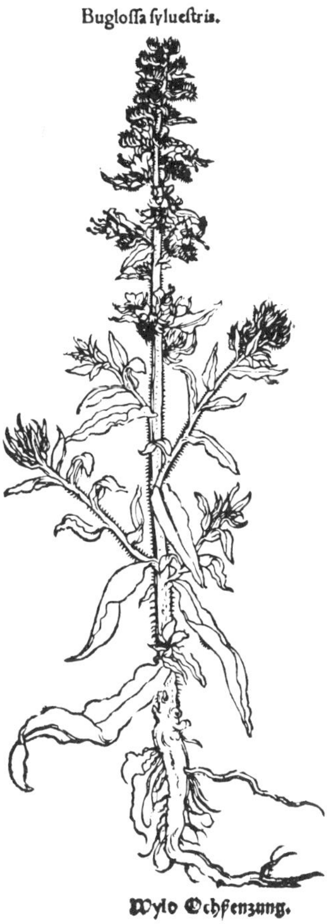

FIG. 46. Viper's bugloss (*Echium vulgare* L.). *Herbarum* (1532–39), vol. 1, p. 111.
FIG. 47. True alkanet (*Anchusa officinalis* L.). *Herbarum* (1532–39), vol. 1, p. 112.

been attained to. Thus may we learn, and thus only, the lesson of the modern development of the very old idea of plant families.

Nomenclature

All plant names are names of groups; and to group things together under a common name is to classify. Nomenclature and classification are therefore so intimately connected that neither topic can be fully discussed apart from some consideration of the other. They cannot be completely divorced; and so it was inevitable that something in relation to Brunfels as nomenclator should be brought out under the heading of taxonomy. It will be useful, however, to epitomize his work as nomenclator, and particularly since he now and then evinces a disposition to amend and improve upon ancient and

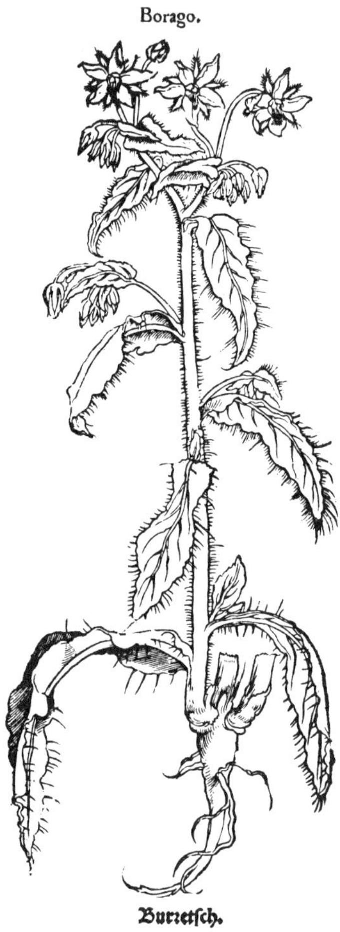

FIG. 48. Borage (*Borago officinalis* L.). *Herbarum* (1532–39), vol. 1, p. 113.

mediæval names and name-making methods; or, it might perhaps better be said, a disposition to return from mediæval to ancient methods; for what I have in mind is something like a distinction which, in a general way, holds between what may be termed the ancient and the mediæval plant naming. It is, however, not much more than a difference between the genius of the Greek language and that of the Latin as to manner of framing distinctive names for things.

In Greek the noun and adjective readily combine to form a single word, such word beginning with the adjective part and ending with the noun, whereas in Latin noun and adjective are kept as distinct words, even with the noun rather than the adjective standing first. To make this as plain as possible let us use a few examples:

Greek	Latin
Leucoion	*Viola alba*
Melanion	*V. nigra*
Chrysion	*V. aurea*
Herpetion	*V. repens*
Chelidonion	*V. hirundinaria*

By many scores of such one-worded Greek plant names which by translation into Latin become binaries, there is revealed one of the misfortunes under which mediæval and early Renaissance botany labored everywhere—for mediæval botany was Latin botany—that of having in its employment hundreds of binary names, some of which were of specific import, while as many more were but the names of monotypic genera.

The continual perplexities involved in this phase of nomenclature seem to have exercised the mind of Brunfels to a degree, so that he made bold to displace here and there some binary generic name, substituting one of a single word. A few examples of such action on his part were brought forward under the heading of his taxonomy. A more considerable exemplification of this practice is given below, in a selection made from the first volume of the *Vivæ Icones*:

Early binary generic names		Brunfelsian substitutes
Fumus terræ *Fumaria herba*	=	*Capnos*
Ferraria major *Consolida minor*	=	*Sanicula*
Testiculus canis *T. vulpis*	=	*Satirion*
Lingua bubula	=	*Borrago*
Sacra herba *Verbenaca supina* *Cincinnalis herba* *Lacryma Junonis* *Herba sanguinalis* *Sanguis Mercurii* *Mustelæ sanguis* *Crista gallinacea*	=	*Verbena*
Trixago minor *Quercula minor*	=	*Chamædrys*
Scrophularia minor *Chelidonium minus*	=	*Ficaria*
Herba Apollinaris *Faba suilla*	=	*Hyoscyamus*[48]

The credit of having reformed the nomenclature of genera by the exclusion of names made up of two distinct words has been given to Linnæus, who, in the year 1751, is thought first to have laid down such a principle.[49] But the actual reform had been quietly inaugurated by Brunfels[50] two hundred and twenty years before Linnæus came forward with his *Philosophia Botanica.*

Sprengel, the one nineteenth-century author of a *Genera Plantarum* who has observed the law of priority in the crediting of generic names, ascribes to Brunfels the authorship of the following:

Ammi	*Fragaria*	*Pyrola*
Calendula	*Linaria*	*Sanicula*
Carthamus	*Melissa*	*Scrophularia*
Castanea	*Parietaria*	*Spinacia*
Euphrasia	*Potentilla*	*Valeriana*

To this list of fifteen, credited to this author by Sprengel,[51] I find two more to be added, namely *Hepatica* and *Ficaria.* Sprengel's reason for not taking them into the reckoning was simply this, that he did not admit the types as worthy of generic rank, but held with Linnæus that the former was but an *Anemone* and the latter a *Ranunculus.*

Now when Sprengel and other advocates of priority credit such genera to Brunfels, it is not to be understood as their meaning that in his book these types are for the first time named and defined. The truth is, that all of them had been known before Brunfels, and some of them had been much written about, under different names. For a heading to each chapter in which a genus is discussed, Brunfels selects, out of the several names current for that genus, the one that pleases him best; and, by virtue of the great prestige which his book obtained, the plant names in it were continued in use by other authors. Therefore they who credit *Sanicula*, *Potentilla*, *Fragaria*, or *Hepatica* to Brunfels affirm no more than this, that each such name, as the fixed appellation of a certain generic type, is traceable back to Brunfels.

In his researches upon native German plants he came to know here and there a type which, after the most diligent comparison with all the classical plant descriptions, he felt certain had not been known to the ancients, neither been described by any one. They were new generic types; and to such he never assigns any name at all, other than that by which it is known to German peasants. There is beautifully figured in one place a flowering plant of *Cardamine pratensis.*[52] Above the figure the German name Gauchbluem is inscribed; beneath it the statement in Latin that the plant was unknown to the ancients, though common enough in Germany, and native. One page is occupied by a most accurate and life-like representation of *Anemone nemorosa*, with the legend: "A wildwood herb, the name of which is unknown."[53] Nor is there any other mention of the plant, not so much as a record of its being known by a vernacular name. This is doubtless the earliest publication of the Wood Anemone.

Out of such namelessly figured types there might here be gathered surprising items of plant history. For one instance: any one informed as to how familiarly hundreds of useful plants were known two thousand years before Brunfels would expect to find so common and important a plant as *Trifolium repens* among that number. Nevertheless, it is one of the things which Brunfels presents as new to botanists.[54] He says it is well dispersed throughout Germany, chiefly in meadows, and is known to the common people by the name of Weiss Fleischbluem; also that his engraver brought him the plant under that name. That Old German name of the plant, and Brunfels's brief remark upon it, both printed on the plate page, seem to constitute the earliest publication of *Trifolium repens*.[55] Both the botanist and the artist seem to have agreed in the opinion—a purely philosophic one—that no little weed was beneath botanical notice, and between them they have given us the beginning of the history of *Draba verna*. The plant is ele-

FIG. 19. Wood anemone (*Anemone nemorosa* L.). *Herbarum* (1532–39), vol. 2, p. 84.

gantly figured under the vernacular designation of Gensbluem; but not another word is said about it.[56] It is, however, the first record, and a perfectly definite record, of an interesting though diminutive type, one that within the last century has been much discussed by very able botanists who have investigated it morphologically, taxonomically, and even as to its rightful name; and that Old German popular name Gensbluem—in later German Gansblum—has proven a somewhat fateful appellation.[57] More than two centuries after Brunfels had printed it, Michel Adanson proposed its adoption as being by right of priority the lawful generic name.[58] For two reasons, not calling for mention here, Adanson's movement failed of any public approval. Yet once again, in the end of the nineteenth century, Otto Kuntze renewed the Adansonian proposition;[59] but the attempt to reinstate Gansblum was again fruitless, at least as to gaining public approval. It was not a Latin-made name. Probably it did not occur to Brunfels's mind that a little weed, of no use in medicine or any art, needed to be dignified by any other name at all than that by which the country people of Germany knew it.

To the nomenclature of species it is evident Brunfels gave no thought; nor was there any reason why he should have given it any attention. Most of the genera, with him as with the botanists of antiquity, were monotypic, and the generic name was all that was needed. There was not the shadow of a reason for appending a second name; and he, no more than hundreds of botanical writers before his day, ever thought of such a thing. Sometimes when there are one or more notable modifications of a type—varieties or species of it—the original goes by the generic name only, while the others have each its own cognomen. Of this sort is his nomenclature of three buttercups which he figures and gives account of.[60] In the nomenclature of today they are (1) *Ranunculus acris*, (2) its double-flowered garden variety, and (3) *Ranunculus bulbosus*. The generic name which Brunfels adopts is *Pes corvinus*, *i.e.*, Crowfoot, turned into Latin. With him the first species is simply Crowfoot, its variety of the gardens is Full-flowered Crowfoot, the third plant is Lesser Crowfoot.

This early practice of leaving the one original representative of a genus without any cognomen, even after said genus has ceased to be monotypic, is a practice doubly suggestive in relation to the philosophy of nomenclature; for, in the first place it plainly reveals the antiquity of the idea of generic types, and emphasizes it. In the second place, the failing to assign a cognomen to the type species entails a difficulty, becomes a possible source of ambiguity and perplexity; for, *Pes corvinus* being mentioned, the question may chance to be asked: Which one of the three? That question is virtually a demand, and a most reasonable one, that the type species have also its particular cognomen. That botanists of fifteen centuries anterior to Brunfels had seen this to be desirable, one may infer from the nomenclature of *Plan-*

Gensblům.

FIG. 50. Common whitlow grass (*Erophila verna* (L.) Chevall). *Herbarum* (1532–39), vol. 2, p. 38.

tago. Two species of this genus were known to Pliny; and he had a specific cognomen for the type species as well as for the other. They were *Plantago major* and *Plantago minor*; and Brunfels follows Pliny in this. His type species is not simply *Plantago*; it is *P. major*, which name, as well as *P. minor*, the German father duly credits to the Latin author of the olden time.[61] He uses, then, a specific name for the original representative of a genus when there is classic authority for so doing; but I have not observed him taking the initiative in this course by actually himself assigning to the type species of any genus a cognomen.

These paragraphs on Brunfels as nomenclator ought not to be concluded without our having taken a briefly comprehensive survey of his principles. These principles, such as he was more or less ruled by, will be all the more suggestive to us from the very fact that he did not professedly have any, for doubtless he had never so much as heard, or even thought, the phrase "botanical nomenclature." The laws governing the naming of plants were not different from those observed in the naming of other things. All that we may gather by observing his procedure along these lines may be the course which a cultivated and philosophic mind, unhampered by prejudices, will naturally take. But such a study will be well worth while; because one does not often meet with an author who so nearly antedates all our stereotyped conventionalities, and takes his own course so little influenced by traditions and prejudices. Without having enunciated one of them, he seems to have been more or less under the guidance of principles like the following:

1. That for the science of botany there is an initial book; that is the *Historia Plantarum* of Theophrastus of Eresus. He quoted that work constantly, but never, I think, any earlier book or author. Others of Brunfels's time and a little later we shall find citing Moses, Solomon, and other Hebrew writers as if these had been botanists; but not so Brunfels, who, notwithstanding his training in theology, and the distinction he had won as a Biblical scholar and commentator, does not intimate that he has found botany in Holy Scripture, and never cites an author who antedates Theophrastus. It will not, however, follow that he must adopt Theophrastian generic names in such wise as to make that author's monumental work the point of departure for nomenclature. The existence of an historically first book of botanical science is one thing. The having a starting point for a universal nomenclature of botany is quite another; and the two are both logically separable and historically separate. Brunfels was well informed about the historic beginnings of botany; but the idea of a universal system of nomenclature for groups of plants had not in his day been conceived.

2. Brunfels writes in Latin. The text of his book is for those who knew Latin, and, knowing it, know things by their Latin names. The writer is under the necessity of using the Latin names of plants rather than those by which the same plants are known in Greek or Hebrew, Arabic or Persian. If a man pretending to write in Latin about animals should write *hippos* instead of *equus*, or *alopex* in place of *vulpes*, he would stultify himself, would be writing unintelligibly, absurdly, and ridiculously. It is not imaginable that Brunfels, in a Latin book of botany, should have done so insanely as to write *drys* instead of *quercus*, or *kittos* in place of *hedera*, *ion* rather than *viola*, or *arnoglossa* rather than *plantago*. Nevertheless he might have done so and most reasonably; indeed he must have done so, had the ideas of universality and priority in nomenclature been conceived and approved by him, because botany is of Greek rather than Latin origin, and so the Greek names of plants happen to be older than the Latin names. It was needful here to take a survey of the whole situation; for from Brunfels forward we must be looking for adumbrations of any of those principles which in our time have come to rule—or misrule—biologic nomenclature.

3. Even as a Latin writer, and using none but Latin generic names as headings for his chapters, Brunfels does not pay respect to priority. He readily adopts, out of several Latin names, ancient and mediæval, not the oldest, but the one that best suits his own purpose or fancy. From before the Christian era until six or seven centuries after it the water lilies had been known as the genus *Nymphæa*. Then from the eighth century forward to the thirteenth and later the Arabic name *Nenuphar* had usurped its place in Latin

botany generally. Brunfels adopts *Nenuphar* and writes *Nymphæa* down among the synonyms; this manifestly for the reason that most of the botanists and druggists of his own time knew the plants as *Nenuphar* and would be disturbed if he should restore the classic name. Here, then, we have

4. The principle that the name by which a genus is known to most of one's contemporaries is the one to be taken up, there being no other objection against the name.

5. That a later name consisting of one word only is commonly permitted by Brunfels to supplant a very ancient one made up of two words has been already quite clearly demonstrated.

6. A species name, or cognomen, is not assigned the type which alone represents its genus.

Kuchenschell: Hacketkraut:

FIG. 51. Pasque flower (*Pulsatilla vulgaris* Mill.). *Herbarum* (1532–39), vol. 1, p. 217.

7. While plainly favoring the selection of the best of several names as the one to be perpetuated, Brunfels, as if realizing the inconvenience of having many synonyms, is moved to use the greatest care and caution against creating them, that is, against creating Latin synonyms. This is well shown by his great aversion to assigning Latin names to types which to him appear undescribed. He publishes freely the engravings of such, but is careful to label them with no other than the German vernacular names. I have not found him once deviating from this very conservative practice. And, under his beautiful plate of *Pulsatilla*, in a long paragraph he explains why he holds to such a course. In none of the authors whom he has been able carefully to study has he found any description of this sort of plant. It may have been named and described somewhere nevertheless. He is resolved to print the figure, and leave it to others who have more leisure than he to study it in the light of all descriptions to them accessible.[62] Meanwhile the thing may be known by one or other of those German names by which the common people know it.

8. The student of botanical nomenclature should here note well the distinction which Brunfels tacitly makes between the Latin names used by Latin botanical writers, and those invented in their mother tongue by the common people. It is plain that with him they have not the same status. The vernacular name cannot figure among the Latin synonyms. It is upon no equality with them. His action and his words together bring it out clearly that, in his mind, there is a botanical nomenclature, and synchronously with it a kind of plant naming that is not valid scientifically. The botanists of antiquity had not, and hardly could have had this thought. Is the expression of it new with Brunfels? He who is to answer this question must first learn pre-Brunfelsian and mediæval botany.[63] The prevalence of that opinion is long since become universal, despite its having been ably disputed two centuries after Brunfels. It will be important to the history of nomenclature that one trace its progress from Brunfels forward.

9. In respect to the nomenclature of species it should be observed that what is often spoken of now as the phrase name, or more unadvisedly the "polynomial," and commonly attributed to all botanists preceding Linnæus, is a thing unknown to Brunfels. In genera of several species I have not found him using in a single instance any name that is more than binary. Where there are three words to a name the first two are the generic name.

6. Leonhard Fuchs 1501-1566

An early and a clear vindication to Brunfels of the honor of having made an epoch in the history of botanical iconography is the fact that his *Herbarum Vivæ Icones* inspired a younger countryman of his to embark at once in a still larger enterprise of the same kind, this with the manifest purpose of outdoing the originator of the movement. Brunfels and Fuchs were alike in that they were college-bred Germans and university graduates; their early and also their later academic and professional training had been acquired in Catholic schools, and mostly while they themselves were yet Catholics; and they were witnesses of the beginnings of the Lutheran movement, and both became partisans of the Augustinian, Brunfels with voice and pen actively and zealously furthering the movement, and Fuchs so expressly in sympathy with it as to have forfeited thereby the professorial chair with which his alma mater had early honored him. Both were regularly graduated medical practitioners, and both eminently successful, even famous, as physicians; though this good fortune came to the elder of the two only very late in life, and after he had abandoned theology and polemics.

Life

Fuchs was a native of Memmingen,[1] Bavaria. The father died when the child was in his fifth or sixth year. The small boy must have been precociously intelligent. The care and cost which the widowed mother bestowed on his education,[2] and the academic honors conferred on him in boyhood, youth, and very early manhood attest this. At the age of ten he was sent away to a noted school at Heilbronn, at eleven to Erfurth where, after a year and a half of very special and zealous devotion to what were already his favorite studies, the ancient languages, the university conferred on him the degree of Bachelor of Philosophy; this before he had completed the thirteenth year of his age. During a year and a half, and that while he was somewhere between the ages of fourteen and eighteen years, he was master of a private school in his native town, to the success of which undertaking on the

FIG. 52. Leonhard Fuchs (1501–66) at age 41, holding the medicinal plant speedwell (*Veronica* L.). Fuchs, *Stirpium* (1542), p. ii. (Print at Hunt Institute)

part of one who was a mere boy in years, a precociously large stature, quiet seriousness of mind, and a manly dignity of bearing are said to have contributed.[3] At the age of eighteen he again left home, and this time to enter the university at Ingolstadt, where at first he applied himself to advanced classical studies, and two years later obtained the degree of Master of Philosophy. Entering at once upon the study of medicine at the same institution he won the doctorate in the year 1524, at the age of twenty-three. He undertook to establish himself in the practice of medicine at Munich; but after a residence there of some two years, within which time he had married, he was called to the Professorship of Medicine at Ingolstadt. Here another and more honorable place was soon tendered him, and he became physician to the Margrave George of Brandenburg. During an outbreak, at Anspach, the residence of this prince, of that very fatal epidemic which one reads of as the plague,[4] Fuchs acquired reputation by the success that attended his treatment of the disease. He remained physician to George of Brandenburg some five years, and it was during this period that his career as an author began. He published a Compend of Medicine,[5] then a translation from the Greek of one of the books of Hippocrates.[6] He was now called a second time to the Chair of Medicine at Ingolstadt. The call was accepted; but again the stay was short. This university still remained one of the strongholds of the old faith. Doctor Fuchs let fall expressions of sympathy with Luther's movement. Within less than a year he withdrew, returned to Anspach, where the Margrave George gave him welcome, and reappointed him body physician. The next year witnessed another outbreak of the plague, and this time Doctor Fuchs with his wife and children fled the place.[7]

In the year 1535 he received a call to the Chair of Medicine in the then newly established Protestant university of Tübingen.[8] Here he remained to the end of his life, that is, for thirty-one years; and they appear to have been years of the most arduous and unremitting activity. His lectures on medicine were extraordinarily popular, and the intervals between lectures were occupied by the duties of the practitioner. He declined one offer of a professorship in the celebrated University of Pisa, and another to the office of Physician to the King of Denmark.[9]

As to when, and amid what surroundings Fuchs became interested in botany, I have met with no record. It may, however, safely be assumed that no passion for nature study, or for plants in particular, was congenital with him.[10] There is internal evidence in his book that as a botanist he was not born but made. The curricula of the schools of medicine at that period offered the possibilities, at least, of the making of botanists.[11] The medicines in use were still chiefly plant products, either native or imported from Asia. The names of them were plant names. Each was the subject of a chapter in the standard books of the materia medica. Those books were all ancient, and had been written by the Greeks, Hippocrates, Theophrastus, Dioscorides, Galenus.[12] Their chapters were the texts on which university professors of medicine lectured to their students;[13] and the identification of the plants more or less succinctly described in those classic chapters was a part of the regular work. It was the examination of plants in the light of what purported to be the original and authentic descriptions of them. Critical work of this kind may be done by a student as a piece of drudgery, or it may become an inspiration. To those not too sluggish it must have been stimulating to be able to demonstrate by Greek texts ten or fifteen centuries old that the vendors of drugs were selling important medicines under wrong names, that what they sold under the Greek name *aristolochia*, for example, was not in the least like what the great Greek physicians had used under that name either morphologically or qualitatively. And if such questions took them to the drug gardens, or led them afield into wild places in search of medicinal plants in their fresh and growing condition, all this would tend to the fuller development and the deepening of a sincere interest in botany.

There is every reason for believing that Fuchs's interest in botany was thus awakened. His earliest botanical publication fully substantiates this view. It was issued by Brunfels as an appendix to the second volume of his *Icones* in 1531, that is, ten or eleven years earlier than the appearing of his principal botanical work. Its title translated is "Leonard Fuchs' Notes on certain Herbs and Simples not yet rightly understood by the Physicians."[14] It consists of thirty-four long chapters upon more than as many plants and plant products then in use; dealing mainly with the right application of ancient names; often quoting the language of authors whom he charges with having brought in this confusion under which the pharmacy of his time is laboring and denouncing the errors of such authors with scathing sarcasms. The aim of the essay is the elimination of gross errors from the pharmacopeia, and all the subjects are plants that have been in use for ages. Nothing new is added; neither is there any trace of the philosophic investigation of the plant world as such; or the revelation of any interest in plant life and form in themselves considered. But this is Fuchs's juvenile botanical production. Will there be awakened within him later an interest in plants as plants rather than as drugs?

There is no evidence that such an awakening ever came, or that any considerable part of his work with plants had other than utilitarian ends primarily in view. In the last chapter of this earliest piece of his botanical writing he expresses the design of going through the whole of Greek and Latin botany, correcting errors and giving the right identification of everything, after the method exemplified in the treatise he is now concluding, even adding that he has been urged to do this by those fully convinced of the great need of such a work. But this promise remained unfulfilled. The two-fold duties incident to a professor's chair and an extensive medical practice claimed his energies, and the two-fold emoluments enabled him to undertake a line of botanical work—botanical recreation, rather—which it is improbable he ever would have thought of but for the great success which had promptly attended the publication of Brunfels's *Icones*. He employed two painters, and also the best engraver in Strasbourg,[15] and set them to work figuring plants. Thus within seven or eight years after the appearing of Brunfels's work Fuchs had ready for the press his great volume of the *Historia Stirpium*, though it was not issued from the press until some four years later, that is, in 1542.[16] Its success seems to have been speedily assured, and was

FIG. 53. Heinrich Füllmaurer and Albert Meyer, the draftsmen for Fuchs, *Stirpium* (1542), p. 897. (Print at Hunt Institute)

really wonderful. To a generation that had been accustomed to such books as the *Hortus Sanitatis*, filled with the most wretched caricatures of plants in place of true representations of them, this great book by Fuchs must have appeared as nothing less than luxurious; and the epoch which, ten years earlier, Brunfels had introduced by his 135 good illustrations of as many plants, was strongly accentuated by the appearing of this new volume with upwards of 500 large plates more than equaling, on the average, those of Brunfels.[17] Within a year, or a little more, there was issued an edition in German, this augmented by six more plates. Then in 1545 there came out, as if in condescension to the class of the unlettered, an inexpensive edition of the plates only, and this was so successful that a second issue was made in that same year; both these were in octavo. After that there were not a few small-sized and cheap editions brought out, with Latin text, but with figures so greatly reduced in size as to be of little use. With these, however, Fuchs had nothing to do.[18]

During the long tenure of the professorship at Tübingen, which covered nearly half his lifetime,[19] there was no return to that critical work upon the history of medicinal plants with which he had inaugurated his career as botanical author; nor were there any more than casual questionings of nature even as to the affinities of plants. But the botanical artists were kept at work.

To have more plants figured, and to formulate a page of text to accompany each plate, gave him pleasant respite from professional work, and promised greater fame and fortune. Before his death he had ready for the press the plates and descriptive texts of fifteen hundred plants, a work which, if it had been printed, would have made three folio volumes as large as the *Historia Stirpium.* But when all was done, no publisher could be found who would undertake the issuing of so vast a work without the advance of a considerable sum of money. This the author would not—perhaps could not—accede to, and the manuscript remained unpublished.[20]

In the original Latin edition of 1542, the Introductory Epistle, addressed to the Margrave of Brandenburg, is a document deserving fuller notice than can here be given it. It is a rather lengthy discourse, but withal instructive as to the condition of botany at the time, and exceedingly well written, amounting to something like an abstract of the history of medical botany from the earliest times down to his own date. As a piece of writing it reveals in its author general abilities altogether superior to what I cannot but consider the mediocrity of his gifts as a botanist. It is in these introductory pages that he earns for himself the praise of being a fair and equitable judge and critic of the work of others, of whatever race, religion, or nationality.[21] At a time when it was usual in Germany to depreciate, if not to denounce, all French and Italian efforts to restore botany, Fuchs proclaims that they are all inexpressibly indebted to such great scholars as Ermolao Barbaro, Marcello Virgilio, Ruel, and others who, through corrected texts and Latin versions of the Greek fathers, first placed them at the service of the botanists of every country. He takes pains to defend them one and all against aspersions that have been cast upon their works by men incompetent to criticize them, and recommends the study of them to all.[22]

When, in their turn, his own countrymen and contemporaries, Brunfels, Cordus, Tragus, come up for mention, it is always most respectful and even honorable mention; and this despite the fact that among them there is a rival or two whom he fears, and has good reason to fear, still is he solicitous to be just to each, and to speak out the favorable things concerning their work which may be said.

A representation which he makes in this preface, of the low estate into which the pharmacy of his time had fallen, I must in substance reproduce.

> The times once were when not only great philosophers and poets but kings and princes both investigated plants, and favored others so occupied. But in our day even the physicians are so much averse to that kind of study that you will hardly find one among a hundred of them who has correct knowledge of even a very few kinds. They appear to think that this kind of information does not belong to their profession, and to judge that it would be condescending from their proper dignity to entertain doubts about the accuracy and trustworthiness of those who buy and sell such things. And so it comes to pass that the druggists—God knows that they themselves

FIG. 54. Veit Rudolf Speckle, the wood engraver for Fuchs, *Stirpium* (1542), p. 897. (Print at Hunt Institute)

are for the most part an illiterate set—leave all this to the foolish and superstitious old women who gather herbs and roots. Error is therefore heaped on error, and will be so long as the identification of vegetable medicines is left to rustic and vulgar ignorance.[23]

The superb South American genus FUCHSIA was dedicated to this, the second father of plant iconography, by Plumier in the year 1703.[24]

Vegetative Organography

Fuchs has a very instructive and useful introductory chapter which he styles "An Explanation of Difficult Terms." From the historian's point of view this will be regarded as most valuable. It is the earliest vocabulary of botanic terms that I have met with thus far; and no historian that I know of has made mention of it.[25] One gathers from this vocabulary good information of progress gained—and also of retrogressions made—in descriptive and organologic botany in the seventeen centuries and more between Theophrastus and Fuchs. True to his title, our author omits all easy and familiar terms, does not define anew the root, the stem, the leaf, the flower. By this we know that he has nothing to add to the ancient and classic diagnoses of these organs. He does, however, define a bulb. "Bulbs are roots that are

Colchicum cum floribus.
Zeitlosen mit den blůmen.

Colchici folia & semen.
Zeitlosen mit blettern vnd somen.

FIG. 55. Autumn crocus (*Colchicum autumnale* L.). Fuchs, *Vivæ imagines* (1549), pp. 201–2. Rather than being reproduced from the first edition at greatly reduced size, all these illustrations are taken from the small-sized Basel edition; even smaller illustrations were published at Lyon, 1549.

round and tunicated; such are those of the hyacinth, asphodel and colchicum." There is here a retrogression from Theophrastus, who doubted that the tunicated mass ought to be considered a root, and who also mentioned that the tuft of fibres descending into the ground from underneath are undoubtedly roots.[26] Neither of these considerations affects the mind of stolid Fuchs. Bulbs are roots that are rounded and tunicated. I do not recall having met with an earlier use of the word tunicated as describing certain bulbs. It is very apt, and has now long been everywhere in use as definitive of one kind of bulbs. With him, however, the employment of it is unfortunate; for it makes the tunicated structure to be characteristic of all bulbs, which is a bad mistake, as excluding the scaly kind, like that of lilies, from the category of bulbs; for not the crudest morphologist could call a scale a tunic. And

Fuchs proves his definition fallacious by stating, when he comes to the figuring and describing of the true lilies, that they have bulbs.[27] His referring to the asphodel as an example of a bulbous plant will be misunderstood. He has not at all in mind that plant which in later times has been identified as the famed asphodel of antiquity, the underground parts of which have nothing that is in the nature of a bulb of any kind. That which Fuchs believes to be the asphodel, and figures for it, is a lily, and its scaly bulb is well shown.[28] If his third familiar example, colchicum,[29] illustrates to us what we distinguish now as a corm, it is at least fibrous-coated on the outside, and would therefore answer at least to the letter of his diagnosis of a bulb.

There is one term in use in the sixteenth century in connection with certain bulbous plants which has not survived; that is, the neck (cervix). Fuchs defines the *cervix* as "an elongated and cylindric body intervening between the summit of a bulb and the tuft of leaves, and has the appearance of a neck." From its position, and its external appearance as cylindric and supporting leaves in onions, leeks, daffodils, and their kindred, one might have expected to find it designated as a stem. That it was not, is a circumstance that must convince us of two things: first, that Theophrastus's immortal definition of a stem as made up of bark, wood, and pith[30] was a part of the very alphabet of botany in Fuchs's time as it is in ours; and second, that the stem-like cylinder surmounting onion, leek, and daffodil bulbs must have been examined in cross section and found completely destitute of every characteristic of stems in general, discovered to be made up of nothing else but the compacted bases of the leaves themselves. It looked much like a stem. Investigation proved it wanting every claim to that title; and they named it cervix, the "neck" of a growing bulb. The word was not destined to a permanent place in the vocabulary of the science. When at length it came to be seen that the bulb itself, as well as the cervix, was also but a mass of leaf bases, and therefore no root at all, the term lost its particular significancy and disappeared. And this very fact of the invention of a new word that seemed to be called for, and its later passing into desuetude, is an interesting kind of episode in the history of morphology, and well merits notice in a place like this. We shall meet with other instances.

Respecting that more marked phase of underground stem, the rhizome, or rootstock, Fuchs appears quite securely to rest in a position which Theophrastus had held with wavering. Fuchs denominates them all *radices geniculatæ*. The Greek had realized that they have rather too much in common with stems.

In his treatment of stems in general, one observes in Fuchs some divergencies from, even here and there some little advancement beyond, the status of these things in the minds of the ancient authors. The word *culmus*, modified from the Greek *calamos*, is his term for the stems of grass-like plants. The first and largest divisions of tree trunks are denominated *brachia*, arms, though not unless such diverge from one another rather strongly, suggesting, as he says, arms of the human body when extended. Others had always noted what they called the knots, or nodes, or stems. Fuchs uses, and even defines, the good term *internodium*, internode; though I much doubt his having invented it.[31] He also observes in trees and shrubs the occasional development of long and vigorous shoots from trunks and main stems, points at which branching is unusual if not abnormal. He names such shoots *adnates*. Botany still recognizes this class, and knows them as adventitious shoots, from adventitiously formed buds.

One reads in this author and in others of his time of such things as the *alæ* of stems. The usual meaning of alæ is wings, as of a bird; but in ancient Latin the term also meant the armpits; and quite like this is the sixteenth-century use of it in botany.[32] Fuchs defines the alæ of stems as being a kind of sinuses from which new growth proceeds; they are the axils of modern terminology. It is here that I meet with *scape* as a botanic term;[33] and the application is just that now in use, designating an elongated peduncle arising from under the ground; though neither Fuchs nor his contemporaries so understood it. They regarded it as a true stem without nodes. But Fuchs's "scapus" was not at once adopted; until long after his time it was usually denominated a *stylus*.

Two of the several modes of leaf arrangement are named and defined in this vocabulary, the decussate[34] and the verticillate;[35] but there is yet no one word in use by which to distinguish leaves as opposite. A phrase is required to express that. Any leaf margin that is evenly indented is described as crenate,[36] or as serrate, quite indiscriminately, the terms being treated as synonymous; but if serratures be quite deep and close, as in the nettles, the leaf is fimbriate.[37] Pediculus and petiolus, *i.e.*, peduncle and petiole, are employed as indiscriminately, either one applying to leaf stalk or to flower stalk.[38] The word *stipula* also makes its appearance in Fuchs's vocabulary, but with nothing like its meaning in more recent botany. His definition proves it to have been in his mind merely a special name for the peculiar leaf of grass-like plants, not a part of such leaf, but the whole of it.[39] It is a definite proposal that, since the stems of grains and grasses have the special name of culm, the leaves of the same class of plants ought not to be called leaves, but should have their own special designation—should be called stipules; and this is perfectly logical and consistent, for the leaves that grow on culms are quite as unlike all other leaves as culms are unlike other sorts of stems. It will be recalled that Theophrastus had named this entire group the Calamophylli in allusion to the remarkable characteristics of the foliage.[40] But Fuchs does not seem to have met with success in this endeavor to have grain leaves and grass leaves become known by the name of stipules; and, more than two centuries later, Linnæus picked up the old term *stipula* and applied it anew, and with perfect success.

Fuchs tried also to invest the compound leaf with a name of its own, as a thing too different from the simple leaf. The distinction itself, as we know, was perfectly and for all time made by Theophrastus, who discovered things and left them nameless. The German father would have the compound leaf called a *frons*,[41] *i.e.*, frond, thus restricting the other Latin word *folium* to the simple leaf and the individual leaflet of the compound. But this also fell short of general adoption; and long after Fuchs, frond came into use as designating the peculiar foliage of ferns. It must be said of Fuchs's application of frond that it was the more correct; for frons with the old Latins[42] meant the leaves of trees, or even leafy twigs of trees, such as, anterior to Theophrastus, and by thousands in later times, the compound leaf was believed to be.

Inflorescence

The term inflorescence of course did not appear in botany until long after Fuchs; but the thing had been of necessity both observed and discussed. The using of words definitive of the various clusterings of fruits and flowers must be older than history. Perhaps few if any of those defined by the German father were newly coined, or even otherwise applied than they had been in far earlier times. But here in the *Historia Stirpium* we have a goodly number of them brought together, along with not indefinite statements of what that author understood to be their meanings. And what must vouch for the importance of this paragraph of history is the fact that not one of Fuchs's terms relating to inflorescence bears with him the meaning which the same term has in the botany of our own time.

Take the word *thyrsus*, which at its first origin in Greek and Latin[43] was but a synonym of *caulis*, any stalk or stem; though later, and still in ancient times, it acquired a special significance; while with botanists of our time it means a particular kind of inflorescence. There is with Fuchs no kind of a flower clustering that is called a thyrse; yet he essays to define the term as if in the botanical terminology of his time it had gained some new shade of meaning. From his definition itself nothing of the kind is apparent; but at definition Fuchs is not adept; and when he says a thyrsus is a straight wandlike or arrowy stalk he has hardly departed from the earliest of ancient definitions. But when we make search for his practical use of the term we find that it has with him a meaning which he had not indicated or intimated in his definition. In the description of the hyssop he uses the expression: "Flowers purplish-blue, investing the thyrses like a spike."[44] Here it is plain that the thyrsus is the axis of a spicate inflorescence, that which in much later botany is become the rachis. But it is only now and then that he notes the arrangement of flowers, though the clustering of fruits is much more frequently taken into account.

The term *racemus* occurs not infrequently; but I think only as specifying

FIG. 56. Hyssop (*Hyssopus officinalis* L.). *Vivæ imagines* (1549), p. 486.

the arrangement of some berry-like kinds of fruits. The type of the raceme is the grape cluster; but in his definition of *racemus* he says it is the fruit cluster not of the grape only, but also of the ivy and of any kind of herbs or shrubs that have bunches of berries. In practice we shall find him writing of the compactly spicate berries of the arum as forming a *racemus*.[45]

As with the ancients, so with Fuchs the spike is rather more a taxonomic than morphologic term. His definition of it is the most concise of all. "A spike is that which a culm bears at its summit. . . . It consists of three parts, grain, glume, and beard. A muticous spike is one that is beardless." Plainly, then, the spike with him is the peculiar inflorescence of grains and their natural allies. The typical form of a spike, that in which as in wheat and barley there is a simple rachis—thyrsus, he would have called it—up and down which the several parts are sessile, is not alone a spike. The one prerequisite of a spike is, that it shall crown the summit of, or at least be connected with, a culm. And so we find him naming as spikes the inflorescences of broom corn,[46] and also of maize.[47] In modern botany they are panicles. Only thus far, however, does he abide by his own diagnosis of the spike as the fruiting cluster of grass allies; beyond this point, and here and there, we find him overstepping the bounds which he himself has set to the application of that term. *Lavandula stœchas*, an aromatic shrub common in European gardens of that time, is of the family of the labiates, the flowers of which are congested within a somewhat elongated and cylindric involucre of chaffy and overlapping scales. This involucre vividly recalls the head or ear of some short-spiked kind of wheat; and Fuchs transgresses his definition boldly enough by calling the involucre of this inflorescence a spike.[48] In justice to the botanist it must, however, be admitted that he was following popular precedent. The inflorescences of not only this but a number of other labiates had long been called spikes. There is then traceable in Fuchs a tendency toward a point that was not actually gained until more than two centuries later, of defining terms by morphologic rather than any other characteristic, of naming a raceme from its structure rather than from the fact of its bearing berries and not capsules; and a spike not as the fructiferous terminal of a culm, but as an axis, bearing up and down its length sessile flowers or fruits, this irrespective of the family of plants in which it may occur.

The *panicula* is almost unique among Fuchsian inflorescences in that it is not made the mark of any particular taxonomic group, and is defined in quite strictly morphologic terms. As far as possible from the panicle of modern vocabularies, it consists of almost any very compact cluster that is somewhat elongated and, at the ends, well rounded. The first example given is that of the cones of spruces, which are not conical, and therefore would not have been called cones by botanists of antiquity, nor by their disciples of the sixteenth century. But it is said by our author that the Latins applied panicle mostly to such as, being of the requisite form and density, were also appendaged by some sort of coma;[49] and so, among the Fuchsian panicles, one finds the bristly hairy spikes of millet, and the elongated furry heads of the mouse-ear clover, otherwise called lagopus, *i.e.*, *Trifolium arvense*. Also individual spikes of scirpus, eleocharis, and others of their tribe—the *individual spikes*, I say, and not the whole inflorescences—are panicles.

The umbel, though defined morphologically as a flower and fruit cluster constructed upon the plan of an umbrella, would never be applied to an umbel of berries. The umbels with Fuchs are the inflorescences of the family of umbellifers, or at least of dry-fruited plants, exclusively. The flat-topped clusters of certain anthemideous composites like millefolium he speaks of as umbels; though they are not really umbels, but corymbs, a distinction that had not then been made.

Anthology

In the vocabulary of Fuchs there is vouchsafed a perfectly intelligible definition of what he calls the calyx. It is a kind of "bag within which at first the flower, and after that the seeds are enclosed." Note first of all, that such a calyx as this can be no part of a flower. It cannot be determined to be a calyx until it has shown itself permanent, until the seeds have ripened. A deciduous calyx would be a contradiction in terms. A circle of green sepals behind a flower, even though at first enclosing the "flower," does not constitute a calyx; at least if it fails to persist and to enfold the seeds after the other parts have fallen. Under this definition all mintworts and sages, the borrageworts and other synsepalous things, have a calyx, while the poppies and the buttercups and their allies have none. The pomegranate and all pomaceous fruits are furnished with that organ; the olive and all drupe-bearing trees are destitute of it. This appears to be the earlier idea of a calyx,[50] the first movement toward the bringing in of that green-leafy circle close behind what was

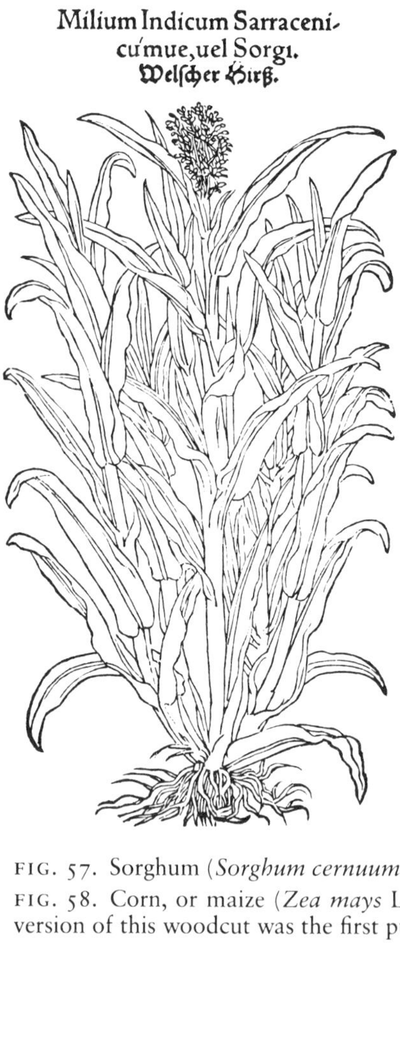

FIG. 57. Sorghum (*Sorghum cernuum* (Ard.) Host.). *Vivæ imagines* (1549), p. 447.
FIG. 58. Corn, or maize (*Zea mays* L.). *Vivæ imagines* (1549), p. 476. The 1542 version of this woodcut was the first published illustration of this species.

FIG. 59. Hare's-foot clover (*Trifolium arvense* L.). *Vivæ imagines* (1549), p. 281.

called the "flower," to where it should be recognized as a part of the flower. But his having technically defined the calyx does not preclude his occasional employment of the term in an untechnical sense, that is, as designating some other organ that happens to be shaped like a cup; thus we find him calling the lily flower a calyx,[51] *i.e.*, a chalice-shaped flower.

His doctrine of the flower in general is that of Theophrastus hardly improved upon. There are two kinds, the leafy and the capillary; but both are united in, for example, the rose. The term petal is still wanting in botany. Its introduction into the vocabulary will not be proposed until two generations after Fuchs. The foliar parts of a flower are still leaves only. Yet what is curiously interesting is, that already as the green leaf is seen to have usually that which they have called its petiolus, or pediculus, so the flower leaf is credited with having its unguis, or claw, the more or less narrowed basal part by which it is attached to its receptacle. Fuchs defines well enough this unguis,[52] even remarking that in the flower of a red rose this claw is white. And so the distinguishing of the two parts, blade and claw, in this organ historically antedates the naming of the whole organ as petal; and Fuchs, so far from affirming this to be a new distinction of his own making, freely attributes it to "the ancients."

In this Fuchsian vocabulary occurs what is perhaps the earliest botanical definition of stamens. There is so much of history in it that one must reproduce it as literally as may be. "Stamens are those apexes that come forth from the middle of a flower-cup; and are so called because they rise up like filaments out of the inmost bosom of the flower."[53] As a definition this is illogical and bungling; for both the anthers and the filaments are separately called the stamens, not by any means the two parts that go to the making up of stamens; either one alone is stamen according to his absurd statement. In the first clause the anthers, *apices*, are the stamens; in the second, the filaments are the stamens. That what he denominates apices are the anthers is clear as day from his description of the flower of the common white lily, where he remarks that in this the apices are yellow; for no other parts of the lily blossom are yellow but the anthers.[54] The term *filamenta* by its very meaning applies to no other organs but the filaments and styles. Let us note here that there is yet no description of a *stamen*. The author neither thinks nor speaks of the thing but in the plural. What he has seen, and what all the botanical authors before him had seen, is a tuft of delicate things standing up from the midst of the circle of colored flower leaves. The ancients had written of them as *flocci*, or as *capillamenta*,[55] still only in the plural. No one had ever looked into the individuality of one of those flocci or capillamenta. To have done that would have been to lead the way to the discovery that the members of the filamentose tuft are not all alike, that at least the one central member—if not the whole central membership of the tuft—is different from those that stand between them and the circumference of the flower. There would soon have been two kinds of stamens to describe if one stamen had been defined, because of those—the styles and their stigmas—which would not have answered to the definition.

Now Fuchs in practice writes of these tufted things in the middle of a flower as the apices. This is a distinct departure from the terminology of antiquity, and is withal a departure in the right direction; for the ancients had seen and written of flocci, capillamenta, and the German had seen—perhaps by some unknown mediæval botanist had been taught to see—the little knots that surmount the outer set of the flocci, and from these little apical knots the whole stamen-tuft had been named anew, "apices." This term, whensoever it made its appearance, came in like a kind of prophecy that the terminal knots were one day to be received as the only essential parts of the tuft. One would willingly concede to Fuchs the invention of this term which shows that anthers are being noticed; but he was in no sense a botanical discoverer, and he availed himself of many an old book and manuscript of which we have no knowledge.[56]

For the staminate tassels of hazel, walnut, and oak trees he has also now a name; whereas the ancients seem to have had none. But he has no more idea what these tassels are for than the ancients had; though he ventures the guess that they are instead of flowers, thereby proving that he had never seen the pistillate or real flowers of any of them. He calls the pendents *nucamenta*, nut-tassels, and describes them in language borrowed from Theophrastus who, as we have been learning, had a much better knowledge of nut-tree flowers than Fuchs ever attained to.[57]

Fruit and Seed

The Greek philosopher's comprehensive and classic definition of a fruit[58] is either unknown to Fuchs, or else he purposely condescends to the popular notion; for he says that a fruit is something made up of flesh and seeds. This is quite in keeping with his definition of a raceme as being a cluster of fruits, *i.e.*, fleshy fruits, rather than a particular mode of clustering in both flowers and fruits. Seed coverings other than fleshy are disposed of by him under two or three distinctive terms. There is the *silique*, the *vasculum*, and the *caput*, as well as its diminutive, the *capitulum*.[59] The legumes he says are siliques, and yet many other herbaceous plants and even shrubs bear siliques; and we infer his silique to be any kind of an elongated and two-valved pod. He ventures upon no definition of what he calls vascula, beyond this, that they are the coverings of seeds. His account of head and capitulum is, that they apply to any well-rounded and solidified part of a plant, whether basal or terminal. An onion bulb, resting wholly above ground and therefore not a root, is a head, a *caput*; and so is the head of a cabbage, and also the round indurated capsule of a poppy. If he knew anything of Theophrastus's technical definition of a fruit as composed of pericarp and seed, he does not appear to me to have made use of it, or of the term pericarp.

The seed is not defined in our author's vocabulary, or even mentioned there otherwise than incidentally.

Phytography

There has already been cited this opinion of phytography from Brunfels,[60] that the best way to reform and improve upon mediæval plant description would be to restore word for word the descriptions of the ancients. At a later time it was thought—and the thought was carried into action—that the only possibility of improvement in phytography lay in wholly disregarding the classic texts, and writing all plant diagnoses anew. Such a proposal as this last would have filled either Brunfels or Fuchs with amazement; and also not unreasonably, for neither of them had in view the reformation of botany in general. Both were aiming at the correction and improvement of that which we of today would speak of as the botanical part of the pharmacopeia. They were interested in phytography, because it is one of the necessities of medical botany. The most trying part of their work was that of the identification of ancient remedies; and their only clues to the identity of any of them were in the ancient descriptions. We have followed Brunfels in his giving, for some plant newly figured, page after page of different descriptions literally quoted from those whom he regards as standard authors, not willing to divert the attention of the student by a single line of his own, willing that they who study his book shall judge for themselves whether what he has figured under a given Theophrastian, or Dioscoridean, or Plinian name has been rightly identified.

The ideal of Brunfels is high. His book is for scholars; but Fuchs plans to be more popular. He will publish twice as many plates as the former had done, and for the sake of economy must reduce the number of pages of printed matter. The average is not much more than a half-page to a plate; yet the descriptive part of Fuchs's volume is not in all respects insignificant when compared with that of Brunfels. There is a falling off in the bulk of interesting and useful matter; but some good phytographic distinctions are made which do not appear in Brunfels.

Every chapter of Fuchs is divided into separate and separately named paragraphs, of which always the first, headed *Nomina*, is devoted to the name and synonymy of the genus, and contains nothing else. In the case of every monotypic genus the second paragraph is headed *Forma*. In this we read always the morphologic marks of this type, at least such of them as Fuchs can copy out of standard authors. When, however, a generic type is known to be made up of two or more species or varieties, then this morphologic paragraph is not second in order but third, the second being now given to the naming and defining of the species or varieties; and this occasional second paragraph is always under the caption *Genera*. In an earlier chapter an explanation has been given of the primitive use of the term genus, and its plural genera, as meaning nothing more nor less than the kind or kinds of a thing.[61] It is plain that Fuchs's "genera" are the species and varieties, while under *Forma* he gives the morphology of the genus as a whole. The placing of the descriptions of species and varieties first, and that of the genus next below is illogical in the extreme; but there are still other intimations that a logical mind was not among the learned Fuchs's natural endowments.[62] But this segregating of the morphology, the ecology, and the properties of a plant, and the relegating of each to its own paragraph is definitely an improvement in phytography, and is perhaps an invention of his own.

Taxonomy

We are learning that there was in Fuchs nothing of the plant anatomist or physiologist, something of the organologist, but that he was in no wise given to philosophizings about plant life and form in general, that he was a medical botanist, dealing with plants from the utilitarian point of view. He would not have appeared as a taxonomist had not taxonomy been inevitable wherever more than one individual plant is treated of.

Concerning the larger and more comprehensive groups, the *genera majora* of the botanists of a somewhat later period, the orders or families of today, there is little to be read in Fuchs; for his sequence of genera is alphabetic, a kind of arrangement which precludes the grouping of genera into families. It does not, however, in any way influence the definiteness of the genera themselves, or of the species composing them. The work is divided into 343 chapters, and each chapter is devoted to one genus and no more, so that the number of genera treated of is the same as the number of the chapters. Many of the genera are represented by only a single species, many by two or three, several by four or five, and one has seven species, all described and figured. But one must not pass to the study of his genera without noting certain lapses from the alphabetic arrangement which are made in deference to what are held to be the affinities of a genus. These are not numerous, but they are enough to show Fuchs as susceptible of being influenced by the idea, even in his time an old one, that some genera are interrelated. Here it must also be observed that the alphabetic order he follows is that of the Greek rather than Latin names of genera. For medical botany, all through the ages and down to Fuchs's time, Greek rather than Latin was the preferred language of nomenclature, at least with those best educated, because all three of the most venerated authorities upon that part of botany, Hippocrates, Dioscorides, and Galen, had been Greeks and had written their treatises in Greek. Now while in his Latin text Fuchs uses the Latin names of things, quite as he ought, the generic names heading the chapters are the Greek generic names, and the sequence is that of Greek letters of Greek names; so that the chapter on the genus of the elder trees or bushes is headed not *Sambucus* but *Acte*, and that of the plantains not *Plantago* but *Arnoglosson*.[63]

Now for a sequence of genera in a book of sixteenth-century botany, the choice of the Greek alphabetic order left its author certain liberties. All Latin names are naturally exempt from such a rule; and there were now in Fuchs's time not a few plants holding places in the pharmacopeia which had not been known to the Greeks of old, and which therefore had only their Latin and their vernacular appellations. Since the Latin names of these may not consistently head chapters where the headings are professedly Greek, Fuchs is apt to use these as occasions for giving expressions of opinion about natural affinity by placing them in the line of what he conceives to be their real kindred. For one instance, take his placing of *Datura Metel*, a plant then newly introduced into Europe, where it was known only by the Italian name of *stramonia*. Since it has no Greek name he locates it where he will; and we who have been taught that there was no natural classifying of plants until well toward the end of the eighteenth century may well be surprised that Fuchs places this at the end of a line which begins with *Solanum nigrum*, such perfectly solanaceous types as *Solanum melongena*, *Physalis somnifera*, and *Atropa Belladonna* intervening.[64]

That very early in the sixteenth century there was already in exercise the taxonomic skill to put together as under one comprehensive natural genus such diverse-looking consanguinities as the five plants here named is something that merits more than passing notice. It is another of those forceful intimations that much of the botanical history we once learned must be unlearned. Let us remember that we are now at a point more than one hundred and fifty years anterior to those great lights, Morison, Ray, and Tournefort, and some two hundred anterior to Haller, Jussieu, and Adanson, to which latter trio is usually accorded all the glory of having first outlined such natural groups as this. But Fuchs himself intimates that from very ancient times there had been a somewhat familiar knowledge of four out of these five plants, and says that both Dioscorides and Galen had held them to be kinds of *στρύχνον*, *i.e.*, *Solanum*.[65] But the referring of so marked a new type of plant as *Datura* to the anciently recognized group of the *Solanum* allies by men of the sixteenth century will hardly fail to suggest to some that there must have been, after all, some appeal to anthology. Certainly the conventional nineteenth-century botanist finds no stronger links uniting these two plants to one family than the symmetrical pentamery of the flower, coinciding with a plicate prefloration and superior ovary. It is nonetheless well beyond question that not a single point anthological had influence in determining the affinities of *Datura*. The corolla has not yet at this period obtained its name. It is still the "flower" simply. Neither stamen nor style had been taken note of as an individual organ, much less had the two been distinguished, or the members of either set been counted. Floral symmetry was yet unnoticed; and, marked as is the æstivation, or prefloration of *Solanum* and all its allies, the very topic of prefloration was not yet heard of in Fuchs's time, nor did it begin to figure in taxonomy until some two centuries after his demise. According to Fuchs, his best reason for locating this type in the line of the nightshades is the fact that its herbage exhales the narcotic odor characteristic of all other nightshade allies. Plant classification, then, in Fuchs's time has not yet emerged from the period of antiquity; and botanists are still considering agreement among plants as to their vegetative and qualitative characteristics, giving little heed to those of flowers or fruit. And a fair measure of success not rarely attended taxonomic effort guided by these criteria. But the success, however marked in the case now in mind,[66] was not quite complete. *Capsicum* is undoubtedly a solanaceous type, but this fact Fuchs failed to apprehend; yet the failure is not unaccountable, notwithstanding that, viewed morphologically, *Capsicum* is much more plainly

Nux Methel.
Rauch öpffelkraut.

Capſicon rubeum & nigrum.
Roter vnd brauner Calecutiſcher Pfeffer.

FIG. 60. Hindu datura (*Datura metel* L.). *Vivæ imagines* (1549), p. 399.
FIG. 61. Pepper (*Capsicum annuum* L.). *Vivæ imagines* (1549), p. 425.

akin to nightshade than is *Datura*. The plant was a new one in German gardens of the sixteenth century, as was also *Datura*. It had been as unknown to the ancient Greeks, and had no Greek name; therefore Fuchs could have placed it next to *Solanum*. The fact is, no one had yet seen its relationship. The plant lacks the narcotic odor. Every part from leaf to seed is of a peppery odor; and this quality, amounting to a burning pungency of taste, joined to the peculiar medicinal qualities, blinded every one to that affinity for *Solanum* which every one now sees in *Capsicum*.

It must not, however, be inferred that this principle of qualitative agreement is held an inviolable rule in this early taxonomy. With botanists of that time, quite as with those of every later generation, the endeavors to form groups are dominated sometimes by one principle, sometimes by another. It is easy to bring forth out of Fuchs instances of putting things together as of the same genus regardless of odors and flavors and of almost all other marks save those of roots and leaves and general mode of growth. Under the genus *Verbena* he figures two species. There is the most deceptive likeness between the two as to roots, stems, leaves, and a slenderly spicate inflorescence of small flowers; but one of these verbenas is *Verbena supina*, the other is *Sisymbrium Læselii*, a true verbena and a crucifer made congeneric, in the Fuchsian binary nomenclature *Verbena recta* and *Verbena supina*.[67] It will not be easy for the systematist of today to imagine so natural an alliance as that of the crucifers remaining unrecognized by those who had already recognized the Solanaceæ; yet as to roots, foliage, and habital characteristics the crucifers differ among themselves a hundred-fold more widely than do the Solanaceæ. Despite what we of later times perceive to be the naturalness

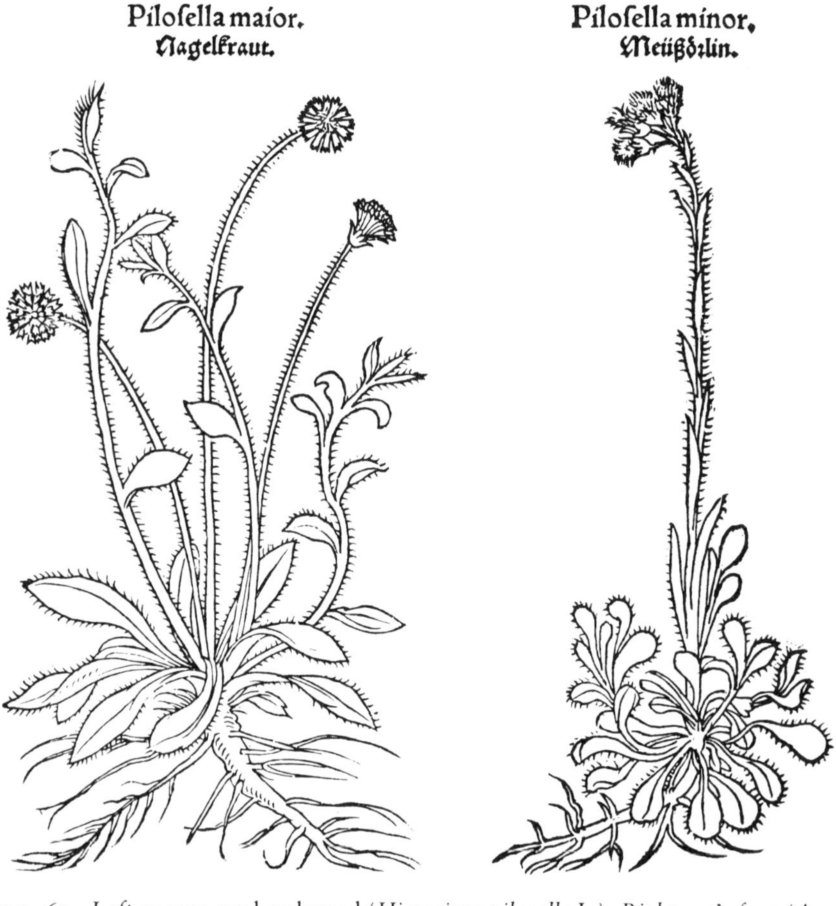

FIG. 62. *Left*, mouse-ear hawkweed (*Hieracium pilosella* L.). *Right*, cat's-foot (*Antennaria dioica* (L.) Gaertn.). *Vivæ imagines* (1549), pp. 346–47.

of the cruciferous group, its recognition had to await the development of anthology; and I have met with no evidence that up to the time when the career of Fuchs was ended there had yet lived a botanist who had known the floral structure of a crucifer. Even the Theophrastian elements of anthology seem to have been sunken in oblivion ages before the birth of Sprengel's and Meyer's and Sachs's "German Fathers of Botany," one or two of whom were to renew the investigation of floral structures; though Fuchs was hardly one of these.

The contemplation of these quaint herbalistic genera based on vegetative characters and ignoring flowers and fruits is both entertaining and instructive. It is as if one had discovered in these antiquated tomes a fossilized and now extinct system of plant classifying; and the reader will not fail to be interested in glancing at the outlines of other such genera. Fuchs has one little genus for which he brings forward into print, from out of an unpublished manuscript, the name *Pilosella*.[68] Here is the composition of the genus:

Fuchsian	Recent
Pilosella major	*Hieracium Pilosella*
P. minor	*Antennaria dioica*[69]

He has tried to make one of the plants answer to the *Myosotis* of Dioscorides. It does not well agree, and he is confident that neither of the two was known to the ancients. Both are well known in Germany, and of repute as vulneraries, on which account he must not omit them. The German herbalists of his time know one of them by the name *Pilosella*. That will suffice for a formal generic name, and with it he has already headed his Chap. CCXXX.

Two kinds of it are found, differing in nothing but the flowers. One of them has leaves that are larger, and do not lie flat upon the ground. Its flowers are yellow, and it is named Pilosella major. The other has smaller leaves that lie flat upon the ground, and purple flowers which disappear with a pappus. The Germans have for this the names Little Mousear and Rabbitsfoot. In a manuscript herbal I find these plants disposed of as Pilosella major and minor.

I thus present a literal version of what our author had to relate respecting the components of this his genus *Pilosella*, even to the interesting admission that the whole chapter had been borrowed from an old herbal in manuscript—illustrated by drawings, no doubt—which had been available to him. Even the specific names *P. major* and *P. minor* were from said manuscript. Contemplating these two plates, whether the proposed genus consisting of a hieracium and an antennaria seem rational or absurd will depend entirely upon whether one view the types with the eye of the modern botanist, or with that of him of Fuchs's time. The modern, taught by the traditions of not more than five or six generations of his more recent botanical

FIG. 63. Wormwood (*Artemisia absinthium* L.). *Vivæ imagines* (1549), p. 2.

ancestry, looks at the flowers only, and in consequence realizes only that the two must be regarded as of widely separate genera; for one has the flower head of a hawkweed, the other a congested bunch of cudweed heads; and as to the structure of their individual flowers they represent something like two extremes. But it will be both illogical and unfair to test the consistency of the earlier classification by presuming to hold it amenable to modern taxonomic principles, though this seems to be about what the readers of old botany, and even the historians, have always hitherto been doing. The consistency of the Fuchsian *Pilosella* is readily seen if, blinding ourselves as he and his forbears were blind to small floral structures, we look at and compare those parts of the two plants which they looked upon as bearing marks of consanguinity. Both the antennaria and hieracium are small perennial herbs of one and the same mode of growth, and that mode rather exceptional. Each has its leaves mostly in a basal tuft, and bears its flowers at the summit of scapiform stems. A number of depressed stolons leafy with a smaller foliage radiate from the base of the stem of each, so that both in the same fashion propagate vegetatively and form colonies. Add to these and other points of morphologic agreement the consideration that both were received as possessing the same remedial virtues, and we have a rational genus according to all the leading principles of sixteenth-century taxonomy.

Thus comprehending the situation—realizing that these groups that look so strange and motley have not been formed at random, but rather under guidance of definite principles—every such group acquires a new and even a lively interest. Let us open the book at its initial chapter. The name of the first genus is *Absinthium*. Three species are described, two of them figured well. They are:

Fuchsian	Recent
Absinthium vulgare	*Artemisia Absinthium*
A. Seriphium	*Sisymbrium Læselii*[70]

Thus a cruciferous plant, common in German vineyards and hedgerows, is made congeneric with wormwood because it has similar foliage, and also is thought to answer to *Seriphium* in several other particulars. It is a good example of futile effort to identify with the *Seriphium* of Dioscorides a very different plant of central Europe which Dioscorides never saw, and which, by the way, if he had seen it, he would never have thought of as a kind of wormwood; for, as we have seen, the Greek did not quite disregard floral structures, but could distinguish genera by anthologic marks.[71]

But Fuchs's disregard of flowers in these generic groupings is manifest again close by *Absinthium*, in his *Anthemis*. The name of the genus and those of all three species are taken up from Dioscorides, as he says, and the following is his identification of them, the names at the right being those now in use.

Fuchsian	Recent
Anthemis leucanthemon	*Matricaria chamomilla*
A. chrysanthemon	*Anthemis tinctoria*
A. eranthemon	*Delphinium Consolida*

Judged by the modern and improved standards, the locating of a larkspur as congeneric with chamomile is the worst of conceivable taxonomy; and it is impossible that in this Fuchs interpreted Dioscorides otherwise than erroneously and even almost stupidly. The Greek had habitually looked at the flowers of things, and had shown clearly a tendency to regard the forms of flowers as taxonomically significant. Concerning *Anthemis* in particular he declares that the flowers in the three species differ only as to the color of the little leaves that encircle the centers of the flowers, which central part he says is yellow in them all.[72] Now concerning the identity of the white-rayed and the yellow-rayed anthemis species of Dioscorides there never had been any doubt; therefore as to these two Fuchs was but reiterating the expression of a common opinion. The third member, however, that is, the purple-rayed anthemis, was problematic; for there is not known in those regions where the ancients botanized a purple-rayed composite having the foliage of an anthemis. But there need not be; for any ranunculus-flowered or anemone-flowered branching herb, if it had the foliage of anthemis, might have been relegated to that genus. It is only to the trained eye of the modern botanist that the rayed head of a composite and a ten-petalled flower of an anemone or a buttercup seem very unlike. By the ancients, and by everybody down to Fuchs's time and much later, they were not regarded as being different. The superficial likeness between the two is great. In either type there is a yellow center made up of little things compacted together, and this is encircled by rather many narrow leaves apt to be different in coloring from those of central tuft, or mat, or cone. The different constituency of those yellow centers in anthemis and anemone no one had yet perceived, even in Fuchs's time. With him, as with all botanical antiquity, the yellow middle part was made up of the "stamina," the "capillamenta," the "flocci," in anthemis as in anemone; and that the circle of colored leaves is a circle of ray-flowers in anthemis, and of petals or sepals in anemone and buttercup—that is a refinement anthological of which neither Dioscorides in his day, nor Fuchs fifteen centuries later, had ever dreamed. But in the grain fields of Greece and Italy there grew in abundance one anemoneous herb with perfectly anthemideous habit and foliage, and flower leaves dark-red—easily within the wide range of the purples of the ancients. Must not this have been the *Anthemis eranthemon* of Dioscorides? Its modern name is *Adonis æstivalis.*[73] Certainly this, rather than *Delphinium Consolida*, is the third anthemis of the Greek physician. And the fault of Fuchs is his utter disregard of those floral marks in respect to which the Greek had said that all three species of anthemis were at agreement; though as to mere foliage, and the annual root as well as more of growth, the larkspur-anthemis of Fuchs answers well enough to the other kinds; and this would have been the apology for Fuchs's erroneous determination of the plant, had he not virtually disclaimed it for himself by attributing it to some unknown earlier botanist whose anonymous manuscript has been at his disposal. In this manuscript he says, "there is an exquisite drawing of this plant, which is commonly called consolida regalis," and then he proceeds to quote, from this unknown author, the following: "Some call the herb Monachella or else Capuciaria, doubtless in allusion to the hood of the monks, which the flower recalls. Dioscorides calls it Eranthemon, and it is one of the kinds of Anthemis, having the foliage of the chamomile, though of a darker green; but the flower is rather like that of a violet."[74]

Evidently the author of that unpublished commentary had been too superficial in his knowledge of the wild plants of Mediterranean fields to well interpret Dioscorides. But his identification of the third anthemis appears to

FIG. 64. Foxglove (*Digitalis purpurea* L.). *Vivæ imagines* (1549), p. 515.

have suited the fancy of Fuchs and he approved it. His former rival, Brunfels, a dozen years earlier had maintained that the *Consolida regalis* was *sui generis*, and did not think it had been known to any of the botanists of antiquity.[75]

There seems to be evidence within the book of Fuchs itself that the work was long in preparation, and the middle and later chapters printed so much later than the earlier as to bear intimations of his mind's having changed somewhat in the direction of an appreciation of floral structure as having some value in the determining of plant relationships. While his contemporaries Brunfels and Tragus hesitated to distinguish generically between the real nettles and the labiate-flowered dead nettles, Fuchs separates them widely, and under the names of *Urtica* and *Lamium*;[76] this, however, not as an original proposition, but as adopted from Dioscorides and Pliny.

In his description of the genus *Pisum*, the garden pea, he says that its flower is shaped like a butterfly;[77] but I do not find him using the expression

FIG. 65. Grape vine (*Vitis vinifera* L.). *Vivæ imagines* (1549), p. 47.

in describing other plants of that family; and while this is the earliest mention of the papilionaceous corolla form that I have met with, I still think it improbable that it was original with Fuchs.

Among several new genera proposed by this author there is one, namely *Digitalis*, which he establishes almost upon the form of the corolla alone.[78] This genus of two species, which he names respectively *D. purpurea* and *D. lutea*, practically concludes this volume of more than 900 pages; and so a course that began in almost total disregard of anthological considerations ends in the admission that floral structure may upon occasion be of such high taxonomic import as to furnish the most essential character of a genus. Contrasted with the beginnings of the volume, this conclusion of it is taxonomically very significant, even prophetic. It forecasts the time a hundred and fifty years later when Tournefort, running to another extreme, would essay the systematization of all petaliferous plants, almost by the corolla alone.

NOMENCLATURE

All the unconventionality, simplicity, and brevity of a primitive and even a classic nomenclature marks the plant-naming of Fuchs. As already indicated he heads each chapter[79] with the Greek name of the genus, unless perchance the plant was unknown to the Greeks and has no Greek name; then the Latin name is used. For every one the Latin and the German synonyms are given, and that in the first paragraph of the chapter, under the caption "NOMINA." Still more carefully considering the convenience of pharmacists and the untutored collectors of simples, he causes to be printed on each plate the Latin name of the species at the lower left-hand corner, and at the right-hand corner the German. Quite as we expect to find it, we read on plates representing monotypic genera no name at all but the generic. Commonly such names consist of but one word. Such are *Adiantum*, *Althæa*, *Anethum*, *Asarum*, *Asparagus*, and scores of others as familiar and as ancient. But there is also no dearth of purely generic names for monotypic genera that are of two terms. There are *Acorus officinarum*, *Tagetes Indica*, *Plantago aquatica*, *Aster Atticus*, *Barba Capri*, *Vitis vinifera*, *Vitis alba*, *Vitis nigra*, *Viola purpurea* (= *Viola*), *Viola alba* (= *Matthiola*), and very many more like these. There is no ground for questioning that every one of these names is purely generic. There is no warrant for denominating so much as one of them "a pre-Linnæan binominal," that is, a name of which the first word is, in the accepted sense, generic, the second specific. Names binary, and of just this last-named quality, do also abound in Fuchs; but these now cited, and also a host of others like them, have nothing in them of the generico-specific meaning which all binaries in use today convey, and are understood to convey. Now, that Fuchs does not mean by *Plantago aquatica* any species of the genus *Plantago* is put beyond all doubt or cavil by the facts, first, that he takes it as the topic of a chapter apart from that in which *Plantago* proper is discussed;[80] and second, that the officinal, or more properly scientific (Greek), names heading the chapters are totally distinct, the one being *Arnoglosson*, the other *Alisma*. Concerning *Vitis vinifera*, now long in the status of a generico-specific binary, it is as readily demonstrable that with Fuchs it is purely generic. Assuming that in primitive times the one word *Vitis*—in Greek ἄμπελος—was, as it now again is become, the generic name for grape-bearing shrubs, there came in later such two-worded generic names as *Vitis alba*, *Vitis nigra*, *Vitis Idæa*, etc., each standing for a completely and widely different genus. It is easy to see that such multiplication of generic names beginning with *Vitis* rendered it needful that the original *Vitis* should receive also a second and modifying term in order to avoid confusion; and thus there came into existence, and of necessity, the generic name *Vitis vinifera* instead of the simple and primitive *Vitis*. Let us see what Fuchs's genera of which this term is a part of the name really are:

Fuchsian	Recent
Vitis vinifera	*Vitis*
Vitis alba	*Bryonia*
Vitis nigra	*Clematis*[81]

Specific names, with this German father, quite as with all his forbears that I know aught about from Theophrastus forward, are strictly binary. One simple word, almost always adjective, constitutes the specific term of such binary, and is the cognomen, as some called it. It happens, indeed, with many a such binary name that it is composed of three separate words; but that is always for the reason that the generic name is of two words. Not much less than a hundred of Linnæus's trivial names of plants are of three distinct words; but this is because he makes the specific part of such names to consist of two words, and never the generic part. This is precisely the difference between the Fuchsian and the Linnæan binary nomenclature; and there exists no other difference. We amend Linnæus by connecting by a

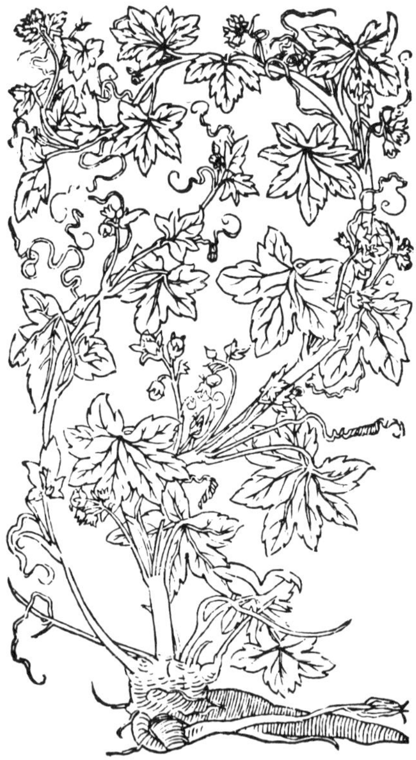

FIG. 66. White bryony (*Bryonia alba* L.). *Vivæ imagines* (1549), p. 52.

FIG. 67. Traveler's joy (*Clematis vitalba* L.). *Vivæ imagines* (1549), p. 53.

hyphen his two-worded specific names. This is done in order to preclude if possible any questioning of the fact that the two words which we have hyphenated are to be thought of as one. It were equally in place to hyphenate the terms of Fuchs's two-worded generic names. This done, he who can might read the truth that all Fuchsian plant names not those of monotypic genera are as strictly binary as those of Linnæus; with even this one difference in Fuchs's favor, that he has no two-worded specific names.

As to general principles of botanical nomenclature, those of Fuchs seem few, and easily ascertained. Those principles appear to be convenience, etymological suitability, brevity.

If all his names are binary, and, as being the mere names, hold places entirely apart and distinct from the descriptive paragraphs, as they always do, then there is not even the suggestion in Fuchs of those "phrase names," so called, which became a burden upon the phytography of two centuries later; and it may be said of his work that it is quite a model of brevity in nomenclature.[82]

Inasmuch as the Greek names of genera are older than the Latin, and he gives to them a kind of precedence, making them the headings of his chapters, there might arise a question whether the Greek names were preferred on account of their right of priority. But there is other evidence that the principle of priority in nomenclature obtained no recognition with him, was perhaps never once thought of. The standard works on the materia medica in use everywhere in Fuchs's day had been written in Greek, which fact alone would give the highest prestige to Greek plant names. Their Latin names, needful to be used for the convenience of the many who knew Latin but not Greek, were in every way as valid as the Greek names. Both were used, chiefly as a matter of the greater convenience to readers and students as a body. That he never thought about rights of priority as worth contending for comes out as clearly as possible in his presenting a new genus with the new generic name *Digitalis*. "We make use of this name, until we ourselves or some one else shall have invented a better one."[83]

The above remark attests its author's opinion that there could be appropriate names and inappropriate, and that names either bad, or even not very good, might well be suppressed in favor of new ones more suitable. Even the principle of convenience, which always favors the retention of an established name whether bad or good, may be overruled for the rejection of a name that is ill constructed, and the substitution of a new and better one. There is one generic name that had held good for some fifteen centuries, a Greek name too, which he declines to adopt as the heading of its chapter, evidently because etymologically distasteful to him. The name in Greek is *Ocimoides*, formed by the addition of *oides* to the generic name *Ocimum*. Instead of the ancient and established *Ocimoides* he writes for a heading to the chapter the new name *Ocimastrum*,[84] an initiative in the reform of generic nomenclature which Linnæus two centuries later was to carry forward with universal approval.

He who thinks that nomenclature, like the science itself, should be subject to advancement and improvement must be believed to have his reasons, though Fuchs does not appear to have declared his. One thing, however, we observe, and that is that all the names he used have their meanings. A genus is named in allusion to some morphologic or qualitative characteristic; or else in honor of some personage who had to do with botany; or rarely, from some foreign country whence the plant or tree had been introduced into Europe. But there is nothing of the meaningless and cabalistic in any of the 343 generic names that head as many Fuchsian chapters. The meaning is not in every instance plainly to be read in the name itself. That which it once had when newly coined has now and then become obscure, if not quite lost, through the lapse of ages. But there are not yet in botany anagrammatic names, nor any that had been framed by the putting together of two or three meaningless though perhaps euphonious syllables. The beginnings of such an epoch in nomenclature we shall not look for until after the time of Linnæus.

In specific nomenclature, however, we note in Fuchs a free use of cabalistic names. In genera of two species one is very apt to be named *mas*, the other *fæmina*, and since at that time nothing was known about sexuality in plants, such names had little meaning. But quite as frequently the first species is called *prima*, the other *altera*; and in the case of his fine plates of six species of *Geranium* they are named *G. primum*, *G. alterum*, *G. tertium*, *G. quartum*, *G. quintum*, and *G. sextum*.[85] Though without a trace of diagnostic significancy, and purely cabalistic, we shall find that this kind of specific adjective came into use very extensively in the works of botanists of a generation later than Fuchs.[86]

7. Hieronymus Tragus (Jerome Bock) 1498-1554

An original and even eccentric character, singularly gifted as a botanist, was Hieronymus Bock, whose names in literature are several; for his earliest publications were made in Latin, under the name Hieronymus Herbarius. Brunfels in his first volume publishes a number of paragraphs the manuscript of which had been furnished by his friend Bock, and these are all credited to him as Hieronymus Herbarius—Jerome Botanist, or Jerome Herbalist, as you like. So again, in the Appendix to Brunfels's second volume, there is a document of some length, entitled *Apodixis Germanica.*[1] This is in German, and is by Tragus, but still under the name Hieronymus Herbarius. In the numerous German issues of his principal work he figures as an author whose name is Hieronymus Bock. When the great botanical merit of this work had been intimated abroad, and a Latin edition of it was thereby called for, in this he appeared under the Græco-Latin name Hieronymus Tragus. By this name therefore is he known in the botanical world in general. The genus that was dedicated to him by Father Plumier therefore necessarily took the form of TRAGIA.[2]

LIFE

No adequate biography of this interesting character seems to be known.[3] We trust that we have to a certainty the year of his birth; we have the date of his marriage, the maiden name of the bride, the names of her parents, and even the number of guests that were in attendance; and all this out of Tragus's own diary.[4] Neither is there any disputing the date of his death, the place of his burial, or the name of the preacher who delivered the funeral panegyric. But all these are matters which, in the biography of a reputed scholar, a practicing physician, and the beneficiary of a lucrative parochial endowment, are of subordinate interest. What one most wishes to know are the names of the colleges and universities at which the man studied, the schools whence he had his degrees in medicine and in theology. In other cases there is ordinarily no doubt of the date of the conferring of a degree, or of the admission to sacred orders. Upon these several points, each of prime importance to even the briefest sketch of the career of a noted scholar, physician, and clergyman, Tragus's biographers from Melchior Adam away back in the year 1620, down to Ernst Meyer in 1857, are silent. Not one of them names a school of any grade, or of any profession, which the man was known to have attended; and if any should suspect Tragus of having been a medical practitioner without a diploma, and of having enjoyed as a Protestant layman the income of a large Catholic parochial endowment, there is not a line in the most authentic biographies, as I have read them, by which to allay either suspicion.

Tragus was born at Heidesbach, not very far from Heidelberg, in the year 1498. Concerning the estate of his parents as poor, or as in easy circumstances, nothing is known. As a youth he was found uncommonly well educated, whence Meyer inferred that the parents were well to do. That inference may as easily be wrong as right.[5] At the monasteries of the period there was free education for boys of intelligence and promise, if their parents were poor. Meyer, writing in the middle of the nineteenth century, should have bethought himself of this, that in the Germany of Jerome Bock's boyhood Luther and the reformation had not yet appeared, and monasteries were everywhere. The biographers all affirm the disappointment of the parents at the young man's final and very fixed determination not to become a monk; and the disappointment seems to imply that he had been placed under monastic influence, and on their part hopefully.

The considerable views which we obtain of Tragus's life here and there do not show him always happily circumstanced, but rather more commonly as acquainted with adversity. Physically he was a consumptive. Even out of the ten children born to him eight died young; and he survived also the mother of them. It is in the year 1523, when he is twenty-five years old, that we find him settled in Zweibrücken, under the appointment of the Palgrave Ludewig, as a school teacher, and also as in charge of gardens of this prince, which he is said to have enriched with many plants.[6] This favorable incumbency Tragus seems to have held for some nine years, and came to an end in 1532 by the death of the Palgrave Ludewig. Quite before this date we should have found Martin Luther a central figure before the German public, the ecclesiastical revolution in progress, and Jerome Bock a zealous and outspoken Protestant. Prince Ludewig's successor, Friedrich II., was either still firmly a Catholic, or else unwilling to declare himself a partisan of Luther; and Jerome Bock, botanist, lost his position and became penniless with a young family on his hands.

A plan inaugurated by some sympathizing friend for his relief is one that ought to have been regarded as somewhat hazardous, and indeed may have seemed so to them; but it was attempted. A beautiful country place called Hornbach was the site of a Catholic church dedicated to Saint Fabian and,

FIG. 68. Jerome Bock, or Hieronymus Tragus (1498–1554), drawn by David Kandel, who also drew many of the plant illustrations for *Stirpium* (1552), preceding p. 1. (Print at Hunt Institute)

long before Tragus's time, richly endowed by private munificence. The parish was vacant. Luther's influence had been felt there and the membership was divided between allegiance to Rome and sympathy with Luther. It was hoped, so says Melchior Adam—himself strongly a Lutheran partisan—that the "old superstition" had been banished to such a degree that the new incumbent would be able to complete the "reformation" of St. Fabian's. The event proved otherwise, and the incumbent was obliged to retire; not, however, very promptly, nor until after he had done much botanizing in the wild rich regions roundabout, such journeyings being made, as he informs us, disguised as a peasant.

On his enforced retirement from this anomalous and only quasi-official incumbency he was again in the straits of poverty extreme, until relief came in the form of a bidding to make his home in the castle of Count Philip of Nassau, who remembered our botanist as having once brought him safely through a dangerous illness. His sojourn with this friend seems to have continued through several years, though for precisely how long cannot be ascertained. During this interval, however, the affairs of St. Fabian's Church at Hornbach had undergone a change, probably that of the elimination of the Catholic element, for a way was now open for Tragus's return to the enjoyment of that benefice. He is said to have received a hearty welcome back by those remaining, and that he ended his days there not very long after his return is certain. He died of consumption early in the year 1554, at the age of about fifty-six years. A half-century or so later when Melchior Adam was engaged upon his biographies of German worthies, a search appears to have been made for Tragus's tomb at Hornbach. On the site of St. Fabian's Church nothing remained but a mass of ruins, but forth from under the fallen rocks they were able to bring a memorial tablet bearing this inscription: "Anno Dom. 1554, 21 Febr. Hieronymus Tragos, animæ corporisque medicus, et canonicus huius ædis, in Domino Jesu obdormivit; cuius anima in consortio beatorum quiescit. Amen."[7]

Phytography

The third in order of time among the more noted German botanists of his century,[8] Tragus is the first in their line to actually describe plants. Brunfels had caused pictures to be made, and then by way of comment on each picture had gathered together all that creditable authors in times past had written about the plant, repeating their language to the letter, and citing religiously the volumes and the pages. Fuchs had also pictured every plant, and then to lessen the cost of printing, had presented short descriptions, compiled from other authors, and for the most part given forth as his own.

Two circumstances, both beneficent, united their influences to make Tragus's work peculiarly and distinctively a book of plant description. First of all, the man was by nature an ardent lover of plants. He began to pursue botany for the mere love of it, without thought of thereby acquiring either fame or profit. He who loves things will see much in them to which the indifferent calm and cool observer is blind. If the born botanist—not the machine-made one—write of plants he will find language wherewith to enable his readers to see what he has seen in a plant; and this is phytography. The second favorable circumstance was that of poverty. Tragus at first had no money with which to employ draftsmen and engravers. When he yielded to the entreaties of friends of botany that he would prepare a book, he wrote it in the German language, and with the intent that unfamiliar plants at least should be recognizable with German readers by his verbal descriptions of them alone. In passing at once from the descriptions in Fuchs to those of the

Latin edition of Tragus, one is impressed by the originality and the vastly superior excellency of the latter; it should be recalled that the high quality of them is largely due to their having been first written and published in German,[9] and without thought of their ever being accompanied by pictorial aids to the identification of the plants, and that it was their great success as descriptions that called for the republication of the book in a Latin version, so that the whole might be available to scholars everywhere outside of Germany.

While easily to be ranked among the fathers of phytography, and even as the first of them in point of time after Theophrastus, Tragus does not systematically describe all the plants which he finds occasion to present for discussion. The most familiar things of field and garden, such as every one has known from time immemorial, he neglects, as to their organography, assuming, as did the ancient authors, that the names will recall the images of the plants to the mind of almost any reader. It is the wild things, the native growths of Germany, together with the more recent introductions into German gardens, which engage his powers of verbal delineation.

To the vocabulary of descriptive terminology Tragus does not make any considerable additions. He very commonly employs the old established system of comparisons, conveying an impression of the root, the stem and mode of growth, the leaf outline, the inflorescence of a less known plant, by comparing them with the like organs in very familiar types. Sometimes he does what every one besides had always done, that is, he repeats old comparisons that had been in steady use unaltered for a thousand years and more, and occasionally he makes bold to suggest a new one which he thinks better. Dioscorides had described in the following terms the foliage of a certain plant: "Cyclamen has the leaves of the mature ivy bush, but colored purple, and variegated with whitish markings."[10] Of course Tragus had pondered well this classic line, but thought it might be improved. Here is his own description of the leaves of what he takes to be the same plant. "The leaves of cyclamen are spread over the ground in a circle, are very similar to those of the ivy, or, as I think, of the asarum rather, of a dark green, underneath somewhat reddened, above more brownish, and marked with whitish spots."[11] This is liable to be promptly adjudged a better description than the other, at least for cyclamen leaves as most people are used to seeing them; and the cyclamen which Tragus knew in Germany has foliage assuredly more like that of asarum than like that of old ivy. But what renders the attempted amendment of the Greek author's diagnosis infelicitous is this, that Dioscorides never saw the cyclamen species that Tragus knew, and that at least some of the several Mediterranean cyclamens have leaves of more nearly ovate outline, and therefore better likened to bush-ivy leaves than to those of asarum. The audaciousness of Tragus is not diminished by the consideration that Dioscorides had not only also known asarum, but in describing its foliage had definitely indicated wherein it differs from that of ivy;[12] so that if the cyclamens of Greece and Italy as he had known them had shown asarum leaf outlines he would have been likely to have said so. Tragus was deceived by his own gratuitous supposition that the cyclamen of Germany was the only cyclamen existing.

The common mullein had been used in medicine from the time of Hippocrates; but in the books not much more than one line of description, and that of the leaves only, had been accorded it. Nothing even remotely approaching the following by Tragus had been written about it:

A very notable thing in this plant is the long straight thick root, of a woody hardness. Its leaves, especially the earlier, lie close to the ground, are rather broad and long, of a whitish aspect and woolly, more so than those of helenium (that is, elecampane, *Inula Helenium*). Not until the second year does it send up its stem, full of a white pith within, like the elder, and sometimes attaining a man's height, clothed with leaves which gradually become smaller and narrower as they approach the summit. The flowers, yellow, woolly, and most sweet smelling are of five distinct leaves, and

FIG. 69. Common cyclamen (*Cyclamen purpurascens* Mill.). Bock, *Stirpium* (1552), bk. 2, ch. 134, p. 906; after Fuchs, *Stirpium* (1542), p. 451.

Verbaſcum.
Wullkraut.

De Lilio conuallio.
Meienblümlin.

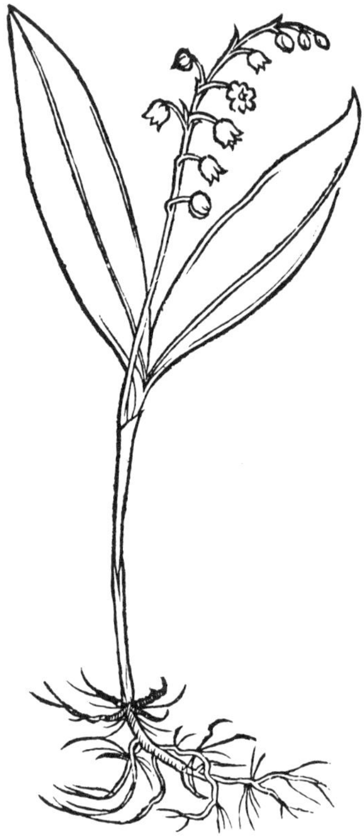

FIG. 70. Mullein (*Verbascum phlomoides* L.). Bock, *Stirpium* (1552), bk. 1, ch. 72, p. 217; after Fuchs, *Stirpium* (1542), p. 848.

FIG. 71. Lily of the valley (*Convallaria majalis* L.). Bock, *Stirpium* (1552), bk. 1, ch. 197, p. 572; after Fuchs, *Stirpium* (1542), p. 240.

completely cover the stem from where they begin up to the very apex of it; which falling away are succeeded each by a woolly globe crowded full of seeds not unlike those of a poppy. When the plant is in flower it well resembles a beautiful torch, whence the name King's Torch has been given it.[13]

It is the earliest botanical account given of the mullein. The writer of it is manifestly a botanist, for he has busied himself with the investigation of this plant as a plant, not as a thing either useful or useless. The subject of this piece of research is but a weed, but he has followed it through its life history, examining its root, dissecting its stem, noting the norm of its foliage, and also the deviations from it, has counted the segments of the corolla, discovered that from within there is exhaled a pleasant odor, has inspected with care the seed vessel and its contents, likening the seeds within it to other seeds that every one is familiar with. There is even added an item of the folklore of the plant. The like of this comes very near to being something new in the history of botany; and the book abounds in plant descriptions of this new and original type.

For an example of his diagnosis of a plant never before described by any one, take that of the Lily of the Valley:

> From a fibre bearing an extensively creeping root somewhat like that of lolium there arise in the month of April green sprouts not so unlike those of asparagus, which turn out to be nothing else than a pair of green leaves closely wrapped together, and supported on a common footstalk. Now when these two begin to separate from each other they assume the appearance of a pair of leaves of the white lily; at the same time there is seen to arise as from between them a triangular peduncle ornamented by five or six little globes placed one after another, of about the size of a chick pea, which in the latter part of April expand to the little cymbal-shaped flowers of a snowy whiteness, round, hollow, serrated around the lower part (*i.e.*, that part looking toward the ground), in the middle within marked with a purple spot, the whole flower most fragrant, but of a bitter taste. After the falling away of these they are succeeded, at the end of June, by a coral-red fruit not unlike that of the asparagus.[14]

The production of this fine word picture had been prompted by something more than mere admiration for a beautiful plant which, although unknown to descriptive botany, had risen to high repute in medicine, at least in Germany. A spirit of friendly rivalry seems to have added zest to those morphologic investigations from which the description followed. Brunfels, who had been first to figure the plant, had found no description of it anywhere, and left it without any. Concerning the medical authorities of the time, whose chapters he had ransacked in search of some account of it, he says he has found them "as silent as fishes."[15] We learn from Tragus that Brunfels afterwards changed his mind insofar as to doubt whether this might not be the *Hemerocallis* of Dioscorides;[16] and Tragus the more carefully studies lily of the valley from earliest spring to the end of the season, and describes it every part from root to fruit, that he may successfully controvert the view of that "man of pious memory, Otto Brunfels." The argument, given at full length, occupies a separate paragraph appended to the description.

On his frequent excursions to the woods and other wild lands, Tragus had discovered many growths hitherto unknown; and while never until late in life having thought of availing himself of the engraver's art, he published a long list of such by verbal delineation only; and so well that plates, when at last in a new edition he made use of them, were well-nigh superfluous, at least to all who were competent to read the descriptions. No greater praise can be given any man of that epoch, than will be rightfully accorded to Tragus in adjudging him to be, for the whole era of modern botany, the first father of phytography after Theophrastus.

Anthology

It was now quite time that some one should resume the investigation of floral structures, a part of botany in respect to which no advancement had been made for fifteen centuries. Neither Brunfels nor Fuchs had even so much knowledge of them as had been attained to by Theophrastus eighteen centuries before. The Greek had said concerning nut trees and oak trees that, over and above those vermiform tassels whose use and nature he could not explain, they have real flowers, things out of the very heart of which nut and acorn were developed. Fuchs, the famous professor of medicine, physician to the titled and the affluent, and moneyed employer of draftsmen and engravers to figure plants, had boldly ventured the unwarranted opinion—contradictory to that of the great Theophrastus—that the tassels of oak and hazel take the place of flowers, and that such trees have no other kind at all.[17]

Tragus, the poor schoolmaster, the apparently unlicensed country doctor, the unordained preacher of the new evangel, welcomes this kind of an opportunity to assail the errors of the dogmatists who sit in exalted station, for, whatever else Tragus may or may not be, he is a botanist well worthy of the name. The combined botanical knowledge of all the Brunfelses and Fuchses of Germany is but a small fraction of what he has seen and taken note of in the book of nature. He has studied the hazel bushes both wild and cultivated in their several species, from the time of the lengthening of their aments in February, all through the spring, and the summer, and the autumn. Theophrastus had averred that, whatever the tassels might not be, the bush has flowers, demonstrably such by the fact that the fruits develop from them. Of the form and coloring of such flowers, and the time of their appearing, he had given no hint; and since at the period with which we are dealing, quite as during two thousand years before, the first idea of a flower was exactly our idea of a corolla, and none had ever seen such things on hazel bushes at any time of year, it was not so wholly inexcusable to deny that there were hazel flowers. And yet, the most cursory reader of Theophrastus's chapters on flowers must have seen that he recognized flowers as either petaliferous or apetalous, even though he did not use those terms; and thus, if corylus failed ever to deck itself with corollas, the next thing in order would be to scrutinize the shrubs in quest of the Theophrastian "capillary" flowers. In such a quest as this Tragus was successful; and since he is the first to actually describe the fertile flowers of *Corylus Avellana*, his words have such historic value that we must quote them:

> All kinds of corylus have diminutive red flowers, resembling the very short stamens of a crocus, which they display just before the unfolding of their leaves. It is in the month of February that those aments which some erroneously suppose to be the flowers, acquire their yellow coloring. Theophrastus in the sixth chapter of his third

book of the *History of Plants* makes mention of the proper flowers of corylus; nevertheless Leonard Fuchs and John Ruelle, besides some others, persistently deny that corylus has ever any flowers; a thing which beyond doubt they would not have done, had they ever once looked into the book of nature on the subject.[18]

After so signal a contribution to the anthology of the hazel, and especially after such public castigation of Fuchs and Ruel, it was fateful that Tragus should err, and that ridiculously, in his philosophy of alder blossoms. Here, having denied that hazel aments are flowers, he is confident as can be that alder aments are true flowers; and he proves it in this wise:

The alder, at least when full grown, adorns itself in spring with brownish flowers, almost like those of the birch and the beech; but they do not fall off, as we have observed that they do in the beech, but are permanent, and come to be drawn together so that they at length acquire something like the configuration of an olive. And this is the fruit of the alder, which matures at the end of summer, and then falling sows itself, so that by this means new alder trees are produced.[19]

FIG. 72. Hazel (*Corylus avellana* L.). Bock, *Stirpium* (1552), bk. 3, ch. 67, p. 1096; a composite figure, after two in Fuchs, *Stirpium* (1542), pp. 398–99.

FIG. 73. Corn poppy (*Papaver rhœas* L.). Bock, *Stirpium* (1552), bk. 1, ch. 39, p. 122; after Fuchs, *Stirpium* (1542), p. 518.

Afterwards in giving account of the flowering and fruiting of birches he betrays the same innocence of the real origin and history of the small seed-bearing cones,[20] inferring that they had been produced by a final contracting and thickening of the pendulous staminate tassels. A less erratic and more logical mind would not have been content with affirming that the loose pendulous aments of birch and alder are real flowers, and at the same time denying that the very similar ones of hazel have anything at all of the nature of flowers. Yet when it comes to the oaks, Tragus must again be given a credit mark as having detected, next after Theophrastus, their very obscure fertile flowers. "The leaves of all kinds of oaks when young are very small and delicate, and there appear with them at this stage long yellow aments. . . . After the aments there come forth very small red flowers which subsequently transform themselves into acorns."[21]

Sexuality in plants never yet having been apprehended, Tragus had no conception of the fact of diœcism. He studied the catkins of willows, but without discovering that those of some individuals are promptly deciduous, and that only those of certain other individuals remain longer. He therefore wrongly attributes to willow catkins indiscriminately the quality of remaining on the tree until they have developed a kind of wool which sails away on the passing breeze. In just this connection, however, he makes a remark which reveals him for once in the character of a truly inductive philosopher, unwilling to venture a broad general conclusion from an isolated fact. "Whether this wool of willows be their seed or not I do not know, except as regards the fourth species, in the case of which it at least takes the place of seed; for in this I have caught the floating wool, have sown it, and have seen willows of this same species spring up from the sowing.[22]

As to certain particulars in the structure of petaliferous flowers the chapters of Tragus seem still more clearly to herald the coming of a new era in anthology. Not that he has any new doctrine of the flower. So far from it, he follows the universal and time-honored practice of calling, in the case of petaliferous flowers, that and that alone the flower, which only long after his day came to be known as the corolla. But he observes and takes note of things outside of this "flower," and of other things inside it, the tendency of which notes and observations is to raise a question as to whether there are not other things which, taken together with the circle of colored leaves, should all collectively be thought to constitute the flower. He does not formally propound any such question. Even that suggestion of it which his language carries, it is more than likely he himself did not perceive. Let us attend to a few instances of anthological comment that seem to be new and original with Tragus.

As subjects of brief description and of various comment the poppies are very ancient, and were so in Tragus's time; and he is the first man in all history to describe the plants as with the pen of a botanist, the pen of a man who had looked at them with his own eyes, and had seen points of morphology never mentioned before him. One of these points relates to the "flower" before its unfolding; or as we should now say, to the flower buds. No such expression as flower bud, however, was yet in use; and Tragus names the thing a *capitulum*, a little head. Poppy plants have what have been called from time immemorial their heads, that is to say, their capsules, or seed vessels; and Tragus now attributes to them a second kind of head, one never before mentioned; and this is his language: "The head from which the flower is to break forth is covered with a pair of skins (cuticuli), green as to color, and also hairy. When the flower itself is ready to make its exit the two integuments separate and promptly fall away."[23]

He has several genera, and in all some eight species and varieties of poppy allies, and to the group as a whole he attributes this pair of caducous integuments. It is a significant item in the history of anthology. He has recorded the discovery of an organ, and has given at least the hint of its possible availability in taxonomy; but, as is usual with discoverers, he is in advance of the time in which his work will be appreciated.[24] It will be yet two hundred and forty years before this pair of integuments to the poppy bud will acquire their name as sepals, and about as long before their caducous nature will be recognized as a good character for the family of Papaveraceæ.

In respect to the forms of flowers, *i.e.*, corollas, there are evident traces of attempts—perhaps half-unconscious efforts—to generalize. As if the wild rose might have been looked upon as the most perfect flower form, or at least as a most representative type, he is given to speaking of other broad-petalled and subrotate flowers as being rose-like, or even as being roses. In describing *Pæonia*, which he figures in a single-flowered state, he twice refers to its "roses," hardly using the word flower at all. "In all our Germany you will hardly find a more elegant rose than that of pæonia"; and again he

FIG. 74. Peony (*Pæonia officinalis* L.). Bock, *Stirpium* (1552), bk. 1, ch. 200, p. 581; after Fuchs, *Stirpium* (1542), p. 202.

FIG. 75. Stock (*Matthiola incana* R. Br.). Bock, *Stirpium* (1552), bk. 1, ch. 191, p. 561; after Fuchs, *Stirpium* (1542), p. 315.

says that "toward the end of April a round head at the summit of each stem all at once breaks into a broad red rose."[25] In like manner he speaks of the flowers of hollyhocks only as roses,[26] and is wont to denominate any large flower of five petals widely spreading as rose, or at least as rose-like. Smaller ones, with petals equally spreading, and especially if they be acutish petals, are his "stellate" flowers. But if the five spreading petals be unequal in some degree, and especially if one of them be at all prolonged at base into a hollow protuberance or spur, such a corolla is not with him a flower but a "violet." The proper violets, that is, the acaulescent kinds and the pansies or tricolor kinds, are regarded by Tragus, as by all others before him, as of two genera with different generic names; yet flowers of both are spoken of not as flowers but as "violets"; and the same term is applied to the flowers of larkspurs.[27]

But there is a second and very different corolla type which is also called by Tragus a violet flower, namely, the cruciform. This came about by reason of the familiarly known fact that in the nomenclature of that period there was a genus *Viola purpurea*—sometimes called *Viola Martia* (= *Viola*)—and also a genus *Viola alba* (= *Matthiola*), besides even a *Viola lutea* (= *Cheiranthus*). It is with this cruciferous viola type in mind that he describes the flower corymbs of *Viburnum Opulus* as adorned with "an outer circle of large white violets consisting of four leaves."[28] But among the crucifers particularly, the flowers of all that have rather large and showy petals he habitually speaks of as violets, for example those of the mustard,[29] the cabbage,[30] and the turnip,[31] with also that of celandine.[32] That it was not merely the cruciform arrangement of spreading petals that caused a flower to be called a violet of this type becomes apparent when one has observed that Tragus never applies it to the blossom of any small-flowered crucifer. In the case of the lepidiums,[33] and *Bursa Pastoris*,[34] and all others having diminutive petals, it is his custom to say only that the flowers are small and of such or such a color, never speaking of them as even small violets. For the crucifer-violet to be called a violet there must be some approximation to the size and showiness of the wall flower and gilliflower.

But now, the learned philologist may interpose that, after all, both rose and violet, in the speech of primitive peoples, are terms indicative of not particular kinds of flowers, that each, in its origin, is but the synonym of that other and later word flower. Nor may this be successfully controverted. The very nomenclature of scores of familiar flowers today attests the truth of it. There is Christmas Rose (= *Helleborus*), China Rose (= *Hibiscus Rosa Sinensis*), Guelder Rose (= *Viburnum Opulus*), Rock Rose (= *Cistus*), Rose of Sharon (= *Hibiscus Trionum*); also there is Dame's Violet (= *Hesperis matronalis*), Dogtooth Violet (= *Erythronium*), and a great number of other such, for plants that have no affinity to either the rose genus or violet genus. Doubtless with some of the peasantry of several countries today the showy flowers they know may be found classified in their languages more or less definitely as roses, lilies, and violets. And something like this was most certainly true four hundred years ago in rural Germany where Tragus was born, and where he did all his botanizing. And although I find him twice using the term violet for small flowers that do not readily fall into either of his two definable categories of violets, I do not think that either one instances a lapse into that primitive usage under which the term is synonymous with smallish petaliferous flowers in general. If he calls the flower of the catsfoot, or ground ivy,[35] a violet, the color of the flower and the irregularity of its corolla limb may have suggested it; and the corolla of a bryony[36] is not unlike that of the sterile outer row of those in viburnum[37] corymbs. That he in truth goes far on the way toward a convenient morphological classification of corollas is evinced by his giving diagnostic names to still other forms. Among the polypetalous he even distinguishes the rotate from the rosaceous. The flower of nigella he says is "round like a

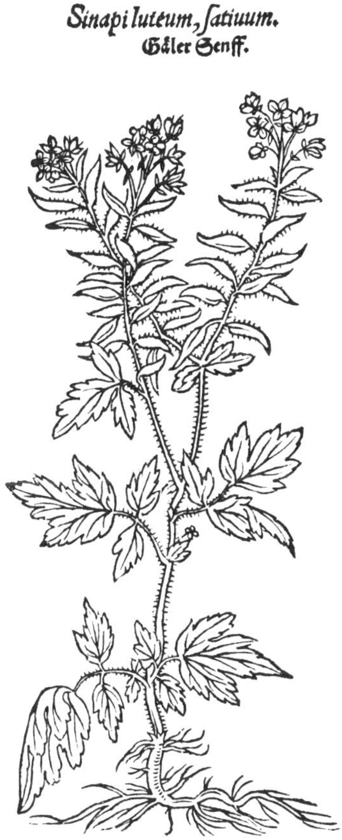

FIG. 76. Guelder rose (*Viburnum opulus* L.). Bock, *Stirpium* (1552), bk. 3, ch. 25, p. 1002.
FIG. 77. White mustard (*Sinapis alba* L.). Bock, *Stirpium* (1552), bk. 1, ch. 32, p. 101; after Fuchs, *Stirpium* (1542), p. 538.

wheel,"[38] that is, the petals spread away in a flatly horizontal direction from their axis, a thing which cannot be said of either rose or pæony petals. It is an excellent distinction, not noted even by Tournefort, the great corollistic systematist of a hundred and fifty years later; for he describes the nigella flower as rosaceous.[39] The campanulate, or campaniform, was also named by Tragus. In his description of the plant hyoscyamus he says that it has little bells for flowers.[40] The corollas of campanula,[41] of digitalis,[42] and of vaccinium[43] are said to be campaniform. The funnelform is also alluded to, though under the term cymbaliform. He attributes that configuration to lily[44] and morning-glory flowers.[45] In a word, he is so definitely the first forerunner of Tournefort in the matter of observing and carefully noting a num-

FIG. 78. Love-in-a-mist (*Nigella damascena* L.). Bock, *Stirpium* (1552), bk. 1, ch. 37, p. 117; after Fuchs, *Stirpium* (1542), p. 504.

ber of marked corolla forms, that it seems due him that a list of them should here be given synoptically:

1. Rotate, typified in *Nigella*
2. Roseform, typified in *Rosa*, *Pæonia*
3. Violetform (proper), typified in *Viola*, *Delphinium*
4. Violetform (cruciferous), typified in *Cheiranthus*, *Matthiola*
5. Stellate, typified in *Sedum*, *Solanum*
6. Campaniform, typified in *Campanula*, *Digitalis*
7. Cymbaliform, typified in *Convolvulus*, *Oxalis*

Gifted with keen perceptibilities in the matter of floral structures, it is rather remarkable that Tragus did not assign names to such very strong corolla types as the bilabiate and the papilionaceous, for I find in him no trace of any term by which he would designate either one. Even the later Latins seem to have denominated certain very nettle-leaved labiates as "nettles with a lip"; and this most lynx-eyed and original inspector of flowers seldom names anything more than the mere color of labiate flowers, once only—the case of glechoma above referred to[46]—giving perhaps a hint of the form. In respect to the flower of leguminous plants he does not so much as second the suggestion of Gesner that the pea blossom has the form of a butterfly, but attributes, for example, to the pea vine the flower of genista, and to genista in its turn the flower of the pea, and so on to the end of the series.[47] There was nevertheless in Tragus's time a German school boy who had already coined the term papilionaceous for these flowers, though only in manuscripts that were not published until after Tragus's demise.[48]

Among those scattered anthological notes which, to his contemporaries and to himself, may have seemed of least moment were his various observations upon stamens. Their function had not yet been guessed at, or even by Tragus himself so much as thought about. The organ had not been even morphologically contemplated, in its individuality, up to the time when Tragus began to examine it in different flowers comparatively. From time al-

FIG. 79. Henbane (*Hyoscyamus niger* L.). Bock, *Stirpium* (1552), bk. 1, ch. 41, p. 133; after Brunfels, *Herbarum* (1532–39), vol. 1, p. 224.

most immemorial they had been mentioned only collectively, in Latin writing, as the capillamenta, the flocci, the stamina, the apices; mostly a tuft of thready things, with or without knotted tips. It is altogether a new thing in botany for a man to write as follows concerning the white-lily flower: "From out the bottom of it there stand forth six apices, and these yellow; then prominently from the midst of them a kind of thickish stamen that is green and has a triangular head, the whole being shaped something like a walking stick."[49]

In a flower of the size of a lily such diagnosis of less obvious parts is easy, insomuch that one need not have been surprised if many a writer before Tragus had indicated them as well; but for similar inspection and diagnosis a gooseberry blossom was small and difficult for a botanist whose researches antedated the invention of hand lenses. The following is our botanist's account of it, and it is the oldest one extant:[50] "In March the bush puts forth its flowers, small, concave, purple, which if you examine them one by one you find to contain within five diminutive apices each supported on a hair, the whole resembling a little bell."[51] Another flower which he is first to describe is that of the hawthorn. This is not much larger than the gooseberry blossom, and of less simplicity in its structure; but he brings out its characteristics quite as clearly. "The individual flower is made up of five small white leaves, from out the middle of which there stand many white things like hairs supporting rose-colored apices, such as one observes in all pomaceous flowers."[52] Among his "pomes" Tragus includes also the drupe-bearing trees, and his account of their floral structure is given once for all under the caption of the Wild Plum.

> The individual flowers of the wild plum consist of five bright-white leaves; and in the midst are seen about eighteen as clear white capillamenta or stamina each supporting its small yellow apex. Just about this is the structure of the flowers of nearly all the pomes, cherries, plums, pears, and apples, except that some of them exhibit a greater number of capillamenta and apices than do those of this wild plum.[53]

From this point forward throughout the line of the pomaceous and drupaceous trees, he has little to say of the flowers beyond this, that "they are those of all the pomes," sometimes remarking that the "apices" are red, or that they are yellow.

In thus taking a census of individual stamens, recording the number of them as being constant in each different kind of flower, even distinguishing by name each of their two parts, it is evident that Tragus on his own part was but entertaining himself and some of his readers with mere curiosities, that he had no notion whatever of the real importance, to the future of botany, of that which he was doing. Let us, then, take a still more careful survey of the things which he accomplished in this particular direction which were for the advancement of anthology, albeit he himself was all unconscious of their import. To begin enumerating the stamens of particular flowers, and to record in each instance the count, was to make a decided innovation upon the immemorially established anthology of Tragus's time, which had always treated of them in the aggregate only, as tufts of hairs or threads. If he announces that in every gooseberry blossom there are just five such hairs or knotted threads, that is announcing that he has made two different comparisons of them: first, a comparison of the five one with another, and then of these with two other hairs quite different from them which occupy the very center within their circle. He does not mention these, the styles, or style branches, yet cannot have failed to perceive them. But he now proceeds virtually to define a stamen. It is the little apex, and the hair on whose summit the apex is sustained. Elsewhere he repeats the definition more explicitly,

FIG. 80. Gooseberry (*Ribes uva-crispa* L.). Bock, *Stirpium* (1552), bk. 3, ch. 14, p. 977; after Fuchs, *Stirpium* (1542), p. 187.

FIG. 81. Blackthorn (*Prunus spinosa* L.). Bock, *Stirpium* (1552), bk. 3, ch. 33, p. 1016; after Fuchs, *Stirpium* (1542), p. 404.

and, avoiding the use of any new terms, he calls that the "capillamentum" which we now know as the filament, and "apex" is his application of an old term to what has become the anther of modern anthology.

Tragus is, then, the discoverer of the stamen as a definable organ made up of two separate and different parts, each part with its own name. The same is very near being true of him as regards also the pistil. We have just seen how accurately he could describe the style and stigma of a lily as things quite apart from the stamens; and as to the flowers of common fruit trees, we must feel assured that he saw the one style of each plum and cherry blossom that he inspected, and the five of them that are conspicuous in the flowers of pear and apple trees. He did not mention them; but then, they are as plainly visible there as the stamens themselves, and both his enumeration and his definition of these prove that he did not confuse the styles with them. Curiously enough, after having located and well outlined the large lily style and its stigma, the very next account he gives of such central thing is in connection with a flower even smaller than that of a gooseberry, namely, of the little ericaceous undershrub then called Myrtillus.[54] Here the stamens seem to have escaped his notice, as they easily may have done, being both minute and hidden away within the partly closed hollow of the corolla at its base, falling away with it; but the style is very obvious, and in describing this hollow bell of a corolla he says: "It has in the middle a red-brown something like the clapper of a bell, which, after the falling of the flower grows into a round reddish berry."[55] The Latin term used for this thing in the middle of the flower bell is *pistillum*. It is the first instance of the employment of that term to designate what has long since become universally known by that name.

On the whole, then, as by suggesting the classifying of corollas this author is the first herald of Tournefort; much more conspicuously as the first investigator, even fairly the discoverer, of stamen and pistil, is he the first forerunner of Linnæus. It was anthology which created the new botany of the eighteenth century and the nineteenth; and the beginnings of the modern anthology are with Tragus.

Fruit and Seed

Well in advance of classic antiquity in the knowledge of floral structure, Tragus not only added nothing to carpology, but had never learned either from Theophrastus or by research of his own anything like all that Theophrastus had attained to along this line. One may doubt if any other book was ever printed, or even written in any age, in which there find expression so many whimsical and superstitious fancies about seeds. In a general way they seem to fall short of having that value in the economy of plant life and plant distribution, in Tragus's opinion, which even a remote antiquity accorded to them. The transmutation of the seeds of cereals into germs of chess and darnel he accepts without the expression of a doubt. Abiogenesis, the doctrine of the origin of living things from lifeless matter, he accepts unwaveringly,[56] and with a plenitude of faith probably surpassing that of the Greeks who aforetime invented the theory. He even defends it by theologic arguments;[57] and by it he explains the coming into existence not only of low and simple flowerless things, but also of some of the highest types, even of trees. The time seems to have been when he had thought that all willows had at first come forth spontaneously from the mud of river banks; for he knew them to be propagated by cuttings always, and no one had ever seen or heard of a willow seed. Even when with such truly scientific inquisitiveness he had planted some of the "white down" that he had gathered as it was about to float off from a willow, and had afterwards the satisfaction of seeing willows spring up as from that down, this neither convinced him that all the other kinds of willow might be raised in that way, nor even that any willow had ever borne a real seed. He says that one out of the four species of willows which he describes can be propagated in this way; but he vouches for none of the others, does not infer that the other three may be found producible from "willow down." Nor did he have any idea that he had planted willow seeds. "The down takes the place of seed."[58] There are certain kinds of poplar, and even of maple, which he thinks never bear seed at all, and which he therefore thinks came into existence abiogenetically. These were diœcious trees, none of them indigenous to Germany, and perhaps then existing there in only the male sex. He has seen the tassels of *Populus alba*,[59] and knows them to be always deciduous; also no one knows of any seed as consequent to the flowering of *Acer pseudo platanus*. They are seedless, and therefore, to the faith of Tragus, abiogenetic in their origin. It even seems to be his opinion that certain plants plentifully seed-bearing may upon occasion spring up and mature in places where no seed of them had fallen. It is upon just this theory that he accounts for the occurrence of individual plants of many kinds on high walls of solid masonry and upon the roofs of buildings. No orchids, in his understanding of them, produce seeds at all. He is familiar with the fact that in autumn, as the plants are withering away, a very fine dust falls from where the flowers were, but he affirms that this perishes together with the season's growth of stem and leaves.[60] He presents to his readers a strange fancy about the primal origin and the perpetuation of this class of plants. He is the first author to mention, and may or may not have been first to observe in the flowers of orchids, resemblances to birds and other flying things; but he writes much as if he had been the inventor of the theory that this kind of seedless plants originates from certain excretions of birds. It is in the chapter that is devoted to the birds' nest orchis that he explains this belief. Such plants, he remarks, abound chiefly in and about thickets where small birds mate and nest.

The belief in all kinds of spontaneous generation, of even seed plants as well as the seedless, did not preclude all research on Tragus's part along these lines. There is a picturesque account of nightly vigils in search of the problematic seeds of *Osmunda regalis*.[61] While it was generally held that all fern-like plants are seedless, this one seems to have been popularly credited with shedding seeds regularly on one particular night of the year. Thus runs Tragus's own story:

> Inasmuch as all writers about herbs have said that ferns produce neither flowers nor seeds, I have thought it worth while to place on record, for the information of botanists, an account of my own experiences, which prove the contrary. For four years in succession I kept vigil all the night preceding the Feast of St. John Baptist,[62] and always found in the very early morning, before the break of day, very minute black seeds, not so very unlike poppy seeds lying on the pieces of cloth, and the mullein leaves which, in order that I might not miss the seeds, I had placed under the plants beforehand. Some of the ferns had shed no seeds at all; others had deposited them by the hundred. Moreover, in these experiments, I employed no cabalistics, no conjurings, no incantations, no superstitious observances of any kind, nor did any one of

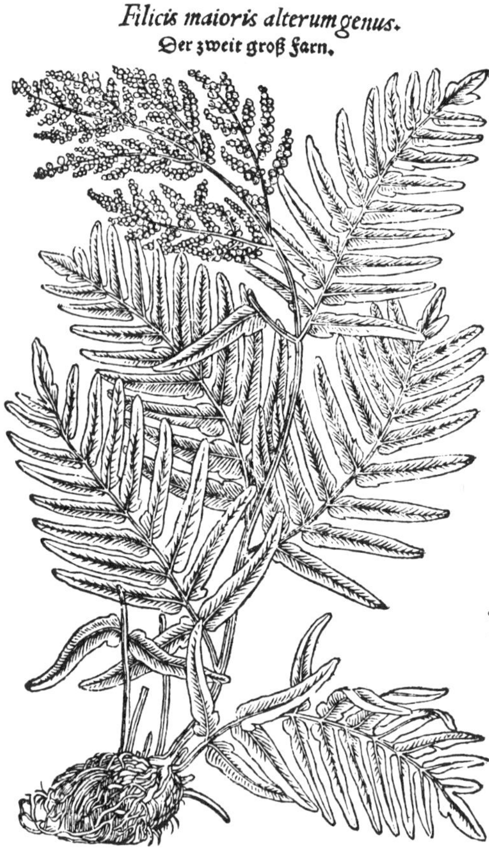

FIG. 82. Royal fern (*Osmunda regalis* L.). Bock, *Stirpium* (1552), bk. 1, ch. 186, p. 543.

the three companions of my vigils; but having made a fire we watched and waited, sometimes finding none, at other times a few here, and many there. Why there should be such a diversity in the yield of seed, and what the purpose of nature may be in all this matter, I do not understand.[63]

This account incidentally reveals that in middle Europe in the sixteenth century there still flourished the ancient profession of the root and herb gatherer, in the practice of its old-time superstitions; that men believed that under the sacred spell of the summer-solstitial midnight such flowerless and seedless herbs as ferns, by help of solemn incantation, could be made to scatter seeds, these presumed to be efficacious in medicine or magic. Tragus, the inquirer and reformer, half believing and half disbelieving, investigated the matter, proving to his own satisfaction that ferns bear seeds, that they produce them naturally, without the promptings of conjuration; yet it seems not to have occurred even to our reformer botanist to look for fern seeds in the day time, or at night except on that immediately preceding St. John's Day!

The distinction, indicated so long ago by Theophrastus, between plants with one seed leaf and those with two, though never again brought forward prominently until long after Tragus, had not been ignored by him. It had been the cereals and their kindred to which the Greek had ascribed the one seed leaf as a universal characteristic. There was now common in Germany one bread-stuff plant which the ancients had not known, namely, buckwheat. Tragus describes and figures it among the proper cereals,[64] but with the remark that it differs from all the rest of them in that it comes up from the seed with a pair of leaves instead of with one alone, in this respect more like a turnip or cabbage.[65] And although he was a student of wild plants rather than of cultivated, there are many instances of his concluding a description with the statement that the seedlings of such a plant come up with two leaves. Now and then he mentions the outline of such seed leaves in some particular plant as contrasted with those of some other, showing that he not only observed but compared them in different plants. But this does not seem to have had any purpose beyond that of gratifying his own curiosity and stimulating the like in his readers. There is no indication of his having apprehended the taxonomic significance of these distinctions between monocotyledonous and dicotyledonous seeds. The time for the birth of this great thought lay distant from Tragus a hundred years and more.[66]

Taxonomy

In his Preface Tragus abjures the alphabetic arrangement of genera as unscientific, bringing in confusion where natural order ought to be. He is clear in expressing the determination to adopt a natural sequence. "In describing things, I come as nearly as I can to keeping by themselves such plants as nature seems to have linked together by similarity of form."[67] This was no new proposition. Ever since plants had first been observed philosophically, and written about, various groups, varying severally as to their extent and inclusiveness, had gained recognition as natural groups through resemblances in morphology. Tragus knew this well, and was only indicating his choice of natural method, in preference to the purely artificial alphabetic arrangement of genera, such as Fuchs and Gesner saved themselves labor by adhering to. Neither does he contemplate considerable innovations upon the long-established method of grouping and arranging things. By his own frequent peerings into the curiosities of floral structures and recording what he saw, he has vividly suggested a new anthology. He has even begun it; and in the course of its future development it is going to revolutionize taxonomy completely and that twice over; yet nothing of this is even dreamed of by Tragus; and his superior knowledge of floral morphology has little real and almost no appreciable effect upon his own classifyings.

Tragus's taxonomy, like that of all his forbears, is established upon characters of the vegetative organs, with no strong appeal to flower or fruit; and the primary grouping of things as trees, shrubs, and herbs is in his estimation natural and valid. Nevertheless a few exceptions are taken by him against this rule. An underlying principle of such a rule, or at least a logical deduction from it, is that no one genus embraces both trees and herbaceous plants. And really there was but a solitary genus known in early botany that was troublesome to those who regarded the distinction between woody plants and the herbaceous as being taxonomically fundamental. There was the tree *Sambucus* and the herb *Ebulus*. These are their classic Latin names, and the import of this nomenclature is, that they are of two genera. *Ebulus* is not even in the least degree woody or shrubby in any part. In texture and duration it is as perfectly a perennial herb as the common asparagus or rhubarb. Dioscorides, the great Greek physician and medical botanist, having regard to pharmaceutical principles as well as taxonomic, received *Ebulus* as a kind of *Sambucus*. They are alike, he says, in foliage, flower, fruit, and medical properties.[68] Tragus, consistently adhering to the fundamentals of classification there accepted, describes and figures *Ebulus* among the genera of herbs; for even his readers, every one, would look for it among the herbs and not among the trees; and *Sambucus* is treated of far away, in the third book, under the general topic of trees.[69] In both places, however, he ventures the opinion that the two are naturally of one genus. Under *Ebulus* he says, "If you consider its foliage, flower, and the heavy somewhat sickening odor of the herbage, you must regard it as nothing else but a smaller and herbaceous kind of *Sambucus*." Under the latter he says again, "As to its leaf, flower, fruit, and odor this is so exactly like *Ebulus* that the ancients were wont to receive them as of one and the same genus."

From a period too remote for precise limitations, only *Sambucus* and *Ebulus* seem to have militated against the taxonomic validity of that old dis-

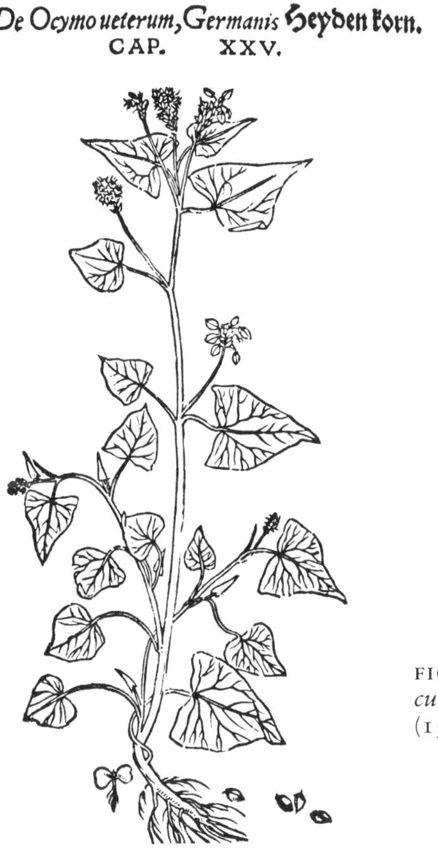

FIG. 83. Buckwheat (*Fagopyrum esculentum* Moench). Bock, *Stirpium* (1552), bk. 2, ch. 25, p. 648.

tinction between herb and tree; and we shall be interested in following the subsequent history of this taxonomic puzzle to its final solution.

That old philosophy of the three grand divisions of Tree, Shrub, and Herb unquestionably carried with it the opinion, certainly not altogether unreasonable, that trees are of highest rank, and herbaceous plants of the lowest. Under this system it will be seen that some authors began with the highest and proceeded to the lowest, and that other authors, beginning with the herbaceous genera, ended with the ligneous, the largest and most enduring trees coming last of all. This last is the order followed by Tragus; not, however, as one philosophically viewing the plant world from lowest type to highest as a genetically connected whole. It is, on the contrary, quite certain that that Aristotelian idea, only now very lately reinstated, never entered Tragus's thought at all. If he takes up herbaceous plants first in order, it may well be because they are both the most numerous as to the genera and species, and of the highest importance to man. Nevertheless we shall find him very much given to running like things together and thus forming groups within groups, lesser ones within the more comprehensive, whether he be dealing with herbs or with shrubs or with trees. This is taxonomic work; and this is the way in which he fulfils the promise made in his Preface about natural arrangement. We must follow him now for some distance, and very carefully, if we are to arrive at an understanding as to what botanical system really was, in Germany, in this first half of the sixteenth century.

In no author as early as Tragus is there given any introductory synopsis or tabulation of the system. Such convenient and helpful skeletonization is a later invention; and here one gathers information about the system, even to the principles that underlie it, only through following the author chapter by chapter from the beginning of the volume to its end.

For a work like this, of 1,200 pages,[70] the selection of a hundred for such analytic study must suffice. They might be taken at random from any one of the three divisions of the treatise; but we shall select the first hundred. Within these there are embraced figures and descriptions of some 74 species, distributed to about 31 genera. Now this proportion of something like two and a half species to a genus was something new in botany and is therefore one of the very significant features of the book, so much so that we must give it a moment's consideration before passing to a study of the sequence of the genera. In order to realize the meaning of my statement that, for the time at which he wrote, the ratio of two and a half species is so great as to amount to an innovation in taxonomy, a brief comparison must be instituted. The five or six hundred plants that the ancients had dealt with[71] represented, in the great majority of cases, what we of today are accustomed to speak of as monotypic genera. Not any very considerable number of their genera are defined as consisting of two or more species, so that they had but one and a small fraction species to each genus. Fortunately for our comparisons the volume of Tragus deals with not very far from the same number of species as that of Dioscorides, and seems to have rather more than twice as many species to each genus; or, to express it differently, an equal number of species is distributed to only half as many genera by Tragus as by Dioscorides; and this must be regarded as having been somewhat revolutionary, especially in view of the fact that Brunfels and Fuchs had much less perceptibly, if at all, departed from time-honored usage in such matters.

There may be reason to doubt that such a movement originated in the mind of Tragus himself. There are intimations elsewhere that as a new departure it may have been suggested if not advised by a mind more philosophic than his own. But this is a matter the investigation of which may be deferred. What should here be remarked is, that almost every generation of active and leading systematists during now nearly four hundred years has been divided upon the question of whether plants are more philosophically disposed in few genera of many species, or in many genera of correspondingly few species; and in Tragus's book we are at a kind of starting point in this perhaps endless controversy about the delimitation of genera. As perceptibly inclining to reduce the number of them he is again the forerunner of Linnæus.

Surveying now somewhat closely 62 consecutive pages of the first 100, we enumerate 45 species all in one line, and with a solitary exception, all at agreement as to certain very obvious characteristics. All of them exhibit fibrous roots, quadrangular stems, and opposite leaves. In 41 out of the 44 the leaves are simple and in no wise divided or even cleft. Here is proof that things had been selected and brought together under the guidance of definite morphologic principle. Also much time and toil must have been bestowed upon the getting together of so considerable a number of square-stemmed opposite-leaved herbs all at agreement in a general characteristic of leaf outline. Some of them are much-branched herbs in which no trace of quadrangularity is seen on the maturer branches, but only on the growing twigs; and it was the judgment of something like a botanical expert that brought such into line among the square-stemmed. It is also worth noticing that 41 of the 45 are plants more or less aromatic.

The number of the genera among which the forty-five plants are distributed is twenty. First in order stands *Urtica* with three species;[72] then seventeen genera of the family of the Labiatæ;[73] then the genus *Valeriana* with three species.[74] As to the nettles proper and the labiate dead nettles, Tragus has order where in Brunfels there was confusion,[75] for the figures of the two nettles have *Urtica* printed over them, while the two dead nettles are respectively indicated as *Galeopsis* and *Lamium*. To these succeeds *Marrubium* with an array of four species. Only one of them is of that genus; two belong to a genus that had not at the time been proposed, that is, *Stachys*; the fourth is a plant that was destined to stand as prototype of the genus *Lycopus*. The figure well represents *Lycopus Europæus*; and this appears to be the first publication of the species, though he writes as if it were already familiarly known in the pharmacy of the time under the name *Marrubium palustre*.[76] It is in the course of his definition of this too-amplified *Marrubium* that he describes the fruiting calyx with its four naked nutlets, one of the most important characteristics of the whole family of the labiates; though as we have said before he is far enough from realizing the taxonomic value of what he has thus been the first to discover and describe.[77] The presence of this pouch, as he calls it, with its four naked seeds occurring as it does in some herbs with dissected foliage, does not induce him to place any such in the same line with the nettle-leaved labiates. His mind upon these matters is the mind of all antiquity, and of his contemporaries, dominated by the idea that likeness as to foliage and stem and root, together with agreement in sensible qualities, more surely indicate consanguinity than do similarity in respect to seed vessels and seeds. We shall meet with plenty of proofs of this.

Next after *Marrubium* the first considerable genus is *Mentha*. Tragus is aware that it is difficult, and says that quite a number of plants which some have regarded as mints others have referred to other genera. He seems to have a new view of his own, namely, that no plant is properly a *Mentha* that has not an upright mode of growth, with flowers separate from the leafy part, and borne in naked pedunculate spikes at summit of the stem. It is a group of mints that is sufficiently natural, and has been recognized as such by all special students of the genus of later periods. Tragus, as his four plates show, limits the genus to this group, disposing somewhat variously of the equally numerous species that have all their flowers in the axils of the leaves. As good a species of *Mentha* as that called *Pulegium* is excluded, and held as a monotypic genus, doubtless partly on account of its peculiar odor and its efficacy as an insectifuge, and partly because its stems are almost or quite prostrate, for in this antique classification by vegetative characters the very posture of stems was regarded as a weighty consideration, as we shall see later.

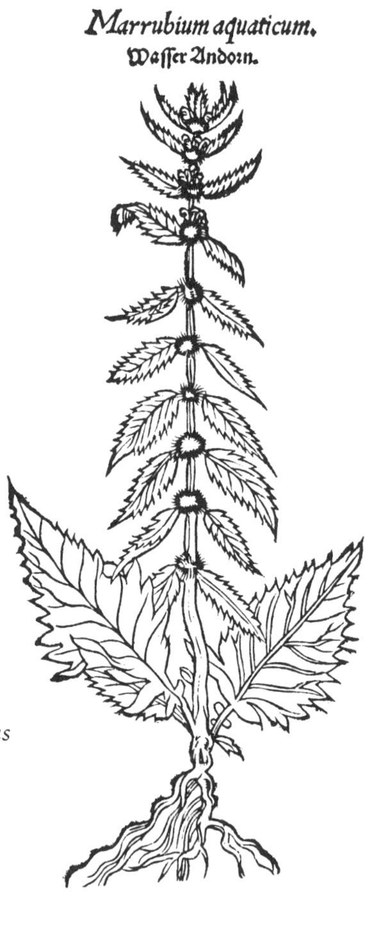

FIG. 84. Gipsy-wort (*Lycopus europæus* L.). Bock, *Stirpium* (1552), bk. 1, ch. 2, p. 9.

In the treatment of these plants now long known as the family of the Labiatæ, Tragus, quite as if he had recognized the family and wished to keep the members of it in as close juxtaposition as possible, again does violence to one of the very fundamentals of the old system to which he has professed fealty. Rosemary and lavender are genera of labiates, square-stemmed, opposite-leaved, aromatic, and have the flowers and fruits of labiates, but they are shrubs; at least rosemary is, and the lavenders are strongly suffrutescent; therefore the proper place for them is away in Tragus's Third Book, among the woody growths, where also we find plenty of growths that are both smaller and less woody than either of rosemary or lavender; but he has both these here in the First Book, at the end of the line of the labiates, all the rest of which are herbs.

With this ending of the series of the labiates we are brought to about the sixtieth of our one hundred sample pages. The number of genera embraced within the fifty-nine pages is eighteen. All are genera of Labiatæ except the first, and that is *Urtica*. We have already seen how nettles and dead nettles were primitively regarded as of one and the same genus. The conceding that the two were generically distinct did not necessitate any wide sundering of them. The close resemblance between them as to vegetative organs, and the clustering of the flowers in the leaf axils, betokened still a close consanguinity. To Tragus and his contemporaries the transition from *Urtica* to *Lamium*, so far from seeming to be abrupt, was a perfectly easy and natural one.

That close against the aromatic labiates of herb and drug gardens *Valeriana* should be located is not so difficult to explain, now that we have Tragus's point of view; for the valerians, at least as to their basal and underground parts, are notably odoriferous; they are not indistinctly square-stemmed, their leaves are opposite, their inflorescence is of the verticillastrate type, and their flowers are bilabiate.

At this juncture the series of the square-stemmed and opposite-leaved is briefly interrupted. The intercalated genera are *Asarum*, *Geum*, and *Ruta*. But in the primitive classifyings aromatic properties were much deferred to; and inasmuch as this whole series, all the way from *Lamium* to *Ruta*, is a line of aromatic plants, the three above named do not interrupt it save only as to stem and leaf morphology, which, at least for the moment, is held subordinate to the qualitative.

After *Ruta* there is at once a return to the line of the quadrangular stem and opposite leaf. The genus is now *Hypericum*, and five species of it are described and figured. These harmonize with the main series as to an undivided foliage set oppositely and at regular intervals up and down square stems, and also as being notably odoriferous even if not distinctly aromatic like labiates. At the same time they have manifest points of contact with rue, that genus which they immediately follow, for their leaves are dotted, and

FIG. 85. Rue (*Ruta ?graveolens* L.). Bock, *Stirpium* (1552), bk. 1, ch. 22, p. 69.
FIG. 86. Common Saint-John's-wort (*Hypericum perforatum* L.). Bock, *Stirpium* (1552), bk. 1, ch. 23, p. 73; after Fuchs, *Stirpium* (1542), p. 831.

small flatly opening yellow flowers crown the stems and branches. Next after hypericum stands the undershrub *Santolina*, an anthemideous composite; and, viewed in the light of twentieth-century taxonomy certainly rue and hypericum and santolina placed in line together make a motley order; but, what we are here in quest of is the set of principles on which sixteenth-century taxonomy was grounded. We are certain that Tragus had his taxonomic reasons for locating santolina where he did, for he states them. Every one who knows the plant is aware of its being notably odoriferous; and he gives as one reason why it may well stand next hypericum the fact that its aroma is that of hypericum intensified. Now on the other side we shall find santolina flanked by two labiate plants; and in respect to its mode of growth, and its aspect as clothed thickly with small grayish foliage, Tragus says that in these things it well resembles lavender and hyssop and thyme. So then, judged by the criteria employed at that period, this was not a motley arranging of things.

From santolina there is a return to labiates. Two species are figured and three described. We wish to know why he thus separated them from the rest of their line. It would be interesting if we could learn his reasons for intruding almost into the midst of the line of the mintworts *Asarum*, *Ruta*, the whole series of the *Hypericum* species, and *Santolina*. There is one thing which gives to these last members of the line of labiates an aspect very unlike that of the others; for their leaves are much dissected, while in the line of more than thirty that precede rue and hypericum there is not one that displays any other than simple leaves. This short concluding series consists

FIG. 87. Lavender cotton (*Santolina chamæcyparissus* L.). Bock, *Stirpium* (1552), bk. 1, ch. 24, p. 77; after Fuchs, *Stirpium* (1542), p. 874.

of compound-leaved species of the genus *Teucrium*. Did Tragus, blinded by foliage so exceedingly different, fail to see that these are true allies of that simple-leaved series that has been interrupted by rue and hypericum? We have the most positive proof that he did perceive the relation; for he says that the stems are quadrangular, that the flowers are verticillastrate around the stems, and that each comes forth from what he calls the seed-pouch, in which particular he likens them to those of hyssop and satureja.[78] But in other instances besides this he is seen to pay such deference to the distinction of compound and simple foliage as to make use of it taxonomically. Of even this genus *Teucrium* one simple-leaved species is not only held generically distinct from those with dissected foliage, but is located 124 pages away from them, where, by the way, on account of its veronica-like habit and foliage it is associated with several veronicas to constitute a genus *Chamædrys*. In his Third Book, in taking up the natural series of the pomaceous and drupaceous trees, the compound-leaved genus *Sorbus* in three species

FIG. 88. Wild service tree (*Sorbus torminalis* Cr.). Bock, *Stirpium* (1552), bk. 3, ch. 30, p. 1010.

heads the series, quite as if by virtue of its compound foliage it had been regarded as the highest or most advanced type of its alliance.[79] Again, the bulk of the umbellifers, all having pinnated or more dissected foliage, Tragus adopts as a natural alliance, following of course the botanists of remote antiquity; but *Bupleurum*, vested as it seems to him in a perfectly simple and even entire foliage, he on that account excludes from the family. The genus *Achillea*, of the anthemideous composites, quite imitative of the umbellifers as to foliage and inflorescence, intervenes between *Bupleurum* and its compound-leaved affinities.[80] Other proofs need not be adduced, for we must return to that group of dissected-leaved labiates that close the line of their cognates. They have brought us to number eighty of our one hundred pages.

With page eighty-one, and thenceforward, one notes an abrupt change, at least respecting the morphology of things, for within the next one hundred it will be rare to meet with a plant square-stemmed and opposite-leaved. The stems are now terete, and the leaves alternate; and in place of aromatic odors there is now everywhere a peculiar pungency of flavor to the herbage. The genera and species are, for a time, those of the family of the crucifers. There occurs at first an unbroken line of seven of these. The student to whom the book is not available will be helped by a list of the names of the species:

	Tragus	Modern
1.	*Nasturtium hortense*	*Lepidium sativum*
2.	*N. aquaticum*	*Nasturtium officinale*
3.	*N. pratense*	*Cardamine pratensis*
4.	*Thlaspidium*	*Lepidium ruderale*
5.	*Alliaria*	*Alliaria officinalis*[81]
6.	*Thlaspi*	*Thlaspi arvense*
7.	*Lepidium*	*Lepidium latifolium*

It will be seen at a glance that no taxonomic account is made of the pods in drawing up the line of those crucifers. Siliculose and siliquose genera are completely intermixed. It is just what Sachs, from his mistaken point of view, calls a "motley order."[82] Yet there is a system in the arrangement of these seven crucifers, and it is plainly apparent. Draw a line between the numbers 4 and 5 and all above the line are compound-leaved, all below it simple-leaved. It is a natural arrangement, according to Tragus's notions—certainly very crude, sometimes—about the tests of affinity. From this point, however, the succession of members of this family is broken by the intrusion of eight species belonging to three other natural groups. The intruders are two persicarias, five buttercups, and a rather peculiar plantain. Only at the end of such a line as this does he again take up the crucifers.

If in this procedure the German father seems chargeable with having abandoned the principle laid down in his Preface, I see no clear defense for him. If he had been intent upon collocating things by the criterion of "form," as he calls it, he must needs have kept the line of the crucifers uninterrupted by plants polygonaceous and ranunculaceous. He was botanist enough easily to have seen the marked contrasts of form between these plants and those crucifers that flank them in considerable file on either hand. Being such, these intrusions were not made by chance or whim. Apparently it is a case of the subordinating of taxonomic principle to the convenience of a multitude of those to whom morphologic marks would go for naught, and with whom a grouping according to useful qualities would be welcome. Let us understand how, with such a purpose before him, Tragus would not unnaturally break the succession of crucifers and interpose a line of buttercups.

Where the succession of square-stemmed, opposite-leaved aromatic herbs ends in a fragrant *Teucrium*, and the terete-stemmed alternate-leaved series begins with the garden peppergrass, I have characterized the transition as abrupt; but this was done with a mental reservation. Morphologically it is abrupt, extremely so. Qualitatively considered, however, the entire series, from the first labiate to the last crucifer has a common character; and Tragus realized this character and allowed it to influence his classifying. The whole line, or group, is one made up of herbs either aromatic or pungent, or both. Let us accentuate this fact by denominating the mint-lavender division of the series the spicy plants,[83] and the peppergrass-buttercup end of it the peppery plants. Now if one should catalogue and enumerate all the aromatic-scented and all the pungent-tasting herbs that are, the aggregate of them all would be but a fraction of the whole number of herbaceous plants. Such constitute, I suppose, not much more than a tenth of all the herbs known to Tragus. But his thought is that such a qualitative thread as this, pervading many species, may be used to line them up by, even so as to include within the line here and there a few which in the particulars of their morphology are not at agreement with the others.

Now this first series of seven crucifers is made up of species the tender stems and leaves of which were eaten raw as salads, or else the crushed seeds were used as condiments.[84] They are particularly pungent or peppery crucifers. Even the name nasturtium, which half the species bore, the etymologists derive from that irritation of the nasal passages experienced during mastication of these things. It must here be stated that, after the interpolation of the buttercups, the resumption of the line of crucifers is made at the genus of the mustards, plants the ground seeds of which had been employed in medicine as counter-irritants from time immemorial.[85] And there will be readers to whom the information will be new that the seeds of the buttercups are as pungent as the seeds of mustard and were long used for the same purpose of raising blisters on the skin. Yet this acridly pungent quality of them is expressed in the very names by which the commonest species are known in botany, that is to say *Ranunculus sceleratus* and *Ranunculus acris*.[86] And Tragus was so familiar with all this, as to have been constrained to locate the *Ranunculus* species in the midst of the counter-irritants, for his whole volume was indited to a great extent in the interest of those who practiced, even rudely and primitively, the healing art. These would expect to find remedial equivalents treated of in contiguous chapters, and he was willing to meet their expectations.

The resumption of the line of the crucifers has brought us to the limit of the one hundred sample pages which we were to examine somewhat thoroughly in quest of the man's mind and method. And now, perhaps in no way may one more easily arrive at a still fuller comprehension of his method than by following him in his further treatment of the cruciferous plants. The list of seven species which, set in line, precede the buttercups, together with the line of five species that succeed the buttercups, amount to about half the number of this family which he figures and describes. The other twelve or fifteen are discussed in two groups, each widely removed from the present series and also from one another. This is because the author must conform to the ancient usage of treating ornamental plants all by themselves in one place, and the edible plants of the kitchen garden also apart from all others. Under these divisions, however, we find the crucifers in each well kept together. The ornamental kinds, to give them by later names, are *Cheiranthus*, two species of *Matthiola*, and *Hesperis matronalis*.[87] The last is a little separated from the line, two other plants, both remote from the crucifers, but popularly called violets, being intruded. I entertain no doubt about Tragus's having perceived the real consanguinity subsisting between these wallflower-gilliflower ornamental plants—all known as violets—and the other crucifers, for in describing the wallflower he remarks that it belongs to the four-leaved—that is, the four-petalled—group of the violets, rather than to the five-leaved sort. Then again under *Hesperis* he describes the seeds of it as being enclosed in elongated and terete siliques like those of the cabbage.

For a glance at his final series of crucifers we must pass to the Second Book, where the topic is that of culinary herbs and roots in general. The series begins with cabbage, which is at once followed by the kales, plants the herbage of which and not the roots is the useful part. At the opposite end of the line occur in order the turnips, the radish, and lastly horseradish;[88] all these being so-called root crops, the sum of the members of the family assembled at this point being seven, the line being divided according to nature of the roots as fibrous or fleshy. More than that, the cruciferous series is here again slightly interrupted; for just after the cabbage-kale series, and before that of the real turnip-radish series, two campanulaceous plants are intruded, both of them with fleshy roots so turnip-like in form, and in such frequent use in cookery as substitutes for turnips, that people call them wild

FIG. 89. Turnip (*Brassica rapa* L. emend. Metzger [=*B. campestris* L.]). Bock, *Stirpium* (1552), bk. 2, ch. 61, p. 728; after Fuchs, *Stirpium* (1542), p. 212.

turnips, or little turnips, so that this vernacular name became turned into Latin as *Rapunculus*,[89] the earliest Latin name for the genus now long known as *Campanula*. There are three crucifers described and figured, each of which stands isolated from all the others. One is *Bursa pastoris*.[90] Its having been located so far from its allies of the peppergrass kind was but accidental. At the time of the printing of those, the shepherd's purse does not appear to have been well known to the author, and his first expression concerning it is that its rightful place is next to what he has called *Thlaspidium*,[91] *i.e.*, *Lepidium ruderale*.

Another of the isolated members is *Camelina*.[92] In certain districts it is common, he says, in fields, especially among flax, to which he likens the plant, except as to its having small yellowish flowers. He reports that its seeds ground with grain impart a certain sweetness to bread, also that the

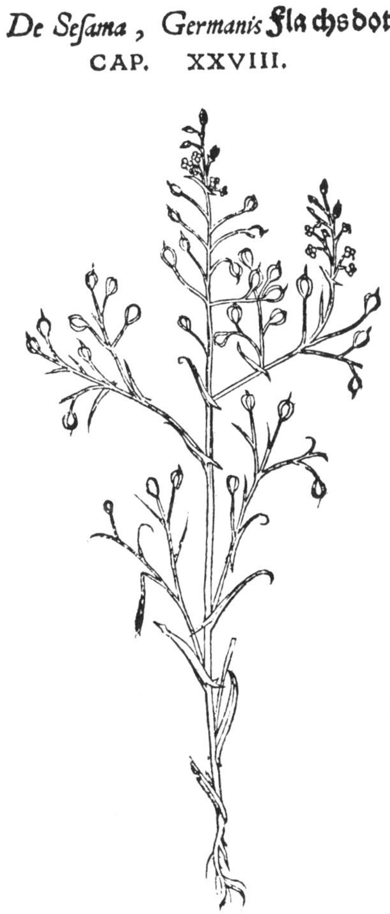

FIG. 90. Gold of pleasure (*Camelina sativa* (L.) Crantz). Bock, *Stirpium* (1552), bk. 2, ch. 28, p. 655.

oil expressed from this seed is, in his opinion, of a flavor superior to that of olive oil. From many a passage in Tragus it is evident he was accustomed to identify plants as of the cress and mustard alliance by a pungent flavor of the seeds. Possibly in *Camelina* they lack this quality. Possibly also our author, variously misled, never tested them in this regard, or thought of such an experiment, for he nowhere intimates that the plant is of that alliance.

The genus *Isatis*[93] is a second member of the crucifers whose relationship Tragus shows no sign of having recognized. Possibly he did not know the plant but by hearsay. His draftsman copied Fuchs's plate of it, and in so doing made the mistake of representing most of the flowers as either five-petalled or six-petalled. It is also to be noted that the fruits of *Isatis* at first sight are sadly bewildering, pendulous like the samaras of the ash tree, which they also much resemble. In the next generation after Tragus, and by one of the most illustrious of all botanical systematists, *Isatis* was indeed referred to the crucifers, though even a century after that its right to a place there was disputed. Tragus will, then, be excused for not having guessed this thing to be a cress-mustard ally; nevertheless, in describing the fruit he proved that a German father of the sixteenth century in his going "straight to nature" might well have taken with him the old Greek father Dioscorides more often than he did; for Tragus, having described the pendulous pouches or bags that succeed the flowers avers, that "this pouch is the seed of the plant." Dioscorides in describing the fruit of *Isatis* had said that "within the pouch there is a seed";[94] and here Tragus's powerful rival, Fuchs, by plagiarizing Dioscorides' whole account of *Isatis*, had been right about its fruit and its one seed.[95] It seems not improbable that a man as keen of botanical vision as Tragus had been relying on other people's statements when he wrote that pouch and seed are here one and the same.

With the exception of these two, *Camelina* and *Isatis*, Tragus's comprehension of the group of crucifers appears to have been complete; and this will become still more manifest by an item of his taxonomic procedure still to be adduced. Fuchs, as was related in the preceding chapter, guided by superficial resemblances in purely vegetative characters, and wholly inattentive to their small flowers and fruits, had received a hedge mustard and a small-flowered vervain as members of one genus which he called *Verbena*.[96] To Tragus this misplaced plant is so plainly of the mustard alliance that he becomes impatient of his rival's blunder.

> This thing is about as much like a verbena as a nettle is like a rosemary bush. I could wish that none should be displeased with me for saying this; but I am aware there are some who will take it much to heart that I have transferred their *Verbena fœmina* to the category of the mustards, and judge me rashly for having done so. But it was reason that compelled me to this course, when I perceived the plant by its whole substance and flavor to be at accord with *Sinapsis*.[97]

Though he names the texture and flavor of the herbage as the reason, that is because it is the one which will appeal to most people, those into whose minds anthological considerations do not enter. It is nonetheless presumable that the diminutively mustard-like flowers, along with the pods and seeds so concordant therewith, first led Tragus to investigate the qualities of the plant.

Some twenty-six species of crucifers all told are described by Tragus. That is probably three times as many as may be found in any author earlier than he, except Fuchs,[98] who allows them to be scattered about according to the alphabetic order of their Greek names, and nowhere gives expression to a thought about their affinities. And Tragus, inasmuch as he plainly discloses his recognition of their consanguinity, except in the case of two species both of them anomalous, ought to be accredited as the first discoverer, so to speak, of this important and taxonomically interesting family. He did not,

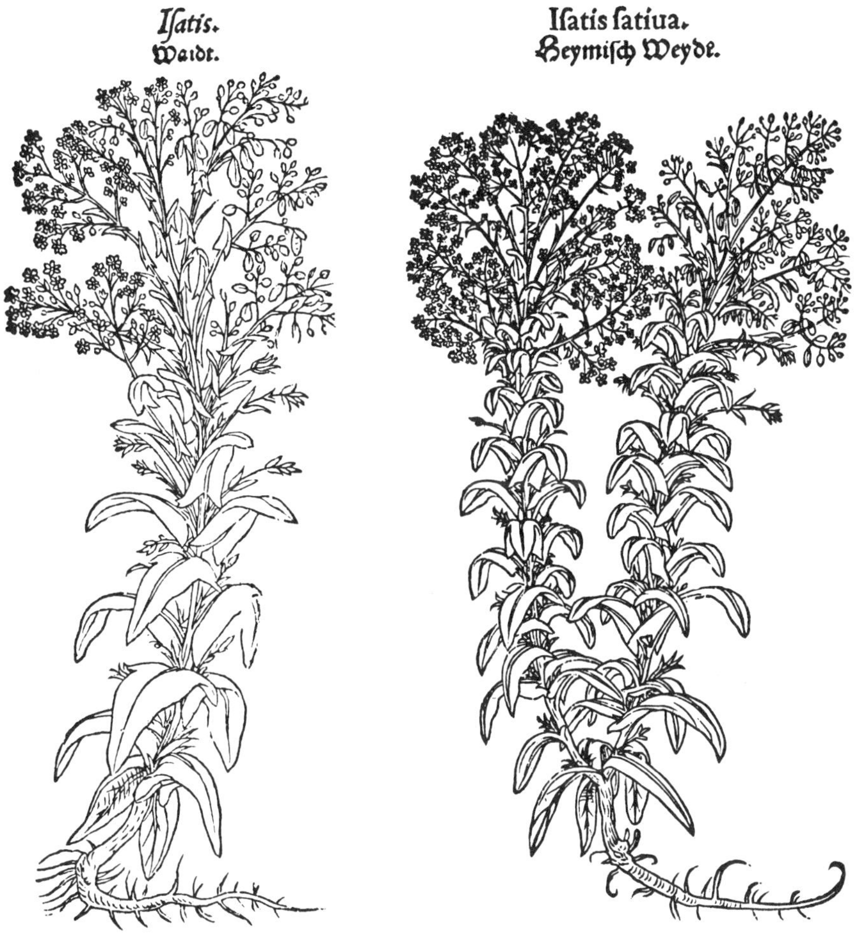

FIG. 91. Woad (*Isatis tinctoria* L.). *Left*, Bock, *Stirpium* (1552), bk. 1, ch. 85, p. 256. *Right*, Fuchs, *Vivæ imagines* (1549), p. 188. The flowers in both figures have four to seven petals, though in the original drawing the flowers have four petals only; Fuchs, *Stirpium* (1542), p. 331.

indeed, name the family. He was not, like the philosophic Theophrastus of old, given to using significant names for such groups of genera. But let us not here perpetrate a fallacy too common with historians, of attributing the discovery of a thing to the man who did but name it, after it had been discovered by another.

There are to be noted in Tragus not a few other instances of decided movement in the direction of a better grouping of genera; but only two or three more may here be allowed even a passing mention.

In our study of Fuchs we had observed that, despite the alphabetic artificiality of his arrangement, by dint of stretching to the uttermost the application of the Greek generic name *Strychnos* (= *Solanum*), he had brought almost all solanaceous plants into one line:[99] but that while a thing as anomalous as *Datura* had thus gained admission to the company of its cognates, *Capsicum* had not been at all apprehended by him as a member of that group. He does not appear to have seen in it any likeness thereto. Tragus, while also doubtless like Fuchs finding the peppery properties of all parts of the plant too foreign to those of other solanaceous genera, nevertheless observes that as to the form of its leaves, and especially of its flowers, it recalls *Solanum*.[100] This was giving that serviceable hint by which later taxonomists were to be led to place *Capsicum* within the lines of the Solanaceæ.

The borrages are a group all the then known members of which are first brought together in unbroken line by Tragus;[101] and he has seven genera of them, embracing something like twice that number of species. All much

FIG. 92. Alkanet (*Anchusa italica* Retz.). Bock, *Stirpium* (1552), bk. 1, ch. 77, p. 232; after Fuchs, *Stirpium* (1542), p. 343.

alike in habit and inflorescence, but differing one genus from another very notably as to nature of the pubescence, and still more so as to form of the "flower," they again come to almost one and the same thing as to the calyx and its . . . three or four naked seed-like nutlets; and all these peculiarities of the flowers, together with the aspect and character of the fructification, Tragus is the first botanist to describe; and he describes them for each genus. In all except the naming of it he is the founder of the family of the Borraginaceæ.

While there is evidence enough that this man's perceptions of plant affinity were keener than those with which any earlier author had been endowed, yet there was never with him any such thought as that natural groups of genera are very numerous; still farther from his mind was such an idea as that all genera may be reduced to a line of natural families. The time was not ripe for the engendering of that thought; nor was it to find expression until three generations later. The list of genera which Tragus cannot systematize, as he had done those of the labiates, the crucifers, and the borrages, is a long one. But he must needs bring them all into some kind of grouping, or succession; and the principles upon which he collocates monotypic genera are various. We must take note of several of them.

There is one quite extended series made up of the following: *Convolvulus*, two species, *Nummularia*, *Cuscuta*, *Humulus*, *Smilax*, *Dulcamara*, *Clematis*, *Bryonia*, and *Lonicera*.[102] There is one character, at least a negative one, by which all these are connectible. Not one of them has an upright stem. All are in some manner climbing; and twining and prehensile plants are much more exceptional, at least in cool-temperate latitudes where Tragus botanized, than one would suppose. Outside of this series now in hand not many were known to him; for the series does not end with *Lonicera*. To that there immediately succeed all the genera of cucurbits that are rough-leaved and are grown in gardens.[103] The smooth-leaved and unvigorous or delicate genera *Bryonia* and *Momordica* had not yet gained recognition as members of the Cucurbitaceæ. And if neither the twining leguminous plants nor the tendril-bearing are placed in this succession, it is for the reason that in their case, and ages before Tragus, the principle of stem posture had been subordinated to the higher one of their agreement with the upright and bushy kinds in a peculiar and distinctive morphology of flower and fruit. Tragus was perfectly aware of all this, and cannot have had so much as a thought of including in this present line the weak-stemmed and prehensile peas and beans.

Sometimes one finds him placing two generic types in juxtaposition for no botanical reason, but only for what may be called a literary motive. The vine, the fig tree, the palm, the olive, and the bay are types not genetically interrelated;[104] neither to Tragus's knowledge was any one of the five related to any other tree. But in ancient history and poetry all had often been associated. They form a group, and that most historic and distinguished, but on the basis not of botany but of literature. And yet, as regards the after part of this line, comprising palm, olive, and bay, one would not dare either affirm or deny that these three had been botanically connected in the author's mind by their drupaceous fruits.

Repeatedly does he bring together two types in every way dissimilar for the sole reason that their names are practically the same; this of course in condescension to those who, in looking for a given plant under a certain name, may find that and its homonyms all in one place. Thus at the end of the buttercups does he locate the herb called *Coronopus* (= *Plantago Coronopus*). It would be irrational to require of Tragus that he should have referred this to the genus *Plantago*. The floral structure which connects them could never have been seen until after the invention of hand lenses at least. The leaves of this *Coronopus* are cut into narrow and remote pinnated seg-

FIG. 93. Honeysuckle (*Lonicera periclymenum* L.). Bock, *Stirpium* (1552), bk. 2, ch. 96, p. 822; in part, after Fuchs, *Stirpium* (1542), p. 646.

ments and beset with bristly hairs, on the whole as far from plantain leaves in form as imaginable. The form of these leaves had procured for the plant the name *coronopus*, Greek for crowfoot, and that very anciently.[105] *Ranunculus* had also for a second Latin name *Pescorvi*, the exact equivalent of the Greek *coronopus* and English crowfoot. Because they had the same name our author placed side by side these different plants. He did the like with the labiate that was usually called *Hedera terrestris*, that is, ground ivy, locating it next the true ivy, *Hedera Helix*,[106] as well aware as any one ever was that there is no consanguinity between them; but this disposal of it would suit the convenience of those untaught in better classification.

Nomenclature

No special attention is given to nomenclature by this author. He follows the usages of antiquity and of his own period, yet in ways of his own by

FIG. 94. Composite illustration of five species of Cucurbitaceæ. Bock, *Stirpium* (1552), bk. 2, ch. 99, p. 831; adapted from five figures in Fuchs, *Stirpium* (1542), pp. 697–701.

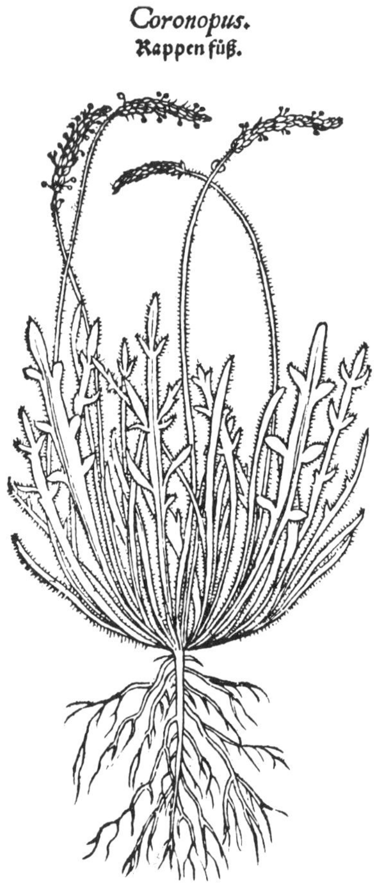

FIG. 95. Buckhorn plantain (*Plantago coronopus* L.). Bock, *Stirpium* (1552), bk. 1, ch. 31, p. 99; after Fuchs, *Stirpium* (1542), p. 449.

FIG. 96. Lesser knapweed (*Centaurea nigra* L.). Bock, *Stirpium* (1552), bk. 1, ch. 72, p. 219.

which it comes to pass that he illustrates those usages uncommonly well. We have already observed that such family names as Umbelliferæ, Cichoriaceæ, Carduaceæ, Legumina for the leguminous plants, and Malvæ, for the malvaceous, had been in familiar use time out of mind.[107] Tragus essays the addition of a few new terms of that kind to comprehend other groups of genera, but these have not been successful; and the cause of their failure will readily be seen. He proposed the name Serpentariæ[108] for that group of trailing, twining, and climbing herbs referred to above as embracing *Convolvulus*, *Humulus*, *Clematis*, and others, both the outlining and the naming of it being made, curiously enough, just at a time when taxonomists—even Tragus himself foremost among them—began to depend less upon the texture and posture of stems, and more upon characteristics of flowers and fruits. And his small group of the Lappæ was equally futile; for while this series might at first view seem to be connected upon a thread of fruit characters, a more attentive inspection brings it out that nothing more significant than the hooked character of the prickles investing the fruits holds the genera together; for they are *Lappa*, *Xanthium*, *Trapa*, *Caucalis*, and *Agrimonia*.[109] Thus Tragus's two new family names, Serpentariæ and Lappæ, were both destined to suppression, because the grouping in either case was little better than fanciful, being based on agreement as to certain peculiarities that are of no general taxonomic value.

Upon the then settled principles of generic nomenclature—principles approved by all antiquity—Tragus attempts no inroads. It does not enter his thought to question the perfection of the established methods in naming things. A generic name of two words, noun and adjective, suits him as well as one of a single term and that substantive, perhaps even better, as signifying somewhat more; for there is more of meaning conveyed by a noun qualified by an adjective than there is in a noun standing alone; and the time is yet distant when meaningless and cabalistic names will be tolerated. So when he becomes the discoverer of a new and nameless generic type that is an ally of *Cyanus*, the common cornflower, or bluebottle, though not of its genus exactly, he assigns the new genus the compound name *Cyanus silvestris*;[110] and we, well aware that half the generic names in sixteenth-century botany are thus made, must read his whole account of the plant in order to assure ourselves that he does not, after all, mean simply a new species of the genus *Cyanus*.

Fuchs, convinced that the genus *Plantago aquatica* is identifiable as the *Alisma* of the Greeks, had taken up the latter name;[111] but Tragus shows a preference for the two-worded appellation and restores it, taking pains also to inform the untaught that, although the plant's name is *Plantago aquatica*, it does not belong to the genus *Plantago*.[112]

Even for Fuchs's new genus aptly named *Digitalis*[113] Tragus thinks that such a two-worded name as *Campanula silvestris* would be better; and he formally proposes this as a substitute, writing *Digitalis* down as a synonym. His remarks at this point illustrate well the idea then prevailing, that the nomenclature of newly proposed genera should be freely open to amendment and improvement: "Let any one name this plant what he will. We, in consideration of the form of the flower, shall name it *Campanula silvestris*, at least for the time being, and until a still more appropriate name shall arise. There are those who call it *Digitalis*."[114]

While Tragus, like others both before him and long after, leaves the representatives of monotypic genera without specific names, yet up and down the

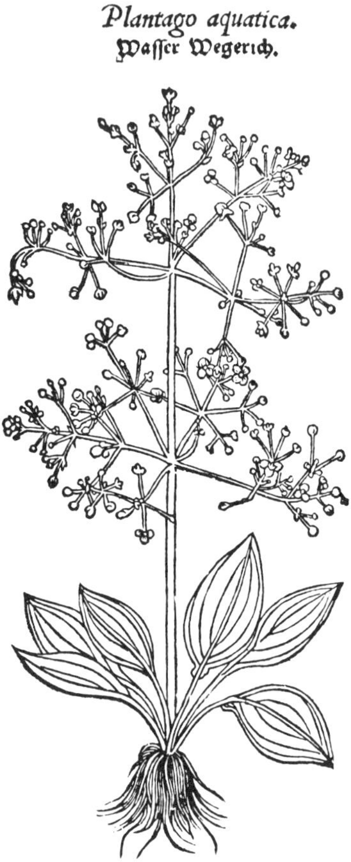

FIG. 97. Water plantain (*Alisma plantago-aquatica* L.). Bock, *Stirpium* (1552), bk. 1, ch. 75, p. 226; after Fuchs, *Stirpium* (1542), p. 42.

margins of a great majority of his 1,100 pages[115] are the binary names of species. If many of these seem to consist of three terms, it is usually because two of them constitute the generic name. Occasionally the third word indicates that what is in hand is a mere variety of the species preceding; and now and then it will be seen that a fourth word is introduced to indicate the variety. In case the generic name itself is binary, the fourth term becomes needful for the indicating of a named variety. Still there is no trace in this author of those phrase names that became a burden upon the botany of the seventeenth and eighteenth centuries. The marginal placing of these binary names is as universal with Tragus as one finds it in Linnæus; but these same names are often repeated as headings to the chapters, and again over the plates. The author employs less freely the numeral adjectives for specific names, and has not many that are geographic. The personal names for species are less rare, though most of these are of earlier mediæval origin, commemorating saints of the Roman calendar; and he is perhaps the first of botanists to have dedicated a new species to himself.[116]

Ecology, Phenology

Tragus is far from emulating in any general way the endeavors of Theophrastus to indicate groups of plants ecologically considered; but there is one piece of such work that ought not to be allowed to pass unnoticed. The Third Book of his volume is to be devoted to the trees and to other lesser but strictly woody growths.[117] Accordingly, in the first chapter at its beginning there is introduced the figure of a large tree, a spruce tree, as we are able to determine from a branch or two of small dimensions which are all that remain alive; for the tree is moribund. All up and down its trunk there are fungi and lichens of several kinds; then upon the ground beneath are as many more. The text of the whole chapter, and it is a long one, covering

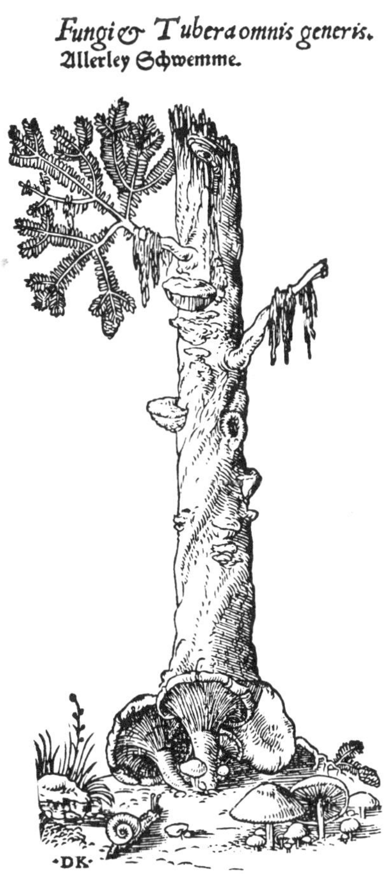

FIG. 98. Several kinds of mushrooms shown in natural habitat. Bock, *Stirpium* (1552), bk. 3, ch. 1, p. 940.

seven pages, relates exclusively to these plants, and a considerable number of them of various genera, embracing an aggregate of some thirty species. The next chapter treats of mosses; the third, of mistletoe. Thus the whole assemblage of German saprophytes and tree parasites is made one ecologic group,[118] as of things growing together, many of them upon trees, most of the others upon the ground beneath trees.

Ecology, however, forms an item and a very distinct one in the account of almost every wild plant which he describes, a fact that will be best impressed by a few citations:

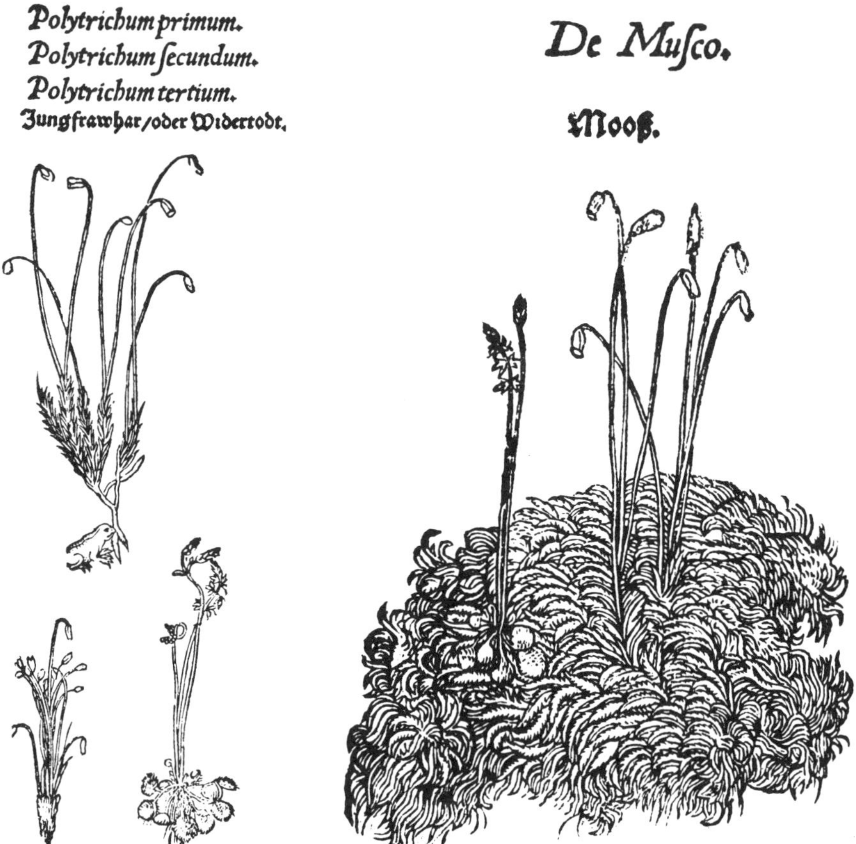

FIG. 99. Moss, as understood by Bock. *Left*, sundew (*Drosera rotundifolia* L.), hair moss (*Polytrichum commune* L.), and either *Bryum* sp. or *Mnium hygrometricum* L. The two moss figures are after Fuchs, *Stirpium* (1542), p. 629. *Right*, three different species: *D. rotundifolia* L., *Polytrichum* sp., and probably *Sphagnum* sp. This figure contains probably the first illustration of a calyptra (on the capsule of the only vertical sporophyte). *Stirpium* (1552), bk. 1, ch. 181, p. 528; bk. 3, ch. 2, p. 946.

Viſcum.
Miſtel.

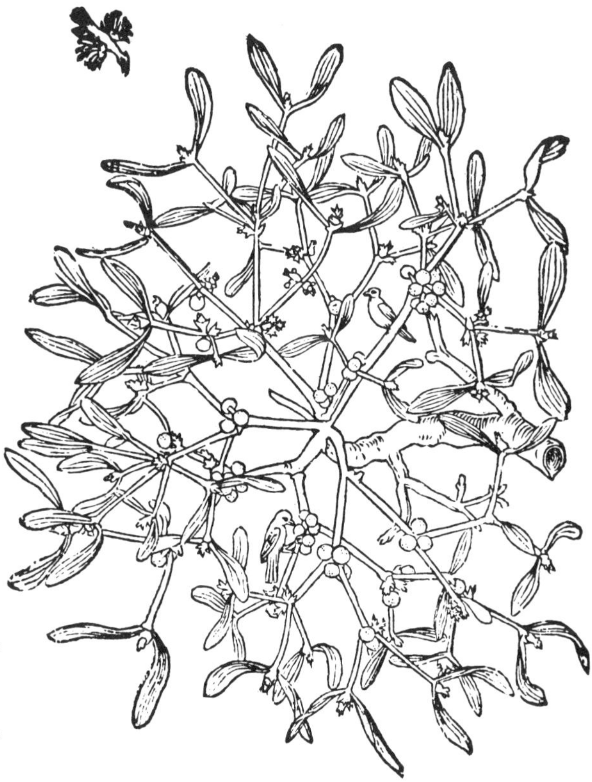

FIG. 100. Mistletoe (*Viscum album* L.). Bock, *Stirpium* (1552), bk. 3, ch. 3, p. 949; after Fuchs, *Stirpium* (1542), p. 329.

Asarum affects shady places where the soil is rather moist, and is usually found under thickets of hazel, but sometimes also in deep damp woods.[119]

Alliaria is an elegant plant which in the month of April is found in certain waste places, under walls, along the bases of hills, by hedges and in cavernous places which are the abode of lizards and other vermin.[120]

The chickweed is located thus: "This most common of herbs is found throughout the whole year in gardens and vineyards; and the richer the soil, the more large and tender the herbage."[121]

Ranunculus sceleratus "grows in low swamps, especially if the soil be sandy, and preferably where there are frogs; but occasionally in very rainy years it will be found in wet lands that are more elevated."[122]

"Fumaria grows in gardens, fields of rye, and also among flax, onions,

FIG. 101. Garlic mustard (*Alliaria petiolata* (M.B.) Cavara & Grande). Bock, *Stirpium* (1552), bk. 1, ch. 27, p. 86; after Fuchs, *Stirpium* (1542), p. 104.

and cabbages, where it flowers in May, and again in autumn it reappears in turnip fields."[123]

The almost omnipresent knotgrass, *Polygonum aviculare*, he thus descants upon:

> Polygonum among common plants is the very commonest of all, at the same time a useful one also. What part of the country is there where one does not meet with it? What roadside is there where it does not abound? What fields (for in cultivated fields it particularly delights), what hedgerows, and what by paths are not covered with it?[124]
>
> Aquilegia, mostly a garden plant with us, also grows wild in elevated woodlands, on rocky hills, and sometimes in the crevices of precipitous rocks.[125]

In phrases like these does Tragus almost always particularize about the habitat, the soil, the exposure of the wild plants of Germany which he describes; and it would not be difficult to gather out of these antiquated and yet living pages definite outlines of the plant associations of every part of the country with which he was familiar.

The fungi, lichens, and mosses, already alluded to under this heading, are not the only plants in connection with which he permits ecological considerations to influence his taxonomy. He collocates in an unbroken series, as plants nearly allied, broad-leaved houseleeks, and sedum species the leaves of which are small and terete. Such a series is of course a faultless one in the judgment of modern botany, because the structure of flower and fruit is the same in all; but the case was otherwise four hundred years ago, when anthology was hardly yet in embryo, and even leaves were more generally received as furnishing the criteria of affinity. Tragus had to defend the position he had taken when placing certain small plants regarded as leafless in the same line with live-forever and houseleek as their next of kin. He himself could not claim that *Sedum acre* and its cognates had leaves at all. They exhibited, he said, in what seemed to be the places for leaves, grain-like things which he preferred to call acini; and an acinus may be a seed, a grain, a germ, or even a berry. He has but one argument to offer in defense of this line of broad-leaved things and things green though leafless, as being a natural series, and that argument is purely ecological. All of them inhabit peculiarly the roofs of buildings, and thrive there much better than elsewhere. Even such of the species as now and then establish themselves on the ground are never seen but in the most open exposures. All of them everywhere avoid all protection from extremes of temperature, retaining their fresh verdure unimpaired under the rigors of the severest winter.[126] Such ecologic groupings are of course traditional, having come down from earlier times; and under such defense as Tragus makes of this one, his contemporaries would perhaps admit its validity despite the great diversity among the members of it as to foliage.

Another instance of this kind of procedure to which I wish to call attention is the reverse of the above as to the result attained. German species of the rather ample genus *Veronica* are placed in widely sundered groups on principles as purely ecologic. In one place, under the generic name *Sium non odoratum*, he describes what is now *Veronica Beccabunga*, appending to his excellent diagnosis the following: "It grows around springs which never freeze, or in such ditches as are equally immune from frost during the whole winter."[127] Then the habitat of a second species of the genus—*Veronica Anagallis aquatica* is its Linnæan name—is given in similar terms: "Throughout the whole winter season this keeps its verdure quite untouched by frost, growing as it does in the water of warm springs." Now six chapters away from this which treats of the two aquatic veronicas, and with more than as many plants not allied to *Veronica* intervening, he describes the dry land members of this same genus, but under the generic names *Chamædrys* and *Teucrium*.[128] These have retained in more recent botany those generic names as specific under *Veronica*.

Thus do we find that our familiar genus *Veronica* was all unrecognized as a whole by Tragus, its members being ranged in two rather widely separated groups, bearing different generic names; and all this in deference to mere ecology, as it were, for, if those of the aquatic group have a tender subsucculent and glabrous herbage holding its freshness all winter, whereas those of the dry land are thin-leaved, soft-hairy, and die down to the ground in autumn, and if these differences may have helped to keep the groups apart, yet are they anatomical differences rather than morphological. And the case cannot fail to convince us of the weight which ecological considerations carried in sixteenth-century classifyings. Neither Tragus, however, nor any of his contemporaries had invented these ecologic distinctions. They were already an old, old story. Contemplate the mere name for those aquatic speedwells, *Sium non odoratum*. It is a generic name, because there are two very clearly distinct species of it. There is somewhat of early botanical history concentrated in that very name. It implies beyond mistake the existence of a genus named *Sium odoratum*. Still further it suggests as almost certain that the name *Sium odoratum* is less ancient than the other. Searching old records now, we shall find that things happened exactly after the manner which the name *Sium non odoratum* seemed to indicate. *Sium odoratum*, the original of all siums, was at first *Sium*, simply, that is, a monotypic genus. More than a thousand years before Tragus, and maybe two thousand or three, the Greeks had known, and had used medicinally, an aquatic of springs and spring runs that never freeze. It had the foliage and the pleasantly odorous quality of certain umbellifers, as Dioscorides had intimated;[129] and this was *Sium*;[130] and centuries later when up in middle Europe men versed in the materia medica looked in wild springy places for the *Sium* of Dioscorides, and more often found there those different things, unknown to the ancients, which though green in midwinter were odorless, and therefore not the real thing, and named them *Sium non odoratum*, they were proceeding upon the principle that both and all such plants, being generated as most of them believed by spring water and the earth at the bottom, were naturally allied, and might all be named so many kinds of *Sium*. Furthermore, the establishment of such a name as *Sium non odoratum* rendered it needful that original *Sium* should be invested with a cognomen in order to avoid misunderstanding and confusion. Hence its later generic name *Sium odoratum*.

That this sium, constant inhabitant of springs and warm drainage ditches, is classed not ecologically but morphologically by Tragus argues no inconsistency. It would be one thing for a sixteenth-century botanist to fail to recognize by morphologic marks the membership of the Scrophulariaceæ, and

Sedum.
Haußwurtz.

Sion non odoratum.
Bachpungen.

FIG. 102. Houseleek (*Sempervivum tectorum* L.). Bock, *Stirpium* (1552), bk. 1, ch. 123, p. 376; after Fuchs, *Stirpium* (1542), p. 32.

FIG. 103. Brooklime (*Veronica beccabunga* L.). Bock, *Stirpium* (1552), bk. 1, ch. 61, p. 188; after Fuchs, *Stirpium* (1542), p. 725.

quite another thing to miss the family characters of any umbellifer. Sium at first glance, as well as by its properties, is unmistakably an umbellifer; and the time is not to be found in the annals of botany when this family had not obtained general recognition, marked as it is both morphologically and qualitatively. The family, so called, to which the veronicas belong is not so. The Scrophulariaceæ have never yet been circumscribed otherwise than most arbitrarily and unsatisfactorily. Tragus understood well the superiority of morphologic over ecologic criteria, and that the latter are to cede to the former when the former are manifest. The anthologic harmony between hardy undying water veronicas and the tender perishable kinds of dry meadows and uplands it was not given him to see, nor, indeed, to any one until long after Tragus's day. And yet, an umbellifer to him was an umbellifer whether hydrophilous or xerophilous. But in the arranging of his umbelliferous genera it will be observed that the two aquatic genera *Sium* and *Apium* are placed side by side.[131]

To the botanist of the fields, the plains, the marshes, and the mountains it is easy to recall the special habitat of almost any plant, but not so the average time of its flowering year by year. To be able to say that this shrub will be found in bloom about the middle of May, that tree in the early part of May, this flower appears late in March, the other in the last days of April, is not likely to be a matter of unaided memory. To know the average time of flowering for everything that grows involves the keeping of written notes through years. Tragus tells this time of the annual flowering of things in almost every chapter of his book; and he is the first of botanical authors to have done this.

Transmutation

Though much given to diversifying his botanical pages by bits of invective against superstitions that are of theologic type, Tragus has never doubted the easy transmutability of wheat and rye into chess. In a long chapter he demonstrates to his readers how this may and does come to pass, under various conditions. And here some experimentations of his own are recorded: "That it is possible for seeds of one species to degenerate and become so changed as to come up as another species is something which I have learned by experience; for from very old cabbage seed sown by my own hands I have raised a crop of turnips."[132] At another place he has the following upon the same subject:

> There are those who think that a sowing of turnip seed upon very dry and sandy ground, especially if the seed be very old, will come up as wild mustard; or at least in that which is as much of the nature of mustard as of that of turnip. In the same fashion cabbage seed very commonly changes into that of a poor and stunted kind of turnip, as I myself have often proven by experiment.[133]

Again in his dissertation upon wheat he reports a certain dark-grained kind as apt to appear intermixed with the other in the low moist parts of the fields, so dark—blackish is his word—as to render flour and bread from such admixture dark-colored. He in perfect confidence accepts this admixture of dark-grained as another instance of transmutation, has never a suspicion that it is another variety of wheat, the seed of which was mixed by chance, in the sowing, with the other.

The case is one entirely apart from that of the melampyrum, or black wheat, of a totally different plant alliance, the seed of which, accidentally harvested with the grain and ground with it, also caused a darkening of the bread; for on this weed he has a chapter apart,[134] as well as another upon smut[135] in grains.

From our view point of four centuries later it may not be easy to understand how such childish fancies could hold their places firmly in gifted minds which, after all, were seriously bent on cold scientific enquiry into all nature's mysteries and were often successful; but of the fact itself there is abundant proof.[136]

8. Euricius Cordus 1486-1535

One of the most gifted and scholarly men among all who figured in German botany in the early sixteenth century is Euricius Cordus, who, though a cultivator of plants, and also a zealous field botanist, wrote no great book, and is chiefly interesting here as having been the father and the educator of that most brilliant of early German botanists, Valerius Cordus.

Life

Henricus—the name was altered by himself in his school days to Euricius—was the thirteenth child of a pair of honest and worthy Hessian peasants, and was born at Siemershausen in the year 1486.[1] His parents died when he was a child, and in some way he became for a time the inmate of a collegiate school at Frankenberg. Here he formed a strong and lasting attachment to a youth, his junior by two years, who afterwards under the adopted name of Helius Eobanus Hessus became celebrated as a philologist and as one of the most elegant Latin writers of the period.[2] On account of a treatise upon dietetics favoring vegetable foods,[3] which in its day was well received and passed through several editions, Haller has enrolled the name of this Eobanus Hessus in his list of botanical authors. What influence he had upon botany was more indirect. It was evidently by virtue of this strong attachment between Hessus and Cordus that the latter was brought finally to devote himself to intellectual pursuits.

After those first school days at Frankenberg, and while Cordus was very young, he married and was settled at his native Siemershausen; in what occupation no records tell; but when in the year 1515 a son was born to him, the event appears to have stimulated him to renewed endeavors to attain distinction in scholarship, for before the son was two years old Euricius Cordus had won the Master's degree at the university of Erfurth, his special studies having been the ancient languages and philology, the same which his particular friend Eobanus Hessus had been all the while pursuing. The year following, *i.e.*, 1517, we find him a student at Leipzig, where also he gives lectures on pastoral poems in Latin of which he is himself the author.[4] Here also he makes a lasting friendship with the young Joachim Camerarius—father of the botanist of that same name—a much younger man than Cordus, and at the time a student at Leipzig, and who subsequently became distinguished in philology. Camerarius soon removed to Erfurth, and Cordus returned thither with him. Eobanus Hessus was still there, and the three determined to open there a select school of their own. That Cordus's lectures

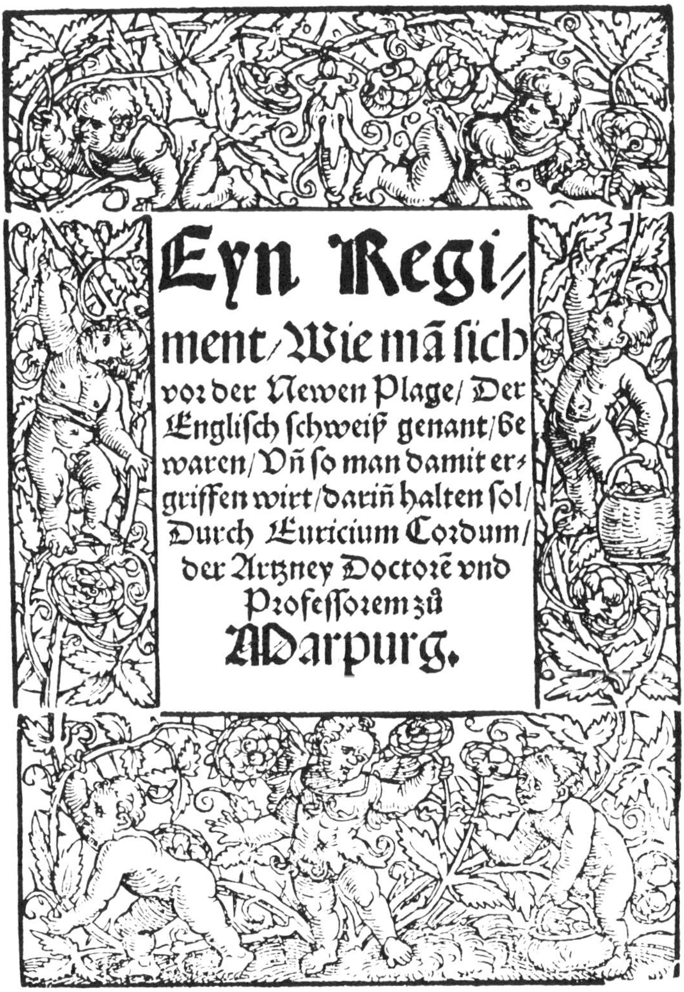
Eyn Regi/
ment/ Wie mã sich
vor der Newen Plage/ Der
Englisch schweiß genant/ be
waren/ Vñ so man damit er-
griffen wirt/ dariñ halten sol/
Durch Euricium Cordum/
der Artzney Doctorẽ vnd
Professorem zů
Marpurg.

FIG. 104. Title page of Euricius Cordus, *Ein Regiment wie man sich vor der newen Plage* (1529). (From National Library of Medicine, Bethesda, Maryland)

and poems had already earned for him a reputation is evinced by this, that his opening a seminary of learning brought him a letter of congratulation and good counsel from so great a celebrity as Desiderius Erasmus.

The time at which this new school enterprise was undertaken proved inopportune, a time when, in Germany, even the oldest and most renowned seats of learning were realizing the influences of that ecclesiastical and civil upheaval commonly called the Reformation; and Cordus and his companions closed their school in the year 1521. And now, as if in hope of thereby gaining a better, or at least a surer, living for himself and his family, he entered the medical profession. Without the means of journeying to Italy and maintaining a year's residence at the most celebrated school of medicine in Europe, that of Ferrara, a physician at Erfurth, one Doctor Sturtius,[5] offered him financial aid; and at Ferrara, in 1522, Cordus received the Doctorate in Medicine at the hands of the venerable Leonicenus, then ninety-four years of age and still active in the discharge of his professorial duties.[6]

Returning now to Erfurth, Cordus practiced medicine during four or five years, and then in 1527 accepted an appointment to the chair of medicine in the newly founded Protestant university at Marburg, from which movement the historians infer, and not unreasonably, that Cordus had abandoned the Catholic faith and become a Lutheran. In this new position he found leisure for study and authorship, for he translated into Latin verse both the *Alexipharmaca* and the *Theriaca* of Nicander, both published by Egenolph of Frankfort, in 1532.[7] Here also his own Latin pastoral poems were published, but without date, or name of publisher. Here also he wrote and gave to the public his one botanical work, the *Botanologicon*, which closed his career as an author. This was published at Cologne in 1534. In this he complains of oppositions and persecutions which he has had to endure at Marburg, such as had compelled him to accept a proffered appointment as City Physician at Bremen, to the Senate and Citizens of which city the book is dedicated. He died at Bremen in 1535, at the age of forty-nine years.

The 'Botanologicon'[8]

The book is in the form of a colloquy between Cordus and a few friends of his, most of them away back in their younger years fellow students at Erfurth, all now men of middle age, physicians, pharmacists, or men otherwise interested in plants, at least the medicinal. As having been university students of medicine every one of them is assumed to be somewhat familiar with all the ancient line of Greek and Roman authors who had written on the materia medica, and whose books were still the standards of study and reference.

Euricius Cordus, even while young, and as yet aiming at nothing else but distinction in languages and philology, had been a great lover and cultivator of plants, training his child Valerius from infancy to know and love them.

EVRICII
CORDI SIMESVSII MEDICI
Botanologicon.

HEVS MEDICE

Vis uarias aliter quàm doctus es hactenus herbas
Scire, nouus multas iste libellus habet.
Vt retinax primum sibi testa reseruet odorem,
Sex nisi quadrantes & breuis hora perit.
Quæ si quàm nostris lusorum perdere chartis
Malis, tunc aliquid doctius æde tuum.

COLONIAE
Apud Ioannem Gymnicum Anno
D M XXXIIII.

FIG. 105. Title page of Euricius Cordus, *Botanologicon* (1534). (From National Library of Medicine, Bethesda, Maryland)

Then when through mad religious partisanship the universities of Germany began to suffer disruption and depletion,[9] and Cordus with a family on his hands was obliged to prepare for a remunerative calling, he was trebly prepared to make a mark in botany. He was a genius. He was intensely a lover of plants. He was uncommonly well skilled in those ancient languages in which the old standards of the materia medica had been written.

The useful purposes which the *Botanologicon* has in view are several, and are essentially reformatory.[10] Prominent among them is that of demonstrating that, through sheer ignorance, a considerable proportion of the jars and drawers and packets in the drug shops are falsely labelled. They are marked with the names of Dioscoridean and Galenian roots and herbs, while commonly filled with things which cannot be the same as those which the ancients knew and made much of under those names. If this was really the case it would follow that the lives of those in illness calling for a certain powerful remedy were apt to be endangered by the administration, either of some drug wholly inert, or else with properties aggravating rather than remedial to that ailment. It was a strong arraignment of the whole united company of the doctors and pharmacists of the time; and Euricius Cordus complains bitterly of the oppositions and persecutions that had followed him and driven him from place to place, while lecturing and writing in endeavors to correct this ignorance. He was not philosopher enough to comprehend that just this abuse and this feigned contempt are the very highest encomiums: the only attestations of his learning and genius which the envious horde of the criticized and the offended know how to pay.

In this colloquy, the *Botanologicon*, there are given many particular instances of errors on the part of the physicians and druggists as to plants. We have already noted, in our study of Brunfels, how that author mistook German species of *Corydalis* for the classic *Aristolochia*, thus, at once agreeing with, and confirming in their ignorance, the whole array of the German doctors and druggists of his time.[11] Euricius Cordus takes up this case as one which easily establishes that for which he contends. One member of his party reads from Dioscorides that *Aristolochia* has leaves like the bush ivy, *i.e.*, well rounded and entire.[12] This which the Germans call by that name has leaves dissected like those of rue. The leaves and even the flowers of ancient aristolochia were described as having an odor somewhat sharply aromatic, a quality of which there is no trace in these fumariaceous herbs. The root of these, it is confessed by all the party, are rounded and turnip-like, as Dioscorides and all the others of olden time had described those of *Aristolochia rotunda*; but that was the only point at which the *Corydalis* answered to the *Aristolochia* description. The fact was plain that people in comparing the plant with the ancient diagnosis of aristolochia, finding that the root agreed, became at once blind to all the points of disagreement. Among the many instances of this kind of error the author presents that of the druggists' having mistaken the common wild plum of German woodland borders for the acacia of the ancients.[13] The acacia had been described as a thorny tree, yielding a mild gum. In these two points the wild plum was at agreement with the old acacia description. They gathered this native German gum and made the accustomed uses of it, believing all the while that in this thorny tree they had the real gum-bearing acacia. Cordus invites their attention to an important part of that classic diagnosis to which they all the while have been blind, that which says that acacia produces its seeds in a pod, after the manner of the lupine; and Cordus asks his party how this

FIG. 106. Birthwort (*Aristolochia pallida* Willd.). Dioscorides, *Codex Aniciæ Iulianæ* (1906), fol. 18ᵛ, tracing in Singer, Herbal in antiquity (1927), p. 9.

gum-bearing thorny tree that yields plums instead of legumes can possibly be believed to be the acacia? It would be easy to multiply by twenty the examples of this kind which find mention and full demonstration in the *Botanologicon*; but the above must suffice. They are representative.[14]

Almost every one who has written a few chapters of botanical history has made record of the fact that at first the botanists of middle Europe wrongly expected to find, and as wrongly believed themselves to have found there, most of the plants that had been described by ancient Greek and Arabian authors. The discovery of the error and the correction of it have usually been credited to botanists of later centuries. But Euricius Cordus is the man who at the very outset, and himself a German, saw the magnitude of this mistake, and so clearly exposed it, that despite the rage of an incensed multitude of doctors and apothecaries, the reform began at once. If Fuchs in his great folio of pictures had in many an instance corrected Brunfelsian errors as to the identity of plants, it was largely if not altogether due to his having studied and been guided by the *Botanologicon*. He pays full tribute to the importance of this work in his Preface, which was not written until after the demise of Euricius Cordus. In this Fuchs says:

In this work of restoring botany Brunfelsius was succeeded by Euricius Cordus, a man of high integrity, great industry, distinguished as a poet, and a man of varied learning. What he accomplished for the elevation and advancement of botany so abundantly appears in his *Botanologicon* as to need no further commendation from us. But this I wish to say; that one so singularly qualified seemed worthy of a longer life, in which to have contributed to medical botany much more matter of the same high import.

At the time when the *Botanologicon* appeared, the illustrated folios of Brunfels were quite new, and the work is often referred to. On one of the earlier pages of this botanical colloquy there is such a record of contemporary opinion about the merits of Brunfels's book as cannot fail to be interesting, and for the sake of which I shall attempt a reproduction of the whole passage:

MEGOBACCHUS. Since Gallus desires it, may we not all go out and botanize a while?

RALLA. Please grant us this favor.

CORDUS. I am willing, and the more so as hoping to learn something from all of you.

NIGER. *You* learn from *us*?

CORDUS. Why not? I have sometimes learned botany from illiterate women and peasants. I am not ashamed of gaining information from a child even, much less from you, men of learning. So whenever it pleases you let us go forth. I will not keep you back; nevertheless I, just as if none of you were here, shall follow my usual practice of taking along a little book or two. I take the greatest delight in these sallyings into the country, where I can have before me fresh and growing those herbs which I have read about at home, and may compare them with the pictures of others which I carry in memory; also taking such note of their names and reputed virtues, as I may gather such from old women whom I meet upon the way. By the use of all these means I am the better able to arrive at a sound conclusion, or at least a more probable opinion, about the identity of a thing.

GALLUS. I wish that Brunfelsius had followed your course; for concerning a good part of his plants it will have to be said that he named them not according to their descriptions, but after the opinions of the ignorant vulgus.[15]

CORDUS. How do you know that?

GALLUS. I have read his two volumes, and have compared them with Dioscorides.

CORDUS. That was advisedly done; and if there are errors, we must overlook them, and stand by the things that are well and rightly said.

GALLUS. What if, in the meantime, by virtue of that authority which a new and plausible work must carry with the unlearned, those old errors which ought to have been eradicated, are only made to strike root still more deeply? I do not think this a matter to keep silence about.

CORDUS. Speak out, then; and if you wish to, cry aloud.

GALLUS. This should be your province rather than mine.

CORDUS. Cordus himself is too unlearned to undertake that piece of criticism. Let him address himself to the task who may choose to; I shall only indicate, with all candor and open-heartedness, some things which I do know and stand by; yet all the while ready at any moment, and without a blush, to be taught by any of you who may know better than I.

RALLA. What are we waiting for?

CORDUS. Young man, lay aside that genteel cloak, and carry this little volume of Dioscorides.

NIGER. You advise well; this author alone suffices.

GALLUS. I have the two volumes of Brunfelsius.

CORDUS. If it please you, we will first enter this my little garden by the house.[16]

The conversation carried on between the five friends while on this botanical excursion constitutes the body of this rare booklet of 185 pages, and gives a clearer insight into the state of medical botany in middle Europe in the time of Brunfels, Fuchs, and Tragus than could be gathered from the most exhaustive study of those authors' folios themselves.

9. Valerius Cordus 1515-1544

Hitherto Valerius Cordus remains almost unknown except by name. Not one of the four of his own countrymen who wrote botanical history within the nineteenth century ever looked into Cordus's writings—all of them published after his early death—far enough to see whether he had been least or greatest among German botanists of the sixteenth century.

Sprengel says that "Valerius Cordus, son of Euricius, if he had lived longer, might have given to his works a certain maturity which is conspicuously wanting to them,"[1] after which anything but laudatory opinion he proceeds to give the young man full credit for having travelled widely in many parts of Germany, and for having discovered and published a goodly number of new plants.

Of a very different tenor is the language of Ernst Meyer: "A splendid and all too transitory phenomenon was Euricius Cordus' son Valerius."[2] While in my view this opinion of Valerius Cordus's merits is not extravagant, still Meyer, as it seems to me, fails to show reason for such high praise. He does much better than Sprengel; yet I am obliged to infer that he borrowed the opinions of sixteenth- and seventeenth-century authorities on Cordus, rather than as having made the young man's writings a subject of careful examination. The estimate of the best botanists of two generations after Cordus was an exalted one, as to his merits; and that outside of Germany.

Sachs in his volume of history seems to have found it easy to adopt Sprengel's tone of indifference to this youth of rare genius. "For the present," he says, "we pass by Valerius Cordus, Conrad Gesner, Matthioli, and several others of no importance,"[3] etc., etc.; and this is the only reference to him which I have been able to find save mention of the fact of his having observed the sensitivity of certain leaves.[4]

Such conflict of opinion on the part of nineteenth-century writers has seemed to make it incumbent on me to determine if possible, out of Cordus's writings themselves, why it was that for nearly two centuries before the

FIG. 107. Valerius Cordus (1515–44). Sambucus, *Icones . . . medicorum* (1574), no. 51. (Print at Hunt Institute)

nineteenth he was held in such esteem by accomplished botanists of every nationality in Europe.

LIFE

Valerius Cordus was born in the year 1515 at Siemershausen, where also his father Euricius Cordus had been born. There is a beauty and a certain pathos in the story of how the destined father while a boy had made fine progress on the road to higher learning, and then by an early marriage, evidently in poverty, seemed to have extinguished all hope of a scholar's career. Yet when this son was born, the young father's zeal for learning returned, for to the child there must be given every advantage of high education; and he himself would be the educator of his Valerius. In some way he managed, as we have already seen, to go to the university of Erfurth, where he very soon obtained his first academic degree. Thenceforward he supported his family by teaching and lecturing until other degrees had been gained, and he was settled in the profession of medicine, meanwhile training his boy carefully in the ancient languages, philosophy, and the sciences, among them botany in particular. We have a biographer's testimony to this.

> Valerius Cordus was imbued with an incredible zeal for learning thoroughly not only medicine, but also the right recognition of plants, to which latter his father Euricius, a physician and also an illustrious poet, urged him by both precept and example; for he had reared the child even from the cradle in the midst of herbs and flowers.[5]

Cordus took his degree in medicine at the university of Wittemberg,[6] and there shortly after gave regular lectures on Dioscorides, and with such marked acceptance that to the audience of medical students certain professors joined themselves.[7] It must have been earlier than this that he prepared his *Dispensatorium*, or manner of preparing all medicines, the only work of his that was printed in his lifetime, and which was a great success. He had not even intended it for publication, but appears to have written it as a pastime while with his uncle, Joachim Ralla, an apothecary at Leipzig. But the uncle thought so highly of the manuscript that he placed it before the magistracy of the city of Nuremberg, and they ordered it printed.[8] It was first published in 1535, was often reprinted during the next 150 years, and was even translated out of the original Latin into French.[9]

The *Annotations on Dioscorides*, being a kind of abstract of his Wittemberg lectures, were not published until five years after his death, and were never by Cordus himself prepared for the press; perhaps not even so much as once written down by him at any time, for the printer's copy, when it came to the printing, consisted of the notes of a student who had been his auditor, whose note book was found available.[10]

In this kind of work young Cordus is before us in but the ordinary role of the early sixteenth-century botanical scholar, a master of the ancient languages, delving deeply into the medico-botanical works of Greek and Roman antiquity, and laboring to correct, amend, and in some degree perhaps augment the ancient pharmacopeia. To have been able to accomplish so much in this direction, and that while yet hardly having attained to manhood, was in itself a proof of genius. To understand the exalted character of this genius it is only necessary to canvass what the youth had also attached to along other and different lines at the same time.[11]

In field work in Germany, for botany alone—not to speak of geology and mineralogy, in both of which he was, for his time, an expert—he had wrought out more results than had his older contemporaries Brunfels, Tragus, and Fuchs combined. In his repeated journeyings to the great forests and wildest mountain districts, it is estimated that he discovered several hundred new plants.[12] Sprengel has given the Linnæan names of some twenty-five of these new discoveries of Cordus;[13] and that is perhaps double or treble the number of novelties gathered in by the whole three above named; and they both were men of longer life and more or less extensive travel.[14] At the time when Cordus's field studies of German botany were in progress nothing had ever been published bearing on Germany in particular as to its plants. The investigator of the botany of its forests, fields, and mountains had no other descriptive resources than the folios of Theophrastus, Dioscorides, and Pliny. Even Brunfels's book, which had appeared when Valerius Cordus was sixteen or seventeen years old was, as to its descriptive text, nothing but the reiteration of ancient botany. Soon after the publication of that work Euricius Cordus had publicly cautioned men against expecting to find all the trees, herbs, and flowers of Germany described in the botanies of the ancients, who had known but the plants of the very different region of the Mediterranean Sea.[15] This was nothing like an intimation that the books of the ancients could not be serviceable to students of German botany, and might therefore well be closed and laid aside. That would have been the proposition of an ignorant man and a charlatan, never of one of reason and erudition. The fields, the gardens, culinary and ornamental, the orchards, waysides, and hedgerows abounded in plants cultivated or naturalized which, in part purposely and in part fortuitously, had been brought into Germany from the South and from the East; and the discussion of just these formed no small part of the phytography of antiquity. All this had been clear to the elder Cordus, and was as easily comprehensible to the younger. But there must now be conveyed a better notion than we have yet gained of the rare subjective equipment wherewith young Valerius went forth to the botanical conquest of the great German forests and unexplored mountain districts. On this we have the following from a contemporary once before quoted.

> To the best possible education of an intellect naturally keen, there was united in him that happy temperament to which nothing is impossible, or even difficult of attainment. To these gifts he added a truly marvellous industry and assiduity in research; and above all a most wonderfully retentive memory for everything he either saw in nature or read in books. In this he so greatly excelled as to be able to carry in mind in their entirety descriptions of things which he had not seen but was looking to find; thus having the descriptions always available whenever occasion called for the use of them.[16]

Probably it is not unusual in modern botany for one who is afield to carry in mind, as gathered out of books, the essential characteristics of certain plants not yet met with, so that he is able to recognize such the instant he first sees them; but it will come as a revelation to most botanists of the present, that just this thing was being done almost four centuries ago, by a German boy in his teens, and while as yet the only plant descriptions extant for him to memorize were those of ancient Greek and Roman physicians, philosophers, poets, and historians.

In the year 1540, when Cordus was twenty-five years of age, he had in manuscript among other things four books which he had entitled *Historia Plantarum*. The four books contain an aggregate of 446 chapters, each devoted to one species, so that in this manuscript he had described that number of species.[17] The work embraces a part of the results of his travels at home in Germany. Finished, as to these four books, as early as 1540, the work was not published until 1561,[18] or seventeen years after Cordus's death. We shall get an idea of the wealth of these pages in botanical matter entirely new if we but glance at the contents of Book I. The first four types described are, in modern nomenclature, *Drosera*, *Gratiola*, *Sagittaria*, and *Bistorta*.[19] Every one of them, at the time when Cordus here described them, was new to science.[20] *Bistorta* had been figured by Brunfels, though guessed by him to be one of the dracontiums of Dioscorides, and not described.[21]

In the year 1542 Cordus went to Italy, dividing his time for nearly two years between the universities of Padua, Ferrara, and Bologna, where he made the acquaintance of Luca Ghini, reputed the most accomplished botanist of his time, but of whose greatness only the tradition remains, because he published nothing;[22] thence he proceeded to Florence, Pisa, Lucca, and late in the summer of 1544, to Rome. The misdirected and ungoverned zeal with which he prosecuted his summer journey in a southern climate cost the young man his life. In the company of two friends and a servant, while tending Romeward, he ranged everywhere from the cool mountains down to the heated and malarious marshes of the seaboard. Almost immediately on reaching Rome, Cordus fell ill of a fever and died there in September 1544, at the age of twenty-nine years. His body found its resting place there in the Church of S. Maria de Anima, where there is a long Latin epitaph, ending with the lines:

> *Ingenio superest Cordus; mens ipsa recepta est*
> *Cælo; quod terræ est, maxima Roma tenet.*[23]

The botanical outcome of these Italian journeyings was a Book V of the *Historia Plantarum*, consisting of descriptions of twenty-five plants which he had studied in that country, as not having been met with by him in Germany. This work was first published at Strasbourg in 1563,[24] then again reprinted as late as 1751, in the large folio entitled *Opera Botanica Conradi Gesneri*.

Cordus left no dissertations on the philosophy of plants, but only his descriptions of some five hundred species; and it is out of these fragments, all posthumously published, that we gather proofs of his resplendent genius.[25]

Phytography

When some years after Cordus's death the manuscript of the *Historia Plantarum* had been sent to the erudite Conrad Gesner at Zürich, in hope that he would approve the work and procure its publication, this worthy, in

FIG. 108. Bistort (*Polygonum bistorta* L.). Brunfels, *Herbarum* (1532–39), vol. 1, p. 62.

his letter of acknowledgment—subsequently printed—says that the four books are "truly extraordinary because of the accuracy with which the plants are described."[26]

Almost a century and a half later Tournefort named Valerius Cordus as having been "the first of all men to excel in plant description."[27] Then coming down to the time of Linnæus we shall find the very learned botanist and historian Haller still more pointedly crediting Valerius Cordus with having been "first to teach men to cease from dependence on the poor descriptions of the ancients, and to describe plants anew from nature."[28]

This, then, appears to be Cordus's title to special distinction among German botanists of the sixteenth century. He is the inventor of the art of phytography. This is saying very much, and the warrant for it must be shown. In our study of Tragus we observed that he, writing in German, and for popular reading, also without thought that his writing would ever have the helpful accompaniment of pictures, used an originality and a minuteness of detail in his descriptions of many plants that were quite new in botanical writing. It is one thing to write popular plant descriptions for every class of readers, and quite another to institute a set form of describing them, and that in the common language of the world of learning, and as if for learned botanists only. Just this is what Valerius Cordus did, thereby actually creating a phytography of a new type. And this new phytography had in view the philosophic end of doing away with the need of pictured illustrations. A leading purpose of Cordus was to demonstrate that every species could be so characterized in words as to be identifiable by description alone. It is, indeed, the only reason there ever was at any time in botanical history for describing plants; and the remotest ancients, when one of them undertook to describe a plant at all did it not so badly, but often very well. The trouble they made was through giving but the names and medical properties in the majority of instances. Euricius Cordus had expressed himself in print as to the unreasonableness of hoping to find the names of all German plants in Dioscorides and Pliny; and the logical sequence was, that German types unknown to the ancients ought to begin to be named and described. Knowing the intense devotion of the father to the son, and recognizing the zeal and ability of the latter, it is not possible to think of Valerius Cordus's work of describing German plants as having had other than this origin. It was likewise of deliberate purpose that the help of the engraver's art was to be dispensed with, as being unnecessary where the verbal descriptions are what they ought to be, except to the untaught, to whom descriptions are useless; for whom, however, Cordus did not pretend to write.

That which I here affirm is a fact which became obscured, and was in effect contradicted, by the editor and the publisher of Cordus's posthumous works; for the folio appeared almost throughout bedizened with woodcuts of plants, to the number of some 280 figures, a condition of things which Cordus could not have dreamed of as possible, and to which, living, it is most improbable that he would have consented. It was by urgent demand of the printer and publisher that figures were inserted. He evidently considered them to be indispensable to the financial success of the undertaking; and most probably he was right in that. The proposal to publish Cordus's works came at the time when the new iconographic movement that had been inaugurated by Brunfels thirty years before was at its high tide of public favor; for Fuchs's larger and more specious volume with doubly numerous plates had followed, and even Tragus had at last come out in an edition with 567 figures. It was not a time when the publisher would look for the success of a volume of plant descriptions in Latin unaccompanied by figures. The Strasbourg printer Rihelius, prospective publisher of Cordus's *Historia*, was in possession of the plates of Tragus's work and desired Gesner as editor to make use of them in as far as possible to illustrate the text. Gesner acceded to the proposition, and did as well as he could, yet not altogether very well, for there were some of Cordus's plant discoveries that were quite unknown to Gesner; moreover, the latter sometimes erred rather sadly in his interpretation of Cordus's diagnoses. Altogether, the attempt to illustrate by those old woodcuts the beautiful texts of Cordus has led to much misunderstanding and many errors in the interpretation of his chapters, the errors being in the main such as have arisen from incautiously concluding the identity of a given plant from the figure which Gesner by mistake associated with the description, rather than from the description itself.

This is not the place for indicating severally the errors made, as editor, by Gesner;[29] but it may be well just here to call attention to the serious mistake made by Tournefort when, justly lauding Cordus's merit as a phytographer, he said also that he "did not disdain to make use of plates."[30] He had not read Gesner's letter prefatory to the History, with its apology for the introduction of the plates; and many another since—even Linnæus among them—has been chargeable with the same oversight, to his own humiliation.

I suppose Valerius Cordus is the first in all history to have formulated for himself a definite plan or model of botanical description. There is a plan which he follows with such uniformity as to leave not the least room for questioning that he had studied it out for himself; and he presents it by example only, without formal announcement, without explanation, defense or apology, and on its obvious merits. In this plan of his we have the first foundation, and the actual beginning, of modern phytography; therefore we must analyze it carefully.

1. There must be a plant before him, a living one; for, while in his day herbarium specimens, especially of uncommon plants, were in the possession of some botanists and pharmacists for purposes of identification,[31] Cordus would not have had the temerity to offer the diagnosis of a dead fragment, or even of a more complete dead specimen, for a plant description. That innovation on phytography was not attempted until two centuries later.

2. The subject must be mature, or at least in flower, the fruit to be waited for if it must be, and described later; for Cordus describes flowers, fruits, and seeds invariably if at all available.

3. He begins with those parts of the plant that are most obvious as it stands living before him. If the foliage is most conspicuous, and the stem insignificant, as in a dandelion or a sundew, he begins with the foliage, proceeding thence to the stem; otherwise the stem is first described, then the foliage.

4. The flower is taken up next in order, the actual diagnosis of it being preceded by a mention of the time of year when the flowering occurs. As to the floral organs, while neither calyx nor corolla has its name as a separate part, he manages to describe both, and always very accurately.

5. Fruits and seeds are described with great precision. In the case of cap-

sular fruits if the cells are several he tells the number of them and notes the lines of dehiscence, often giving not only the color and the form of the seeds, but the number of rows they form within the capsule.

6. The root is always the last part of the plant which Cordus describes. This, as has been remarked in an earlier chapter, is the inconspicuous or at least the hidden organ, the one that has to be unearthed often with difficulty; and if physiologically and biologically speaking the root is the first of plant organs—and Cordus must have known this, or at least might have learned it from Theophrastus[32]—phytographically it is perhaps well enough regarded as the last; for it is perfectly natural that a man in describing a plant should begin with those parts that are obviously before his eyes, and then proceed in a search for those that are hidden away.

7. In describing herbaceous plants Cordus never fails to state the natural duration of the species if he knows it. Everything with him is annual, or biennial, or perennial. He is far more careful about this than most botanists of the present.

8. To the morphology of things he adds faithful reports of the odors and flavors, whether of foliage, flowers, or roots; and he is so distinctively phytographic as to make the briefest possible mention of medical qualities. This all the more clearly reveals in him a purpose to separate between descriptive botany and economic botany, for as a young physician he was particularly distinguished in pharmacy.

The boldest innovation that was made by any botanist of the whole sixteenth century, in whatever part of Europe, was that of Valerius Cordus when he proceeded to describe anew, and according to his own phytographic scheme, some of the best-known and even best-described plants of Dioscorides. And yet the ultimate success of the innovation might have been foreseen, doubtless was foreseen, by the young author. The reader must here be given one example of Cordus's new description of a very old medicinal plant, contrasting his with that of Dioscorides.

DIOSCORIDES, Book II, Ch. 162:[33]

ARUM, called Lupha by the Syrians, sends up leaves like those of Dracunculus, but larger and less spotted; stem purplish, nine inches high, bearing something like a pestle, upon which the red seeds grow; root like that of Dracunculus white.

VALERIUS CORDUS, *Hist.*, Book I, Ch. 50:

ARUM in early spring sends up its leaves each rolled together like a cloak and the roll slenderly pointed; these gradually expand and assume the outline of an ivy leaf, though they are much larger, sometimes attaining the length of nine inches, ending in a point, but widening below, yet receding into a sinus where joined to the petiole. In certain localities the leaves are purple spotted. At the same time of the year it sends up another rolled up cloak[34] which rests at the summit of a short upright stalk, and which about the middle of May opens to something like the form of a rabbit's or a donkey's ear, and shows within that which may be likened to the pestle of a mortar, is of about the length of the little finger, erect, of a dull purple or ashy color, and rests on a kind of roughish tubercle, beneath which there is another tuberculation of the same size, but paler as to color. This last-named tuberculation, after the one above it and the pestle have withered, grows to the size of a walnut and takes on the aspect of a bunch of red berries, each berry containing a seed or two a little smaller than a lentile. This thing ripens at about the summer solstice, and the knot of shining berries and its stalk are all that remain visible of the plant at that time; and when these have fallen away everything disappears. The plant is from a bulbous perennial root of the size of the first joint of the thumb, white, delicate, which is found in a shrunken and withering state under the growing herb, yet after the withering of the herbage is found increased in size and form. It sends out many eyes or tubercles by means of which the plant is propagated. Every part of the herbage exhales a heavy odor, and is so acrid in flavor as to affect the tongue and palate of him who tastes it as if he had swallowed thistles or briers. The plant inhabits shady places in deep woods, or old and shaded drains and ditches, or along hedges. Some cultivate it in gardens.[35]

Cordus's descriptions of new types discovered by himself do not differ in plan from the above. Some such are longer, others shorter, according to the requirements of the plant itself as examined by him in minute detail. One sees that in them all the same attention is given to the morphology, and also to the life history of the plant in as far as this is known to him. In his practice of describing each species both morphologically and biologically, he is a herald of our late nineteenth- and early twentieth-century writers who, now that we have the microscope, give life histories with minuteness of detail before impossible.

It will also be observed that Cordus's descriptions of plants are lengthy. The writing them over again in the descriptive terminology of eighteenth- and nineteenth-century phytography would reduce the number of words by at least one-half, if not by two-thirds, yet would not greatly improve them as to their definitive quality. But the same thing may be said of hundreds of more than equally lengthy descriptions of species published very recently.

Vegetative Organs

One meets with a new term or two in Cordus relating to vegetative organs; and new terms, with a man of his mentality, mean new thoughts. Foremost among such new terms of his is *coliculus*, *i.e.*, cauliculus, or caulicle, diminutive of caulis, a stem. This caulicle appears as the very first word in the descriptive account of an herbaceous plant newly discovered in Cordus's day, now known as *Calla palustris*. The only organ this plant has which Cordus could possibly have identified as any kind of stem is that part now for some time known as its rootstock, or rhizome. The account given of these caulicles is, that they are spread about over the ground, are about a foot long, of the thickness of one's little finger, are green, glabrous, and jointed, that from each joint white fibrous roots descend into the ground, and broad-bladed leaves arise above them,[36] etc., etc. Every other botanical writer all through the long ages had called every such thing a root. Theophrastus alone, and that seventeen centuries earlier, had gone so far as to register a doubt about the propriety of classing them with roots. Cordus is the first to publish a decision that they are primarily of the nature of stems. Note also that the name he assigns this organ is one that accentuates its stem characteristics, veiling those conditions which had led to its having been denominated a root. Such prudence in the selection of a name for the organ evidences a philosophic mind. He might have named the thing as rootstock or a rhizome, thereby, however, veiling those stem characteristics which he wished to impress, and accentuating the very things which had led to their being mistaken for roots; and so the adverse critic would have made light of the whole affair, pointing out that one says it is a stem, yet names it by a name which seems to say that after all it is a kind of root. It is very possible by the injudicious naming of a scientific discovery to retard the acceptance of the discovery itself.

Cordus must not be thought of as having regularly defined these horizontal root-like stems, or as having even recognized them in all their phases as being of the nature of stems. Their distinctly articulated and leaf-bearing character is in many cases far from being obvious. Such plants as iris and acorus are credited by him,[37] as they always of old had been, with having thick fleshy horizontal roots. But he is nonetheless first to name as stem any form or phase of rootstock or rhizome, and all the merit of such fine organologic discovery belongs to him.

The occurrence of adventitious roots at the lower joints of large culms Cordus remarks upon admiringly. He names them *fulcra*, not perceiving that at their very earliest starting forth from aerial joints they are roots. He makes the subjoined comment on them as he has studied them in the Indian millet:

> The plant has many obliquely descending and quite firm roots, but is not content with these; for when the culm obtains its growth and begins to be top heavy with leaves and the growing spike, it sends down from its inferior joints certain braces which, when they reach the ground put forth roots and fibres. Through Divine Providence, by means of these braces the plants more securely maintain themselves erect against the force of winds.[38]

The same is stated more succinctly and briefly in his account of Indian corn.[39]

There are intimations that Cordus is not content with the notion that leaves may spring from roots immediately, but that leaf bearing should be the prerogative of stems, or of that which represents them. When having in view a plant the leaves of which form a rosette hardly raised above the level of the ground, he seems purposely to avoid writing them down as radical leaves, or root leaves, and is wont to describe them as radiating "around" the root. It is an evasion, certainly; but it subserves its purpose, for he thereby escapes the necessity of saying that they grow from the root. But again, in describing some garden biennials, like the carrot, the part to which the leaves are attached is visibly distinct from the fusiform root, though it is extremely short, too short to be called a stem, and he denominates it the caput,[40] the head above the root. Only tardily has botany come to approve, formally, this one of the improvements in organography which it owes to Cordus. It is now taught that such apparently radical leaves grow from what is called a crown, and is understood to be but a shortened and thickened stem; but the teaching is still ineffectual to altogether prevent our occasional speaking and writing about stemless plants and root leaves.

To the fundamentals of leaf morphology I have not found this author contributing very much that is new; but his descriptions show him always carefully discriminating between the equally and the unequally pinnate in compound leaves. Not that these leaves have yet been formally designated as "compound." They have not been, nor will they be so named until after the lapse of more than a century after Cordus's time. The epoch is one of discovery, not of always naming the thing discovered.

But Cordus is the man who first reaches the conclusion that an organ need neither be green colored nor horizontally explanate in order to be a leaf. It seems to have fallen to his lot to describe but two orobanchaceous types. In the case of the first of them it is evident he had not yet seen that the colorless scales investing the base of the stem have the nature of foliage; for he says "the plant is destitute of leaves."[41] Some years after this, traveling in Italy, he becomes acquainted with another and much more scaly member of this alliance. His opinion respecting what the scales are now undergoes a change. "The dense investiture of scales upon the stems of this plant, all of them pointing vertically, are to be interpreted as being its leaves."[42]

Anthology

Among botanical authors of his time Cordus alone gives some attention to inflorescence. He is the first after Theophrastus to have noted the distinction of the centripetal and centrifugal in anthesis, or, to state it otherwise, of the indeterminate and determinate in inflorescence; and every historian of botany appears to have overlooked this. Meyer writing on this topic little more than a half-century since says that, in as far as he is aware, all the way from Theophrastus down to the times of Link and Robert Brown, no mention was made of these distinctions.[43] Cordus's writings antedate those of the worthies last named by almost three centuries; and Meyer cannot have taken the time to read them, for it is a very common thing with this German youth, in describing plants spicate or racemose as to their flowering, to say that the expanding of the individual flowers proceeds from base to summit of the axis, and that thus the succession of the flowering is prolonged indeterminately as it were, and may continue indefinitely, through a long season.

In the case of loose inflorescences such as the corymb and the umbel he does not make note of such differences. But in the crowded, though quite spherical flower cluster of *Echinops sphaerocephalus*, he observes and records it that the first flowers to open are those at the summit of the cluster and that the succession is from that point downward to where the globose head is joined to its peduncle.[44]

It will be taken for granted that Cordus had received from Theophrastus the suggestion of the centripetal and centrifugal in inflorescence; but there is one excellent definition of a particular kind of inflorescence which modern botany receives from this German youth of the sixteenth century, that of the word umbel. The word itself as a botanical term is as old as botany, and was in the first place suggested by the mechanism of an umbrella or sunshade, which in more than one way certain inflorescences recall. There are two things essential to an umbrella, and these not equally conspicuous. The

FIG. 109. Yarrow, or milfoil (*Achillea millefolium* L.). Bock, *Stirpium* (1552), bk. 1, ch. 160, p. 477; after Fuchs, *Stirpium* (1542), p. 726. The illustrations for this chapter on V. Cordus follow Gesner's practice of using those from Bock in Gesner, *Collecta* (1561).

more obvious part is the rounded and expanded externally more or less convex surface; and there is also the framework beneath supporting it in expansion. Now while the modern botanist is taught to look not at the expanded surface, but at the structure of the framework beneath for the evidence that a flattish-topped inflorescence is an umbel, it was quite otherwise with the botanists of antiquity, and with all of them before Valerius Cordus. Brunfels, Fuchs, Tragus, and this young botanist's wisest contemporaries held the broad flat clusters of the elder and the viburnum to be umbels equally with those of carrot and caraway, parsnip and fennel, no heed being given to the complexity of the supporting framework in the elder, or to the simplicity of it in caraway and fennel. Even the small heads of those composites the yarrow and the tansy were said to be arranged in umbels. It is in the midst of his own new and improved description of the common yarrow that Cordus suggests the need of distinguishing between umbel and corymb, and that nothing should be called an umbel the stalklets of which do not all arise from one and the same point.[45] This is done with the utmost modesty, the revolutionizing proposition being enclosed in a parenthesis. He sees the important distinction between umbel and corymb, names briefly the characteristic of the original and true umbel, and quietly passes on, leaving it to be inferred that other flattish-topped inflorescences not answering to this clear definition of the umbel may be called the corymb; and his practice proves this to have been his purpose. We shall be interested in observing later how long after Cordus botanists in general saw the need of distinguishing between umbel and corymb. Just here, however, it may be observed that when the vast family of the composites first gained clear recognition, Corymbiferæ became at once the family name, and remained steadily in use for a century and more; and Cordus made the first beginning of calling the inflorescences of many commonest composites corymbs.[46]

Those modified leaves now known as bracts, and collectively as an involucre subtending inflorescences and flowers, first obtained mention and description by Cordus. In *Daucus* they so conspicuously embrace the umbel, and are so very unlike the proper foliage that he is careful to bring out in writing all their peculiarities,[47] the first step in the direction of giving taxonomic significance to the involucre in the family of umbellifers. Recalling that that comparatively modern and well-advanced anthologist Tournefort wrote down the involucre of every araceous plant as its "petal,"[48] it is interesting to note that this German youth more than a hundred and fifty years before Tournefort denominated it an "involucrum,"[49] as do we of today, for the spathe is but one type of involucre.

The calyx under Cordus makes noteworthy though by no means great progress toward gaining recognition as a part of the flower. Quite in conformity to the definition which we found in Fuchs's vocabulary[50] Cordus calls that a "calyculus" which is synsepalous and either becomes a part of the fruit, or at least a protection to the growing and mature seeds. All labiates and borrageworts will readily be accredited as having calyxes. But how will it fare in this regard with the Solanaceæ? In the some half-dozen genera of this family then known in German gardens there was not one, *Hyoscyamus* excepted, which could exhibit an organ answering to the then accepted definition of a calyx; and curiously enough Tragus and even Fuchs, when they name the "calyx" of *Hyoscyamus*, mean nothing but the operculate, and therefore literally chalice-shaped, capsule.[51] The real calyx goes with Tragus for a "vasculum," while Fuchs makes no reference to it whatever. One cannot but regret that Cordus did not describe *Hyoscyamus*; it would have been so interesting to have seen what he with his perfect originality of view would have made of the anthology and carpology of a type so marked. But the youth was much engaged in bringing from German meadows, river banks, and woods representatives of genera unknown to the ancients, and describing these. The presentation of his views about calyxes in solanaceous plants will be helped by a list of the genera which at that period, over and above *Hyoscyamus*, were in German gardens. They were *Solanum*, *Atropa*, *Mandragora*, *Melongena*, *Physalis*, *Capsicum* and *Stramonium*, or *Datura*. The last had been credited with a calyx by Tragus, though not technically, or as being an organ regularly so named, but only as a cup-shaped green thing forth from which "the flower proceeds."[52] The corresponding part of the plant *Physalis* Fuchs had described collectively as "rounded pouches resembling bladders,"[53] and Tragus calls them simply "*bladders*."[54] These I think are all the references made by Fuchs and Bock to anything like a calyx in their eight or nine solanaceous genera. Of course the small, flat, hardly more than disk-like calyxes of proper solanum, mandragora, capsicum and their like would not be noticed by them in any way. They were no part of the flower, nor had any significancy. With Valerius Cordus every one of them—the tubiform cup of *Datura*, the globiform-inflated and closed bladder of *Physalis*, the prickly saucer-shaped thing holding the mere base of a melongena fruit, and the almost flat green disk visible at the basal end of a pepper pod—is a "calix" or "caliculus." This was a most significant innovation, the employment of an old term in such wise as to make of it a strictly scientific term. Hitherto "calyx" had been used in botany from the remotest times as signifying any organ that happened to be cup-shaped, the ovary of a pear or quince or pomegranate in their early stages, a cup-shaped corolla, the cup-shaped corona within the perianth of a jonquil or daffodil, or even an operculate dry seed vessel. Any one of these was a "calyx" in pre-Cordian terminology.[55] Now, according to this use of the term, which makes its first appearance in Cordus's descriptions of solanaceous plants, calyx is not calyx by reason of its shape, but because it holds always one and the same place-relation to "flower" and to the fruit that succeeds the flower. This principle of the location of an organ, the place which it visibly occupies as next some other organ, rather than its form or coloring or texture, is one which, long after Cordus's day, came to be received as absolutely fundamental in anthology. Morphologically considered the whole doctrine of the flower, as almost every reader will know, rests on this basis, and most securely. Therefore Cordus's mere application of this principle to the identifying of calyx in all its extremes of form throughout the one family of the nightshades must rank among the most important contributions ever made to anthology; for it was the first hint ever given of a truly scientific classification of floral organs.

The truthfulness of this last statement will remain unimpeached notwithstanding that Cordus did not name the calyx as forming a part of the flower. It was too early for the expression of anything so iconoclastic as that. From the earliest dawn of botany down through uncounted ages, a circle of green-colored leaves, no matter what their size or form, or where placed, had been a circle of leaves, and a whorl of leaf-like organs colored otherwise than green had been a flower. Even in the mind of young Cordus this appears to have been a prejudice too deeply seated to fade away even before the light of his own brilliant discoveries. He never admitted the calyx to the rank of a floral organ, though he seems upon the verge of doing so. In describing the "flower" of the white water lily he actually sets apart, as too different from the others, those four outer members which are green externally and which also, as he says, completely enfold all the others, and he gives to them collectively a name by which they must be distinguished from the many and narrower white "flower leaves"; the name is not calyx but "tunica."[56] He does not, however, look on them as partaking so much of the nature of a calyx as of ordinary flower leaves. He observes that they are not altogether of that green color which at first place they seem to be, but that at summit and marginally they are of the same texture and whiteness as the others. They impress him as being modified flower leaves, whereas the green things at the back of buttercups and others like them are but reduced and modified ordinary leaves. The terms calyx and caliculus I have failed to find Cordus making use of at all except for such as are synsepalous. If such a circle is quite chorisepalous, or even approximately so, he calls it a circle of leaves simply. He does not overlook the fact that such chorisepalous circle has a particular function, and that in immediate relation to the flower; and here again he seems as on the verge of extending the use of the term flower so as to make it include the calyx; but he never quite does that. His manner of expressing himself in such cases is exemplified in his description of what he calls *Hepatica alba*, which is his name for a new genus of his discovery, its equivalent in modern nomenclature being *Parnassia*; the type, *P. palustris*. Here is his account of what is now called the calyx in that species. "Beneath the flower itself there are five very small leaves within which the flower before it opens is enclosed."[57] Certainly that which envelops a flower before its expansion can be no part of that flower. There is here a more than whispered call for a new extension of the use of the word flower, a new definition of that organ, by the terms of which the calyx shall be recognized as a part of it. All this will come, but by no means speedily.

There is one use of the term calyx frequent with Cordus which practically implies chorisepaly. I refer to his habitual writing of the involucres subtending the flower heads of all cichoriaceous plants and some of the true composites as the calyx. This was a complying with the terms of the then accepted definition of a calyx as that which enfolded at first the flower, after that the fruit. As a somewhat special application of the term this, perhaps introduced by Cordus, seems strongly to have commended itself to future generations, for, long after the "flower" of chicory and its cognates had been seen to be an inflorescence, this term calyx remained in use instead of involucre. With Tournefort in 1700 it was still a calyx. Linnæus a half-century later modified the expression insofar as to write it "calyx communis"; but this needful modification was afterwards ruled out. A number of prominent botanists, even down to the middle of the nineteenth century, wrote down the involucre even of the sunflowers as a calyx, and the bracts composing it, as sepals.[58]

To the morphology of the corolla—if one may use that term in writing about a time which far antedates the term's invention—Cordus adds a few items of high import to phytography. Botany has now not many expressions which it could as hardly do without as the terms papilionaceous and bilabiate, for they at once recall, and respectively designate, two large and important families of plants. Fuchs in one instance speaks of a certain "flower" as having the form of a butterfly.[59] In as far as I have been able to discover, the idea of comparing the pea blossom to a butterfly originated with Fuchs's brilliant contemporary Conrad Gesner.[60] Neither of these appears to have used the expression "flore papilionis forma" in connection with any more than the single species *Pisum sativum*; but this evidently suggested to Cordus the possibility of something better than the antique usage of describing a vetch blossom as being like that of a lupine, and the lupine flower in turn being compared to that of a pea. And his appears to have been the botanical eye that was first to perceive in the flowers of the whole long line of beans, peas, and lupines, and of broom and cytisus, and laburnum tree, only more or less marked deviations from—mere modifications of—that pea-blossom type which Gesner had compared to the insect called a butterfly; and by and by the adjective term papilionaceous was coined by him. Such readers as may not have access to Cordus's works may be interested in following the gradual genesis of that useful adjective in the young master's mind; and this can be shown by citing his expressions in precisely the order in which they occur in the succession of his pages. I therefore give these in the footnote.[61]

Of equal service to all botany was his invention of a term that should at once indicate and describe the corolla of the Labiatæ. His term is "flos hians," the gaping flower, or corolla as we now say. Nor does it require but a moment's reflection to become convinced that the expression gaping, or yawning, corolla is more perfectly and exactly descriptive of the most common and typical corollas of labiates than is the term bilabiate, for certainly that which one sees clearly in the form of such is, not a pair of lips, but a wide-open yawning mouth, exposing even the throat itself. Terminology certainly lost something of the accurately definitive when later authority displaced Valerius Cordus's "yawning" flower and substituted the less fitly chosen bilabiate.

It was yet far from the time when any taxonomic use would be made of the different ways in which members of floral circles are enfolded in the bud. Cordus was first to observe some such differences, and to name them in his plant descriptions. In describing *Iris* he notes that the parts of the flower are convolute in the bud.[62] Similarly he describes the corolla of *Datura* as being laid in folds before expansion.[63] It will not be assumed that in first writing these things down he had any thought of their important bearing upon affinities. He may have had; but whether so or no the terms he chose are so perfectly correct for the kinds of prefloration which they indicate, that no reformer of terminology has sought to displace them, and they remain in common use today.

The word petal was still unheard and unwritten in our science. It will be proposed by an Italian botanist two generations later.[64] It is evident Cordus has realized to some degree the desirability of some term by which flower leaves shall be distinguished from ordinary foliage, for where flowers are choripetalous he uses the diminutive foliolum (leaflet) instead of folium; this, however, not as an inviolable rule, nor in such wise as to preclude the application of the term to ordinary green leaves that are of very small size. Also such elongated and more or less strap-shaped flower leaves as radiately encircle the middle of a flower, or head of flowers, Cordus repeatedly describes as resembling rays, "radii," the earliest adumbration of the term now long in use for this kind of corolla.

The disk corollas in the composites that have rays were still undiscovered. They were seen in the mass only and were always written of as the "stamina," therefore quite undistinguished from the central parts of a buttercup or anemone blossom. Cordus alone I find in one instance so writing of the "stamina" of one particular composite as to prove that he had seen the individual "stamen" as he could have called it, and had found its form peculiar. The plant is *Tussilago*; and having described the rays, he says that the stamina in the midst of them are "in form like minute lilies."[65] No matter what he calls the thing, it is plain that he is the discoverer of the disk corolla in composites.

If both stamens and pistils are numerous, small, and slender in the same flower, Cordus makes no distinction between them in name. In this case all are "stamina"; thus in *Pulsatilla* all the threads in the midst of the flower are stamina; but those occupying the very center he observes as "changing into a tuft of long hairs at the base of each of which there is a seed as in clematis."[66] If, however, in flowers at once many-stamened and many-pistilled the pistils, being without styles, have no capillary aspect in the aggregate, then the two sets are not confused and the inner ones are not "stamina." This is apparent in the account which he gives of thalictrum flowers. He says they are "very small, and consist of pendulous stamens only; but after the falling away of these, very small oblong black seeds remain."[67] In the case of larger flowers, where the stamens are fewer and easily counted, it becomes clear that by "stamen" Cordus means primarily the filament; this doubtless partly because in very many instances it was all he could find. There are no "apices" to the inner "stamina" of pulsatilla or clematis, none to the forked styles of

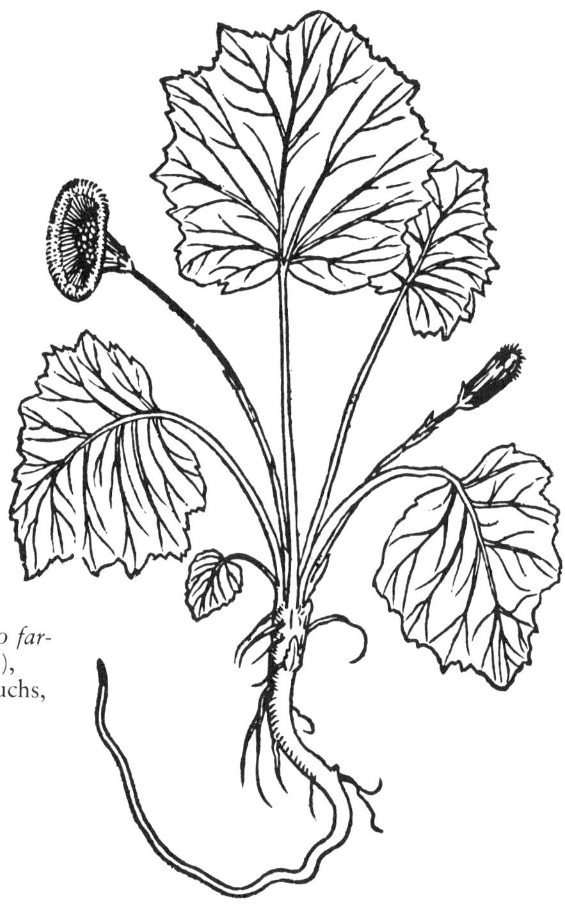

FIG. 110. Coltsfoot (*Tussilago farfara* L.). Bock, *Stirpium* (1552), bk. 1, ch. 139, p. 418; after Fuchs, *Stirpium* (1542), p. 140.

the many Cichoriaceæ and Compositæ which he examined, and he always calls these the stamina. Even in the solanum type of floral structure, where five stamens, almost all anther, form a central column forth from whose summit one slender style-thread protrudes, Cordus, seeing them all, denominated all six indiscriminately "stamina."[68] Evidently his mind was exercised by these small things, in the morphology of which he saw enough to prevent him at least from calling them indiscriminately stamina and apices as others of his time were doing. In the large and therefore convenient flowers of lilies he saw and took note of the transverse position of the anthers, but would not name the things by any name at all. The stalks which these rested on were what he called stamina. The term apex he seems to have wished to transfer to the style and stigma and to have it apply to that part of the flower only.[69] He takes particular notice of dust—*rubiginosus pulviusculus*—which the lily anthers shed before collapsing. Again in describing the an-

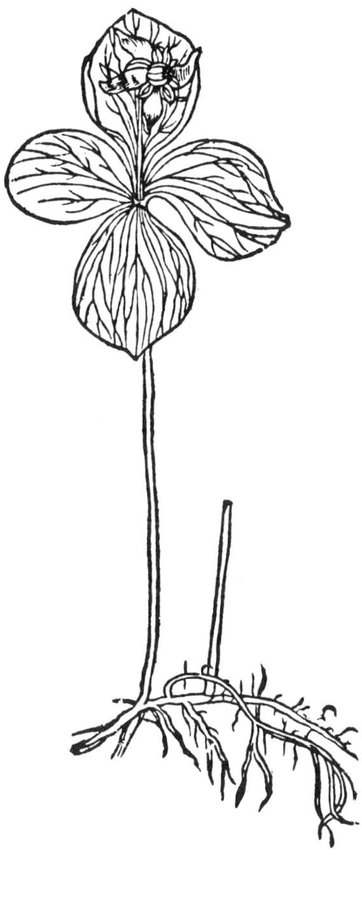

FIG. 111. Herb paris (*Paris quadrifolia* L.). Bock, *Stirpium* (1552), bk. 1, ch. 52, p. 308.

thers of *Paris* he sees this same kind of dust, and there proceeds to assign it a name, even the name which it has always since borne; for he describes these parts as being "luteo *polline* conspersa."[70]

Since the most ancient times what they knew as a fruit they had recognized in its germinal state resting at the bottom of the flower, or else below it; but in every stage from the tender germinal condition forward to its maturity they had called it the "fruit" simply. Cordus makes the first inroad upon this time-honored usage. That which we know as the ovary he uniformly declines to write of as the fruit, as if perceiving an absurdity in making this small and tender mere promise of fruit, identical with the future actual fruit, even in name. He does not, however, formally propose a universal name for this hitherto really nameless thing. Where, as in the water lilies and in some others, it is rather large and conspicuous within the flower, he applies to it that already much and variously used term *capitulum*;[71] but in the more numerous cases of smaller flowers, where it appears as only a small protuberance in the bottom of the blossom, he writes of it as the *tuberculum*. The ovary, then, like the pollen, is one of Cordus's discoveries, so to speak, in anthology. Furthermore, this tuberculum, as he calls it, is openly recognized as an integral part of the flower as a whole. The proof that he so received it is found in the language he employs in describing the pæonia. Having given the characteristics of the flower leaves, he passes from these to the two large ovaries occupying the center. These he calls a pair of horns, *cornicula*, mentioning that they are hedged about by the yellow stamens, then proceeding to state that these two cornicula "remain after the other parts of the flower have fallen," then grow to be an inch and a half long,[72] etc., etc. The idea that corolla, stamen, and pistils collectively constitute a flower could not be expressed more unmistakably; and it is the earliest record I have met with of such a proposition. Theophrastus, as we have seen, first classified flowers as leafy and capillary,[73] so that a mere tuft of stamens only, unattended by leaves of any kind, was a flower. Then, with a wild rose blossom before him he construed the yellow stamens as being, not a circle of floral organs, but a capillary flower within a leafy one. Thus the rose had two flowers; the very central axis, globosely enlarged below, was the "fruit" even then, and no part of the flower. This Theophrastian anthology of two flowers in one, rather than one flower made up of two sets of organs, was everywhere accepted in Cordus's time. Tragus had reproduced it, though with augmentations, even to calling the petals "rose leaves," and the stamens the "rose flower."[74] Now in Cordus's procedure we have an illustration of how the making of one little distinction, and the invention of a word that accentuates that distinction, may revolutionize a science. The man had seen an inaccuracy in the practice of calling by the one name of fruit both the little tubercle lying at the bottom of a cherry blossom and the subsequent ripe cherry. That mere floral sign and promise of a fruit he determined to name a tubercle. Logically a fruit is no part of any flower; as logically this "tubercle" *is* a part of it. In rose and pæony there are not three flowers one within another, but one flower made up of three different kinds of parts. We shall not be likely to meet with any other one botanist in the whole history of our science who so greatly advanced the morphology of the flower as Valerius Cordus.[75] To the finishing of morphologic anthology only two things remained to be done. The calyx was to be added to the category of floral organs, and the "flower leaves" awaited a distinctive name, for the "foliolum" proposed and used by Cordus did not satisfactorily meet the demand. It was not sufficiently different from folium. The discovery of the functions of certain floral parts, though of the very highest importance, awaited not the coming of a more astute botanist, but the invention of mechanical aids to natural vision.

Among a number of Cordus's signal new discoveries in anthology, one must not omit to mention the flowers of the genus *Ficus*. Every botanical authority from Theophrastus forward had averred that fig trees have no flowers, and that the fruits are only fruits and nothing more from their first small appearing to their ripening. Cordus says: "When the figs are half grown they develop in their interior (what you may be surprised to know) what appear to be crowded stamens, of a pale purplish color, standing forth from the fleshy part, and all pointing toward the central hollow, to each of which there succeeds a flattened yellowish seed."[76] It is most remarkable that, two hundred years after this clear description of fig flowers, Linnæus should have placed *Ficus* among the genera of cryptogams.[77]

Fruit and Seed

Not Brunfels, nor Fuchs, nor Tragus shows sign of ever having read Theophrastus's scientific definition of a fruit.[78] As far as my careful reading has gone the Greek's term pericarp first reappears in Cordus; and he uses it frequently. Then, in his full descriptions of various plants it comes out that he has made sections, longitudinal and transverse, of many pericarps, so as to be able to record the number of the seeds when they are not too numerous, or the number of rows in which they arrange themselves when the number of the seeds is too great to be told. While the elder German authors have quite minutely and well described the curious and beautiful seeds of cardiospermum,[79] this youth, as if belonging to a later generation and to a later century than the sixteenth, makes note of the fact that these same seeds are attached to the central part of their pericarp,[80] thus suggesting long before the botanical world had apprehended its taxonomic usefulness, the several modes of placentation.

He is accustomed to peer into and take note of every aspect of the various dry dehiscent fruits that he meets with in the gardens or in the wilds. In describing species with capsular fruits he tells whether its capsule is one-celled or several-celled, naming the lines of dehiscence, the commissures, the partitions, if there be any, the septa, and the compartments themselves, the loculamenta. Then the seeds are reported on, not only as to the method of arrangement but as to every item of their form, the color and texture of their testa, and the color and flavor of the nucleus—so he names it—when they are large enough to be tested by the sense of taste.

Cordus has not, like Tragus, followed up the suggestions of Theophrastus about the cotyledons, for that belongs to the garden student, who plants seeds and watches their germination and first appearing above ground; and Cordus is more zealously devoted to wild botany. To him, however, must be conceded priority in the matter of distinguishing between spores and seeds on the one hand, and between spores and pollen on the other; this also without his ever having seen either an individual spore or an individual pollen grain. For the clearer understanding of Cordus in this particular field of enquiry we must recall Tragus's having so studiously and so laboriously gathered what he and others believed to be the seeds of the fern osmunda.[81] That he called them seeds implies the belief on his part that young ferns could be produced from them. But then, the superstitious Tragus seems also to have believed that trees could be reproduced both by their proper seeds,

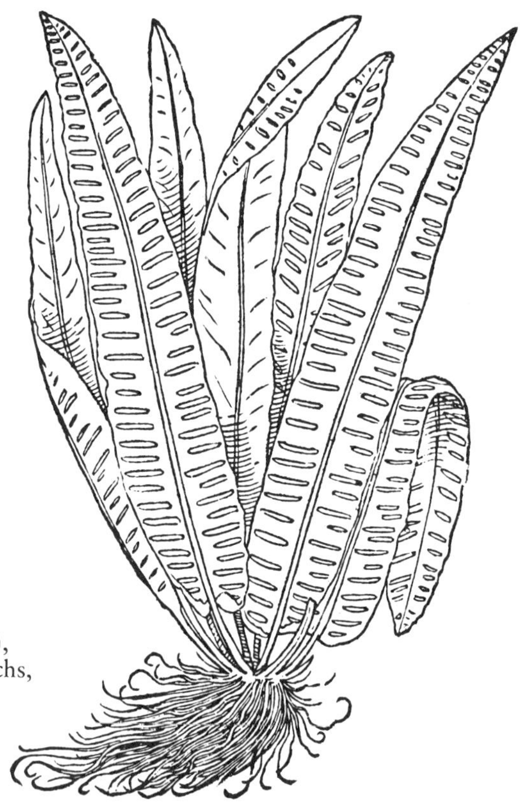

FIG. 112. Hart's-tongue fern (*Phyllitis scolopendrium* (L.) Newm.). Bock, *Stirpium* (1552), bk. 1, ch. 188, p. 549; after Fuchs, *Stirpium* (1542), p. 294.

FIG. 113. Male fern (*Dryopteris filix-mas* (L.) Schott). Bock, *Stirpium* (1552), bk. 1, ch. 187, p. 546.

and also by that flower dust which Cordus afterwards named pollen.[82] Among nature students of four hundred years ago I know not who else is so far from accepting things on other people's guess or hearsay as Valerius Cordus, in whom I have not yet read a line that savors of the fabulous or superstitious; and that, for the period, is much to say of any author. Concerning the propagation of scolopendrium he says: "Phyllitis has no stem, or flower, or seed; nevertheless, from the vermiform patches on the back of the leaf, when these are resolved into powder and are scattered abroad, it is propagated."[83] Substantially the same proposition is repeated with emphasis at the end of his description of *Aspidium filix mas*: "Although it has neither stem nor flowers nor seeds, it nevertheless propagates itself by means of the yellow and hairy powder that is produced on the back of the leaf and is blown away by the wind."[84] Again, of trichomanes it is said: "It grows copiously on moist shaded rocks, although it produces no stem, or flower or seed. But it reproduces itself by means of the dust that is developed on the back of the leaves, as do all kinds of ferns; and let this statement of the fact once for all suffice."[85] From this point forth he proceeds to describe a half-dozen other ferns in close succession, carefully bringing out the form and arrangement of their fruit dots or lines not omitting even the indusium, but not again mentioning their seedlessness or their means of propagation.

This positive and reiterated assertion that ferns have no seeds, yet propagate by organs so infinitesimal that he has never seen one,[86] implies that what ferns shed from the back of their fronds is understood to be of a different structure from that of seeds. It is an easy matter for one frequenting the native haunts of certain ferns to see their prothallia both with and without the first diminutive fern leaf; and it cannot reasonably be doubted—indeed Cordus by his strong language compels us to think—that he had seen these things,[87] and had assured himself that the germination of ferns is most different from that of seed plants, thence inferring to a certainty that those particles, invisible except in mass, are things different and distinct from seeds.

Taxonomy

Cordus's Books I and II are devoted to herbaceous plants, Book III to woody growths great and small. He accepts, then, and without objection, these anciently established first Grand Divisions of the plant world. The primary division of the herbaceous plants is not made from the point of view of affinities, but is ordered historically rather, as one may say; for while the heading of Book I is "Concerning Divers Herbs," that of Book II reads, "Plants whose History was either inexactly transmitted by the Ancients, or else altogether omitted." Such headings do not seem to promise much of taxonomic doctrine, or of the tacit expression of it by grouping. Members of the same natural family, and species of the same genus will almost inevitably be distributed partly to Book I and partly to Book II.

Despite these seeming obstacles to ready expression, and while there has never yet been any attempt to relegate all genera to families, or even formally to characterize any of those several families that have always been recognized, still Cordus advances well beyond all his precedessors in this significant part of botanical systematizing. When, as sometimes happens, his general plan has led to the placing of some type away from its real cognates, he is apt to give the hint that such genus thus isolated in his book belongs to a certain family. An example of this occurs in connection with his new description of the old genus *Lupinus*. His first word is "Lupinus is a leguminous plant."[88] He seems to be offering this as a piece of taxonomic information that is needed. He is not presenting his readers with an empty platitude. When writing of *Faba*[89] and *Cicer*,[90] and *Phaseolus*[91] he does not tell that either one is a leguminous plant. All the world knows that these are so

FIG. 114. Broad bean (*Vicia faba* L.). Bock, *Stirpium* (1552), bk. 2, ch. 14, p. 618; after Fuchs, *Stirpium* (1542), p. 389.

FIG. 115. White lupine (*Lupinus albus* L.). Bock, *Stirpium* (1552), bk. 2, ch. 15, p. 622; after Fuchs, *Stirpium* (1542), p. 309.

classed, and have been immemorially. The family of Leguminosæ of antiquity consisted of such Papilionaceæ as yielded edible seeds and were therefore food plants. Important though they were, they could not be harvested and threshed after the manner of harvesting the frumenta, or cerealia. The individual pods had to be collected by hand; hence the very name legumina. *Lupinus* was not one of the leguminous plants with the ancients. Its seeds were bitter, but endowed with active medicinal qualities. It was by virtue of botanical principles quite new in Cordus's day that he dared to say *Lupinus* is a leguminous plant. The family was now receiving new accessions, not of plants newly discovered, but of such as had been known very anciently but not admitted to the group, but now ready to be received because new and truer criteria of relationship were being exploited, and gaining recognition. Cordus, as we have seen, had been the discoverer of the papilionaceous flower[92]—the expression tells but the truth—and one immediate result of the discovery was, the augmentation of the ancient group of the legumina by the referring thereto of every genus the mere form and plan of whose corolla was at agreement with that of beans and peas. Anthology, dormant if not dead for fifteen centuries, has come to life, is developing with some rapidity, and the ancient taxonomy that had been constructed largely according to vegetative characters, culinary uses, and medicinal properties, is being steadily but quietly displaced. The chief agent of this radical taxonomical reform at this period is Valerius Cordus. We must follow him in his bringing in of one other new accession to the family of the leguminous. The type had been well known to the ancient Greeks. They had given it the name which, as a genus, it has always borne from their day to ours, *Glycyrrhiza*, Sweetroot.[93] No use was known for any part of the plant but its root. Anciently no one would have called it a leguminous plant; and for this very reason Cordus argues the case of its proper admissibility to membership in the family. In the place of a concluding note, supplementing a fine description of the plant in all its parts, in the course of which he has announced having found the flowers, though small and crowded, to be precisely papilionaceous, he says:

> The roots have a very sweet flavor, moderately astringent, and a trifle acrid, to which there is added the mere trace of a certain bitterness that belongs to all leguminous plants; for even this plant is as certainly referable to the leguminosæ as the peas and beans themselves; in view of which decision it seems fitting that I should make mention of this one sensible quality that it has in common with all the leguminosæ; for perhaps not every one would prove to have the sense of taste sufficiently cultivated to be able to perceive it and attest its presence.[94]

One is obliged to pause a moment in admiration of the calm, judicious conservatism of this youthful botanical genius, as it reveals itself in the last of those lines. He has discovered, and as modestly as possible he announces the discovery, that by their floral structures alone the family relationships of hosts of plant genera may be determined, and unmistakably. The detection of a principle so manifestly valid, and so surely destined to revolutionize completely all botanical system, cannot but have fired with enthusiasm a man only some twenty years old—a year or two more, possibly, and quite as possibly a year or two less—but in his writing how completely does he repress all enthusiasm. Profoundly respectful toward venerated authorities of two thousand years before who had held that agreement as to properties was required in order to establish the fact of interrelationship, he investigates licorice root to find that also by qualitative criteria as well as by floral structure the plant proves itself a member of the Leguminosæ. In this repression of excessive enthusiasm for the new, and continuing to respect the old principles, some later celebrities are in unfavorable contrast to Cordus, for they so greatly magnified the value of the new anthology, as to write intolerantly and even in derision of the old ideas that vegetative characters and sensible qualities have taxonomic weight.[95]

We must follow Cordus a few steps further in this path of the discovery of relationships, for he is making distinct and lasting landmarks in the history of plant families.

FIG. 116. Bitter apple (*Citrullus colocynthis* (L.) Schrad.). Bock, *Stirpium* (1552), bk. 2, ch. 98, p. 828; after Fuchs, *Stirpium* (1542), p. 372.

Tracing backwards the history of the Cucurbitaceæ, we reach no point, however ancient, at which gourds, pumpkins, squashes, melons, and cucumbers were not recognized collectively in their status of a family, or larger genus, as such a group was at first named. In their mode of growth, their coarse, rough herbage, and even as to the structure and qualities of their familiar large firm-fleshy fruits, they were in a comprehensive way at one. Any cultivator of them, however untaught, would be botanist enough to see that. Meanwhile there were two or three other types, long and familiarly known, which had never been thought of by even the most skillful botanists as cognates of gourds and cucumbers. They were the bryonias and the momordicas. According to such signs of consanguinity as availed with the ancients, these could never have been thought of as possible cucurbits. They were not large, coarse, harshly almost hispid plants, but were small, comparatively smooth and delicate in texture; and the bryonias, so far from yielding any fruit the least like gourds and melons, put forth bunches of

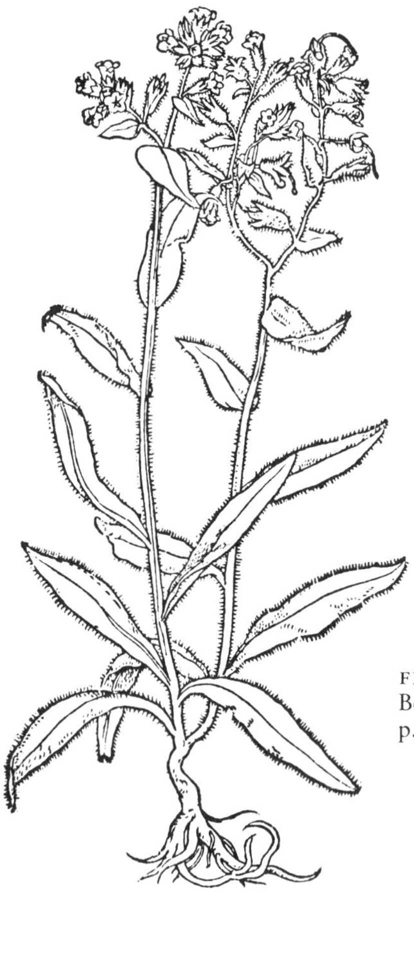

FIG. 117. Bugloss (*Lycopsis arvensis* L.). Bock, *Stirpium* (1552), bk. 1, ch. 77, p. 234.

small, soft, pulpy berries, red or black, more like those of nightshades. In the second place there were no mild qualities in their plants, no parts were edible. They were actively medicinal, and some of them powerfully narcotic-scented. It was no less than a master stroke of the new taxonomy, Cordus's bringing in of momordica and bryonia to augment the series of genera of the Cucurbitaceæ.[96] He had examined their small flowers, dissected the little bryony berries, and compared their plan of structure with that of their robust innocuous neighbors of the fields and gardens, and by these tokens had found them all to be of one lineage.

Singularly the little soft-leaved weed *Lithospermum arvense* had never been thought of as in the least degree akin to the coarse, rough, stinging borrage and bugloss and anchusa; but Cordus describing the plant in every part with a minuteness and accuracy unapproached by any earlier writer concludes it all with the proposition that its affinities are with the anchusas and echiums. "It is of their kindred; something which the ancients did not apprehend."[97] It had been through a comparison of their inflorescences, and by a recognition of the same floral plan and fruit characters that he had become able confidently to add lithospermum to the family of the borrageworts; and yet not by these alone, for he avers that as to properties also it is much like the others.

In his delimitation of the genus *Ranunculus* Cordus defers unwontedly to floral characters, and slights those of root, stem, and foliage. This is very interesting as proving that the trend of his mind has been strongly in the direction of what has come to be the established and settled first principle of classification.

Upon this point the subjoined list of his *Ranunculus* species will be instructive. They are given in the order in which he places them. The equivalents of them, that is to say the identifications of them in modern nomenclature, is somewhat doubtful in only one or two cases.

	Cordus	Modern
1.	*Ranunculus palustris*	*Ranunculus sceleratus*
2.	*R. Sardous*	*R. Sardous*
3.	*R. tertius*	*Anemone ranunculoides*
4.	*R. quartus*	*A. nemorosa*
5.	*R. quintus*	*Ranunculus arvensis*
6.	*R. flammula*	?*R. repens*
7.	*R. coronarius*	*R. acris*, double-flowered
8.	*R. octavus*	?*R. acris*, single-flowered
9.	*R. nonus*	*R. bulbosus*
10.	*R. leptophyllus*	*Pulsatilla vernalis*
11.	*R. undecimus*	*Anemone silvestris*
12.	*R. platyphyllus*	*Ranunculus flammula*[98]

A genus *Ranunculus* of such composition as that perfectly illustrates taxonomy in its stage of transition from where, as at the first, it was almost wholly dependent on vegetative characters, to where it relies almost as entirely upon those of flower alone, and this at a time when as yet the calyx has received but little attention and has not been admitted to the category of floral organs; and the wholly petaloid sepals of *Anemone*, *Sylvia*, and *Pulsatilla* form as good a "flower," *i.e.*, corolla, as do the flower leaves of a rose or pæonia. It is also a stage at which the fruit is not yet accorded the taxonomic weight that was allowed it fifty or sixty years after Cordus by Cesalpino. In the eyes of a twentieth-century botanist the above is a curious medley to be called a genus *Ranunculus*; and what is more, no one anterior to Cordus has done as badly as that. The remotest Greeks seem to have admitted to *Ranunculus* no species not thereto referred by most botanists of the nineteenth century. Here there are added to the genus, and admirably indeed, a considerable list of true buttercups unknown to antiquity; but over

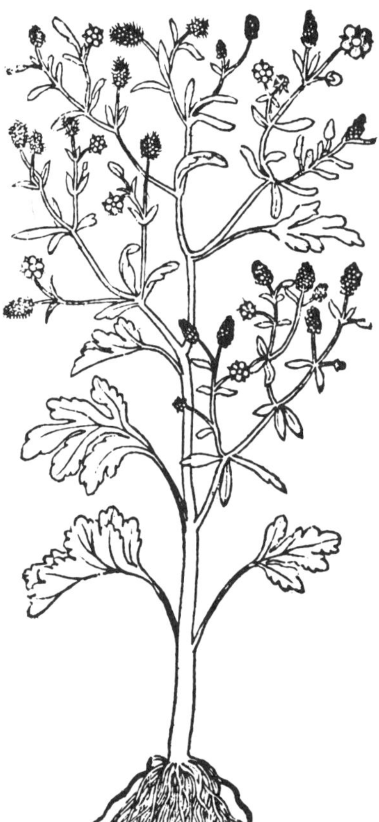

FIG. 118. Celery-leaved buttercup (*Ranunculus sceleratus* L.). Bock, *Stirpium* (1552), bk. 1, ch. 30, p. 93; after Fuchs, *Stirpium* (1542), p. 159.

and above these a trio of representatives of anemoneous genera that differ much among themselves. This kind of a genus *Ranunculus*, for that period, explains itself readily. It is plain that Cordus has yielded for once to enthusiasm for the newly rising anthology, that he has attempted the abandonment of the old reliance on vegetative characters, and is putting things together more with reference to the structure of the flower. Throughout this series there is always a five-leaved or six-leaved "flower," a circle of indefinitely numerous stamens within that, and in the middle a compact head of many "seeds." If the time had then arrived for the formal statement of the generic characters of such a group, we know that it would have been on this wise that Cordus's *Ranunculus* would have been characterized by him, to which morphologic diagnosis it would have been appended that certain acrid properties pervade the entire line of species. When I say that we know he would have done this, I have before me this proof, that the clear mention of just such common characteristics forms a part of the description of each

FIG. 119. Yellow wood anemone (*Anemone ranunculoides* L.). Bock, *Stirpium* (1552), bk. 1, ch. 30, p. 95; after Fuchs, *Stirpium* (1542), pp. 161–62.

leading species. One may take Cordus's separate diagnoses of any one line of related species and cull from them to a certainty that which to his view is the generic character. Not that this is peculiarly true of Cordus; for with equal certainty does one gather out of Brunfels's and Tragus's groups of square-stemmed opposite-leaved axillary-flowered aromatic herbs—or those otherwise vegetatively marked—the characters of their more crudely conceived genera. What distinguishes Cordus is—we must once more insist—this one rather steadily maintained appeal to floral organs, while for the time quite too much neglecting the vegetative, and also too little heeding very marked differences in the fruits of things; for he is perfectly aware that one of his ranunculi has "seeds" with long feathery tails, that another has them densely woolly-coated and compacted into the closest kind of a head, and that in a third they are rather few, and more like those of a *Thalictrum* than they are like the seeds of buttercups.

On the whole, and as thus studiously looked into, this *Ranunculus* of Cordus is one of the most significant chapters of taxonomic history ever written, for herein is illustrated as nowhere else the transition from an old taxonomy to a new one. By its making too little use of fruit characters it calls for the carpologically established genera of a Cesalpino;[99] but by the very pronounced corollism of such a chapter Cordus most clearly presages Tournefort.

We shall fall short of a fair comprehension of all that is in this chapter of taxonomy, unless we as carefully consider his disposal of two other types not referred by Cordus to *Ranunculus*, but so placed as immediately to succeed that genus. These are:

Cordus	Modern
Chelidonium minus	*Ficaria ranunculoides*
Chelidonium palustre	*Caltha palustris*

These bring the number of ranunculaceous species in this unbroken line up to fourteen. That, in as far as it goes, is very good; yet there will seem quite a glaring inconsistency in the man's having excluded *Ficaria* from a genus to which he has already admitted, over and above many good ranunculi, three anemones and a pulsatilla. If the last four could go into *Ranunculus*, why not—and much more easily—*Ficaria*? There never is in any age any other so ponderous a dead weight upon scientific progress as so-called "authority," and the prejudices it entails. Those particular anemones and the pulsatilla were plants in Cordus's time newly discovered, northern types of which no ancient Greek or Latin authority had ever heard. They hardly yet had well-established Latin names, or fixed places in the new books of botany. It was easy for Cordus to name these what he would, and to place them where he would, without risk of seeming to set himself superior to ancient and revered authority. The case of *Ficaria* was as different as possible. All antiquity had known this plant. From Dioscorides forward through the ages its name had been *Chelidonium minus*. Cordus, as a university lecturer on Dioscorides, by the acuteness and originality of his genius, had made himself famous at the age of twenty-three years. To have given any Dioscoridean plant a new name would have been venturesome. He had been bold enough when, afterwards, in his own *Historia Plantarum* he had indicated the plants' intimate relationship to *Ranunculus* rather than to true celandine. This, to my mind, seems the explanation of the inconsistency referred to. The inconsistency is, however, more than condoned by the neat item of constructive taxonomy with which it is intimately connected.

I have never been able to comprehend the view point of the professedly natural systematist who fails to perceive a most intimate relationship as subsisting between *Ficaria* and the plant called *Caltha palustris*. With Cordus the type last named was new. The German peasantry had always known the

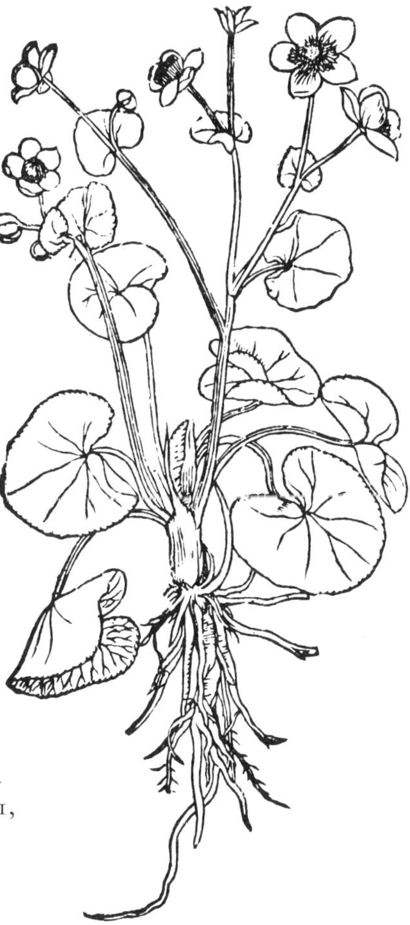

FIG. 120. Marsh marigold (*Caltha palustris* L.). Bock, *Stirpium* (1552), bk. 1, ch. 44, p. 142.

plant, and had their several vernacular names for it; but the botanist became convinced that botanically it was unnamed and undescribed; and then, like an accomplished expert in the detection of affinities—a master in taxonomy—locates it close against *Chelidonium minus*, *i.e.*, *Ficaria*, and to accentuate the expression of this relationship, names the new type *Chelidonium palustre*,[100] this to be understood, I doubt not, as a binary generic name, just as *Chelidonium minus* was understood to be; for no one now thought of this as a mere species of the celandine.

Even as to the conception of a species Cordus is so far in advance of his own time as to appear quite abreast of Tournefort, who flourished a hundred and fifty years later. Among the botanical fables that have passed for history, none is more familiar than that Linnæus was first to clearly recognize varieties; but Cordus seems to realize the difference between species and variety as well as if he had lived in the nineteenth century. The instances of his mentioning varieties are not numerous, but they suffice to show that he discriminated them readily enough. The double-flowered buttercup which he calls *Ranunculus coronarius*[101] he describes with the utmost brevity, remarking that it is not a proper species, but is a "factitious" thing of the gardens.

In giving full account of the *Viburnum Opulus* and its ornamental variety the Snowball Bush which appears so different he is careful to say that there is really no difference between the two save this, that the latter, instead of having both kinds of flowers, the fertile and small and the sterile and showy, has but the sterile kind. Plainly, he regards it as but a freak, a variety of the other. It is also worth noting that his is the earliest mention in history of this universal favorite, which he says grows wild along with the other, but is rare, except in gardens and pleasure grounds to which they transferred it from its native wilds.[102]

The question of the degree of relationship subsisting between the two kinds of *Dipsacus*, or teasel, has exercised the minds of successive generations of taxonomists early and late.[103] Cordus appears to have settled it at the outset in the right way, or at least in a manner to satisfy the requirements of the modern evolutionary. He almost fills a folio page with his fine description of the original wild and useless *D. silvestris*, and then disposes of the cultivated and singularly useful *D. sativus* in five lines; even these five relating in the main to the mechanical serviceability of the hard, prickly heads, presenting as the only morphological distinction between this and the former, the harder, firmer texture and convenient curvature of the prickle-like bracts investing the head.[104]

A philosophic botanist, writing for the philosophic, need not more explicitly avow his belief that *Dipsacus sativus* is but a usefully variant offspring of *Dipsacus silvestris*, nor need he more clearly express his understanding

FIG. 121. Teasel (*Dipsacus sylvestris* Huds.). Bock, *Stirpium* (1552), bk. 2, ch. 106, p. 847; after Fuchs, *Stirpium* (1542), p. 224.

that the wild thing, being the type of the species, is the thing for the botanist to describe in full.

Illustrations have been given already of Cordus's superior skill in bringing into line related genera, as if members of a natural family; but these taxonomic notes must not be concluded without allusion to one of the most striking manifestations of his ability to segregate, amend, and improve larger groups. Perhaps the best example of all is one that occurs in his early lectures on Dioscorides. The pharmacists of the time have a group of plants which they know as the Lactariæ, that is, milky-juiced herbs. Those best informed understand the species of the euphorbiaceous genus *Tithymalus* to be meant by this name Lactariæ, so says the lecturer; but then, he adds the suggestion that so many other herbs besides these have the faculty of shedding drops of milky juice when their stems are cut or broken, and the milks of these different herbs are so dissimilar as to their properties—some being innocuous, others poisonous—that the plants ought clearly to be distinguished in groups:

Let us, then, indicate three groups by their [morphologic] differences. Let the first be the Tithymali, the second the Intybaceæ. Then we shall have a third group that is of neither the Tithymali nor the Intybaceæ. We shall now proceed to demonstrate how these two may be distinguished. Tithymalus is a genus of many species every one of which has elongated leaves devoid of any kind of indentation or division. All intybaceous herbs have their leaves more or less erose or incised all around the edges. The greater proportion of the tithymali have an umbellate inflorescence and all of them, even those not umbellate, have each flower and fruit subtended by a pair of opposite leaves. But the flower stalklets of the intybaceæ bear each a flower composed of small leaves compacted together, yellow as to color, *Chondrilla minor* and *Cichorium* excepted, in which they are either blue or white. The flowers of intybaceous plants resolve themselves into a kind of wool called a pappus. Never so the tithymali, for all of them have a three-celled fruit with a single seed in each cell. The seeds of intybaceæ with their pappus sail away on the breeze. The tithymali eject their seeds forcibly and with a sound. The milk of the intybaceæ is bitter at first taste. That of the tithymali is at first taste mild, even not so unlike that of cow's milk, but after that it begins to burn the mouth and throat, and if applied to the skin may blister it; which the mild juices of the intybaceæ never do. Now these herbs, which I have thus distinguished by their proper marks from the Intybaceæ, are called the Tithymali and may be known by that general [*i.e.*, family] name. To the Intybaceæ belong all the species of *Lactuca*, both kinds of *Chondrilla*, *Intybus sativus*, and *silvestris*, commonly known as *Cichorium*, also *Hieracium*, and whatever other plants are like them [*i.e.*, organologically], and have a milky juice.

My third assemblage of lactiferous plants can not be distributed among these two groups, neither do they of themselves constitute an alliance; for they are not united by any resemblances which they have in common. But they are not very numerous. *Ficus*, *Erinus*, *Scammonia*, and *Cissampelus* are among them; each seeming to belong to an alliance of its own.[105]

Cesalpino, of the end of the sixteenth century, will be praised in future millenniums for having founded Systematic Botany. But had Valerius Cordus lived to only twice his nine-and-twenty years, it is easy to conceive that the great Italian might have missed his laurels.

Nomenclature

Euricius Cordus having demonstrated that many a plant type native to Germany and well known to Germans was entirely unknown to the Greeks and Latins of antiquity, his son Valerius proceeded to name and describe such types, so that they might take their places—so that each might become the subject of a chapter—in books of botany. Valerius Cordus was the first man in history to establish many new genera of plants. I do not know that any one author between Dioscorides and Valerius Cordus proposed more than one or two new genera. This man all at once proposed so very many that his book is a great landmark in the history of genera. What we are here interested in finding out is whether in the naming of his new genera Cordus either tacitly or openly subscribes to any particular principles of generic nomenclature. From what we know of his antecedents as the carefully instructed favorite son of a celebrated Greek and Latin scholar, we shall reasonably expect to find him constructing his new names after classic models, making no inroads upon ancient usage. Let us make a selection of his new names, that the student of nomenclature may observe the manner of their construction.

Cordus	Modern	Cordus	Modern
Anblatum	*Lathræa*	*Moschatella*	*Adoxa*
Balsamella	*Impatiens*	*Helianthemum*	*Helianthemum*
Coralloides	*Dentaria*	*Oxycoccus*	*Oxycoccus*
Drosion	*Alchemilla*	*Pneumonanthe*	*Pneumonanthe*
Isophyllon	*Bupleurum*	*Rorella*	*Drosera*
Limnesion	*Gratiola*	*Sagitta*	*Sagittaria*
Lycostaphylos	*Opulus*[106]	*Thamæcnemum*	*Vaccaria*
Millegrana[107]	*Herniaria*		

It will be seen that by far the greater proportion of these fifteen generic names is regularly Greek-made; some five are as plainly Latin, while one only, and that the first of the list, is very barbarous Latin; for *Anblatum* is said to be the old German name *Ohnblatt* (leafless) written over into Latin and supplied with the convenient terminal. It may be remarked that there is not one in this list of fifteen generic names that ought ever to have been displaced. It will hardly be possible to point out one of them which is not as good as its modern equivalent. Yet *Helianthemum* is the only one of the fifteen that never has been set aside; and the suppression of the rest is mainly to be attributed to Linnæus's recklessness in the matter of priority in nomenclature.

While most of Valerius Cordus's new generic names are of the universally acceptable one-worded type, it is clear from the following list that he did not find the binary sort objectionable.

Cordus	Modern
Chelidonium palustre	*Caltha*
Chelidonium phragmites	*Corydalis*
Hepatica alba	*Parnassia*
Pentaphyllum palustre	*Comarum*
Trifolium palustre	*Menyanthes*

Here again I apprehend a difficulty on the part of many a reader to see that these two-worded names are purely generic; we are so perfectly accustomed all our life long to read every such name as being half generic, half specific; and so I affirm again that not one of those second and adjective terms is a specific name at all. Colloquially, and in our vernacular, we make use of two-worded generic names most freely without thought of impropriety. When we speak of Rose (=*Rosa*), Christmas Rose (=*Helleborus*), Rock Rose (=*Cistus*), and Guelder Rose (=*Opulus*) we have no such thought as that all these are so many different kinds, or species, of Rose. No more had Valerius Cordus any such thought as that the four types which he knew as *Chelidonium majus*, *Chelidonium minus*, *Chelidonium palustre*, and *Chelidonium phragmites* were four kinds, or species, of celandine. To his mind they were as clearly four distinct genera as their respective modern equivalents *Chelidonium*, *Ficaria*, *Caltha*, and *Corydalis* are different genera to the mind of the botanist of today.

Nevertheless there were people in Cordus's time and before that—untaught, unbotanical botanists—to whom such binary generic names were a stumbling block, people who supposed that the above four were but so many kinds of celandine, being misled by the reiteration of the substantive part of the name; and Cordus, while himself creating some such double generic appellations, suppresses several of the older ones in favor of new ones of one term; and this as if to correct in each instance a deep-seated popular error. The labiate plant known now as *Glechoma* had long been known in Latin as *Hedera terrestris*, a name which caused it to be thought of as really a kind of ivy. Cordus proposes that botanists shall call it by a new name *Chamæclema*.[108] Similarly what had always been called *Trifolium acetosum*, or sour clover, seeing they really mistake it for a kind of clover, Cordus proposes shall be known as *Oxys*.[109] The genus, by the way, was so known for two centuries after Cordus had proposed it, and until Linnæus without the shadow of a reason for so doing, changed it to *Oxalis*.

In respect to the construction of new generic names Cordus represents the most rigidly classic type of nomenclator in this, that he creates no meaningless names. Every appellator of this kind that he makes is framed in allusion to some characteristic, either organologic, or ecologic, or qualitative, of the type itself, and is therefore full of meaning. The nomenclature of genera in even the remotest antiquity was not universally so; for they had in ancient times the genera *Artemisia*, *Eupatorium*, *Euphorbia*, *Gentiana*, *Pæonia*, *Valeriana*, and several others which, like these, were named in honor of eminent medical botanists. Not one, however, of Cordus's many new genera is dedicated to any person whether historical or mythical. Even Theophrastus, Nicander, Dioscorides, and Pliny were not to be commemorated in generic nomenclature until after the lapse of more than a century and a half from the date of Cordus's death.[110]

As regards the principle of priority, it is to be observed that he stands by it and contends for it only in the case of names that are of great antiquity. For instance, he finds that the name *Eupatorium* has been displaced from its ancient type so that the plant is everywhere in his time known under the name of *Agrimonia*. He will enter into no compromise with this kind of error. While as he says, "Almost all physicians and pharmacists of today call it *Agrimonia*,"[111] he declines to head his chapter with that appellation, and writes *Eupatorium* instead, regardless of the convenience of the erring multitude. It is taking boldly the ground that the scientific man must be true to history in his plant naming, and that it belongs to the doctors and druggists to correct their errors according to the light of history. This is nothing less than the most tenacious adherence to the principle of priority, the restoring of an ancient name, where the whole concourse of those in a business way interested will be opposed to the restoration, and he knows it.

We have observed Brunfels suppressing so ancient and classic a generic name as *Chelidonium minus* in favor of a new name, *Ficaria*;[112] this presumably on the ground that the latter is short, one-worded, and free from ambiguity. Cordus, as if intolerant of any change that should relegate to synonymy so old a name, continues the old binary in use, not deigning to cite Brunfels's *Ficaria* even as a synonym.[113]

Specific nomenclature, as far as Cordus is concerned, remains in *status quo*. All specific names are binary. There is not in him the trace of a tendency to combine name and description in one phrase. No botanist of the nineteenth century was any further from that than he. But he is no more careful than his contemporaries to avoid the use of those meaningless things, the numeral adjectives, as specific names. He has an even dozen species of *Ranunculus*, and the names of three are *R. octavus*, *nonus* and *undecimus*. The use of these cabalistic appellations, however, perfectly demonstrates that he had no idea of any other than binary names for species in genera that were more than monotypic.

Anatomy and Physiology

Almost every page of original plant description in Cordus bears evidence of his having been accustomed to examine the interiors of organs. We shall hardly expect him to lay any foundations of a science of plant anatomy; for he knows nothing of any artificial aids to vision. The vegetable cell will not reveal itself to him, nor anything else that is too small to be seen with naked eye. But he makes sections of roots, stems, leaf stalks, fruits of all kinds, and even of seeds, and records the anatomical aspects of them all. Neither Grew nor Malpighi, had he lived at the time of Valerius Cordus, could have done more.

One meets with these records of anatomic and physiologic observation only as distributed up and down the pages of the whole volume and forming part of the regular descriptions of genera and species; and they are so very numerous that one may here reproduce but a small selection of them.

That stemless aquatic, the common European sagittaria, is a large plant noteworthy on account of small size of the leaf blades and the flowers, in contrast to the great dimensions of the leaf stalks. The bulk of the plant as a whole consists of mere leaf stalk. It is certain that by way of enquiry into this matter Cordus has dissected those leaf stalks, for the first clause of his description of the plant is this: "*Sagitta* has triquetrous petioles, very thin and spongy as to the inside, a yard high or sometimes longer, or shorter, all according to the depth of the lakes in which it grows."[114] In like manner he has examined the interior of the flower stalk in its turn, ascertaining that while smaller than the petioles it is also of "spongy texture, and of remarkably light weight."

The water lilies are plants large in all their parts, so that viewed in cross section they exhibit clearly their structural characteristics, and he notes them fully. What he takes for the main root of *Nymphæa alba* he reports to be as thick as one's wrist lying horizontally along the mud, knotted on at the places where the leaves of former seasons were inserted, are quite black on the outside, but of a clear white within, and of a very spongy and porous substance. The fibres that descend from these into the ground he reports to be outwardly greenish-white, quite white within and also porous. The stalks that support the flowers and leaves in this species he finds to be terete, and to have open tubes extending throughout their whole length within. The yellow water-lily rootstock he finds less knotted on the outside, and white both without and within, otherwise like the former, but with coarser fibrous roots and these still more spongy; the petioles, obtusely angled rather than terete, show smaller tubes in the middle.[115]

There was a coarse rank dry land herb, the teasel, *Dipsacus silvestris*, which lent itself as readily to Cordus's anatomical inspection. He says it is of but biennial duration, that its root, of a finger's thickness, during the first year of its life is fleshy; the second it is of a ligneous hardness, and in that condition dies after the plant has flowered and fruited. The tall stem of the plant he remarks is heavy enough to warrant the inference that it is solid; nevertheless he finds it hollow throughout in the very center. Even the large egg-shaped head that bears the flowers and the seeds on its surface he has cut across in each direction to find that within it is filled with what he calls a woolly pith.[116] And this inspection of the interior of stems is made not only in easy and inviting cases, but everywhere. Seldom is there wanting to any of his plant descriptions the clause that tells of the internal structure of its stem, as being solid, or hollow, or as filled in the center with pith; and the color of the pith is also often given. Manifestly the cutting of stems across is as much a part of his work with herbaceous plants as it is to note the exterior of them as terete or angled, smooth or rough, even-surfaced, or striate, or fluted; and he is particular about mentioning every such characteristic. Of still deeper interest are his dissections of all manner of fruits, fleshy and capsular, and of seeds; but on account of the taxonomic significancy of their results, these have been related under another heading.

FIG. 122. Sickle-fruited fenugreek (*Trigonella fœnum-græcum* L.). Bock, *Stirpium* (1552), bk. 2, ch. 4, p. 597; after Fuchs, *Stirpium* (1542), p. 798.

The intimations of what seemed like a sense of feeling in some mimosaceous foliage had been one of the wonders of plant phenomena as observed by ancient Egyptians and Greeks. Cordus had seen no mimosa, or sensitive acacia, but in two plants frequent in German gardens he observed the foliage remarkably sensitive to varying meteorological conditions. In his very full and minute description of the licorice plant the first remark about the foliage, after having stated its character as a compound leaf, is that its leaflets exude some substance by which they adhere to one's fingers when handled; then he proceeds to describe the several different degrees of enfoldment or of expansion which the leaflets exhibit according to the altitude of the sun on clear days, and how they keep themselves folded together all day long if the sky be cloudy or the weather wet or cold, concluding with the observation that all this is true of the leaflets only, and that the leaf stalk itself does not alter its position at any time of the day, and that in all kinds of weather its attitude is the same.[117] This last remark is called for by what he had said in an earlier chapter about nyctotropic movement in the common fœnugreek.[118] In this he says the nightly folding together of the leaflets is accompanied by a deflection of the whole leaf, petiole and all.

In every description of a twining species of herb Cordus mentions the direction of the circumnutation, whether as following the course of the sun or as taking the contrary direction. Of these phenomena he is the first of all writers to make record, I think. He was the first to describe the plant called sundew,[119] and its physiology interested him. Its description occupies the first chapter of the *Historia*, and as a new genus he would like to call it *Rorella*.[120] He seems to have perfectly established it, that what seems like dew on the leaves of the plant is really an exudation, for he says that in the very driest weather the plant is still sprinkled all over with those minute drops that have the appearance of dew. He has done what was in his power

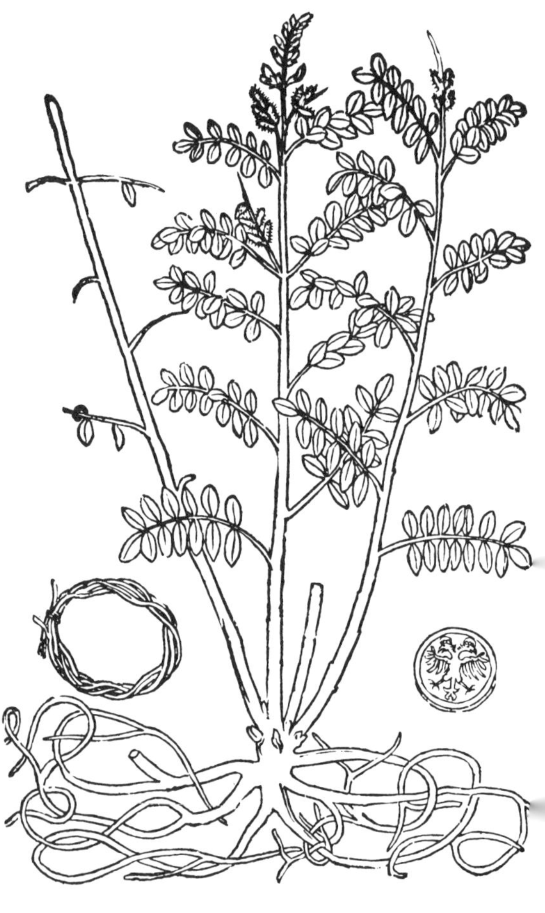

FIG. 123. Licorice (*Glycyrrhiza glabra* L.). Bock, *Stirpium* (1552), bk. 2, ch. 150, p. 935; after Fuchs, *Stirpium* (1542), p. 192.

to ascertain at least the qualities of those minute glistening drops. He has tasted, and reports the flavor to be a little bitterish, with also a hint of the acidulous, and slightly acrid.

The plant physiologist of today, interested in the functions of the root tubercles of leguminous plants may find in Valerius Cordus the earliest mention of these organs. I do not find him taking note of them except as occurring in the cultivated lupine of Europe. Accustomed to give a full account of every kind of root, even to its medicinal usefulness or uselessness, he says of that of the lupine that it is "slender, woody, white and without useful properties, parted into a few slender fibres upon which there sometimes grow small tubercles." [121]

Ecology

We have already been learning that even from the most primitive times every botanist was an ecologist, at least to the extent of observing and recording the special environment which every kind of wild plant affects, and sometimes to the mentioning of some of its associate species. Valerius Cordus, being well skilled in both chemistry and mineralogy, goes beyond all his predecessors in that he names the petrography of a plant's habitat, or otherwise indicates the constituency of the soil in which it is to be looked for. We can in no other way so well present this, his own new aspect of matters ecological, than by the translation of a few of his passages.

The fern called hart's tongue, best known as *Scolopendrium*, but which Cordus knew as *Phyllitis*, he says, "grows on shaded and rocky declivities of mountains; loves a rich soil, though not springing from the soil directly, but from the moss that covers the rocks, especially limestone." [122]

To *Saxifraga Aizoon* he attributes the habitat of "Limestone cliffs, especially where they are wet and overgrown with moss." [123]

Describing two species of *Sanguisorba*, that which he calls *S. major* "inhabits low clayey pasture lands that are subject to inundation from rivers," while *S. minor* also "grows in clayey soil, or even gravelly, but on open sunny slopes and along roadsides." [124]

He seems to take a special interest in the ecology of such plants as he has himself first discovered and described as new. The cranberry is one of these. He says he found it growing "in wild and very wet and soft mossy bogs; often with the sundew not far away." [125] Then, under his description of the sundew itself—also new—we learn that he found that, though as a neighbor to the cranberry, yet in soil distinctly of another character, that is, "in very wet sandy places." [126]

In the case of that new generic type which he denominates *Moschatella* (= *Adoxa* L.) he mentions its most interesting associates. "It grows in shady places, under trees, in soil very rich, along with the fumariaceous kind of Celandine (= *Corydalis cava*) and also the ranunculaceous Celandine (= *Ficaria ranunculoides*)." [127]

FIG. 124. Great burnet (*Sanguisorba officinalis* L.). Bock, *Stirpium* (1552), bk. 1, ch. 157, p. 470; after Fuchs, *Stirpium* (1542), p. 788.

Cordus knows at least one specific type which, as he observes, has a way of establishing itself upon some diversity of soils, and adapting itself to several different kinds of locality; and the phases which it assumes according to its altered environment so much interest him that he gives a particular account of them. The subject is the shrub *Spartium scoparium*.[128] He says:

> It inhabits rough places on mountains in a hard reddish soil somewhat sandy, as in Hesse, etc.; but sometimes occurs on the lowest plains in mere sand and gravel, as about Nuremberg, etc. Nor should it be left unmentioned that while on sandy plains it is a low bush seldom exceeding a yard in height; on the mountains, where the soil is better, it approaches the dimensions of a tree, with a trunk from seven to nine feet high and so thick that one can not span around it, supporting a head of virgate branches so dense as to intercept and hold all the snow of a considerable storm, so that the traveller, passing through such a wood in winter, may walk on almost bare ground under arches of snow overhead.

There is no indication that the author would distinguish even as varieties these rather strikingly different phases of the shrub. He regards them as the natural products of different conditions as affecting a simple species. It is the well-skilled botanist's view, whether of the sixteenth century or of the twentieth.

Pomology

The recognition of marked varieties in the same species of cultivated fruits—varieties originating under cultivation—is so very ancient that there is no hope of one's ever tracing it to its beginnings. Also before the dawn of history men had learned that valued varieties of fig, olive, grape, and other fruits could not be depended on to come true to seed. Seedlings of these were apt to prove degenerate, as they called it; and the propagating of them by layers, and especially by grafting, had been invented as the sure means of preserving and perpetually reproducing choice varieties.

They, who wrote of fruit culture two thousand years ago and more, mention by name great numbers of varieties, not only of fig, olive, and grape, but also of peaches, cherries, and other fruits,[129] sometimes favoring the reader with a few hints of the differences subsisting between two varieties; but I have met with nothing like descriptive lists of the varieties of even such common and variable fruits as figs, olives, and grapes, in the writers of antiquity, nothing that was written for the purpose of enabling the reader to identify the varieties. I cannot discover that any one anterior to Valerius Cordus engaged in this kind of an enterprise.

There are long chapters in Cordus's book which so read as to make it certain that in the course of his botanical expeditions to many parts of Germany, as well as at home, he made everywhere very special studies of the different varieties of apple and pear which were under cultivation in the orchards of the time, and that he wrote a careful description of each on the spot, and that so full that the properly qualified reader would be able to identify the different kinds by the description alone. I say the qualified reader, meaning of course the educated, for every line of Cordus's pomologic writing, like all the rest of his botany, is in Latin; and a knowledge of the Latin terminology of descriptive botany is essential to the full understanding of him here.

With the intention, then, of interesting the botanical fraternity in this diversity of cultivated varieties, he describes as many as fifty named varieties of pears, and thirty-one of apples, all of which he has found in one part or another of Germany.[130] The original German names are always given, then the name is turned into Latin as if for the convenience of the botanists, all of whom in the time of Cordus, like the other educated people, find Latin the only adequate medium of scientific converse. The excellency of these pomologic diagnoses can more readily be seen than described, and I therefore present English translations of two or three of them; and since it is his practice to describe one variety partly by comparison with another, I shall take up three descriptions that are consecutive. Also inasmuch as most of the pears which one sees in the fruit marts of today are quite pear-shaped, I shall begin with one which Cordus knew that was globose as an apple:

KAULBIRN, that is, Globe Pear, is almost as round as a globe, except that at base it ends in a blunt and scarcely manifest protuberance. It is hardly two inches long, seldom at all exceeding that, and the diameter is a trifle greater than the length. The color is altogether pale green; the flesh very tender, melting in the mouth, and of a mild delicious flavor, and by its abundant juiciness satisfying thirst, delightfully fragrant when pared. This matures at the beginning of autumn, and is very perishable. It is cultivated about Eisleben.

HANFFBIRN, that is, Hemp Pear, is similar to the Globe, but a little larger, in color green, with darker spots and dots; in flavor answering to the Globe, but the flesh not as tender and juicy; matures at the same time and soon decays. This also is grown at Eisleben.

GLOCKENBIRN, that is, Bell Pear, is abruptly narrowed below into a narrow neck, this again at the very base widening into a blunt head-like protuberance, the fruit as a whole having the configuration of a bell,[131] whence it has its name. The color is yellow, well speckled with green, the length a trifle less than three inches, the diameter not more than two inches. Rather fragrant when pared; the flavor that of the Hemp Pear; matures at the same time with that and is perishable. Grown in quantities at Eisleben.

KÜNIGSBIRN, that is, Royal Pear. Large and ventricose, sometimes four inches long, the diameter somewhat less; color blue-green, but on the side exposed to the sun faintly reddening. Flavor a trifle astringent, flesh somewhat juicy and vinous, assuaging thirst. Matures at the end of September and is not very perishable.

It is perhaps less difficult to describe apples well, than pears; and all those familiar to Cordus are very vividly depicted. We present two or three:

HARTLINGE WEISS, that is, White Harding. Somewhat depressed-globose, the height about two inches, the diameter two and a half or somewhat more. Colored reddish-yellow on the sunward side, elsewhere greenish-white, dotted with specks that seem to lie beneath the transparent epiderm. Flesh tender though firm, juicy, of an acid-vinous and excellent flavor; the fruit fragrant for a considerable period after having been stored. Ripens in early autumn and keeps until the end of winter.

Grown almost throughout Germany.

SAFFRANECKE. Saffron Apple. In quality scarcely second to any apple, this is also nearly globose and but an inch and a half in diameter, the breadth not rarely a little exceeding the length. Color something between green, yellow, and pale saffron; in warm and dry seasons adorned with dots and narrow streakings of scarlet and orange; in moist and cloudy summers devoid of any such markings. Flesh tender, sweet slighty acidulous, luscious, and with a kind of spicy pungency which commends it to any palate. Moreover, it is delightfully fragrant. Begins to ripen in early autumn, and keeps until the vernal equinox, and frequently even longer.

Cultivated at Hildesheim, particularly in the orchards of the monks at the monastery in Sulta, located outside the city.

HARTLINGE ROT. Red Harding. Either much flattened at the ends and orbicular, or now and then quite globose; in size also as variable, but on the average perhaps about two inches high, and two and a half in breadth. Color rose-red, marked with long blood-red stripes. Flesh rather solid and dense, yet very tender, juicy, acidulous, and with a peculiarly luscious flavor. Rather pleasantly scented. Ripening in early autumn, and the most enduring of all apples, sometimes keeping until the middle of the next summer, and according to the statements of some, even longer than that.

It is found in almost all orchards of nearly every part of Germany.

I apprehend that a skillful pomologic artist should be able to draw and color these apples and pears so vividly word-pictured by this perhaps the very first of descriptive pomologists, and for the history of pomology it may be doubted that there are extant any more important chapters than these written by Valerius Cordus three hundred and seventy years ago.

Commemorating Valerius Cordus, Plumier established the genus CORDIA in the year 1703.[132]

Appendixes A and B

APPENDIX A

Pre-Grecian Knowledge of Plants

Greene's hypothetical discussion of preclassical knowledge of plants does not adequately convey the extent of this knowledge; and an underestimation there can lead to an overestimation of the Greek achievement. The Greeks did develop scientific theory to an impressive extent, and their achievement was hardly foreshadowed by earlier cultures, but the information that their theories interpret was to a large extent the legacy of earlier peoples.

Prehistory

From our modern perspective, the domestication of plants and animals seems one of the greatest and most enduring achievements of early peoples. For this reason, farming has often been viewed as a brilliant discovery. It seemed so because farming was assumed to provide a more secure living than did hunting and gathering. However, when anthropologists investigated this assumption, they found that hunter-gatherers have a more secure food supply obtained for less work than is the case for their neighbors whose subsistence comes from primitive agriculture; Lee & De Vore, *Man the hunter* (1968). If this is true, then agriculture may have been not so much a brilliant discovery as a desperate expediency, resorted to in overpopulated regions where hunting and gathering could no longer feed the people. According to this scenario, the essentials of farming were not so much unknown, before farming began, as they were unnecessary; Cohen, *Food crisis* (1977). See also E. Anderson, *Plants* (1952); Brotherwell & Brotherwell, *Food in antiquity* (1969), chs. 6, 7; Heiser, *Seeds to civilization* (1973); Simmonds, *Evolution of crop plants* (1976); Smit, *History of life sciences* (1974), cols. 181–230; Tannahill, *Food in history* (1973), chs. 1, 2.

Historical Times

The use of plants for medicines in ancient India, Persia, China, Egypt, and Israel is discussed in Berendes, *Pharmacie* (1891), vol. 1, pp. 1–124.

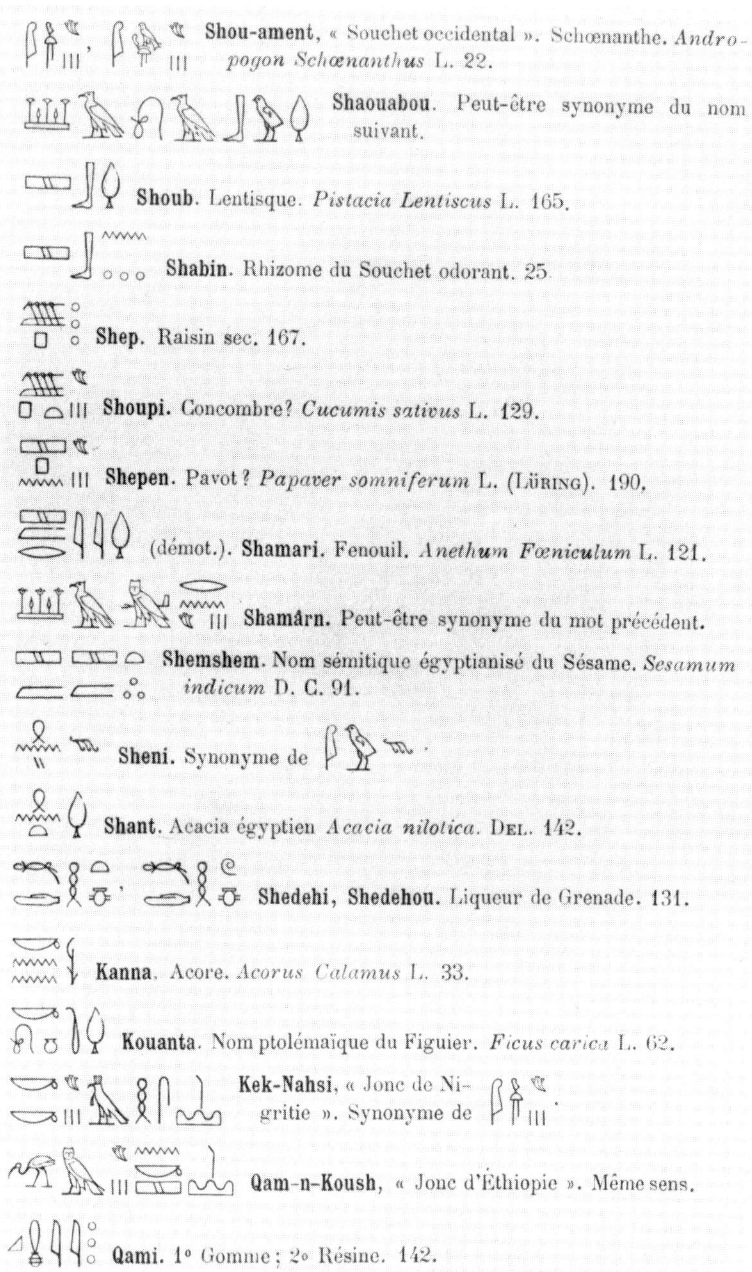

Shou-ament, « Souchet occidental ». Schœnanthe. *Andropogon Schœnanthus* L. 22.

Shaouabou. Peut-être synonyme du nom suivant.

Shoub. Lentisque. *Pistacia Lentiscus* L. 165.

Shabin. Rhizome du Souchet odorant. 25.

Shep. Raisin sec. 167.

Shoupi. Concombre? *Cucumis sativus* L. 129.

Shepen. Pavot? *Papaver somniferum* L. (Löring). 190.

(démot.). **Shamari**. Fenouil. *Anethum Fœniculum* L. 121.

Shamârn. Peut-être synonyme du mot précédent.

Shemshem. Nom sémitique égyptianisé du Sésame. *Sesamum indicum* D. C. 91.

Sheni. Synonyme de .

Shant. Acacia égyptien *Acacia nilotica*. Del. 142.

Shedehi, Shedehou. Liqueur de Grenade. 131.

Kanna. Acore. *Acorus Calamus* L. 33.

Kouanta. Nom ptolémaïque du Figuier. *Ficus carica* L. 62.

Kek-Nahsi, « Jonc de Nigritie ». Synonyme de .

Qam-n-Koush, « Jonc d'Éthiopie ». Même sens.

Qami. 1° Gomme; 2° Résine. 142.

FIG. 125. Some of the 202 kinds of plants named by the Egyptians in hieroglyphics. Loret, *Flore pharaonique* (1892), p. 143. (Courtesy Tozzer Library, Harvard University)

EGYPT

By the time writing had been invented, various cultures had knowledge of 100 to 200 kinds of plants; the Egyptians, for example, had hieroglyphic names for 202 kinds of plants (see Fig. 125). Egyptian use of herbal medicines undoubtedly began before 3000 B.C.; Budge, *Divine origin* (1928), ch. 5, and Dawson, Studies in medical history (1929). The Egyptians cultivated gardens for food, medicines, and ornaments for their surroundings. An excellent illustration of an Egyptian garden comes from a tomb of the reign of Amenhotep III (*ca.* 1411–1375 B.C.), reproduced in color in Berrall, *Garden* (1966), p. 10. See also Gothein, *Garden art* (1928), vol. 1, ch. 1; Hartmann, *L'agriculture dans l'ancienne Égypte* (1923); Joret, *Plantes* (1897–1904), vol. 1, pp. 1–326; Keimer, *Gartenpflanzen im alten Ägypten* (1924); Laurent-Täckholm, *Faraos blomster* (1951).

Among the Egyptian technologies was the making of paper from papyrus,

FIG. 126. Watering a garden in Egypt by the shādūf method, thirteenth century B.C. (Permission of the Science Museum, London)

ancient accounts of which are in Herodotus, *History* (1921–24), bk. 2, secs. 37, 92, 96, and bk. 7, sec. 25; Pliny, *Natural history* (1938–62), bk. 13, secs. 21–23, and bk. 24, sec. 88; Theophrastus, *Enquiry into plants* (1916), bk. 4, ch. 8, secs. 3–4. A recent account is in Lucas, *Ancient Egyptian materials* (1962), pp. 137–40. Egyptian plant foods are discussed in great detail in Darby, Ghalioungui & Grivetti, *Gift of Osiris* (1977), and more briefly in Tannahill, *Food in history* (1973), chs. 2, 3. On drugs and preservatives, see B. Baumann, Botanical aspects of . . . embalming (1960), and Von Deines & Grapow, *Ägyptischen Drogennamen* (1959). Further references on plant use in Egypt are in Smit, *History of life sciences* (1974), cols. 231–50.

The Near East

The Near Eastern knowledge of plants in antiquity is described in Gothein, *Garden art* (1928), vol. 1, ch. 2; Joret, *Plantes* (1897–1904), vol. 1, pp. 327–504, and vol. 2; Laufer, Sino-Iranica (1919). Further information on early plant usage is in Levey, *Chemistry . . . Mesopotamia* (1959), Babylonian botanical gardens (1961b), and *Early Arabic pharmacology* (1973), pp. 1–9.

The plants known and used by the Assyrians are discussed in Bonavia, *Flora . . . Assyrian monuments* (1894). More comprehensive and recent, though less readable, is Thompson, *Assyrian botany* (1949). Other references on the plant lore of ancient Mesopotamia are in Smit, *History of life sciences* (1974), cols. 251–62.

The plant lore of the Bible has attracted the attention of scholars almost since the beginning of Biblical scholarship. From a religious perspective, Maimonides discussed this lore in Book 7 of *Mishnah Torah* (1180, in Hebrew; see Appendix B). Very useful recent discussions are in Harrison, Healing herbs of the Bible (1961); Moldenke & Moldenke, *Plants of the Bible* (1952). Other references are in Smit, *History of life sciences* (1974), cols. 314–17.

India

In India, a knowledge of plants can be traced back many centuries before Christ, though dating is often only approximate. Tied to either agriculture or medicinal usage, this knowledge is discussed by the following authors: Ainslie, *Materia Indica* (1826); Chopra, Nayar & Chopra, *Indian medicinal plants* (1956); Chowdhury, Botany: Prehistoric (1971a); Dutt, *Materia medica of Hindus* (1877); Ghosh & Sen, Botany (1971); Giboin, *Épitomé de botanique . . . l'Inde* (1949); Hammett, Agricultural and botanic knowledge (1950); Levey, *Early Arabic pharmacology* (1973), pp. 10–19; Majumdar, History of botany . . . India (1951); Nadkarni, *Indian materia medica* ([1955]); Smit, *History of life sciences* (1974), cols. 263–82.

FIG. 127. "Winged genii" fertilizing a stylized date palm (*Phœnix dactylifera* L.). From the Assyrian palace of Ashur-Nasir-Pal II at Calah, 883–859 B.C. This bas-relief on limestone is 2.26 m high. (Courtesy Museum of Fine Arts, Boston; Charles Amos Cummings Fund, 35.731)

China

The Chinese began writing about plants in antiquity, from both medical and agricultural points of view, though some of these early writings cannot be dated precisely. *Shih-ching*, the book of odes, is an anthology reflecting Chinese life from the late eleventh century to the middle sixth century B.C. Its discussions of some 150 kinds of plants also provide information about the natural vegetation of ancient China. See Ho, Origins of Chinese agriculture (1969); Li, Vegetables of ancient China (1969).

The earliest extant Chinese agricultural treatise was written in the last half of the first century B.C. There is also extant a comprehensive agricultural encyclopedia of the sixth century A.D., written by Chia Ssu-hsieh, which provides insights into earlier knowledge; an English summary is in Sheng-han, *Agricultural encyclopædia* (1962). The study of materia medica, called *pên-ts'ao*, existed before the year 200 B.C.

The earliest extant botanical treatise is the *Nan-fang ts'ao-mu chuang* [Plants of the southern regions] by Chi Han (*ca.* 263–307 A.D.), a poet and government administrator who was assassinated during a civil war. His botanical report covers only eight pages of Chinese in the recent edition (1979), with English translation by Hui-Lin Li. Chi Han wrote brief accounts of about 80 kinds of plants that have practical or ornamental uses and are found in a region now in southeastern China and north and central Vietnam. The *Nan-fang* text, as now known, may have been altered by additions and losses since Chi Han's time. Li provides both English and scientific names for the species Chi Han discussed, along with notes and bibliography to relevant literature on economic botany in Chinese, Japanese, and western languages. See also Bretschneider, *Botanicon sinicum* (1882–95), vol. 1, pp. 38–39 (on Chi Han), 39–75 (on materia medica), and 75–86 (on agriculture).

The best and most comprehensive contemporary work on Chinese medicine, including medical botany, is Chamfrault & Ung, *Traité de médecine chinoise* (1959–61). The best brief, comprehensive guide to Chinese science, including botany, is Sivin's introductory bibliography of literature in western languages, along with several articles, in Nakayama & Sivin, *Chinese science* (1973). Still useful is much of the information on Chinese science in Sarton, *Introduction to history of science* (1927–48). Joseph Needham and his associates are currently providing a new and authoritative account, *Science . . . China* (1954–). Volumes 1 (*Introductory orientations*) and 2 (*History of scientific thought*) provide useful background for the history of Chinese botany. There is also a one-volume abridgement of these two volumes, Needham & Ronan (1978). Volume 6 of the unabridged treatise is a history of Chinese botany (*Biology and biological technol-*

ogy), being written by Needham with the assistance of Lu Gwei-Djen and Georges Métailié. Needham has already published two sections from this volume: Botanical taxonomy (1971); and, with Lu, Hygiene . . . China (1962). Also helpful is the annotated bibliography in Smit, *History of life sciences* (1974), cols. 281–310.

APPENDIX B

Medieval Knowledge of Plants

The last ancient authority discussed by Greene is Galen, and the earliest Renaissance author is Theodore of Gaza. The gap between them is more than twelve hundred years. Classical authors had firmly established the basic features of botany, and the works of Theophrastus, Nicander, Dioscorides, Galen, and others retained their influence throughout this long period. The transition from "ancient" to "medieval" is therefore difficult to discern, but it was the practical rather than the theoretical writings of antiquity that provided the most widely emulated examples for the Middle Ages. This emphasis was set within the contexts of agriculture and pharmacy, and medieval handbooks on plants became important documents that passed on much of the accumulated data of classical antiquity. But the herbals of the period, though written in Greek and Latin, also evinced a new approach to the study and utility of plants in pharmacy, medicine, agriculture, and veterinary care. Thus, when Greene discusses botanical works of the Renaissance, one cannot fairly evaluate those works without some notice of the writings in botany and related subjects that were produced in that vast intervening period.

Meyer did provide a good survey of this literature from Islam and Europe in his *Geschichte der Botanik* (1854–57), vols. 2–4, and there are bibliographical guides to more recent literature in Sarton, *Introduction to the history of science* (1927–48), and in Smit, *History of the life sciences* (1974). The objective here, then, is to indicate some of the more noteworthy botanical works of the medieval period, drawing on, but not attempting to duplicate, the information published by these three authors, and also providing additional references, especially to the more recent literature. Two other works that contain much useful material are Stillwell, *Awakening interest in science* (1970); Ullmann, *Medizin im Islam* (1970).

Byzantium

Greek civilization endured in the eastern Mediterranean region throughout ancient and medieval times, although its extent was gradually reduced during conflicts with Persians, Arabs, and Turks. The latter finally overran Constantinople in 1453. Since the ancient Greek scientific writings did not require translation to be understood, they continued to exert a strong influence during the medieval period. Byzantine scholars assiduously analyzed ancient Greek scientific materials and wrote commentaries on them. These "scholia" provide modern students with a clear view of what works were available to Byzantine savants, as well as the problems they encountered as they interpreted ancient authorities. Byzantine botany inherited the traditions of nomenclature and classification bequeathed particularly by Theophrastus and Dioscorides, and the medieval Greek scribes and scholars updated these works by adding new names as they were obtained from neighboring Slavic and Arabic peoples. They also invented new methods of arranging the older materials, as is seen in the numerous alphabetical texts of Dioscorides' *Materia medica* that began with Oribasius's quotations from Dioscorides. Medieval Greek students of botany were thus groping for a sensible system of nomenclature.

Notable Byzantine contributions to the history of medicine and science are the numerous compendia that began with Oribasius in the fourth century and continued through Joannes, son of Zacharias, in the fourteenth. The most important of these medical and botanical encyclopedias merit attention here as part of the background for Greene's discussion of Renaissance authors.

One of the most famous and influential encyclopedias, which combined practical agriculture and botany with classical theory in medicine and pharmacy, is the lengthy manual called the *Geoponica.* Early Byzantine physicians and students of natural history had joined the traditional old Roman treatises on agriculture to the science of botany, and as early as 360 Didymus of Alexandria wrote a *Georgica* as well as an *Octatomus.* Didymus's *Georgica* became a primary source of the *Geoponica*, and details of various remedies in the *Octatomus* were combined with the writings of Anatolius of Berytos (*i.e.*, Beirut, Syria, fl. *ca.* 365). This legacy was then employed by Cassianus Bassus when he put together his agricultural manual sometime in the late sixth or early seventh century. Bassus's materials were, in turn, the basic foundation for a massive revision, prepared at the order of Constantine VII Porphyrogenitus, sometime after 944. It is this last handbook that is known as the *Geoponica.* In the estimate of Kenneth White, *Roman farming* (1970), p. 32:

> The *Geoponika* is an astonishing assortment, containing, in addition to a great deal of worthless magic, some good summaries of well-tried practices. There is very little in these pages that is new: most items can be paralleled in earlier writers, and the debt to Cato, Varro and Columella is evident. Yet these authors are never cited; instead there are numerous citations from writers otherwise unknown, or from known authors who were not agriculturists. Critical discussion is entirely excluded, each

recommendation being laid down as a rule to be followed. No less than five of the twenty chapters are devoted to the cultivation of the vine, and to wine-making recipes.

Geoponica was first published in an abridged Latin translation entitled *De agricultura libri viginti* (1538) and in a complete edition in 1704. Beckh has edited the modern critical edition: Bassus, *Geoponica* (1895). On the history and bibliography of *Geoponica*, see Fehrle, *Griechischen Geoponikern* (1920); Meyer, *Geschichte*, vol.2, pp. 258–61, and vol. 3, pp. 344–49; Quinby, *Catalogue of botanical books* (1958), no. 41; Sarton, *Introduction*, vol. 1, pp. 370–71, 452–53; Wellmann, Didymos (1894); Wilson, Greek treatise on agriculture (1939). Difficulties in identifying authorities are illustrated in Rodgers, Varro and Virgil in *Geoponica* (1978).

Closely related to the agricultural manuals that combined the old Roman traditions embedded in Cato, Varro, and Columella is the collection of extracts called *Corpus hippiatricorum græcorum*, also compiled at the order of Constantine VII. Veterinary medicine employed numerous botanicals, and the application of treatment broadened from animals of the farm to the care of the cavalry horses in the fourth century. Quotations from Apsyrtus of Prusa (fl. under Constantine I) in this *Corpus* on veterinary botany describe a large number of botanicals used especially in the treatment of intestinal problems in horses. Hierocles (fl. *ca.* 360) also composed veterinary and botanical tracts, and the writings of Hierocles and Apsyrtus were the major sources for the *Corpus*, which included, in addition, materials from at least 15 authorities. What is particularly interesting about the medieval Greek veterinary literature is the Byzantine translation from the Latin of a number of earlier works on agriculture and veterinary botany. This is, of course, mirrored in the medieval Latin tracts by Vegetius, Pelagonius, and Palladius and in the curious manual called *Mulomedicina chironis*, which sought to update Varro, Columella, and earlier authorities. See Oder & Hoppe, *Corpus hippiatricorum græcorum* (1924–27); Walker, "Roman veterinary medicine" (1973).

Oribasios (fl. 4th century), from Pergamum, became physician and adviser to Emperor Julian at Constantinople. At Julian's suggestion, he made abstracts of the works of Galen, but this epitome has not survived. His surviving *Iatrikai synagogai* (Medical collections) is also mostly a compilation from earlier authors, many of whose writings are known primarily from these extracts. Latin translations of some of his writings began to be published at least by 1533; Durling, *Catalogue of . . . books* (1967), pp. 437–38. Bussemaker and Daremberg edited *Œuvres d'Oribase* (1851–76), the Greek text of which is superseded, but their French translation remains useful. The standard Greek text is now that by J. Raeder (1928–33). For summaries of Oribasius's life and the history of his writings, see Baldwin, Career of Oribasius (1975); Burgess, *Asters* (1902), pp. 163–65; Harig,

Galenschrift . . . Oreibasios (1966a); Kudlien, Oribasius (1974); Meyer, *Geschichte*, vol. 2, pp. 261–73; Sarton, *Introduction*, vol. 1, pp. 372–73; Smit, *History*, cols. 411–12; Stillwell, *Awakening*, nos. 462–65.

A general appreciation of the importance of Dioscorides' *Materia medica* is impressively demonstrated by an edition of the manuscript with color illustrations, *Codex Aniciæ Iulianæ*. "This magnificent manuscript, which dates from about the year 512, displays a standard of excellence in plant drawing that was not to be surpassed for almost a thousand years"; Blunt, *Botanical illustration* (1950a), p. 10. The codex was a gift for the daughter of an emperor. In 1423 it was kept in the monastery of St. John the Baptist in Constantinople, but after the fall of the city in 1453 it came into the possession of Hamon, physician of Suleiman the Magnificent. In 1562, Ogier Ghiselin de Busbecq, ambassador to the court of Suleiman from Emperor Ferdinand I, saw it and reported its price to the Emperor. The Emperor presumably purchased it, since it came into the Imperial Library in Vienna seven years later. There are two facsimile editions, one in black and white (1906) and one in color (1965–70).

Three medical encyclopedists who wrote compendia in Greek that were widely used in the medieval and Renaissance periods were Aëtius of Amida (fl. *ca.* 540), Alexander of Tralles (fl. *ca.* 575) and Paul of Aegina (fl. 640). Aëtius, from Amida, Mesopotamia (now Diyarbakir, Turkey), became a court physician at Constantinople in the reign of Justinian I. His work is entitled *Tetrabibloi*, since it is divided into four sections, or *Sixteen medical books*, since each of the sections is further subdivided into four parts. The entire work was published in Latin (1533–35), and Durling, *Catalogue of . . . books* (1967), lists ten Latin volumes from the sixteenth century, which contain portions of Aëtius's encyclopedia (nos. 43–52). Portions of the Greek text have been published from 1534 down to recent times. Aëtius's accounts of materia medica are discussed by Meyer, *Geschichte*, vol. 2, pp. 374–78; Stannard, Plant called moly (1962), pp. 267–68. On his life and his encyclopedia, see Kudlien, Aëtius (1970a); Sarton, *Introduction*, vol. 1, pp. 434–35; Stillwell, *Awakening*, no. 260.

Alexander of Tralles, in Lydia (now in Turkey), was the most original of these encyclopedists in his own observations. Even so, he "was uncritical of a great deal of the older medical literature that he cited, and was by no means entirely free from superstition"; Kudlien, Alexander of Tralles (1970b). His encyclopedia was first printed in Latin in 1504 and in Greek in 1548. The best Greek text is Puschmann, *Alexander von Tralles* (1878–79). Brunet, *Œuvres médicales d'Alexandre de Tralles* (1933–37), the French translation, provides in Volume I a lengthy introduction to Alexander's life and to Byzantine medicine. Alexander's materia medica is discussed from a botanical perspective by Meyer, *Geschichte*, vol. 2, pp. 379–81. For summaries of his life and bibliography, see Durling, *Catalogue of . . . books*

FIG. 128. Bean seedling. Tenth-century alphabetical codex of Dioscorides. Cheltenham, Phillipps, 21975, fol. 186[v]. (Courtesy The Pierpont Morgan Library, New York City)

(1967), nos. 143–57; Kudlien, Alexander of Tralles (1970b); Sarton, *Introduction*, vol. 1, p. 453; Stillwell, *Awakening*, no. 265.

Paul, from the island of Ægina in the Saronic Gulf, studied and practiced medicine at Alexandria, where he remained after the Arabs overran Egypt. His encyclopedia is entitled *Epitome medicæ libri septem.* "He principally drew on Oribasius and Galen (though he effectively used most of his other predecessors as well) and produced a comprehensive manual for the practicing physician"; Rice, Paulus Aegineta (1980), p. 146. His books on gynecology and toxicology are no longer extant. He exerted a great influence on Muslim medicine, his encyclopedia being translated into Arabic by Ḥunayn ibn-Isḥāq (Johannitius) in the second half of the ninth century. The first printed Greek edition was in 1528, and the modern critical edition is edited by J. L. Heiberg (1921–24). There were several partial and complete Latin translations published in the sixteenth century. The English translation by Adams, *Seven books* (1844–47), is quite satisfactory, and it includes lengthy commentaries by Adams with corollary references from Dioscorides, Nicander, Galen, and others. See especially Books V and VII, which consider toxicology and general materia medica. There are botanical discussions of his materia medica by Meyer, *Geschichte*, vol. 2, pp. 412–21; Stannard, Plant called moly (1962), pp. 268–69. For summaries of his life and bibliography, see Rice, Paulus Aegineta (1980); Sarton, *Introduction*, vol. 1, pp. 479–80; Stillwell, *Awakening*, nos. 472–75; Thomas, Paul of Ægina (1974).

Stephanos Magnetes (*ca.* 1050) wrote a medical dispensatory, a Latin edition of which Caspar Wolf edited, *Alphabetum empiricum de remediis expertis* (Zurich, 1581). A list of his plants is included in Meyer's account, *Geschichte*, vol. 3, pp. 365–79. The very little that is known about him is related in Sarton, *Introduction*, vol. 1, p. 727.

Symeon Seth (fl. 1070–80) was a palace official under Emperor Michael VII Ducas. He was also an encyclopedist and a translator from Arabic into Greek. Of his two works of botanical interest, most important is *Syntagma de alimentorum facultatibus*, on the medicinal properties of foods. It includes the first Occidental discussions of Hindu and Arabic knowledge on camphor, hashish, cloves, nutmeg, and julep; there is a modern edition (1868). He also wrote a botanical dictionary. For accounts of his work, see Burgess, *Asters* (1902), pp. 187–88; Harig, Arabischen Quellen des Simeon Seth (1967); Meyer, *Geschichte*, vol. 3, pp. 356–65; Sarton, *Introduction*, vol. 1, p. 771.

Nicholas Myrepsos (fl. 1222–80), from Alexandria, became physician to Joannes III Ducas Batatzes and reportedly traveled extensively. He compiled an antidotary based on a shorter one by Nicholas of Salerno, and Johann Agricola Ammonius's abridged Latin translation was published in 1541. Leonhard Fuchs then published a more complete Latin translation in 1549.

The Greek text remains unpublished. Some scholars, including Lehmann, Nicholaus Myrepsus (1925), have argued that the man did not exist. Sarton did not accept their conclusion; *Introduction*, vol. 2, pp. 1094–95. For botanical aspects of his antidotary, see Meyer, *Geschichte*, vol. 3, pp. 382–85.

Joannes, son of Zacharias, called Actuarios (fl. 1330's), was a physician at the court of Andronicos III Palæologos. Sarton insisted on calling him Actuarios rather than Actuarius, "not only because he was a Greek, but also because these words have different meanings in Greek and Latin; actuarius means notary, ἀκτουάριος court physician"; *Introduction*, vol. 3, pp. 889–92. Sarton considered him to have been "the last of the great Byzantine physicians." His major work was a medical compendium in six books, *Methodus medendi*, of which Books 5 and 6 are a pharmacopoeia of botanical interest. Jean Ruel translated these two into Latin, *Medicamentorum compositione* (1539), listed in Durling, *Catalogue of . . . books* (1967), pp. 324–25. Meyer discussed Joannes's compendium from a botanical perspective, *Geschichte*, vol. 3, pp. 386–90, and Kudlien discussed it from the perspective of urine analysis, Aktuarios (1973).

General References

Lenz compiled an anthology, *Botanik der alten Griechen und Römer* (1859), which is also relevant for the early Byzantine period. Langkavel did the same for the later Byzantine periods, *Botanik der spæteren Griechen* (1866). Although Langkavel discussed his authors in the introduction, both works organize quotations from their sources according to the species of plants described rather than by author. An introduction and bibliography are included in Thomson, *Textes grecs . . . plantes* (1955), along with manuscripts in French translation dating from the twelfth to the sixteenth centuries. The whole Byzantine period, with many references, is encompassed in Stannard, Byzantine botanical lexicography (1971a). Although now challenged in various details, a still useful summary is Singer, Herbal in antiquity (1927), which emphasizes the derivations of the ancient botanical illuminations in the famous *Codex Aniciæ Iulianæ* (1906) of Dioscorides, and also Eutecnius/Nicander, Oppian, and others. Particularly good for a detailed analysis of the various botanical and pharmacological works is Kritikos & Papadaki, Pharmacie chez Byzantins (1967). Three authoritative essays on Byzantine medicine by Temkin are reprinted in his *Double face of Janus* (1977). Byzantine pharmacology is discussed in Watson, *Theriac and mithridatium* (1966), pp. 94–100.

More general references on Byzantine science, medicine, and culture are listed with annotations in Smit, *History*, cols. 173–74, 421–30. Additional discussions and references are in Vogel, Byzantine science (1967), especially pp. 287–94, 457–62. Extant manuscripts of the works of Nicander and Dioscorides are discussed, with useful references for the history of botany, in Kádár, *Greek zoological illuminations* (1978). Byzantine medical works published in Greek or Latin in the sixteenth century are listed in Durling, *Catalogue of . . . books* (1967).

ISLAM

The Arabs and Persians, Martin Levey asserts in Ibn Māsawaih (1961a, p. 394), "far surpassed the Greeks and Latins in pharmacology and pharmacognosy. Not only did they systematize the pharmacological knowledge of the Greeks, Indians, and Babylonians, but the writers in Arabic made many genuinely original contributions to enhance the materia medica of ancient times."

Abū Zakarīyā' Yūḥanā ibn-Māsawaih (777–857), known in the Latin West as Mesuë Maior and as Johannes Damoscenus, was the earliest notable Islamic author in pharmacology. The son of an apothecary, he became the last great professor at the medical school at Jundīshāpūr and was also the physician to several caliphs. The Latin translation of his *Opera medicinalia* is discussed briefly in Meyer, *Geschichte*, vol. 3, pp. 182–83. Fifteen Latin and four Italian editions of his works were published before 1500; Klebs, Incunabula scientifica et medica (1938), pp. 226–28. Numerous editions published in the sixteenth century are listed in Durling, *Catalogue of . . . books* (1967), pp. 398–403. His *Kitāb jawāhir al-tīb al-mufrada* is translated and discussed in Levey, Ibn Māsawaih (1961a). References in the literature to "Mesuë the Elder" and "Mesuë the Younger" are probably mistaken references to the same person; see Sournia & Troupeau, Mésué (1968). See also Sarton, *Introduction*, vol. 1, pp. 574, 728–29; Ullmann, *Medizin*, pp. 112–15, 314.

Abū Yūsuf Ya'qūb ibn-Isḥāq al-Kindī (*ca.* 800 to *ca.* 866) was perhaps from Kūfa (now in Iraq), where his education began. He completed his education at Baghdad, where he remained to improve on earlier translations of philosophical and scientific works from Greek into Arabic. He also wrote commentaries on earlier works in philosophy and science and his own contributions to these subjects. Of botanical interest are two pharmacological works. His *Risāla fī ma'rifa quwwat al-adwiya al-murakkaba* was translated into Latin as *De medicinarum compositarum gradibus investigandis* (1531). It had been translated by Gerard of Cremona and also by Arnold of Villanova. Al-Kindī's *Medical formulary* (1966) was translated into English with a study of its materia medica by Levey. On his life and work, see Hamarneh, Al-Kindī (1965); Jolivet & Rashed, Al-Kindī (1978); Sarton, *Introduction*, vol. 1, pp. 559–60, and vol. 2, pp. 342–43; Stillwell, *Awakening*, no. 266; Ullmann, *Medizin*, pp. 123, 301.

Although Arabic authors in medicine and agriculture went beyond the contributions of the ancient Greeks, the writings of those Greeks nevertheless constituted an important foundation for their work. Thus, the transla-

FIG. 129. Autumn crocus (*Colchicum autumnale* L.). Arabic translation of Dioscorides, *Materia medica*, by Abdalla ibn al-Fadl, Baghdad, 1224 A.D. (Courtesy Smithsonian Institution, Freer Gallery of Art, Washington, D.C.)

tors of the ancient scientific and medical works were vital to the development of Islamic botanical knowledge. Ḥunayn ibn-Isḥāq, known in the Latin West as Johannitius (808–73) from Hīra, Persia, was the most important of the early translators. He was a Nestorian Christian who went to Baghdad to study medicine under ibn-Māsawaih. Ḥunayn ibn-Isḥāq and his school of assistants also worked at Baghdad. They translated many of the writings of Galen, and Iṣṭifān ibn-Basīl and Ḥunayn ibn-Isḥāq translated Dioscorides' *Materia medica* into Arabic. The work of this group of translators is discussed in Anawati & Iskandar, Ḥunayn ibn-Isḥāq (1978a), and in Sarton, *Introduction*, vol. 1, pp. 611–13. This translation is reproduced in

1952, with scholarly apparatus, in Dubler & Terěs, *Versión árabe de la Materia médica de Dioscorides*, reprinted as vol. 2 in Dubler, *Materia médica de Dioscórides* (1953–59). See also Meyer, *Geschichte*, vol. 3, pp. 135–46; Ullmann, *Medizin*, pp. 258–63.

Abū Ḥanīfa Aḥmad ibn-Dā'ūd al-Dīnawarī (*ca.* 815–95), born in Dinawar, Iraq, was interested in Persian history, lexicography, botany, and astronomy. His botanical work, *Kitāb an-nabāt*, though philological in orientation, is nevertheless important for the history of botany. There are recent editions of two portions of this work (1953, 1974), with introductions, notes, indexes, and bibliography by Lewin. A paragraph on classification of plants from the *Kitāb an-nabāt* is quoted in English in Bodenheimer, *History of biology* (1958), p. 196. On al-Dīnawarī's life and work, see Lewin, Al-Dīnawarī (1962); Meyer, *Geschichte*, vol. 3, pp. 163–67; Sarton, *Introduction*, vol. 1, p. 615; Sezgin, *Geschichte des arabischen Schrifttums* (1970–71), vol. 4, pp. 338–43.

Abū Bakr Aḥmad ibn-'Alī ibn-Waḥshīya (*ca.* 860 to *ca.* 935) was born near Janbalā, Iraq, and died in Baghdad. He practiced astrology and magic, and his veracity is suspect. He was descended from the Nabatæans, and his agricultural work, *Al-filāḥa al-nabaṭiyya*, was allegedly translated from ancient Nabatæan writings. His other work of botanical interest, *Al-sumūm wa'l-tiryāqāt*, is translated and discussed in Levey, Medieval Arabic toxicology (1966). See also Bergdolt, Ibn Wahhschija: Kultur des Veilchens (1932); Fahd, Ibn Waḥshīyya (1971); Hamarneh, *Origins of pharmacy* (1973), pp. 104–6, and ibn Waḥshīyya (1976a); Meyer, *Geschichte*, vol. 3, pp. 43–88; Sarton, *Introduction*, vol. 1, pp. 634–35.

Abū Bakr Muḥammad ibn-Zakariyā' al-Rāzī, or Rhazes (*ca.* 854–925), from Rayy, Persia (now Iran), was one of the leading medical authors of Islam. One of his most important works was a medical encyclopedia, *Kitāb alḥāwī fīal-ṭibb* (*Liber continens*) (12 parts, Hyderabad, 1955–62). In it he drew on Greek and Hindu authors, as well as his own experience. Book II is on materia medica, or simple drugs, and Book III on compound drugs. An account of this work, with a translation of the last chapter of Book I, is in Iskandar, *Catalogue of Arabic manuscripts* (1967), pp. 1–26. Among his other works is one on materia medica, *Al-ṣaidana fi al-ṭibb*, which Hamarneh characterized in Pharmacy . . . of al-Bīrūnī and al-Ghāfiqī (1976b, p. 4) as "a useful innovation" that nevertheless "fell short from the high standards expected from its author, containing very little personal observations." See also Ebied, *Mediæval Arabic and Jewish medicine* (1971), pp. 82–85; Hamarneh, Bibliography (1964), pp. 88–90, and Arabic manuscripts (1977), pp. 86–89; Meyer, *Geschichte*, vol. 3, pp. 167–70; Pines, al-Rāzī (1975); Sarton, *Introduction*, vol. 1, pp. 609–10; Stillwell, *Awakening*, nos. 500–504. For Brunfels's edition of al-Rāzī's work, see ch. 5, n. 19.

Abū 'Alī al-Ḥusain ibn-'Abdallāh ibn-Sīnā, or Avicenna (980–1037), born in Afshana, near Bukhara (now in Uzbek S.S.R.), was renowned for centuries as a leading physician and philosopher. He also held various important government positions. Iskandar judges him to have been a better philosopher, but a less original medical author, than al-Rāzī (in Anawati & Iskandar, Ibn Sīnā (1978b), p. 498). Avicenna's major work was a huge medical encyclopedia, *Al-qānūn* [Canon], Book 5 of which contains accounts of compound drugs, largely of plant materials. Gerard of Cremona translated *Al-qānūn* into Latin, and many editions of his translation were published, beginning in 1473. For a discussion of the influence of Rhazes's *Al-ḥāwī* on Avicenna's *Al-qānūn*, see Iskandar, *Catalogue of Arabic manuscripts* (1967), pp. 26–32. On Avicenna's life and work, see also Campbell, *Arabian medicine* (1926), vol. 1, pp. 77–82; Durling, *Catalogue of . . . books* (1967), pp. 47–53; Gruner, Interpretation of Avicenna (1921) and *Canon of . . . Avicenna* (1930); Hamarneh, Bibliography (1964), pp. 99–101, and Arabic manuscripts (1977), pp. 89–94; Klebs, Incunabula scientifica et medica (1938), nos. 131, 132; Meyer, *Geschichte*, vol. 3, pp. 184–203; Sarton, *Introduction*, vol. 1, pp. 709–13; Smit, *History*, cols. 470–72; Stillwell, *Awakening*, nos. 287–90. For Brunfels's reference to Avicenna's De viribus cordus (1476), see ch. 5, n. 19.

Abū Raiḥān Muḥammad ibn-Aḥmad al-Bīrūnī (973–1050) was a Persian geographer, mathematician, astronomer, and historian born at Khwārazm (now Khiva, Kara-Kalpakskaya, A.S.S.R.). Sarton thought he was "one of the very greatest scientists of Islam, and, all considered, one of the greatest of all times"; *Introduction*, vol. 1, p. 707. The last of the some 130 works he wrote was a materia medica, *Al-saidana fī al-ṭibb*. A recent edition of the Arabic text was edited by Said and Elahie, with English translation entitled *Al-Bīrūnī's book on pharmacy and materia medica* (1973). Volume 2 consists of an introduction and commentary by Hamarneh. (In his review of the work, Batshon (1976) praised Hamarneh's introduction, but found some difficulties with Said's translation.) Hamarneh summarized his introduction to al-Bīrūnī's book in *Origins of pharmacy* (1973), pp. 79–87. Also still important is Meyerhof, Drogenkunde des Bērūnī (1932).

Al-Bīrūnī was interested in the numerical patterns in nature, and on one noteworthy occasion he applied his mathematical outlook to the description of flowers:

> Among the peculiarities of the flowers there is one really astonishing fact, *viz.* the number of their leaves [petals and sepals], the tops of which form a circle when they begin to open, is in most cases conformable to the laws of geometry. In most cases they agree with the chords that have been found by the laws of geometry, not with conic sections. You scarcely ever find a flower of 7 or 9 leaves, for you cannot construct them according to the laws of geometry in a circle as isoscele [triangles]. The number of their leaves is always 3 or 4 or 5 or 6 or 18. This is a matter of frequent occurrence. Possibly one may find one day some species of flowers with 7 or 9 leaves, or one may find among the species hitherto known such a number of leaves, but, on

the whole, one must say nature preserves its *genera* and *species* such as they are. For if you would, *e.g.* count the number of seeds of one of the [many] pomegranates of a tree, you would find that all the other pomegranates contain the same number of seeds as that one the seeds of which you have counted first. So, too, nature proceeds in all other matters. Frequently, however, you find in the functions [actions] of nature which it is her office to fulfil, some fault [some irregularity], but this only serves to show that the Creator, who had designed something deviating from the general tenor of things, is infinitely sublime beyond everything which we poor sinners may conceive and predicate of Him. (*Chronology* (1879), pp. 294–95.)

For accounts and references on al-Bīrūnī, see Boilot, Al-Bīrūnī (1960); Kennedy, Al-Bīrūnī (1970); Sarton, *Introduction*, vol. 1, pp. 707–9.

Abū Manṣūr Muwaffaq (fl. 975) lived at Herāt, Persia. He was the first to write a materia medica in the Persian language. "Abū Manṣur's 'Principles of Pharmacology' is a book of fundamental importance, in that it is the first to reveal what Persian-Arabic medicine and pharmacology owe to India, and how Indian drugs were further conveyed to Europe. The author himself informs us that he had been travelling in India, where he became acquainted with her medical literature." Laufer, Sino-Iranica (1919), p. 580. Of the 585 remedies that he described, 466 are derived from plants. A Persian copy of his treatise made in 1056 survived and was published in 1838 with Latin translation. There is also a German translation, *Pharmakologischen Grundsätze* (1893). Abū Manṣur's observations are quoted frequently in Laufer, Sino-Iranica. Laufer also provides (pp. 580–85) comments and corrections on Achundow's German text. See also Meyer, *Geschichte*, vol. 3, pp. 39–40; Mieli, Abū Manṣur Muwaffaq (1940); Sarton, *Introduction*, vol. 1, pp. 678–79; Ullmann, *Medizin*, pp. 266–67.

Abū al-Muṭarrif ʿAbd al-Raḥmān ibn-Wāfid (*ca.* 1008–75), of Toledo, was a physician who wrote on both pharmacology and agriculture. He worked for 20 years on his *Kitāb al-adwiya al-mufrada*. It was translated into Latin as *De virtutibus medicinarum* (1531) and into Catalán by at least the fourteenth century. The latter was published with introduction, index, and glossary by Luis Faraudo de Saint Germain (1943). A Hebrew translation is discussed briefly in Millās-Vallicrosa, Texto hebraico . . . Ibn Wāfid (1943a). Ibn Wāfid's *Madjmū ʿal-filāḥa* exists in several manuscripts that are described in Millás-Vallicrosa, Traduccion castellana . . . Ibn Wāfid (1943b); Manuscrito árabe . . . Ibn Wāfid (1954a); Nuevos textos manuscritos (1954b). For more on his life and writings, see Meyer, *Geschichte*, vol. 3, pp. 205–8; Ullmann, *Medizin*, pp. 272–73; Vernet, Ibn Wāfid (1976).

Abū ʿAbd Allāh Muḥammad ibn-Ibrāhīm ibn-Baṣṣāl (fl. during reign of al-Ma'mūn, 1038–75), of Toledo, wrote an important treatise in Arabic, which was published under the title *Libro de agricultura* (1955) with a Spanish translation and introduction by Millás-Vallicrosa and Aziman. Since the manuscript of this book was first discovered in 1942, Sarton had not had a chance to discuss ibn-Baṣṣal in Volume 1 (1927) of his *Introduction*, but did mention him in passing in Volume 3 (1948), pp. 1637, 1850. Sarton took the opportunity of his favorable review, Ibn Baṣṣāl, *Libro* (1956), p. 76, to provide a useful summary of ibn-Baṣṣāl's life and work. Sarton found the book to be "matter of fact, practical, didactic, concise; there is no nonsense in it and no digressions on materia medica, no magic, no astrology. He was a man of great practical experience, but it is unlikely that he had paid no attention to his predecessors."

Ibn-Sarābī, or Serapion the Younger (fl. first half of 12th century?) was a physician about whom little is known; Ullmann, *Medizin*, p. 283. His treatise of botanical interest, *Kitāb al-adwiya al-mufrada*, Sarton characterized as being "an elaborate compilation derived from Byzantine and Muslim sources"; *Introduction*, vol. 2, p. 229. Sarton stated that ibn-Sarābī wrote in Arabic, but according to Meyerhof he wrote his treatise in Syriac and it was soon translated into Arabic; Introduction, p. xiii, in Maimonides, *Noms de drogues* (1940). Ibn-Sarābī's *Kitāb* became popular in the Latin translation by Simon of Genoa (Simon Januensis), entitled *Liber Serapionis aggregatus in medicinis simplicibus* (editio prima, 1473), on which see Klebs, Incunabula scientifica et medica (1938), nos. 911, 913; Stillwell, *Awakening*, no. 556. One edition (1531) was edited by Otto Brunfels, and one of his references to it is cited here in ch. 5, n. 19. On this edition, see also Durling, *Catalogue of . . . books* (1967), no. 2524. There was also a vernacular translation into Italian from the Latin, by Frater Jacobus Phillipus de Padua, recently edited by Ineichen and entitled *Libro* (1962–66). For further bibliographic details on ibn-Sarābī, see Smit, *History*, col. 473.

Abū Ja'far Aḥmad ibn-Muḥammad al-Ghāfiqī (died 1165) probably came from Ghāfiq, near Cordova, Spain. Nothing is known of his life except that he was a physician who studied the medicinal uses of the plants of Spain and North Africa. "His description of plants was the most precise ever made in Islam; he gave the names of each in Arabic, Latin, and Berber. He is very often quoted by Ibn Baiṭar." Sarton, *Introduction*, vol. 2, p. 424. Meyer's discussion of al-Ghāfiqī's work was based on quotations by ibn-al-Baiṭār and others; *Geschichte*, vol. 3, pp. 210–15. However, since Meyer wrote, part of an abridged version of al-Ghāfiqī's *Al-jāmi' fi al-adwiya al-mufrada* has been published with English translation and commentary in Meyerhof & Sobhy, *Abridged . . . Book of simple drugs* (1932–40). They discontinued printing it upon discovery of several unabridged manuscripts of *Al-jāmi'*. The complete text can be reconstructed from these manuscripts, according to Hamarneh, *Origins of pharmacy* (1973), pp. 87–92 and n. 87, and Pharmacy . . . of al-Bīrūnī and al-Ghāfiqī (1976b).

Al-Sharīf al-Idrīsī (1100–1166), from Ceuta, Morocco, was one of the most important medieval geographers. From 1138 to 1154 he worked with Christian scholars at the court of Roger II at Palermo. There he wrote *Al-*

kitāb al-Rujarī [Roger's book], which was the most detailed geography treatise of the Middle Ages. There is a two-volume French translation (1836–40), which is not considered entirely accurate. Fragments on particular regions have been translated into various languages. A botanical-pharmacological manuscript ascribed to al-Idrīsī was discovered in this century at Istanbul. It is discussed and partly translated in Meyerhof, Pharmakologie und Botanik des Edrisi (1930). Meyer's discussion of al-Idrīsī in *Geschichte*, vol. 3, pp. 285–301, is limited to *Al-kitāb*. Oman provides a guide to the discussions and partial translations of *Al-kitāb* in Notizie bibliografiche . . . al-Idrīsī (1971). For more detailed discussions, see Ahmad, al-Idrīsī (1973); Hamarneh, *Origins of pharmacy* (1973), pp. 92–93; Sarton, *Introduction*, vol. 2, pp. 410–12; Ullmann, *Medizin*, pp. 277–78.

Abū Zakarīyā' Yaḥyā ibn-Muḥammad ibn-al-ʿAwwām (fl. second half of 12th century) lived in Spain; nothing else is known of his life. He is known only for his agricultural treatise, *Kitāb al-filaḥa*, which drew on classical authors and even more extensively on Hispano-Islamic authors. "Ibn al-ʿAwwām's treatise deals with 585 plants, and explains the cultivation of more than fifty different fruit trees. It contains striking observations on the different kinds of soil and manure and their respective properties, on various methods of grafting, on sympathies and antipathies between plants, etc. The symptoms of many diseases of trees and vines are indicated, as are also methods of cure." Sarton, *Introduction*, vol. 2, pp. 424–25. The Arabic text with Spanish translation by Josef Antonio Banqueri was published in 1802 in 2 volumes entitled *Libro de agricultura*. J.-J. Clément-Mullet translated the work into French, *L'agriculture* (1864–67). Sarton considered the published text and translations unsatisfactory. A previously unpublished part of the text, about pruning grapevines, was published and translated into Italian, Sul Taglio (1889); this edition includes annotations by Moncada. The Banqueri volumes were available to Meyer for his discussion in *Geschichte*, vol. 3, pp. 260–66. Clément-Mullet's translation was used by Olson & Eddy, Ibn-al-Awam (1943). Khan, Ibn al-ʿAwwām's Kitāb (1950), contains a long chapter-by-chapter summary of the work in English; technical matters are discussed in Khan, Further elucidation (1954–56). See also Vernet, Ibn al-ʿAwwām (1970a).

Moses ben Maimon, called Maimonides (1135–1204), was born in Cordova and died in Cairo. He was the most eminent Jewish physician and philosopher of the Middle Ages. Although he wrote a major religious work in Hebrew—the *Mishnah Torah* (1180), a codification of the Talmud—his other important works he wrote in Arabic. The most famous of all, *Dalālat al-hayrīn* [The guide of the perplexed] (1190), examined the limits of human reason and the relationship between science and religion, particularly of the Jewish religion in relation to Greco-Islamic natural philosophy. Four of his works relate to medical botany.

Kitāb al-sumūm wa al-mutaḥarrīz min al-adwiya al-qattāla, written in

1198, was translated during the medieval period into both Hebrew and Latin. A photographic copy of a Hebrew-Arabic manuscript version is in *Treatise on poisons* (1966), accompanying the English translation by Muntner.

Sarḥ asmā' al-'uqqār, a commentary on the names of drugs, was unknown after the Middle Ages before Ritter's modern discovery in Istanbul of a copy made by ibn-al-Baiṭār, which Meyerhof published along with an introduction and French translation, *Noms de drogues* (1940). It contains the names of some 2,000 drugs given in Arabic, Greek, Persian, Berber, and Spanish, along with descriptions and comments. Sarton suggested in his review, Maimonides, *Un glossaire* (1941), that Hebrew names were omitted because they were not in use among the Islamic physicians and druggists. *Glossary of drug names* (1979) is an English translation by Rosner.

Fī tadbīr al-ṣiḥḥa was written between 1193 and 1198, then translated into Latin by Soldanum Saladinum Babylonice with the title *De regimine sanitatis*. For the six Latin editions published from 1477 to 1535, see Durling, *Catalogue of . . . books* (1967), no. 3304; Klebs, Incunabula scientifica et medica (1938), no. 643. Bar-Sela, Hoff, and Faris have published a very useful annotated English translation, Maimonides' two treatises (1964). In this work (p. 19) Maimonides cautioned that garlic, onion, leek, radish, cabbage, and eggplant are very bad for one's health, as are also most fruit trees.

Maqālah fī bayān ba'ḍ al-a'rāḍ wa al-jawāb 'anhā, a treatise elucidating some kinds of accidents and the responses to them, was written after *Fi tadbīr* and is the other work included in Bar-Sela, Hoff & Faris, Maimonides' two treatises (1964). Despite its name, it is not on the repair of wounds, but on the diet that can help one recover from accident or illness.

The *Mishnah Torah* (1180) is Maimonides' longest work, but it relates to the sciences and medicine only indirectly. Of greatest botanical interest is its *Book 7, The book of agriculture*, translated by Klein (1979). Only the *Mishnah Torah* was available to Meyer for his discussion of Maimonides in *Geschichte*, vol. 3, pp. 218–22. The account in Sarton, *Introduction*, vol. 2, pp. 369–80, carries no discussion of the *Sarḥ asmā' al-'uqqār* because it had not yet been rediscovered. The account in Pines, Maimonides (1974), emphasizes Maimonides as philosopher rather than as physician or scientist. The literature on Maimonides as physician and scientist is so extensive that one can find it useful to consult not only the works cited above, but also these bibliographies: Ebied, *Mediæval Arabic and Jewish medicine* (1971), pp. 110–16; Hamarneh, Bibliography (1964), pp. 75–77, and Arabic manuscripts (1977), pp. 96–98; Rosner, Maimonides (1969); Smit, *History*, cols. 324–26; Stillwell, *Awakening*, nos. 444, 445; Ullmann, *Medizin*, pp. 167–69; Whitrow, *Isis cumulative bibliography* (1971–); vol 2, pp. 131–35.

Abū Ḥāmid Muhammad ibn-ʿAlī ibn-ʿUmar Najīb al-Dīn al-Samarqandī (d. 1222) of Herāt, Persia, was killed in the destruction of that city by the Tartars. An eminent physician, he wrote a number of treatises on simple and compound drugs, a dozen of which survive in manuscript. One, *Aqrābādhīn*, has been published in photographic copy of the Arabic manuscript, along with an English translation and introduction by Levey and al-Khaledy, *Medical formulary* (1967). Levey, who also translated into English the *Aqrābādhīn* of al-Kindī (see above), found that al-Samarqandī drew on many more plants and other materials from Persia, India, and China than did al-Kindī, reflecting the fact that the latter lived in Baghdad, whereas the former lived in northeast Persia (now Afghanistan). See also Hamarneh, Arabic manuscripts (1977), pp. 99–100; Ullmann, *Medizin*, pp. 170, 308–9.

ʿAbdallāh ibn-Aḥmad, called ibn-al-Baiṭār (*ca.* 1190–1248), was born in Málaga, Spain, and died in Damascus, Syria; he studied pharmacology and medicinal plants in Seville. Later, he traveled extensively in the Mediterranean lands collecting plants and information on them. On the basis of his personal information and the previous literature, he produced two different herbals.

> His main work is a collection of simples, *Kitāb al-jāmiʿ fī al-adwiya al-mufrada*, which is the foremost Arabic and mediæval treatise of its kind, the greatest from the time of Dioscorides to the middle of the sixteenth century. It is not only a very methodical and critical compilation, but it contains also a good number of personal observations. It does not deal only with simples or drugs, but also with various species of food. Leaving out duplications, some 1400 different items are considered, of which about 300 (including some 200 plants) were novelties. Practically the whole of Dioscorides' and of Galen's knowledge on the subject was incorporated in the *Jāmiʿ*, but many other authors were quoted, some 150 in all, among whom were twenty Greeks. (Sarton, *Introduction*, vol. 2, p. 663.)

Ibn-al-Baiṭār's other herbal, *Kitāb al-mughnī fī al-adwiya al-mufrada*, was "almost a reversion of the first one. It deals largely with the same simples and vegetables, but instead of being arranged in their own alphabetical order as in the *Jāmiʿ*, the new order is therapeutical." Sarton, pp. 663–64.

There were several attempts in the nineteenth century to translate *Al-jāmiʿ* into European languages, the most complete and reliable of which is entitled *Traité des simples*, translated by Leclerc (1877–83) shortly after publication of an Arabic text (4 vols., 1874–75).

Sarton's evaluation was based on this text and Leclerc's translation, but he also mentioned that his evaluation might require some modifying to take into account the allegation in Meyerhof & Sobhy, *Abridged . . . Book of simple drugs* (1932–40), p. 27: "Leclerc calls him 'the greatest botanist of the East.' This is somewhat exaggerated, but he was the greatest and most intelligent compiler of pharmacological works in the Arabic-writing world. We shall prove in the next chapter and by our edition itself, that he took al-

Ghāfiqī's book as a basis for his work and added quotations from later authors with some remarks of his own." Vernet has suggested that intellectual indebtedness was perceived differently in the Middle Ages than it is now, and that ibn-al-Baiṭār should not be considered a plagiarist; Ibn al-Baytār (1970b). Sarnelli has studied some of the extant manuscripts in order to clarify the extent of ibn-al-Baiṭār's indebtedness; Al-Ghafiqui . . . al-Baytar (1956).

Hamarneh has translated into English ibn-al-Baiṭār's organizing principles for *Al-jāmiʿ* and also the account from it of ambergris; *Origins of pharmacy* (1973b), pp. 93–96. Meyer had available a German translation (considered inferior to Leclerc's) of the *Al-jamiʿ* when he discussed ibn-al-Baiṭār's work; *Geschichte*, vol. 3, pp. 227–34. For more recent references on ibn-al-Baiṭār's life and work, see Ebied, *Mediæval Arabic and Jewish medicine* (1971), pp. 117–18; Sarton, *Introduction*, vol. 2, pp. 663–64; Smit, *History*, cols. 456–57; Ullmann, *Medizin*, pp. 280–84.

Abū Yaḥyā Zakarīya' ibn-Muḥammad ibn-Maḥmūd al-Qazwīnī (*ca.* 1203–83), from Qazwin (now Kasvin), Persia, was an encyclopedist who wrote two works on cosmography and geography. The latter contained observations on plants that are extracted from the German translation (1849) in Wiedemann, Pflanzen von Qazwīnī (1918). See also Ahmad, Al-Qazwīnī (1975); Sarton, *Introduction*, vol. 2, pp. 868–70.

Shihāb al-dīn Abū-l-ʿAbbās Aḥmad ibn-ʿAbd al-Wahhāb al-Nuwairī (1279–1332) was from Upper Egypt, the son of a government secretary. He followed his father's profession, and for a time was intendant of the army in Tripolis, Syria. Later he returned to Egypt, and he died in Cairo. Al-Nuwairī wrote a very large encyclopedia, which took him over 20 years, *Nihāyat al-arab fī funūn al-adab* [Aim of the intelligent in the arts of letters]. Part 4, on plants, includes the following topics (Sarton, *Introduction*, vol. 3, p. 621):

I. (1) Origin of plants, and the soil they need; (2) plants proper to each region; (3) food plants, vegetables

II. Trees, distinguished with regard to their fruits—(1) skin or burr not edible; (2) stone or seed not edible; (3) fruits without skin or burr or stone

III. Perfumed flowers—(1) those whose perfume can be distilled, like rose and water lily; (2) those whose perfume is generally not extracted, like violet, narcissus, jasmine

IV. (1) Gardens; (2) flowers; (3) gums, of which there are 18 kinds; (4) mannas (mann, amnān)

V. Aromatics and drugs (subdivided into 11 categories)

An edition of *Nihāyat* was prepared by Aḥmad Zakī (Cairo, 14 vols., 1923–43, incomplete). Portions of Part 4 were translated into German, with commentary, by Wiedemann (1913–18; listed in Sarton, *Introduction*, p. 622).

Qāsim ibn-Muḥammad al-Wazīr al-Ghassānī (fl. 1586), physician to the sultan Aḥmad al-Mansūr, wrote a herbal, *Ḥadīqat al-azhār fi sarḥ māhīyat al-ʿushb wa al-ʿaqqār* [Garden of flowers, or explanation of the characters of herbs and drugs]. It contains 379 articles on plants and drugs from Morocco, but its chief interest lies in its classification of plants, the earliest in Islam. See Renaud, Classification botanique . . . marocain (1928); Ullmann, *Medizin*, pp. 286–87.

General References

The following works provide a general guide to information on medieval Islam and the development of science and medicine of that period: Ebied, *Mediæval Arabic and Jewish medicine* (1971); *EI*; Hamarneh, Bibliography (1964), pp. 117–89; Hassan *et al.*, *Journal for the history of Arabic science* (1977–); Mieli *et al.*, *Science arabe* (1938) and facsimile edition augmented with bibliography by Mazahéri (1966); Millás-Vallicrosa, *Historia de la ciencia española* (1960); Nasr, *Science . . . Islam* (1968) and *Islamic science* (1976); Sarton, *Introduction*; Sezgin, *Geschichte des arabischen Schrifttums* (1970–71), especially vol. 4, pp. 330–45, on "Arabischen botaniker und agronomen"; Smit, *History*, cols. 430–86; Ullmann, *Medizin* (1970), *Natur- und Geheimwissenschaften im Islam* (1972), especially pp. 62–94 on "Die Pflanzenkunde," and *Islamic medicine* (1978).

The most detailed survey of Islamic botany is still Meyer, *Geschichte*, vol. 3, pp. 1–327. More recently, there are the well-documented survey in Meyerhof, Pharmacologie et botanique chez les musulmans (1935), and the survey with generous quotations from modern translations in Mieli, Botanica . . . los Arabes (1941). Useful studies relating to the translation of Arabic plant names into Romance languages are Asín Palacios, *Glosario* (1943); ʿĪssa Bey, *Noms des plantes* (1930); Johnstone, Some early "Spanish" plants (1976); Schweinfurth, *Arabische Pflanzennamen* (1912).

Hardly separable from Islamic botany is Islamic pharmacology, as noted in Meyerhof's article, cited above. Included in Dubler's fundamental study, *Materia médica de Dioscórides* (1953–59), is the Arabic text and information on its translation into Latin (see vols. 1, 2); see also Dubler, 'Materia medica' unter den Muslimen (1959). The earlier literature on Islamic materia medica is listed in the bibliographies by Hamarneh and by Ebied cited at the beginning of this section. Two authoritative recent surveys are Hamarneh, *Origins of pharmacy* (1973); Levey, *Early Arabic pharmacology* (1973). Also useful is Iskandar, *Catalogue of Arabic manuscripts* (1967).

For the appreciation and understanding of Islamic botany, the illustrations that accompanied some of the manuscripts of Dioscorides and other authors on materia medica are also very important. Day reproduced a good sampling of these illustrations in Mesopotamian manuscripts (1950), and two articles in the memorial volume for Ernst Herzfeld provide additional illustrations, discussions, and references: Farès, Herbier arabe (1952); Weitzmann, Greek sources of Islamic scientific illustrations (1952).

The literature on Islamic agriculture is less extensive than that on Islamic pharmacy. Millás-Vallicrosa was for a number of years the most active contributor to knowledge on Islamic agriculture, and he published two bibliographic guides: Bibliografía agronómica hispanoárabe (1954c) and Tradicion de la ciencia geopóneca hispanoárabe (1955). More recently, al-Shihabi *et al.* have written a useful comprehensive survey, Filāha (1964). Also still useful is Laufer, Sino-Iranica (1919).

Latin Europe

Rutilius Taurus Palladius (fl. late 4th century) was a Roman who wrote a farmer's calendar in Latin known as *De agricultura* or *De re rustica.* It is based largely on earlier writings, especially Columella's. It was first published at Venice in 1472 along with the works of Cato, Columella, and Varro under the title *Scriptores rei rusticæ.* Rodgers, *Introduction to Palladius* (1975), emphasizes the history of the manuscripts. His briefer article, Palladius (1976), covers similar ground but is somewhat broader in scope. For discussions of the early editions and also the content of Palladius's work, see Burgess, *Asters* (1902), pp. 173–76; Klebs, Incunabula scientifica et medica (1938), no. 902; Meyer, *Geschichte*, vol. 2, pp. 328–33; Quinby, *Catalogue of botanical books* (1958), nos. 15, 16, 44, 49; Sarton, *Introduction*, vol. 1, p. 355; K. White, *Roman farming* (1970), pp. 30–31 *et passim.* The vernacular translations include one made about 1420 into Middle English. Lidell edited the better edition of it, *Middle-English . . . De re rustica* (1896). Rodgers has edited a modern edition of Palladius, *Opus agriculturæ* (1975). It includes *De insitione*, which is of special interest for showing ancient and early medieval knowledge of plant grafting.

Apuleius Barbarus, Pseudo-Apuleius, and Apulei Platonicus are the various names attached to a very popular medieval herbal, *De herbarum virtutibus*, by an unknown author of the fourth century who may have lived in North Africa. It was originally written in Greek, according to Singer, Herbal in antiquity (1927), pp. 37–43. However, most authorities accept a Latin origin. Although it draws on ancient medical authorities, in this herbal "the virtues of the herbs often operate in an occult manner, or procedure suggestive of magic is often involved in collecting or applying them. Often diseases are cured merely by holding an herb in the hand, wearing it with a string about the neck, or placing it behind one ear." Thorndike, *Magic and experimental science* (1923–58), vol. 1, p. 598. Because this herbal was so popular, many of the manuscripts have survived, and two have been published: a ninth-century manuscript edited by Hunger, *Herbal* (1935), and a twelfth-century manuscript edited by Gunther, *Herbal* (1925). An early English translation, made shortly before the year 950, has also been published, with a modern English translation by Cockayne, *Herbarium* (1864–66). Voigts provides references to other studies and background for the pseudonym, Significance of name Apuleius (1978). This herbal is placed in the context of the history of medical botany by F. Anderson, *Illustrated history* (1977), ch. 3; Arber, *Herbals* (1938), pp. 14–19, 38–41; Meyer, *Geschichte*, vol. 2, pp. 316–28.

Another popular herbal, by "Pseudo-Dioscorides," was written in southern Europe around 550. Its great popularity is partly attributable to its illustrations. Its history has been confused by a tendency of copyists to fuse its chapters with those in the herbal of Apuleius Barbarus. Riddle has clarified the history of this herbal, in Pseudo-Dioscorides (1981), and his assessment (p. 59) follows:

> *Ex herbis femininis* is substantially a new botanical-medical treatise and contains new medical and botanical contributions; it added to, modified,and subtracted from the Dioscoridean core. Of the seventy-one chapters, only eighteen or nineteen are translated from Dioscorides with insignificant changes. In no case is the translation a complete, unmodified Latin version of Dioscorides' Greek. In some twenty-nine chapters, the author has modified the Greek text and added medicinal uses of the plants not contained in Dioscorides' work. In thirteen or fourteen, the text is completely new and totally unrelated to Dioscorides.

Isidore of Seville (*ca.* 560–636), Bishop of Seville, organized the surviving Latin knowledge and presented it in encyclopedic form. He wrote after the disruptions from numerous barbarian invasions, yet his writings have come to symbolize not the ongoing struggle to preserve the ancient knowledge so much as the depths to which learning had sunk. The most important of his writings was the *Etymologiarum sive originum libri xx*, written about 622–33. Of botanical interest are Book 4 on medicine and pharmacy and Book 17 on agricultural and plant terminology. The plants he discussed are listed in Meyer, *Geschichte*, vol. 2, pp. 389–97. The terms Isidore used in Book 17 are given with English definitions in Sprague, Botanical terms in Isidorus (1933b). For further historical and bibliographic information, see Sarton, *Introduction*, vol. 1, pp. 471–72; Sharpe, Isidore of Seville (1964, 1973); Stillwell, *Awakening*, no. 665; Whitrow, *Isis cumulative bibliography* (1971–), vol. 1, pp. 633–34.

Benedictus Crispus (died 725 or 735), Archbishop of Milan, wrote a medical poem known as *Commentarium medicinale* or *Poematium medicum*, based on Latin authors. Although not published until 1833, it is now available in English translation with extensive historical and botanical notes in Stannard, Benedictus Crispus (1966a). See also Meyer, *Geschichte*, vol. 2, pp. 421–23.

Walahfrid Strabo (*ca.* 809–49), from Swabia, Germany, was raised from age eight in a Benedictine monastery on Lake Constance. In the 830's he served as a tutor to the future emperor Charles the Bald. Walahfrid's *Hortulus* was inspired by Virgil's *Georgics*, but it is "in no way a *réchauffé*, or a *pastiche*, of a classical poem which had caught a scholar's fancy; it is a down-to-earth practical but very artful manual by a man who had a garden of his own, who loved it and who himself cultivated it." Blunt, Walahfrid Strabo, p. 12, in Walahfrid Strabo, *Hortulus* (1966). It was first published in 1510, on which see Stillwell, *Awakening*, no. 698. Choulant established a critical edition in 1832. The English edition, *Hortulus* (1966), includes a very good bibliography, illustrations, and a photographic copy of the earliest extant manuscript (Vatican Codex no. 469).

Macer floridus de viribus herbarum is the generally used title for a popular herbal that may have been written around 1100, possibly by Odo de Meung-sur-Loire. Part of this catalogue of 77 herbs and medicines was incorporated into the famous, though anonymous, *Regimen sanitatis salernitanum. Macer floridus* was translated during the Middle Ages into English, French, and Polish. It was published in Latin in 1477, and the critical text is Choulant, *Macer floridus* (1832). *Middle English . . . Macer floridus* (1949) includes an introduction, notes, and a bibliography by Fisk. The following selected studies place this herbal within the historical context and also provide still more references: F. Anderson, *Illustrated history* (1977), ch. 4; Burgess, *Asters* (1902), pp. 196–214; Daems, MNL. Macerglossen (1966); Flood, Tradition of Macer floridus (1976) and Pliny and . . . Macer (1977); Quinby, *Catalogue of botanical books* (1958), nos. 1, 3, 22, and app. no. 29; Stillwell, *Awakening*, no. 442; Thorndike, *Magic and experimental science* (1923–58), vol. 1, pp. 612–15.

The Salerno school of medicine in southern Italy near Naples was the earliest in Europe. The treatises written by its physicians during the twelfth century probably reflect Arabic influences and are often of uncertain authorship. Of botanical interest are an antidotary attributed to Nicolaus Salernitanus, *L'antidotaire Nicolas* (1896), written in the mid-twelfth century and first printed in 1471, and a herbal known as *Circa instans*, probably by Matthæus Platearius (died 1161), which was printed first in 1497. Extracts from both works are translated into English by Michael McVaugh in E. Grant, *Source book in medieval science* (1974), pp. 787–89. The last of the botanically significant professors at Salerno was Matteo Silvatico (died 1342). He compiled a large *Liber pandectarum medicine* (completed about 1317, published in 1474), which contained some original observations on plants and also many mistakes, owing to his ignorance of both Arabic and the correct names of many plants he discussed. Silvatico also "kept a botanical garden in Salerno, which is the earliest garden of its kind known to us, if conventual herb gardens are not taken into account. In that garden he cultivated not only domestic but also foreign plants, such as Arum colocasia; he sowed in it seeds which he had obtained in Greece [see the article "Cantalides" in his work]. He seems to have traveled extensively, and to have observed plants in many localities and collected information on them [see the articles "Condros," "Pusca"]." (Sarton, *Introduction*, vol. 3, p. 816.) For further information on the contributions of the Salerno school, see F. Anderson, *Illustrated history* (1977), ch. 6, and *Circa instans* (1978); Bayon, Masters of Salerno (1953); Beck, *Studien . . . Circa instans* (1940); Burgess, *Asters* (1902), pp. 253–75, 303–4; Meyer, *Geschichte*, vol. 3, pp. 435–513, and vol. 4, pp. 167–77; Ongaro, Antidotari salernitani (1976); Sarton, *Introduction*, vol. 2, pp. 239–42; Smit, *History*, cols. 544–45, 547; Stillwell, *Awakening*, nos. 461, 522.

Henrik Harpestræng (died 1244) was a canon at the cathedral of Roskilde, Denmark. He was probably also physician to King Erik Plovpenning. His writings "show him to have been a remarkable medical author both in Latin and in his native tongue"; Pedersen, Harpestræng (1972). *De simplicibus medicinis laxativis*, evidently his first medical work, survives in a single manuscript, edited with commentary by Johnsson (1917). Harpestræng wrote a herbal, *Liber herbarum*, part of which has been translated into Danish, Norwegian, Swedish, Icelandic, and German. The Latin original was edited for publication by Hauberg (1936). Harpestræng also wrote a Danish herbal based on *Macer floridus* and on *De gradibus liber* by Constantine the African. This *Dansk lægebok* was very popular, and editions have been published by Molbech (1826) and by Kristensen (3 vols., 1908–20). See also Meyer, *Geschichte*, vol. 3, pp. 537–39; Sarton, *Introduction*, vol. 2, pp. 659–61.

The twelfth century was a time of active translation of scientific works from Arabic into Latin, including some treatises that had been translated much earlier from Greek into Arabic. And although the increase of knowledge available in Latin stimulated several authors to write new encyclopedias, these works inevitably contained far more old knowledge than new. The most popular was *De proprietatibus rerum* by Bartholomæus Anglicus (written before 1283), of which Book 17, on plants, was based on Pliny, Dioscorides, Isidore, and *Circa instans*. There were 25 editions published between 1470 and 1500, in Latin, English, French, Dutch, and Spanish. The English version by John of Trevisa was published about 1495; Seymour and others edited a critical text, *On properties* (1975). For a more detailed evaluation and bibliographic information, see F. Anderson, *Illustrated history* (1977), ch. 8; Klebs, Incunabula scientifica et medica (1938), pp. 74–76; Meyer, *Geschichte*, vol. 4, pp. 84–106; Sarton, *Introduction*, vol. 2, pp. 586–88; Stannard, Squill (1974c), p. 700; Stillwell, *Awakening*, no. 595; Thorndike, *Magic and experimental science* (1923–58), vol. 2, pp. 401–35.

Some of the medieval encyclopedias, however, were not published with the advent of printing. Of these, *Shaar ha-shamayim* [The gate of heaven] by Rabbi Gershon Ben Shlomoh (fl. 1250?) is of interest as an example of medieval Hebraic science. Although living in Arles, France, Gershon relied on Arabic rather than Latin or Greek authors when writing his third treatise, on plants. *Gate of heaven* (1953) contains a translation into English by

Bodenheimer (see pp. 54–55 on sources, pp. 127–54 for translation of third treatise).

Albert von Lauingen (Albertus Magnus) (*ca.* 1200–1280) was from a wealthy aristocratic Bavarian family. At Padua he studied liberal arts and joined the Dominican Order. He taught in Germany in the 1230's, in Paris in the 1240's, and then settled at Cologne in 1248, where he lived and taught for much of the remainder of his life. Among his students was Thomas Aquinas. Albert probably became acquainted with the Aristotelian corpus at Paris. His scientific writing dates from about 1245 to about 1275 and includes encyclopedic works on both animals and plants. His *De animalibus* is rather less original than his *De vegetabilibus et plantis*. The latter was the most important theoretical treatise on plants since Theophrastus. Albert did not have access to the Theophrastian *Historia plantarum* and *De causis plantarum*, his point of departure being the writings of Aristotle and the Aristotelian *De plantis*, attributed to Nicholas of Damascus (references to which are given in ch. 2, n. 24). Albert's *Vegetabilibus* (1517) was reprinted in 1661; a critical edition (1867) was begun by Meyer and completed by Jessen. A recent edition without the critical apparatus is included in Albertus Magnus, *Opera omnia* (1890–99), vol. 10, pp. 1–320. Translations and annotations by Grant and Singer from the 1867 text are included in E. Grant, *Source Book* (1974), pp. 689–700.

The studies on Albertus Magnus are so numerous that only a selection can be cited here, beginning with general references: Sarton, *Introduction*, vol. 2, pp. 934–44; Smit, *History*, cols. 522–25; Thorndike, *Magic and experimental science* (1923–58), vol. 2, pp. 517–92; Wallace, Albertus Magnus (1970); Weisheipl, *Albertus Magnus* (1980), which includes Reeds, Albert on the natural philosophy of plant life, and Stannard, Albertus Magnus and medieval herbalism.

References to the botanical works of Albertus Magnus include Arber, *Natural philosophy* (1950), pp. 24–27; Herm. Fischer, *Mittelalterliche Pflanzenkunde* (1929), pp. 34–42, 159–84; Meyer, *Geschichte*, vol. 4, pp. 9–84; Paszewski, Problèmes physiologiques . . . Albert (1968); Sprague, Plant morphology in Albertus Magnus (1933c) and Botanical terms in Albertus Magnus (1933d); Wimmer, *Deutsches Pflanzenleben* (1908).

Rufinus (fl. late 13th century) is unknown except as revealed in his *De virtutibus herbarum* (composed after 1287), and his work was practically unknown in medieval and later times until it was published as *Herbal of Rufinus* (1946). Rufinus was both a medical practitioner and a monk, and at the time he wrote he lived in Genoa. In his introduction to this work (p. xvi), Thorndike emphasizes the originality of Rufinus, seen especially in his

> careful, detailed description of the plant itself—its stalk, leaves, and flower—and an equally painstaking distinguishing of its different varieties or a comparison of it

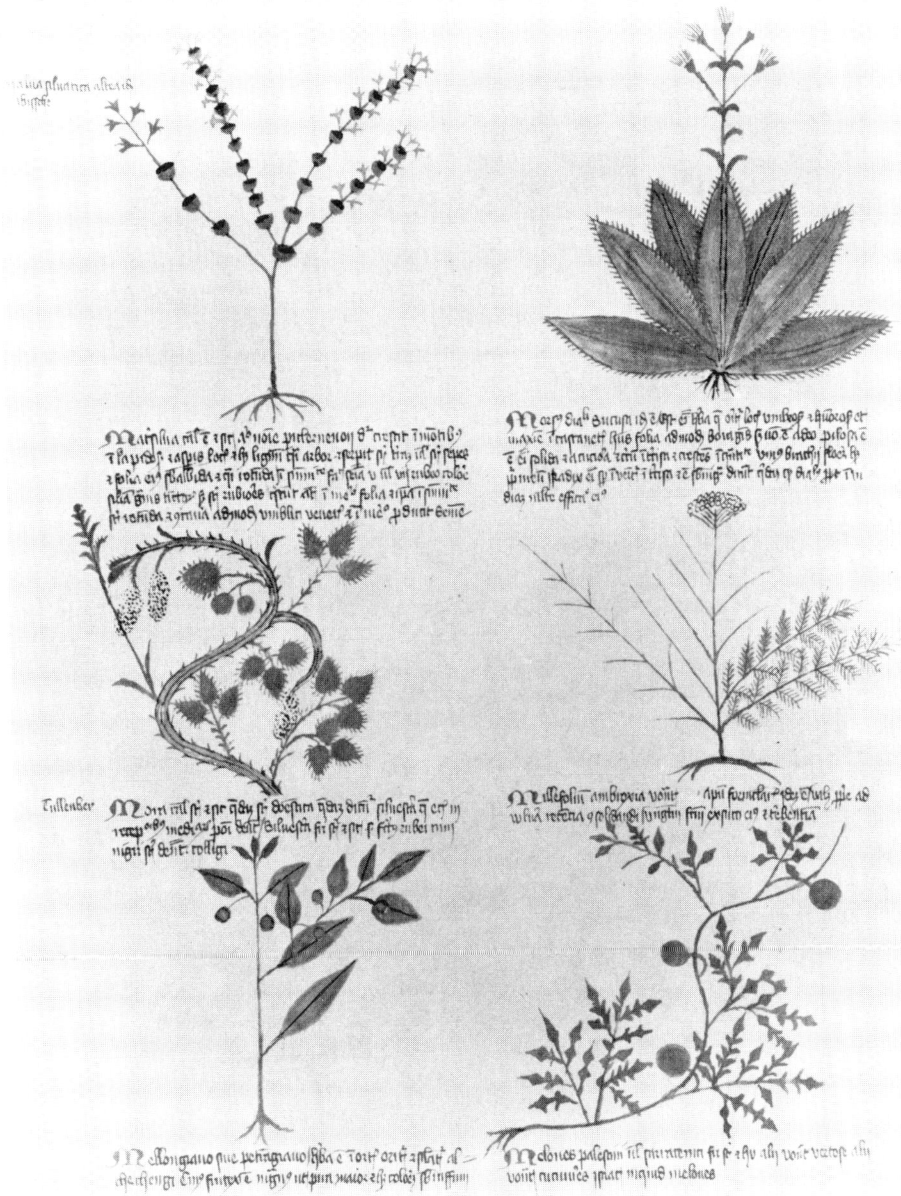

FIG. 130. A page from *De simplici medicina*, a handwritten herbal of the last quarter of the fourteenth century. A facsimile edition was edited by A. Pfister (1961), fol. 21ᵛ. Such illustrations were schematic aids to identification. (Courtesy Universitätsbibliothek Basel)

with, and differentiation of it from, other similar or related flora. He further takes care to inform us as to other names applied to a given herb or other plants indicated by the same name, so that his work should constitute a most useful volume of reference for those having occasion to deal with the vexed question of the medieval names for herbs.

See also Stannard, Plant called moly (1962), pp. 299–301; Stearn, *Botanical Latin* (1966), pp. 24, 26.

Two other late medieval herbals have been published in modern times: Aebischer & Olivier, *L'herbier de Moudon* (1938); Brodin, *Agnus castus* (1950). Brodin discovered (p. 26) that *Agnus castus* was published in 1525 as the first part of an anonymous herbal, printed by Richard Banckes. This herbal is reprinted in facsimile, with introduction and modern English, in Larkey & Pyles, *Herbal* (1941).

Palladius's *De agricultura* was practically a thousand years old when an Englishman and an Italian decided to write their own manuals on the subject. Written first, but not published until around 1510, was the *Boke of husbandry* by Walter of Henley (fl. 1250 or later). There is a recent critical edition, with introduction, modern translation, and ample bibliography by Oschinsky, *Walter of Henley* (1971). Further bibliographic details are in Quinby, *Catalogue of botanical books* (1958), appendix nos. 8, 55; Sarton, *Introduction*, vol. 2, p. 647; Stillwell, *Awakening*, no. 713.

Pietro dei Crescenzi (*ca.* 1230/33–1320/21), of Bologna, studied both medicine and law at the University of Bologna and then held various political and judicial offices in northern Italy. He retired to his estate near Bologna in 1299 and there wrote his *Ruralium commodorum*, which he completed around 1306. It was first printed in 1471. "His book was professedly a compilation, but one made critically by a man deeply versed in the subject. He tried to give the intelligent husbandman as comprehensive an account as possible, reconciling the agricultural point of view of the Romans with the botanical one of Albert the Great, and the practical one of contemporary Italian farming." Sarton, *Introduction*, vol. 3, p. 813; see also pp. 811–15. "The work was published twice in Cracow in 1549 and 1571 and, although it was obsolete at the moment of publishing, it became very popular in Poland as a textbook for gardening according to Italian methods"; Szafer, *History of botany in Cracow* (1969), p. 23. Additional botanical and bibliographical information is in Klebs, Incunabula scientifica et medica (1938), nos. 310–13; Meyer, *Geschichte*, vol. 4, pp. 138–59; Quinby, *Catalogue of botanical books* (1958), nos. 2, 7, 14, 58; Stillwell, *Awakening*, no. 615.

Jan Stanko (1430–93), "one of the most outstanding naturalists of the Middle Ages not only in Poland but in Europe," could as well be called a Renaissance scholar. He is mentioned here, however, because he shared the misfortune of some other medieval scholars whose works were not published in time to be of use to their contemporaries. Sad to say, his has never been published.

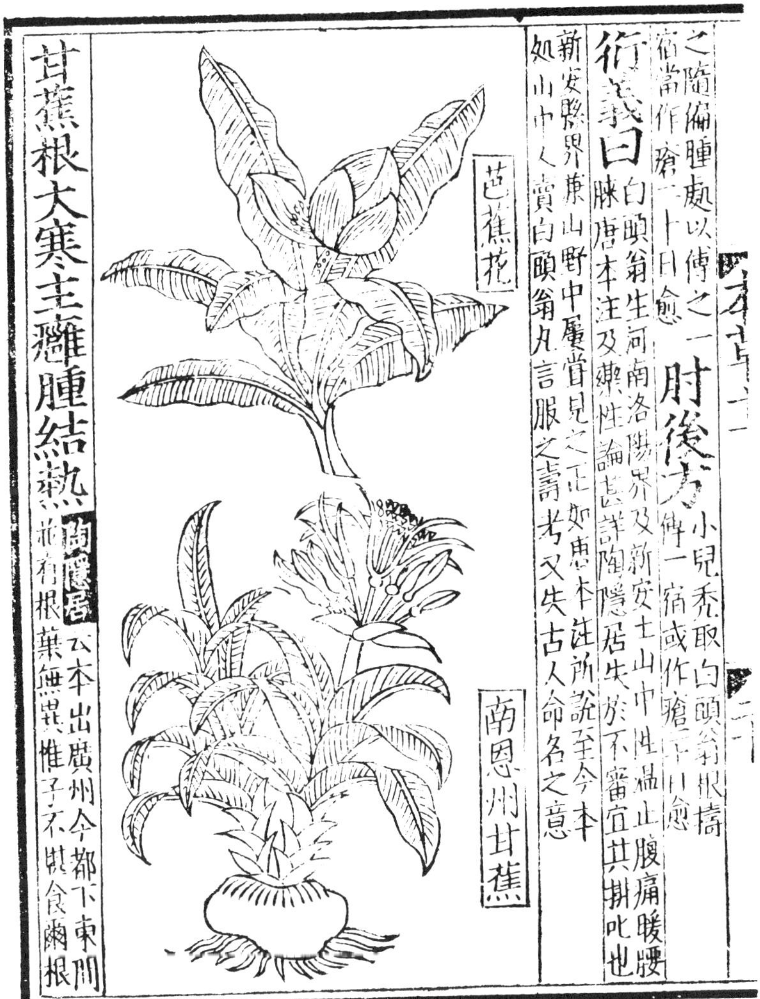

FIG. 131. Woodcuts of banana plants. Above: *pa-chiao* flower (*Musa* sp.). Below: *kan-chiao* of Nan-ên-chou, Kwangtung Province. From *Cheng-lei pên-ts'ao*, 1249 A.D., in the Library of Congress, Washington, D.C.

Stanko was the author of the largest and—in respect of nomenclature—the most critical dictionary of Polish plants and animals, which was known as *Antibolomenum* or *Antidotarium* of 1472. The codex includes about twenty thousand alphabetically ordered words relating to plants not only in Latin, German, and Polish, but also Greek, Arab, Italian and other languages. There are about two thousand Polish words—an astonishing treasure, indeed. (Szafer, *History of botany in Cracow* (1969), p. 7.)

General References

The specific citations given above by no means exhaust the information on medieval Latin botany in Meyer, *Geschichte*, vols. 2–4; Sarton, *Introduction*, vols. 1–3; Smit, *History*, cols. 485–558; Thorndike, *Magic and experimental science* (1923–58), vols. 1–3; or individual studies of restricted scope too numerous for inclusion here. Some other general references are Earle, *English plant names* (1880); Herm. Fischer, *Mittelalterliche Pflanzenkunde* (1929); Fussell, *Classical tradition . . . farming* (1972); Gothein, *Garden art* (1928), vol. 1, ch. 6; Henslow, *Medical works* (1898); Nissen, *Herbals* (1958), pp. 22–42; Riddle, Dioscorides (1971) and Dioscorides (1980); A. Schmid, *Über alte Kräuterbücher* (1939), pp. 3–21; Sigerist, Materia medica in middle ages (1939); Singer, Herbal in antiquity (1927), pp. 20–52 and 10 color plates, and *From magic to science* (1928); Stannard, Natural history (1978) and many of his other articles listed in the bibliography.

The best recent work on medieval botanical illustration is F. Baumann, *Erbario carrarese* (1974). Medieval commentators on Pliny are discussed in Nauert, Plinius Secundus (1980), pp. 302–4, 318, and the medieval use of Dioscorides in Riddle, Latin alphabetical Dioscorides (1974).

India and China

The sources listed in Appendix A for these two regions should also be consulted for the medieval period. In addition, for India, see Chowdhury, Botany . . . medieval (1971b). An interesting example of a medieval Chinese treatise is discussed in Hummel, Printed herbal of 1249 (1941), from which herbal an illustration is reproduced as Fig. 131. The continuing development of Chinese medical botany into the sixteenth century is indicated in Sivin, Li Shih-Chen (1973).

Notes to Part I

Notes to Part I

Full citations of the sources cited below are given in the Bibliography, in Part II of this work.

Preface

1. [Greene nevertheless found it necessary to discuss the contributions of men whose orientation toward botany was practical.]
2. [This point is discussed further in the Editor's Introduction.]
3. [Luca Ghini and Cesalpino are discussed in Part II, chs. 18 and 22.]

The Philosophy of Botanical History

1. Tournefort, *Institutiones* (1700), vol. 1. [Tournefort is discussed in Part II, ch. 26.]
2. [Greene did not discuss their achievements either. He had hoped to write a third volume that might have included these men. Studies on their contributions are cited in Part II, Appendix C.]
3. Adanson, *Familles des plantes* (1763), pt. 1, Preface, pp. i–cliv. [Adanson's Préface istorike de la botanik is discussed in the Editor's Introduction.]
4. Adanson, *Familles naturelles* (1847).
5. [This reference is to Sachs, *Geschichte der Botanik* (1875), which is discussed in the Editor's Introduction. Usually Greene did not cite the English translation (1890), probably because he owned only the German edition.]
6. Linnæus in his *Classes plantarum* (1738) accomplished this admirably for a very limited period, that is, for the time between 1583 and 1738, only a small fraction of the time during which the idea of classes, or families, had been in the minds of botanists and found more or less distinct expression. [On the life and work of Linnæus, see the summary account and references in Lawrence, Strandell & Kiger, *Catalogue* (forthcoming); Lindroth, Linnæus (1973).]
7. Adanson, *Familles des plantes* (1763), vol. 1, pp. lxxxix–xciii. [For references on Adanson, see the Editor's Introduction, n. 3.]
8. [Pena & Lobel, *Adversaria* (1570).]
9. [Porta, *Phytognomonica* (1588). Rienstra, Porta (1975). Neither Porta nor Adanson used the word "ecology," which was coined by Ernst Haeckel in 1866.]
10. [J. Bauhin & Cherlier, *Historia plantarum* (1650–51). For references, see Part II, Appendix C.]

11. [Bachmann (Rivinus), *Introductio . . . herbariam* (1690). For discussion and references, see Part II, Appendix C.]
12. [Boerhaave, *Index plantarum* (1710). Lindeboom, Boerhaave (1970).]
13. [Haller, *Enumeratio methodica stirpium* (1742). Hintzsche, Haller (1972).]
14. [Gleditsch (1714–86), Système des plantes (for 1749, published in 1751).]
15. [Sprengel, *Historia rei herbariæ* (1807–8) and *Geschichte der Botanik* (1817–18). On his training and career, see Risse, Sprengel (1975).]
16. Sachs, *Geschichte der Botanik* (1875), pp. 42–43 [(1890), pp. 39–40].
17. [This void Greene has attempted to fill in the following chapters. Later studies of the history of botanical terminology are cited in the editor's notes to Greene's discussion.]
18. Sprengel, *Historia rei herbariæ* (1807–8).
19. *Ibid.*, vol. 1, p. 3.
20. *Ibid.*, pp. 6–49 [(1817–18), vol. 1, pp. 6–40].
21. [See p. 280.]
22. Genesis 2: 19–20.
23. [See pp. 148–49.]
24. [This word, according to the *Oxford English dictionary*, can be traced back in the English language to the year 800, though its present spelling was uncommon before 1600.]
25. [The earliest record of this word in the *Oxford English dictionary* is from the year 1297.]
26. [Papilionaceæ Giseke = Leguminosæ – Faboideæ or Fabaceæ Lindley.]
27. *Juglans alba* L. [= *Carya tomentosa* (Lam.) Nutt.]
28. *Juglans nigra* L.
29. *Juglans regia* L.
30. Linnæus, *Philosophia botanica* (1751), sec. 54.
31. [Dodoens is discussed in Part II, ch. 24.]
32. [Kaspar Bauhin was the brother of Jean Bauhin (see n. 10). The work to which Greene refers, below, is K. Bauhin, *Pinax* (1671), a concordance to the names that various authors had used for 5,718 species. On both Bauhins, see the references in Part II, Appendix C.]
33. [*Hepatica triloba* Chaix = *H. nobilis* Schreb.]
34. Bretzl, *Botanische Forschungen* (1903), p. 405.
35. [*Armeniaca vulgaris* Lam. = *Prunus armeniaca* L.]
36. [*Amygdalus persica* L. = *Prunus persica* (L.) Batsch.]
37. [*Citrus limonum* (not *limonium*) Risso = *C. medica* L.]
38. [*Cydonia vulgaris* Pers. = *C. oblonga* Mill.]
39. Banister, Catalogo, in Ray, *Historia plantarum* (1686–1704). [On Banister, see Part II, Appendix C.]
40. [Ruel is discussed in Part II, ch. 16.]
41. [Dorsten, *Botanicon* (1540).]
42. [Ray, *Historia plantarum* (1686–1704), vol. 2, p. 1385. For citations on Ray, see Part II, Appendix C.]
43. [One difficulty with this thesis is that in antiquity the Latin names for *Quercus ilex* L. and *Q. suber* L. were *Ilex* and *Suber*, and therefore these words arose among the "unlettered vulgus" in the first place. On the early usage of these words, see André, *Termes de botanique en latin* (1956). Greene retracted the last phrase of the above when he studied the commentaries on Dioscorides by Marcello Virgilio (see Part II, ch. 14).]

Chapter 1: The Rhizotomi

1. [The importance was appreciated, but it was generally assumed that the roots absorbed "food" as well as water.]
2. [Modern research confirms this estimate. See the references under "Prehistory" in Appendix A.]
3. Theophrastus, *Historia plantarum* (1644), bk. 9, ch. 9 [(1916), bk. 9, ch. 8, secs. 4–8].
4. Sprengel, *Historia rei herbariæ* (1807–8), vol. 1, p. 63 [(1817–18), vol. 1, p. 49].
5. [*Toxicodendron vulgare* = *Rhus radicans* L. var. *vulgaris* (Michx.) DC., poison ivy.]
6. [Historians of science have learned to take superstition more seriously than did Sprengel. Recent scholarship does provide some support for Greene's thesis that there was an empirical basis for the rituals of the rhizotomists, although their hopes and fears undoubtedly played a much greater role in determining their actions than did their slight empirical knowledge. The inclinations of ancient peoples to attribute all sorts of magical properties to plants are shown in studies by Budge, Castiglioni, Dawson, Delatte, Frazer, Kanner, Ogle, and Pfister, and the persistence of this tendency into modern times is shown in a whole genre of literature, of which Friend's work is representative. In antiquity, these superstitions were all too often taken over into pharmaceutical botany by men who opposed superstition, but were not always able to recognize it when they saw it. During the Renaissance, superstitions about astrology and the doctrine of signatures were elevated to the rank of scientific theories in the books of Porta, Culpeper, and Coles. Greene discusses Renaissance superstition on pp. 326–27. Arber, Medieval herbalism (1953); Budge, *Divine origin* (1928); Castiglioni, Magic plants (1943a) and Herbs (1943b); Coles, *Art of simpling* (1656); Culpeper, *English physitian* (1652); Dawson, Studies in medical history (1929); Delatte, *Herbarius* (1938); Frazer, *Golden bough* (1911–15), vol. 2, chs. 9–11; Friend, *Flower lore* ([*ca.* 1884]); Kanner, Mistletoe (1939); Leyel, *Magic of herbs* (1926); Ogle, Laurel (1910); F. Pfister, Pflanzenaberglaube (1938); Porta, *Phytognomonica* (1588); Sarton, *History of science* (1952–59), vol. 1, pp. 545–46. See also the works by Marzell, cited in ch. 7, n. 79.

The most notorious superstition concerning plants that arose in antiquity was the belief that pulling up a mandrake root could be fatal. For this reason, it was commonly recommended that a dog be fastened to the mandrake for pulling it out; this might kill the dog, but not the rhizotomi. On the history of ideas about mandrake, see Frazer, Jacob and the mandrakes (1917); Germani, Mirabili virtù della mandragora (1935); E. Gilbert, *Les plantes magiques* (1899), ch. 2; Grieve, *Modern herbal* (1931), vol. 2, pp. 510–12; Hocquette, *Mandragore* ([1963]); LeClerc, Médecine des signatures magiques (1918); Randolph, Mandragora of ancients (1905); Tercinet, *Mandragore* (1950); C. Thompson, *Mystic mandrake* (1934). For photographic copies of some of the ancient illustrations, see Kádár, *Greek zoological illuminations* (1978), pls. 59–63.]

7. [The saying concerns meat, not medicine: "Quod ali cibus est aliis fiat acre venenum." Lucretius Carus, *De rerum natura* (1924), bk. 4, line 637. Francis Beaumont and John Fletcher made it an English saying in their play, *Love's cure* (1625?), act 3, scene 2: "What's one man's poison, Signor, is another's meat or drink."]
8. Theophrastus, *Historia plantarum* (1644), bk. 9, ch. 17 [(1916), bk. 9, ch. 16, sec. 8].

9. [Presumably, Greene means *Historia plantarum* (1644), bk. 9, chs. 17, 18, but it is not clear that all the information in these two chapters came from the three men indicated.]

10. Theophrastus, *Historia plantarum* (1644), bk. 9, ch. 18 [(1916), bk. 9, ch. 17].

11. Aristotle, *Meterologica*, bk. 1, ch. 2 [bk. 2, ch. 9, 370ᵃ10, in *Works* (1908–52)].

12. [Theophrastus, *Enquiry into plants* (1916), bk. 3, ch. 1, sec. 4, and *De causis plantarum* (1976–), bk. 1, ch. 10, sec. 3, and bk. 3, ch. 23, sec. 1.]

13. Aristotle, *De anima*, bk. 1, ch. 2 [405ᵃ1–4, in *Works* (1908–52)].

14. Theophrastus, *Historia plantarum* (1644), bk. 1, ch. 6 [(1916), bk. 1, ch. 3, sec. 5; bk. 3, ch. 2, sec. 2].

15. It was Valerius Cordus, the greatest if not the only botanical genius of the first half of the sixteenth century, who first gave expression to the opinion that, from the morphologic and phytographic point of view, the importance of the root had always been overestimated. He set the example for a reform of descriptive botany in this particular; but, as usual with men of genius, he was a century in advance of the ideas of the multitude. [But see the evaluation of Sprague and Sprague, quoted in ch. 9, n. 28.]

16. [See Oakes & Morris, West Indian weedwoman (1958).]

17. The following are illustrative examples: Alumroot, Bloodroot, Bowman's root, Culver's root, Cancerroot, Cankerroot, Black Snakeroot, Button Snakeroot, Seneca Snakeroot, Indian root, Musquash root, Colicroot, Papooseroot, Pepperroot, Pinkroot, Redroot, Yellowroot, Sheeproot. It was easy to double the number of such names of American plants, not one of which was assigned either by a learned physician or a professional botanist. [The plant knowledge (lore) of various tribes of North American Indians is described in a number of treatises, of which the following are representative: Chestnut, Plants used by Indians (1902); Coville, Plants used by Klamath (1897); Curtin, *Prophet of the earth* (1949); Romero, *Botanical lore of California Indians* (1954); Wyman & Harris, Navajo . . . ethnobotany (1941) and Ethnobotany of Kayenta Navaho (1951).]

Chapter 2: Theophrastus of Eresus

1. Linnæus, *Philosophia botanica* (1751), sec. 9, p. 30.

2. [The plants in the *Hippocratic corpus* are discussed in Sprengel, *Historia rei herbariæ* (1807–8), vol. 1, pp. 35–49; Louis, Dénominations binominales . . . grecque (1971), pp. 2–4; Stannard, Hippocratic pharmacology (1961) and Squill (1974), pp. 689–90.]

3. "Primus verorum botanicorum"; Haller, *Bibliotheca botanica* (1771–78), vol. 1, p. 31.

4. "Celeberrimus autem omnium verus rei herbariæ parens, *Theophrastus* fuit Eresius"; Sprengel, *Historia rei herbariæ* (1807–8), vol. 1, p. 66.

5. [This accusation is a distortion. See Sachs, *Geschichte der Botanik* (1875), pp. 14–18; (1890), pp. 13–16.]

6. [Diogenes Lærtius reported that "about 2000 pupils used to attend his lectures"; *Lives* (1925), vol. 1, bk. 5, ch. 2, sec. 37, p. 483.]

7. Meyer, *Geschichte der Botanik* (1854–57), vol. 1, p. 147. [Diogenes Lærtius, the source of the story, merely stated: "He bore the name of Tyrtamus, and it was Aristotle who re-named him Theophrastus on account of his graceful style"; *Lives* (1925), vol. 1, bk. 5, ch. 2, sec. 39, p. 485. Diogenes Lærtius probably wrote in the early third century A.D., about five hundred years after the death of Theophrastus. R. D. Hicks, his translator, states that "he was uncritical" and "there are also mis-

takes due to careless handling of the vast mass of excerpts"; *ibid.*, vol. 1, pp. xv–xvi. For comprehensive accounts of Theophrastus based on ancient sources, see McDiarmid, Theophrastus (1976); Regenbogen, Theophrastos (1940); Sarton, *Introduction to history of science* (1927–48), vol. 1, pp. 143–44.]

8. Aristotle died at the age of 63. Theophrastus was then 48; and, according to his own statement in his preface to that book entitled *Characters*, he had finished it in his ninety-ninth year. St. Jerome says that Theophrastus died at the age of 107. [The preface to Theophrastus's *Characters* is now considered spurious; Thorndike, Disputed dates . . . Theophrastus (1924), p. 73. Theophrastus died at age 85, according to Diogenes Lærtius, *Lives* (1925), vol. 1, bk. 5, ch. 2, sec. 40. Jerome merely listed Theophrastus in the chronological table in his *Chronicles*, but there is no documentation.]

9. The name of the first wealthy patron of botanical science was Demetrius Phalareus, according to Diogenes Lærtius, *Lives*, vol. 1, p. 350 [(1925), bk. 5, ch. 2, sec. 39. Demetrius of Phaleron, a former pupil of Theophrastus, became dictator of Athens. His assistance was thus a public rather than a private act. Lynch, *Aristotle's school* (1972), pp. 97–99. The garden at the Lyceum would have functioned more as a school campus than as a botanical garden. For gardens in ancient Greek life, see Gothein, *History of garden art* (1928), vol. 1, ch. 3; Grimal, *Jardins romains* (1969), pp. 63–98; Olck, Gartenbau (1910).]

10. Hipparchus, Neleus, Strato, Callio, Demotimus, Callisthenes, and Cresarchus are their names, according to Diogenes Lærtius, *Lives* [(1925), bk. 5, ch. 2, secs. 52, 53].

11. *Ibid.* [(1925), sec. 53].

12. *Ibid.* [(1925), sec. 54. For an analysis of Theophrastus's will, see Lynch, *Aristotle's school* (1972), pp. 99–102.]

13. Diogenes Lærtius, *Lives* [(1925), sec. 55]. Note also the botanical names of the two young slaves, Cymo and Donax.

14. [It need not be assumed that Theophrastus personally collected all the information to which Greene referred. To a large extent (assuming that Theophrastus actually composed most or all of the material contained in the two botanical treatises that bear his name), the contribution of Theophrastus must have been limited to that of synthesizing existing knowledge. Such a synthesis would have made earlier treatises obsolete, and therefore only those of Theophrastus were preserved. This was a common occurrence in Greek science. The disappearance of the earlier treatises makes the existing ones seem to be a more remarkable achievement than may actually have been the case. See Greene's own remarks about this, p. 89.]

15. [These writings were said to contain 232,808 lines; Diogenes Lærtius, *Lives* (1925), vol. 1, bk. 5, ch. 2, secs. 42–50. Recent scholars have reached no consensus about the demarcation between the writings of Aristotle and Theophrastus or the extent to which other members of the Lyceum contributed to the final form of their works. Chroust, Corpus Aristotelicum (1963), especially p. 57; Grayeff, Problem of . . . Aristotle's text (1956) and *Aristotle and his school* (1974), pp. 77–85.]

16. [Johannes Bodæus Stapelius (or van Stapel) was a physician who graduated from the medical course at Leyden University in 1625 and then practiced medicine in Amsterdam. He was reputed to be learned in Latin and Greek and in botany. He died in 1636, before completing his edition of Theophrastus's *Historia plantarum*, which was then brought out by his father, Egbert Bodæus Stapelius, in 1644. A. White *et al.*, *Stapelieæ* (1937), vol. 1, p. 74.]

17. [Theophrastus, *Historia plantarum* (1644). Van Stapel's edition was useful, especially for its botanical annotations. However, it has a number of textual defects, and therefore Greene's citations are supplemented with the critical Loeb edition in

Greek with English translation (1916). Gaza, the translator of van Stapel's text (published first in 1483), is the subject of ch. 10 in Part II.

Because van Stapel's edition was widely used during the seventeenth century, the plant illustrations from it are shown in this chapter. Almost all of them were taken from earlier herbals, but their use here will indicate what seventeenth-century botanists saw when reading *Historia plantarum*. In his notes, van Stapel described species unknown to Theophrastus, among which was the first account of a species of Stapelieæ, called *Fritillaria crassa promontorii bonæ spei* (p. 335), now known as *Stapelia variegata* L. His illustration of it, along with a translation of his account, is reproduced in A. White *et al.*, *Stapelieæ* (1937), pp. 74–75.]

18. See pp. 124–27.

19. [This conclusion is dubious, as Greene himself indicated in his following sentences. Einarson observes: "It was good form not to mention a contemporary by name. So Phanias of Eresus, a contemporary and townsman, who wrote on plants and corresponded with Theophrastus, is not mentioned by him." Introduction, in Theophrastus, *De causis plantarum* (1976–), vol. 1, pp. xix–xx. Another author, Androtion, is cited twice in *Historia plantarum* and once in *De causis plantarum*; *ibid.*, pp. xx–xxi. Theophrastus also apparently utilized the medical herbal by Diocles of Carystus (fl. late 4th century B.C.); Dannenfeldt, Diocles (1971); Jaeger, *Diokles* (1963); Wellmann, Älteste Kräuterbuch (1898a); Scarborough, Theophrastus on herbals (1978).]

20. [The situation seems more complex than Greene indicates. See notes 14, 15, and 24.]

21. [On the earlier Greek knowledge of plants, see Fellner, *Homerische Flora* (1897); Forster, Trees . . . Homer (1936, a portion of which is quoted in Part II, ch. 19, n. 38) and Trees . . . Herodotus (1942); Fraas, *Floræ classicæ* (1845); Kanngiesser, Flora des Herodot (1910); Meyer, *Geschichte der Botanik* (1854–57), vol. 1, pp. 1–146; Murr, *Pflanzenwelt . . . Griechischen Mythologie* (1890); Regenbogen, Forschungsmethode antiker Naturwissenschaft (1930), pp. 165–76; Sprengel, *Historia rei herbariæ* (1807–8), vol. 1, pp. 19–55, and (1817–18), vol. 1, pp. 28–52; Stannard, Plant called moly (1962), pp. 254–58.]

22. [The difference between the following discussion and subsequent analyses of Theophrastus's method resulted from a difference in perspective. Greene wanted to know by what methods Theophrastus attacked problems of botany. Other scholars have asked what ideas about scientific method are revealed in Theophrastus's botanical writings. Senn developed the thesis that in his later writings Theophrastus abandoned Aristotle's teleological principle. To support this idea, Senn supposed that Theophrastus's botanical writings had been rearranged after completion, presumably by someone else. Thompson and others attacked this idea, and Senn's confidence must have been shaken. His posthumous book (1956) on this subject had been withdrawn from publication in 1934. Farrington, *Greek science* (1944–49), vol. 2, ch. 1; Regenbogen, Theophrast-Studien (1934) and Theophrastos (1940); Senn, *Biologische Forschungsmethode in der Antike* (1933), pp. 93–128, Theophraste (1935), and *Pflanzenkunde des Theophrast* (1956); Thompson in Theophrastus, *De causis plantarum* (1941), pp. 6–7.]

23. [Theophrastus did, in fact, often organize his discussion in *Historia plantarum* according to utilitarian principles. Book 5 is concerned almost entirely with the uses of timber, books 7 and 8 are mainly on herbaceous food plants, and book 9 is devoted to economic plants, most of which were of medicinal interest. On his discussion of medicinal plants, see Berendes, *Pharmacie* (1891), vol. 1, pp. 156–70; Scarborough, Theophrastus on herbals (1978).]

24. [What Aristotle wrote about plants is uncertain. The Aristotelian corpus con-

tains a short treatise, *De plantis*, which is generally considered to have been written by Nicholas of Damascus (64 to at least 4 B.C.). Other Aristotelian writings briefly discuss plants and allude to treatises on plants, perhaps to those of Theophrastus. See Arber, *Natural philosophy* (1950), pp. 9–11; Bouyges, Sur le *De plantis* (1924); Hemmerdinger, Le *De plantis* (1967); Lulofs, Aristotle's *περὶ φυτῶν* (1957); Meyer, *Geschichte der Botanik* (1854–57), vol. 1, pp. 81–146; Sarton, *History of science* (1952–59), vol. 1, pp. 546–57; Senn, Aristoteles . . . Schrift über Pflanzen (1930); Thompson in Theophrastus, *De causis plantarum* (1941), p. 73; Wingate, *Mediæval Latin versions* (1931). The Arabic translator of *De plantis*, Hunayn ibn-Ishāq, was a famous Nestorian physician and one of the leading translators into Arabic; see Part I, Appendix B. Recent English translations of *De plantis* are in Aristotle, *Works* (1908–52); *Minor works* (1936), pp. 141–233.]

25. [Questioning the wisdom of scientific investigation might have begun with Socrates, who was skeptical of the possibilities of scientific discovery; Xenophon, *Memorabilia* (1923), bk. 1, pt. 1, secs. 10–17. Aristophanes ridiculed scientific investigation in his play *Clouds*, but he did so inappropriately by making Socrates the scientist who asked: "How many times the length of its legs does a flea jump?" (1924, v. 145). Aristotle defended the study of "ignoble" animals, such as fleas; *De partibus animalium*, 645^a1–25, in *Works* (1908–52).]

26. [Greek science was largely non-utilitarian, and although there were works on agriculture and pharmacy that discussed plants from a practical point of view, the existence of "pure" science before Theophrastus indicates that there must also have existed natural philosophers before him who were interested in plants. Modern scholars have differed in assessing the importance of practical versus theoretical motives for advancing ancient science. The importance of practical motives was defended by Childe, *Man makes himself* (1941) and Farrington's *Greek science* (1950); Farrington, *Greek science* (1944–49). Edelstein argued the opposing view, Motives and incentives (1963).]

27. Theophrastus, *Historia plantarum* (1644), bk. 1, ch. 2 [(1916), bk. 1, ch. 1, secs. 9–12; ch. 2, secs. 1, 2].

28. [The development of Theophrastus's botanical terminology is discussed in Strömberg, *Theophrastea* (1937).]

29. [The idea of classifying the parts of living beings into organs, which are subdivided into something resembling our modern concept of tissues, was discussed in Aristotle, *De partibus animalium*, bk. 2, 646^a7–655^b27, and *Historia animalium*, bk. 1, 486^a5–487^a10, in *Works* (1908–52).]

30. [*I.e.*, evergreens excluded.]

31. Theophrastus, *Historia plantarum* (1644), bk. 1, ch. 15 [(1916), bk. 1, ch. 9, sec. 8].

32. *Ibid.*, ch. 1.

33. [Analogical reasoning is an indispensable part of the scientific method, but it can serve only as a guide, not as a proof. The Greeks, for the most part, failed to realize the limitations of analogy as proof. See Arber, Analogy (1944); Hesse, *Models* (1966).]

34. See p. 120.

35. Theophrastus, *Historia plantarum* (1644), bk. 1, ch. 11 [(1916), bk. 1, ch. 7, sec. 1].

36. *Ibid.*, bk. 8, ch. 2.

37. [Grew, *Anatomy of vegetables* (1672), pls. 1–3. On Grew, see Part II, Appendix C.]

38. Theophrastus, *Historia plantarum* (1644), bk. 1, ch. 5 [(1916), bk. 1, ch. 3, secs. 2, 4].

39. [See n. 29.]

40. [Greene did not, as this statement seems to imply, devote a separate section of this chapter to the botanical terminology of Theophrastus. Rather, terminology is discussed throughout the chapter. For a more detailed discussion, see Strömberg, *Theophrastea* (1937).]

41. Theophrastus, *Historia plantarum* (1644), bk. 1, ch. 6 [(1916), bk. 1, ch. 3, secs. 5, 6, and *De causis plantarum* (1976–), bk. 1, ch. 16, secs. 10–13.]

42. [*agrestis*: relating to the fields or country; wild
silvestris: belonging to a forest
trivialis: common, ordinary
arvensis: growing in a plowed field
pratensis: growing in a meadow
hortensis: growing in a garden
sativus: planted
urbanus: urban]

43. [Theophrastus, *Enquiry into plants* (1916), bk. 1, ch. 3, sec. 6. The list of plants that could not be cultivated was as follows: silver fir (*Abies pectinata* Carr.), Greek fir (*Abies cephalonica* Loud), holly (*Ilex aquifolius* L.), caper (*Capparis spinosa* L.), lupine (*Lupinus alba* L.), and in general those of cold climates. Obviously, Theophrastus had in mind their cultivation in Greece; he might have believed that plants of cold climates could be cultivated in those climates.]

44. [The biology of cultivated plants became an important and exciting field of investigation in the twentieth century, when it was realized that new species or varieties could be found that might be economically important, and when it was discovered that the chromosomes of species long cultivated provide clues to their evolution. The literature on these two subjects is too vast to be cited here, but an introduction to it can be had from the works cited under "Prehistory" in Appendix A.]

45. [Theophrastus, *Enquiry into plants* (1916), bk. 1, ch. 3, sec. 5.]

46. I have inverted Theophrastus's order here. He places fruit-bearing and sterile before flowering and flowerless.

47. Meyer, *Geschichte der Botanik* (1854–57), vol. 1, p. 162.

48. [See, on ecology, pp. 195–98; on geographic distribution, pp. 200–203; on pathology, p. 204; and on transmutation, pp. 205–8.]

49. [For another discussion of Theophrastus's treatment of morphology, see Arber, *Natural philosophy* (1950), ch. 2.]

50. [This question had been explored previously in Aristotle, *Historia animalium*, 588^{b}, and *De partibus animalium*, 681^{a}10–15, in *Works* (1908–52).]

51. Theophrastus, *Historia plantarum* (1644), bk. 1, ch. 2.

52. *Ibid.*, ch. 1. I trust I may not be found to have rendered too freely the words ἤθη and πράξεις.

53. [Linnæus, *Systema naturæ* (1735), p. 10.]

54. *Ibid.* (1744), p. 3. [The attempts of biologists in the eighteenth century to give a precise answer concerning the differences between plants and animals are discussed in Ritterbush, *Overtures to biology* (1964).]

55. [Concerning plant sensitivity, see n. 189.]

56. Theophrastus, *Historia plantarum* (1644), bk. 1, ch. 2 [(1916), bk. 1, ch. 1, sec. 9].

57. *Ibid.*, bk. 3, ch. 18 [(1916), bk. 3, ch. 18, sec. 9. Theophrastus indicated in this paragraph his belief that ivy is a parasite on the trees it climbs.]

58. Theophrastus, *De causis plantarum*, bk. 2, ch. 23 [(1976–), bk. 2, ch. 17.

The passage refers to both *Loranthus europæus* and *Viscum album*. Mistletoe is also discussed in *Enquiry into plants* (1916), bk. 3, ch. 7, sec. 6, and bk. 3, ch. 16, sec. 1.]

59. Theophrastus, *Historia plantarum* (1644), bk. 4, ch. 5 [(1916), bk. 4, ch. 4, secs. 4, 5. See also bk. 1, ch. 7, sec. 3.]

60. Theophrastus, *Historia plantarum* (1644), bk. 1, ch. 10 [(1916), bk. 1, ch. 6, secs. 3–12].

61. [Evidently no study has been made on the development of our understanding of the demarcations between roots and stems. My inspection of botanical treatises leads me to think that this understanding developed between 1790 and 1840. Johann Heinrich Troll (1756–1824) deserves much of the credit for this clarification, according to F. A. Stafleu (pers. comm.). Another student of plant anatomy from that time period who also deserves credit was Pierre Jean François Turpin (1775–1840); see Hocquette, Turpin (1976). For further discussion of this topic, see p. 279.]

62. Theophrastus, *Historia plantarum* (1644), bk. 1, ch. 2 [(1916), bk. 1, ch. 1, sec. 9].

63. *Ibid.*, bk. 7, ch. 8.

64. *Ibid.* (1644), bk. 1, ch. 16, and bk. 7, ch. 7 [(1916), bk. 1, ch. 10, sec. 7, and bk. 7, ch. 8, sec. 2].

65. Excerpts from Theophrastus, *Historia plantarum* (1644), bk. 1, ch. 16 [(1916), bk. 1, ch. 10].

66. *Ibid.* (1644), bk. 3, chs. 11, 12, 13, 16 [(1916), bk. 3, ch. 11, sec. 3; bk. 3, ch. 12, sec. 7; bk. 3, ch. 13, sec. 5].

67. *Ibid.*, bk. 3, ch. 14 [sec. 1].

68. *Ibid.*, bk. 4, ch. 9 [(1916), bk. 4, ch. 7, sec. 8. In the same description, Theophrastus also reported that the leaves of the tamarind close at night and open between sunrise and noon.]

69. [A. Jung, *Isagoge phytoscopica* (1678). On Jung, see Part II, Appendix C.]

70. [*M. polyacantha* Willd. = *M. asperata* L.]

71. Theophrastus, *Historia plantarum* (1644), bk. 6, ch. 2 [(1916), bk. 4, ch. 2, sec. 11].

72. [See p. 280.]

73. Theophrastus, *Historia plantarum* (1644), bk. 6, ch. 2 [(1916), bk. 6, ch. 2, sec. 7. Thiselton-Dyer identified this species as *Ferula communis* L. Silphium, a drug widely used in antiquity, apparently came from a member of the genus *Ferula*. The species has been fairly certainly identified as *F. tingitana* L. by Gemmill in Silphium (1966). The drug and the plant from which it was derived have been discussed by a host of authors cited in Gemmill; see also Andrews, Silphium (1941), and Steier, Silphion (1929–31).]

74. Theophrastus, *Historia plantarum* (1644), bk. 1, ch. 16 [(1916), bk. 1, ch. 10, sec. 2].

75. Namely, by Sébastien Vaillant, *Sermo de structura florum* (1718). [On Vaillant, see Part II, ch. 26, n. 56. On the early ideas of sexuality in plants, see Dzhaparidze, *Sex in plants* (1967); Leroy, Histoire . . . sexe (1959); Lorch, Discovery of sexuality (1966), pp. 212–16; Singer, Greek biology (1921), pp. 86–92; Stearn, *Botanical Latin* (1966), p. 33. For later ideas, see Part II, Appendix C.]

76. [This statement needs qualification: it would have been possible to deduce the function of pollen by removing anthers from flowers and observing that seeds did not develop. But Greene seems correct in saying (p. 166) that a microscope was necessary for making distinctions in some plants between the parts of the seed and the pericarp.]

77. [See Theophrastus, *Enquiry into plants* (1916), bk. 1, ch. 13, secs. 3–5.]

78. Theophrastus, *Historia plantarum* (1644), bk. 3, ch. 6 [(1916), bk. 3, ch. 3, sec. 8]. There is but one record of the discovery of the flower of the fig until after the invention of magnifying lenses, and some 18 centuries after Theophrastus. Even Linnæus, still later by two centuries, had the genus *Ficus* under the Cryptogamia. [See also Theophrastus, *De causis plantarum* (1976–), vol. 1, bk. 2, ch. 9, secs. 5–15. For Valerius Cordus's discussion of the fig flower, see p. 389.]

79. Theophrastus, *Historia plantarum* (1644), bk. 3, ch. 6 [(1916), bk. 3, ch. 3, sec. 8].

80. [Greene gives the impression here that Theophrastus was more explicit than he was concerning flowers. Theophrastus merely raised the question: "There is also a doubt about the flower of some trees. . . ." Then he discussed the evidence noncommittally; *Enquiry into plants* (1916), bk. 3, ch. 3, sec. 8.]

81. Theophrastus, *Historia plantarum* (1644), bk. 1, chs. 20, 21, 22 [(1916), bk. 1, ch. 12, sec. 4, and ch. 13, secs. 1–5].

82. *Ibid.*, ch. 21 [(1916), bk. 1, ch. 13, sec. 1. In this case, Greene's translation of χνοώδης as "capillary" is more appropriate than the more common "downy," which Hort used].

83. *Ibid.*, bk. 3, ch. 7 [(1916), bk. 3, ch. 5, sec. 6; see also bk. 3, ch. 3, sec. 8].

84. Pliny, some three centuries after Theophrastus, refers to the juli of the filbert in terms that prove them still incomprehensible to nature students of the time. He says that they are *ad nihil utiles*, which we remark is a negation, and therefore unscientific. [*Natural history* (1938–62), bk. 16, sec. 120.]

85. [This judgment is substantiated in Bois, *Plantes alimentaires* (1927–37), vol. 2, pp. 515–16; Simmonds, *Evolution of crop plants* (1976), p. 305.]

86. Theophrastus, *Historia plantarum* (1644), bk. 3, ch. 7 [(1916), bk. 3, ch. 5, sec. 6].

87. *Ibid.*, ch. 15 [(1916), bk. 3, ch. 15, secs. 1, 2. Theophrastus discussed both cultivated and wild ones and said that there were "two kinds of each sort; some have a round, others an oblong, nut. . . ." *C. tubulosa* Willd. = *C. maxima* Mill.]

88. Theophrastus, *Historia plantarum* (1644), bk. 3, ch. 5 [(1916), bk. 3, ch. 3, secs. 6, 7. Theophrastus also stated that the flower of the alder "is as large as that of the bay"; bk. 3, ch. 14, sec. 3.]

89. "The flower of the [silver] fir is yellowish red and otherwise beautiful." Theophrastus, *Historia plantarum* (1644), bk. 3, ch. 6 [(1916), bk. 3, ch. 4, sec. 5].

90. *Ibid.*, ch. 6 [(1916), ch. 3, sec. 8. Thiselton-Dyer identified this species as *Juniperus phœnicea* L. In antiquity, plants were said to have different sexes when two kinds were known. Sometimes a dioecious species was so designated, but in this case the fruit-bearing individuals were often called male. *Ibid.* Commonly, however, the reference was to two different, but similar, monoecious species, in which one was arbitrarily called male and the other female. This custom persisted into the Renaissance, and examples from Brunfels's *Herbarum* (1532–39) are discussed in Sprague, Brunfels (1928), p. 87.]

91. Theophrastus, *Historia plantarum* (1644), bk. 3, ch. 6 [(1916), bk. 3, ch. 3, sec. 8].

92. *Ibid.*, bk. 1, ch. 21 [(1916), bk. 1, ch. 13, sec. 2].

93. *Ibid.*, bk. 3, ch. 13 [(1916), bk. 3, ch. 13, sec. 6].

94. *Ibid.*, bk. 3, ch. 16 [(1916), bk. 3, ch. 16, sec. 4. Coats, *Garden shrubs* (1965), pp. 19–22.]

95. Theophrastus, *Historia plantarum* (1644), bk. 1, ch. 22 [(1916), bk. 1, ch. 13, sec. 3].

96. [The morphology of the composite flower was first explained satisfactorily by Cassini (1781–1832) in a series of papers that he began publishing in 1812 and later

republished in *Opuscules phytologiques* (1826–34). However, Ruel in 1536 pointed out that the composite head is composed of many florets (Greene discusses it in Part II, ch. 16). Grew also discussed this (not very clearly) and illustrated it in his *Anatomy of plants* (1682), pp. 38, 170–71, tables 13, 59, 60, 61, 62. The development of this knowledge is summarized in Burgess, *Asters* (1902); Lorch, Idealistischen Morphologie (1958); Jas. Small, *Origin . . . compositæ* (1919).]

97. Theophrastus, *Historia plantarum* (1644), bk. 7, ch. 3 [(1916), bk. 7, ch. 3, sec. 1. On the usage of ocymum, see Grieve, *Modern herbal* (1931), vol. 1, pp. 86–87; Rosengarten, *Book of spices* (1969), pp. 108–17.]

98. Meyer, *Geschichte der Botanik* (1854–57), vol. 1, p. 166.

99. [See pp. 379–80.]

100. [See Theophrastus, *De causis plantarum* (1941), bk. 6, ch. 6, sec. 5; (1976–), bk. 1, chs. 20, 22. See also Aristotle, *De generatione animalium*, 750ᵃ21–750ᵇ1, in *Works* (1908–52).]

101. Theophrastus, *Historia plantarum* (1644), bk. 1, ch. 3 [(1916), bk. 1, ch. 2, sec. 1, and also *De causis plantarum* (1976–), bk. 1, chs. 16, 19, 21.] περικάρπιον was also in use with Aristotle. The invention of it lies between him and Theophrastus. [For a further discussion of pericarp, see p. 389.]

102. Theophrastus, *Historia plantarum* (1644), bk. 1, end of ch. 18 [(1916), bk. 1, ch. 12, sec. 1].

103. [Theophrastus, *Enquiry into plants* (1916), bk. 1, ch. 11, sec. 2.]

104. [*Ibid.* For a discussion of this term, see "lobos" in Carnoy, *Noms grecs de plantes* (1959), p. 163.]

105. [Theophrastus, *Enquiry into plants* (1916), bk. 1, ch. 11, sec. 4.]

106. Theophrastus, *Historia plantarum* (1644), bk. 7, ch. 3 [(1916), bk. 7, ch. 3, sec. 2. Another important discussion of the growth of the plant from the seed is in Hippocrates, On the nature of the child (περὶ φύσιος παιδίου, pars. 22–27), in *Œuvres completes* (1839–61), vol. 7, pp. 515–29. See also Miller, Dynamis and seeds (1966).]

107. Theophrastus, *Historia plantarum* (1644), bk. 7, ch. 1 [(1916), bk. 7, ch. 1, secs. 3–7. Theophrastus discussed the seasonality of buds and fruits and its probable causes in *De causis plantarum* (1976–), bk. 1, chs. 11–16. The understanding of why seeds of different species, though overwintering in the same ground, germinate at different times, is a recent achievement. See Crocker, Dormancy in seeds (1916); Mayer & Poljakoff-Mayber, *Germination* (1963).]

108. Theophrastus, *Historia plantarum* (1644), bk. 8, ch. 2 [(1916), bk. 8, ch. 2, sec. 1].

109. In Latin the clause runs thus: "Id quod in quibusdam partis pudendæ refert formam, ceu in faba et cicere, sed maxime in lupino." ["In some cases there is also a formation resembling the penis, as in beans, chick-peas, and especially in lupines; from this the root grows downwards, the leaf and the stem upwards"; Theophrastus, *Enquiry into plants* (1916), bk. 8, ch. 2, sec. 1.]

110. [Malpighi, *Anatome plantarum* (1675–79). For references on Malpighi, see Part II, Appendix C.]

111. [This statement is true only if one judges botany to have begun with Aristotle. For references on earlier Greek botanical knowledge, see n. 21.]

112. Theophrastus, *Historia plantarum* (1644), bk. 1, ch. 3 [(1916), bk. 1, ch. 2, sec. 1].

113. [This term is the subject of a chapter in Strömberg, *Theophrastea* (1937), pp. 117–21.]

114. Theophrastus, *Historia plantarum* (1644), bk. 1, ch. 4 [(1916), bk. 1, ch. 2, sec. 6].

115. Just as in certain European languages still spoken there is no one word for womb, only the phrase "mother of life." [The term μήτρα is the subject of a chapter in Strömberg, *Theophrastea* (1937), pp. 122–28. It is difficult to see the merit of Greene's argument that a word meaning "womb" is a better word for "pith" than a word meaning "marrow." Here, as in his discussion of the flower, Greene seems to be reading more into Theophrastus's statements than seems warranted.]

116. [Greene seems to be crediting Theophrastus with the discovery and understanding of the cambium, an unwarranted conclusion. On the history of this understanding, see Lorch, Elusive cambium (1967); Schopfer, Débuts de l'anatomie végétale (1948) and Terme "cambium" (1949).]

117. Theophrastus, *Historia plantarum* (1644), bk. 1, ch. 4 [(1916), bk. 1, ch. 2, sec. 6. Aristotle failed to understand the functional difference between nerves and blood vessels, which was why Theophrastus made only a morphological distinction. Souques, *Neurologie dans l'antiquité* (1936), pp. 105–12.]

118. Theophrastus, *Historia plantarum* (1644), bk. 1, ch. 17 [(1916), bk. 1, ch. 10, secs. 1, 2].

119. [*Ibid.*, bk. 1, ch. 10, sec. 10.]

120. [Theophrastus explained the value of analogy and began by applying terms concerning animals to plants; *Enquiry into plants* (1916), bk. 1, ch. 1, secs. 2–4. For discussion and references on the use of analogy in science, see n. 33.]

121. [Ivy is a vine or a shrub, not a tree. On the usage of ivy (*Hedera helix* L.), see Grieve, *Modern herbal* (1931), vol. 2, pp. 440–41. On the usage of sweet bay (*Laurus nobilis* L.), see Coats, *Garden shrubs* (1965), pp. 192–95; Grieve, *Modern herbal* (1931), vol. 2, pp. 464–65; Ogle, Laurel (1910).]

122. See also Bretzl, *Botanische Forschungen* (1903), pp. 8–22.

123. Theophrastus, *Historia plantarum* (1644), bk. 3, ch. 13 [(1916), bk. 3, ch. 14, sec. 1].

124. *Ibid.* [(1916), bk. 3, ch. 13, sec. 5].

125. *Ibid.*, bk. 3, ch. 11 [(1916), bk. 3, ch. 11, sec. 1].

126. *Ibid.*, bk. 1, ch. 16 [(1916), bk. 1, ch. 10, secs. 5, 6].

127. [Tournefort, *Élémens* (1694). He is discussed in Part II, ch. 26.]

128. [Linnæus, *Species plantarum* (1753).]

129. [Adanson, *Familles des plantes* (1763).]

130. This is almost precisely what Tournefort undertook to do in the seventeenth century, and Linnæus again in the eighteenth.

131. [Much more important than these considerations would seem to be the theoretical difficulties involved in establishing a classification system, which Aristotle had discussed; *De partibus animalium*, 642^b–643^a, in *Works* (1908–52).]

132. Theophrastus, *Historia plantarum* (1644), bk. 1, ch. 5 [(1916), bk. 1, ch. 3, sec. 1].

133. See pp. 141–42. [Because Greene's discussion of Theophrastian taxonomy is based heavily on the plant names Theophrastus used, it is useful to consult other works on ancient botanical names and terms: André, *Lexicographie botanique grecque* (1958); Carnoy, *Noms grecs de plantes* (1959); Strömberg, *Theophrastea* (1937); Thiselton-Dyer, Ancient plant-names (1914–18), and in Liddell *et al.*, *Greek-English lexicon* (1925–40).]

134. ["Dendron," "thamnos," "phrygia," "poa"; Strömberg, *Theophrastea* (1937), pp. 150, 155.]

135. [The word "plant" in English usage is traced back to about 825 A.D., when it appeared in the *Vespasian Psalter* 143: 12. *Oxford English Dictionary.*]

136. ["Phyto," "lachanon," "botane," "poa."]

137. There is a familiar sacred verse that attests this: "All flesh is grass, and all the

glory of man is as the flower of grass. The grass withereth, and the flower thereof falleth away."—I Peter 1: 24. At the date of this translation gramineous plants were regarded as flowerless. Therefore in the minds of the learned translators even showily flowering herbaceous plants were a part of "grass." [The statement "all flesh is grass" appeared first in Isaiah 40: 6.]

138. Theophrastus, *Historia plantarum* (1644), bk. 1, ch. 10 [(1916), bk. 1, ch. 6, sec. 6. See also Strömberg, *Theophrastea* (1937), pp. 100, 113.]

139. Theophrastus, *Historia plantarum* (1644), bk. 1, ch. 17 [(1916), bk. 1, ch. 10, sec. 8].

140. *Ibid.*, bk. 1, ch. 16 [(1916), bk. 1, ch. 10, sec. 5. Greene evidently meant that Theophrastus had seen only one kind of palm. Theophrastus discussed others besides the date palm, as Greene indicates on p. 202.]

141. [See Strömberg, *Theophrastea* (1937), p. 86.]

142. [*Ibid.*, p. 113.]

143. [See Yates, Terms acanthus, acanthion, etc. (1846).]

144. Theophrastus, *Historia plantarum* (1644), bk. 1, ch. 22 [(1916), bk. 1, ch. 13, sec. 3].

145. *Ibid.*, bk. 7, ch. 11 [(1916), bk. 7, ch. 7, sec. 1].

146. [Theophrastus, *Enquiry into plants* (1916), bk. 3, ch. 9.]

147. [*Ibid.*, bk. 3, ch. 14.]

148. Theophrastus, *Historia plantarum* (1644), bk. 9, ch. 13 [(1916), bk. 9, ch. 12, secs. 3–5, and ch. 13, sec. 1. Although there are some similarities in both the flowers and fruits of the Nymphæaceæ and the Papaveraceæ, botanists in recent times place these families in different orders, indicating a difference with Greene's judgment on the closeness of their relationship.]

149. *Ibid.*, bk. 7, ch. 4 [(1916), bk. 7, ch. 4, sec. 6. Hort's translation of this sentence differs significantly from Greene's, and the context fits Hort's version better. His rendering follows: "As to the cucumber and gourd, it is said that there are various forms of the former, but of the latter, just as in radish and turnip, the differences are only between better and inferior individuals."]

150. Meyer, *Geschichte der Botanik* (1854–57), vol. 1, p. 162. [In *De causis plantarum* (1941), bk. 6, ch. 16, secs. 1–8, Theophrastus pointed out that wild fruits are more fragrant and less sweet than cultivated fruits. On Hippon, see p. 125.]

151. See his *Characters of men*, a work completed, as he tells us, in his ninety-ninth year. [On the spuriousness of this statement, see n. 8.]

152. [The possibility that Theophrastus withheld such comment out of deference to public opinion seems unlikely. Greene presumably reached this conclusion after reading this statement from Diogenes Lærtius: "And so highly was [Theophrastus] valued at Athens that, when Agnonides ventured to prosecute him for impiety, the prosecutor himself narrowly escaped punishment"; *Lives* (1925), bk. 5, ch. 2, sec. 37. The suppression of science in antiquity is discussed in Farrington, *Science and politics* (1939); Momigliano, Farrington's *Science and politics* (1941).]

153. Theophrastus, *Historia plantarum* (1644), bk. 7, ch. 10 [(1916), bk. 7, ch. 11, sec. 2].

154. *Ibid.*, bk. 9, ch. 11 [(1916), bk. 9, ch. 10, secs. 1–4. Theophrastus observed that these plants "appear to have nothing in common except the name." On the usage of these two species, see Hans Fischer, Helleborus im Altertum (1936); Grieve, *Modern herbal* (1931), vol. 1, pp. 388–91; Kingsbury, *Poisonous plants* (1964), pp. 140, 457–61; LeClerc, Histoire des ellébores (1917); Taylor, False hellebore [*Veratrum*] (1956).]

155. Theophrastus, *Historia plantarum* (1644), bk. 9, ch. 13 [(1916), bk. 9, ch. 13, sec. 1. Greene placed too much emphasis on the fact that *Nymphæa* is discussed

after the poppy. In the 1644 edition, as in earlier Latin editions, *Nymphæa* was under the chapter heading for poppies. However, it is not clear that it was Theophrastus who made the chapter divisions. In Hort's Greek-English edition there is a different division, between the discussions of poppies and water lilies.]

156. [This statement is questionable. Theophrastus mentioned approximately 550 kinds of plants (there is no earlier surviving mention of about 170); Regenbogen, Theophrastos (1940), col. 1467. Pliny discussed 800 kinds of plants, according to Adanson, *Familles des plantes* (1763), p. cxvi.]

157. [The comparable identifications by Thiselton-Dyer in *Enquiry into plants* (1916) are as follows: *Alnus glutinosa*, *Melampyrum*, *Vigna sinensis*, *Ostrya carinifolia*, *Pinus* sp., *P. laricio*, *P. pinea*, *P. halepensis*, *Mespilus germanica*, etc., *Cratægus orientalis*, *Cotoneaster pyracantha*, *Amelanchier vulgaris*, *Quercus ilex* var. *agrifolia*, *Cydonia vulgaris*, *Prunus domestica*, *P. insititia*, *P. avium*, *P. mahaleb*, *Sorbus domestica*.]

158. Linnæus, suppressing the white-water-lily genus, daringly transferred its name to that of a genus of insignificant malvaceous weeds.

159. Sir J. E. Smith, most ardent Linnæan though he was, restored *Cyamus* instead of *Nelumbo*, insisting on its right of priority.

160. [According to Thiselton-Dyer, the species are as follows: *Nuphar luteum*, *Nymphæa alba*, *Nymphæa stellata*, *Nelumbium speciosum*.]

161. [According to Thiselton-Dyer, the species are as follows: *Acorus calamus*, *Dianthus inodorus*, *Castanea vesca*, *Juglans regia*, *Amelanchier vulgaris*, *Vitis vinifera* var. *corinthiaca*. Two recent studies of Theophrastus's nomenclature appear to have been made without awareness of Greene's discussion: Louis, Dénominations binominales . . . grecque (1971), pp. 4–9, 24–25; Váczy, Origines . . . nomenclature binaire (1971).]

162. Burgess, *Asters* (1902), pp. 57 *et seq.* [The species indicated is now known as *Aster amellus* L. Thiselton-Dyer indicates that the τριπόλιον referred to in *Enquiry into plants* (1916), bk. 9, ch. 19, sec. 2, might be *Aster tripolium* L.]

163. Sixteenth-century botanical scholars knew this well. Cesalpino wrote the name always as one word, Asteratticus, or else Asteracticus. [*De plantis* (1583), bk. 12, ch. 30, pp. 495–96.]

164. It was the usage of nearly all authors down to and including Tournefort in the year 1700. A binary name meant either a genus, a species, or a variety.

165. Theophrastus, *Historia plantarum* (1644), bk. 1, ch. 7 [(1916), bk. 1, ch. 4, sec. 2. See also *De causis plantarum* (1941), bk. 6, ch. 13, secs. 3–5.]

166. Theophrastus, *Historia plantarum* (1644), bk. 3, ch. 3 [(1916), bk. 3, ch. 2, sec. 5. See also *De causis plantarum* (1941), bk. 1, ch. 8, sec. 3, and ch. 11, secs. 5, 6.]

167. Theophrastus, *Historia plantarum* (1644), bk. 3, ch. 4 [(1916), bk. 3, ch. 3, secs. 1, 2].

168. *Ibid.*, bk. 4, chs. 7–14.

169. [Theophrastus, *Enquiry into plants* (1916), bk. 4, ch. 7, secs. 2, 3. The discovery that corals are animals was first made by Jean André Peyssonnel in 1723, and was published, in disbelief, in Réaumur, Formation du corail (1729). Plantefol, Peyssonnel (1974). Concerning the controversy over whether corals are plants, animals, or rocks, see Lorch, Nature of corals (1965); Ritterbush, *Overtures to biology* (1964), pp. 128–38.]

170. [Trees are the subject of books 2–5 in *Historia plantarum*. This account is evaluated within the broader context of ancient writings on forestry in Makkonen, Ancient forestry (1967–69), part I.]

171. Theophrastus, *Historia plantarum* (1644), bk. 1, ch. 15. [(1916), bk. 1, ch. 9, secs. 3–7. This account merely explains which trees and shrubs are evergreen and which not, but it does not attempt to explain why.]

172. [The phenological data from *Historia plantarum* are extracted in Stillingfleet, *Miscellaneous tracts* (1762), pp. 321–27.]

173. [Theophrastus, *Enquiry into plants* (1916), bk. 1, ch. 9, sec. 1.]

174. Theophrastus, *Historia plantarum* (1644), bk. 2, ch. 3 [(1916), bk. 2, ch. 2, sec. 10].

175. [Other discussions of the sexuality of plants are on pp. 155, 202.]

176. Theophrastus, *Historia plantarum* (1644), bk. 1, ch. 23 [(1916), bk. 1, ch. 14, sec. 2].

177. *Ibid.*, ch. 1 [(1916), bk. 1, ch. 1, sec. 7. On the early usage of this species, see Galil, Sycamore fruit (1968).]

178. Theophrastus, *Historia plantarum* (1644), bk. 3, ch. 4, and bk. 4, ch. 1 [(1916), bk. 3, ch. 3, secs. 1, 2, and bk. 4, ch. 1].

179. *Ibid.*, bk. 1, ch. 6 [(1916), bk. 1, ch. 3, sec. 6, and bk. 4, ch. 5, sec. 1].

180. *Ibid.*, bk. 2, ch. 8 [(1916), bk. 2, ch. 6, sec. 2. For a more general discussion of the physiological effects of salt on plants, see Theophrastus, *De causis plantarum* (1941), bk. 6, ch. 10, pars. 1–10, and the notes therein by Thompson, pp. 266–74.]

181. Theophrastus had heard that at Babylon there were palms of some sort, the top of which could be made to root and grow again; *Historia plantarum* (1644), bk. 2, ch. 2 [(1916), bk. 2, ch. 2, sec. 2].

182. [Theophrastus, *Enquiry into plants* (1916), bk. 2, ch. 6, sec. 11.]

183. Theophrastus, *Historia plantarum* (1644), bk. 1, ch. 9 [(1916), bk. 1, ch. 6, sec. 2].

184. *Ibid.*, bk. 1, ch. 22, and bk. 2, ch. 8 [(1916), bk. 1, ch. 13, sec. 5, and bk. 2, ch. 8, sec. 4. Apparently Greene means that Theophrastus was less certain and less definite in the other cases; for the sexuality of plants was frequently mentioned in *Historia plantarum*, and at least once in *De causis plantarum* (1976– , bk. 1, ch. 22, sec. 1), but without revealing a correct understanding of sexuality of plants.]

185. *Ibid.*, bk. 4, ch. 2 [(1916), bk. 1, ch. 10, sec. 5; bk. 2, ch. 6, secs. 9, 10; bk. 4, ch. 2, sec. 7. This is the doum palm (*Hyphæne thebaica* Mart.).]

186. [Alexander von Humboldt (1769–1859); August Heinrich Grisebach (1814–79).]

187. Theophrastus, *Historia plantarum* (1644), bk. 4, ch. 2 [(1916), bk. 4, ch. 2, sec. 5].

188. [Theophrastus, *Enquiry into plants* (1916), bk. 4, ch. 2, sec. 8, wherein Thiselton-Dyer identified this species as *A. albida* Delile. See also Theophrastus, *De causis plantarum* (1941), bk. 6, ch. 11, sec. 15.]

189. [Theophrastus, *Enquiry into plants* (1916), bk. 4, ch. 2, sec. 11. *M. polyacantha* Willd. = *M. asperata* L. Naturalists did not begin to investigate the causes of plant sensitivity until the seventeenth century. Ritterbush, *Overtures to biology* (1964), pp. 144–56 *et passim*; Sachs, *Geschichte der Botanik* (1875 or 1890), bk. 3, ch. 3; Webster, Plant sensitivity (1966).

190. Theophrastus, *Historia plantarum* (1644), bk. 4, ch. 8 [(1916), bk. 4, ch. 3].

191. *Ibid.*, bk. 4, ch. 5 [(1916), bk. 4, ch. 4].

192. See Bretzl, *Botanische Forschungen* (1903); also an excellent abstract of the same in Fedde, Uranfängen der Pflanzengeographie (1903). [For further information and interpretation, see Joret, *Flore de l'Inde* (1901) and Recherches botaniques (1904). Although Theophrastus mentioned Alexander in *Enquiry into plants* (1916), bk. 4, ch. 4, sec. 5, and bk. 4, ch. 7, sec. 3, Thorndike pointed out that there is no

evidence to indicate how much of Theophrastus's information on Indian plants came from reports from Alexander's expedition and how much (if any) from other sources; Disputed dates . . . Theophrastus (1924), p. 76. Alexander's expedition did not go beyond the Indus Valley.]

193. [Theophrastus, *Enquiry into plants* (1916), bk. 4, ch. 4, sec. 4.]

194. [*Ibid.*, bk. 4, ch. 4, sec. 2.]

195. [*Ibid.*, bk. 4, ch. 4, secs. 12, 13.]

196. [Theophrastus did refer to *N. oleander* L. in *ibid.*, bk. 1, ch. 9, sec. 3, and bk. 9, ch. 15, sec. 1, but Greene obviously had in mind the plant mentioned in bk. 4, ch. 4, sec. 13, which Thiselton-Dyer identified as *N. odorum* Soland.]

197. [*Ibid.*, bk. 4, ch. 7, sec. 8.]

198. [This phenomenon attracted the interest of botany students in the eighteenth century. However, the periodical movements of plants were confused with irritable movements (on which, see n. 189).]

199. Theophrastus, *Historia plantarum* (1644), bk. 4, chs. 16, 17 [(1916), bk. 4, ch. 14. Although Theophrastus did not discover the cause of plant galls, he did describe several varieties, especially from oak trees. Ten kinds of oak galls that Theophrastus discussed are identified and illustrated in Senn, Theophrastus . . . laubwerfenden Eichen (1928) and Oak galls . . . Theophrastus (1941). For broader surveys of Theophrastus's discussions of insect attacks on plants, see Capelle, Entomologie (1962); Hatch, Theophrastus . . . entomologist (1938).]

200. Theophrastus, *Historia plantarum* (1644), bk. 5, ch. 1.

201. *Ibid.*, bk. 5, chs. 2–5 [(1916), bk. 5, ch. 1, secs 5–10.

202. *Ibid.*, bk. 4, ch. 14 [(1916), bk. 4, ch. 13] and *De causis plantarum*, bk. 2, ch. 11. [See also Aristotle, *De generatione animalium*, 750ᵃ21–750ᵇ1, in *Works* (1908–52).]

203. The conifers were favorite subjects of study with Theophrastus; and if some of our Sequoias are rightly estimated to be 25 centuries old, they were not small trees in our philosopher's time.

204. Theophrastus, *Historia plantarum* (1644), bk. 8, ch. 6 [(1916), bk. 8, ch. 7, sec. 1. Darnel is *Lolium temulentum* L.]

205. *Ibid.*, bk. 8, ch. 7 [(1916), bk. 8, ch. 7, sec. 1].

206. *Melampyrum arvense*, according to Sprengel, *Historia rei herbariæ* (1807–8), vol. 1, p. 96 [(1817–18), vol. 1, p. 79. See also Theophrastus, *Enquiry into plants* (1916), bk. 8, ch. 4, sec. 6.]

207. Theophrastus, *Historia plantarum* (1644), bk. 1, ch. 7 [(1916), bk. 1, ch. 5, sec. 2. Greene has read too much into statements that are not significantly diagnostic. Theophrastus was discussing the qualities of bark and stated that some plants have fibrous bark, others fleshy; some have several layers, others only one. He cited wheat as an example of fibrous bark and darnel as an example of layered bark. However, fibrosity and layeredness are not opposable characteristics. Most damaging to Greene's case is the statement coming later (bk. 2, ch. 4, sec. 1) to the effect that wheat turns into darnel.]

208. *Ibid.*, bk. 2, ch. 3 [(1916), bk. 2, ch. 2, secs. 9, 10. This quotation makes it clear that the argument was not whether one species ever changed into another, but rather, under what conditions it occurs. The instability of species (which is not the same as evolution) was a common assumption in biology down to the end of the seventeenth century. Balme, Biology in Aristotle and Theophrastus (1962); Capelle, Theophrast über Pflanzenentartung (1949); Senn, Species bei Aristoteles und Theophrast (1945); Zirkle, Species before Darwin (1959), pp. 639–40.]

209. See van Stapel's edition of Theophrastus, *Historia plantarum* (1644), p. 78.

[Scaliger, *Historias Theophrasti* (1584). Greene referred to the seventeenth century because he used the reprint of Scaliger's commentary. Scaliger wrote two other commentaries: *De causis plantarum Theophrasti* (1566) and *De plantis Aristotele* (1556); Hall, Scaliger (1950); Rose, Scaliger (1975); Schmitt, Theophrastus (1971), pp. 269–71, 274–75.]

210. [With the advantage of hindsight, Greene's objection is easily raised and sustained, but an ancient proponent of the transmutation might have asked for proof that all tadpoles metamorphose into frogs and all caterpillars into butterflies and moths. Lacking such documentation, one could easily believe that some wheat might grow into darnel. Had Greene's projected controversy developed between proponents and opponents of transmutation in antiquity, scientists might have studied in closer detail the transformations of larvæ into adult animals, but there were few such studies before the seventeenth century.]

211. Theophrastus, *Historia plantarum* (1644), bk. 3, ch. 18 [(1916), bk. 3, ch. 18, secs. 7–10. Pliny stated that "the leaves of the ivy are angular at first but become round"; *Natural history* (1938–62), bk. 16, ch. 35, sec. 86.]

212. Sachs, *History of botany* (1890), p. 16 [(1875), pp. 18–19. Greene rejected this judgment on the basis of his defense of the botanical knowledge of Theophrastus. Often, as in the above discussion of ivy, Greene found a subtle reason for concluding that Theophrastus had conducted many investigations and had, through prudence, withheld his well-founded conclusions. This claim seems implausible. Kirchner also argued that Theophrastus's botanical writings were based on extensive personal investigations, but Regenbogen more recently has argued to the contrary. Kirchner, *Botanischen Schriften des Theophrast* (1874); Regenbogen, Theophrastos (1940), cols. 1467–72.]

213. [Greene had little to say about Theophrastus's physiological writings, found in *De causis plantarum*; see Thompson's discussion in the 1941 edition and Hoppe, *Biologie Wissenschaft* (1976).]

214. Reaffirmed and somewhat improved by Cesalpino in the year 1583. [See Part II, ch. 22.]

215. But, as indicated on p. 166, he did not make this division of vascular plants as it is made today.]

216. [This vague statement attributes more to Theophrastus than seems justified. Greene's own discussion of Theophrastus's writings on the ages of trees (pp. 204–5) does not indicate that he knew the age could be determined by counting annual rings. Pertinent statements from Theophrastus are: wood on the north side of a tree "is closer and more vigorous" (*Enquiry into plants* (1916), bk. 5, ch. 1, sec. 12) and "trees make their growth all at once or nearly so, or at all events we may say that they do so all at one season" (bk. 7, ch. 10, sec. 2). The discovery that annual rings indicate age seems to have been made during the late Middle Ages or early Renaissance. Sarton, Tree-ring analysis (1954b).]

217. [Plumier, *Nova plantarum americanarum* (1703), p. 8. Linnæus first used this name in a catalogue in Royen, *Floræ Leydensis* (1740), p. 528.]

Chapter 3: Greeks and Romans After Theophrastus

1. Meyer, *Geschichte der Botanik* (1854–57), vol. 1, p. 202. [*De plantis* is discussed in ch. 2, n. 24.]

2. This is one of the several valuable papers that are prefatory to Hieronymus Tragus's *Stirpium* [Bock (1552). Gesner is discussed in Part II, ch. 20.]

3. [The evidence for Nicander's floruit is inadequate, but probably the middle of the second century B.C., according to Gow & Scholfield, Life of Nicander, pp. 3–8,

in Nicander, *Poems* (1953); see also Kroll, Nicandros aus Kolophon (1936); Sarton, *Introduction to history of science* (1927–48), vol. 1, pp. 158–59.]

4. [Recent scholars have, on the contrary, concluded that Nicander was a poet who borrowed from other authors the information conveyed in his poems. For details, see nn. 5, 8.]

5. Cicero, *De oratore* [(1942), bk. 1, sec. 16, par. 69. It was a qualified compliment; Cicero observed that Nicander distinguished himself in his *Georgica* because of his poetic ability and in spite of his ignorance of farming. For the extant fragments of Nicander's *Georgica*, see his *Poems* (1953), pp. 144–61.]

6. Athenæus flourished some three centuries after Nicander. [These quotations are from Athenæus, *Deipnosophists* (1927–41), XV, 679–84. For a botanically interesting example of his quotations, see Stannard, Squill (1974c), pp. 687–88.]

7. Haller, *Bibliotheca botanica* (1771–72), vol. 1, p. 54. [See also Buller, Fungus lore (1914).]

8. [The principal source for Nicander's *Theriaca* was a treatise by Apollodorus (fl. *ca.* early 3rd century B.C.), and probably Nicander's *Alexipharmica* was derived mainly from another work by Apollodorus. See Gow & Scholfield in Nicander, *Poems* (1953), p. 18; Scarborough, Nicander's toxicology (1977–).]

9. Meyer, *Geschichte der Botanik* (1854–57), vol. 1, p. 247.

10. [*E.g.*, for snakebite he suggested: "Cast also upon the fire the foliage of the *male fern* with its cloven fronds, or take the heated root of the *frankincense-tree* mixed with an equal measure of *garden-cress*; and mingle the fresh, pungent horn of a *roe* . . ."; *Theriaca*, lines 39–42, in *Poems* (1953). The remedies for animal bites and stings were known as theriacs, from *θηρίον*, a wild or venomous animal.]

11. [". . . in Nicander the difficulties of identification, already formidable in themselves, are enhanced by the doubt whether the poet knew what he (or his authority) was talking about"; Gow & Scholfield in Nicander, *Poems* (1953), p. 24. Apollodorus, Nicander's authority, owed most of his botanical remedies to Diocles of Carystus (fl. *ca.* 300 B.C.). Dannenfeldt, Diocles (1971); Jaeger, *Diokles* (1938); Wellmann, Älteste Kräuterbuch (1898a). See also Burgess, *Asters* (1902), pp. 115–18.]

12. [The centaurea (line 500) was probably yellowwort (*Chlora perfoliata*); aristolochia was mentioned twice (lines 507, 937); the trifolium (lines 522, 907) is *Psoralea bituminosa*, according to Gow & Scholfield in Nicander, *Poems* (1953).]

13. Meyer, *Geschichte der Botanik* (1854–57), vol. 1, p. 248. [See also Louis, Dénominations binominales . . . grecque (1971), pp. 9–13.]

14. Sprengel, *Historia rei herbariæ* (1807–8), vol. 1, p. 129 [(1817–18), vol. 1, pp. 106–10. For the Latin translation of Nicander by E. Cordus, see ch. 8, n. 7.]

15. [Cato (234–149 B.C.) was Roman censor in 184. He is also known as Cato Censorius or Cato Major to distinguish him from his grandson Marcus Porcius Cato Uticensis (95–46 B.C.). He is included in Plutarch's *Parallel lives*, and a recent detailed biography is by Astin, *Cato the censor* (1978).

16. [Cato was noted for his cruelty to defeated enemies and slaves, and his fanatical insistence on the total destruction of Carthage inspired that policy of the Roman senate.]

17. [Indeed, it is the earliest extant prose work in Latin. It is also known by the title *De agri cultura*. The *editio prima* was published with the treatises by Columella, Varro, and Palladius, under the title *Scriptores rei rusticæ* (1472). For further discussions, see Brown, Cato (1980a); Duff, *Literary history . . . golden age* (1953), pp. 183–84, 187–88, 191–93 *et passim*; Fussell, *Classical tradition . . . farming* (1972), pp. 20–25 *et passim*; Mayerhöfer *et al.*, *Lexikon* (1959–), vol. 1, pp. 608–9; Meyer, *Geschichte der Botanik* (1854–57), vol. 1, pp. 338–48; Sarton, *In-*

troduction to history of science (1927–48), vol. 1, p. 186; K. White, *Roman farming* (1970), pp. 19–20 *et passim.*]

18. [Cato, *On agriculture* (1934), secs. 41–52, 133, 151, 161. Ancient grafting practices are discussed in Mendel, Transplantations (1953); Pease, Ancient grafting (1933).]

19. [There is evidence that, by the time Cato wrote, more than the eight cultivars of *Vitis vinifera* L. he listed already existed; some of those he listed are still cultivated. Billiard, *Vigne* (1913); Curtel, *Vigne* (1903); Forbes, *Ancient technology* (1953–64), vol. 3, pp. 106–30; Goor, History of grapevine (1966); Hyams, *Dionysus* (1965); Laufer, Sino-Iranica (1919), pp. 220–45; H. Lutz, *Viticulture and brewing* (1922); Senn, Rebbau . . . Theophrast (1944); Weaver, Grape growing (1960); Younger, *Gods, men, wine* (1966).]

20. *Catonia* P.Br. = ?*Miconia* Ruiz & Pav.

21. [Varro's birth date is now believed to have been 116 B.C. For discussions of his treatise, see Brown, Varro (1980b); Dahlmann, Varro (1935); Duff, *Literary history . . . golden age* (1953), pp. 241–53 *et passim*; Fussell, *Classical tradition . . . farming* (1972), ch. 1; Grimal, *Jardins romains* (1969), pp. 364–67; Meyer, *Geschichte der Botanik* (1854–57), vol. 1, pp. 354–69; Sarton, *Introduction to history of science* (1927–48), vol. 1, pp. 225–26; Thomas, Varro (1976); K. White, *Roman farming* (1970), pp. 22–24 *et passim.*]

22. Quoted in Meyer, *Geschichte der Botanik* (1854–57), vol. 1, p. 356, Symmachi I, epistol. 2.

23. Livy, bk. 22, chs. 34, 35.

24. ["His education was completed at Athens under the philosopher Antiochus of Ascalon"; Duff, *Literary history . . . golden age* (1953), p. 242.]

25. [Varro, *Rei rusticæ* (1472); see also n. 17.]

26. [Varro, *On agriculture* (1934), bk. 1, par. 46. The species Varro meant by "heliotropia" is unknown; Von Erhardt-Siebold, Heliotrope tradition (1938), p. 24.]

27. Varro, *Rei rusticæ*, bk. 2, ch. 5, sec. 13. [Varro referred to Aristotle's discussion of determining sex in cattle; *De generatione animalium*, 765^{a}1–20, in *Works* (1908–52).]

28. [This statement is true only to the extent that Greene qualified it in the next sentence. Certainly, Cato showed more evidence of firsthand knowledge of agriculture than Varro did.]

29. [Virgil's birth date is now believed to have been 70 B.C. Among the many biographical studies of Virgil, those by Frank and Sellar are cited as reliable and easily available: Frank, *Vergil* (1922); Sellar, *Virgil* (1897).]

30. [The village has often been identified with Pietole, but Conway has shown that the most convincing location of Virgil's father's farm is between the Carpenedolo Ridge and the Chiese River about 30 miles east of Mantua and about 5 miles west of Calvisano; *Vergilian age* (1928), pp. 14–40.]

31. [Such was the estimate of John Dryden, who published his translation of Virgil's *Georgics* in 1697. For an analysis of the *Georgics* and a discussion of Virgil's sources, see Duff, *Literary history . . . golden age* (1953), pp. 326–32; Wilkinson, *Georgics of Virgil* (1969).]

32. [A. Stevenson, *Catalogue of botanical books* (1961), no. 517.]

33. [Virgil's *Bucolica* (1810) contains Martyn's translation and an abridgment of his annotations.]

34. [This is incorrect. In a letter to James Edward Smith, dated 12 April 1818, Thomas Martyn stated that his father, John Martyn (1699–1768), had never received a degree. Smith, *James Edward Smith* (1832), vol. 2, p. 387.]

35. Haller, *Bibliotheca botanica* (1771–72), vol. 1, p. 68.

36. [Virgil's poems have also inspired the following commentaries: Abbe, *Plants of Virgil* (1965); Billiard, *L'agriculture . . . Virgile* (1928); Bubani, *Flora virgiliana* (1869); Büchner, Vergilius (1955, 1958); Burgess, *Asters* (1902), pp. 123–33; Consoli, *Neologismi botanici* (1901); Grimal, *Jardins romains* (1969), pp. 379–92; Meyer, *Geschichte der Botanik* (1854–57), vol. 1, pp. 370–75; Paulet, *Flore et faune de Virgile* (1824); Sargeaunt, *Plants of Virgil* (1920); Sarton, *Introduction to history of science* (1927–48), vol. 1, pp. 221–22; Sprengel, *Historia rei herbariæ* (1807–8), vol. 1, pp. 139–48, and (1817–18), vol. 1, pp. 121–26; Thiselton-Dyer, Flora (1938); K. White, *Roman farming* (1970), pp. 39–41 *et passim.*]

37. [These figures are according to Meyer, *Geschichte der Botanik* (1854–57), vol. 1, p. 373. Theophrastus mentioned most, if not all, of the species of which he was aware, but Cato and Varro and even Virgil seldom had occasion to mention any species for which they did not know a practical use. Virgil, when he discussed plants, relied to an important extent on Theophrastus, as shown in Jahn, Prosaquelle Vergils (1903). Virgil's discussion of plants was also influenced by the poetry of Theocritus (fl. 275 B.C.). Two lines from his *Idylls* are quoted in Part II, ch. 19, n. 40. Gow has provided an authoritative text, translation, and commentary in *Theocritus* (1952), see "Virgil" in vol. 2, index; Feé, *Flore de Théocrite* (1832); Lembach, *Pflanzen bei Theokrit* (1970); Lindsell, Was Theocritus a botanist? (1937); Prescott, *Development of Virgil's art* (1927).]

38. Columella, *De re rustica* (1472), bk. 2, ch. 16. [Greene apparently has mistaken Columella's uncle for his father. "The parents of Columella are named nowhere in his works, but he speaks often and with the greatest respect for an uncle, Marcus Columella, an expert farmer of the Bætic province, in whose company much of his youth appears to have been spent"; Ash, Life and works of Columella, in Columella, *On agriculture* (1941–55), vol. 1, p. xv.]

39. [In addition to the writings by Greek and Roman authors, Columella had access to the important treatise on agriculture of Mago the Carthaginian, which was translated by D. Silanus from Punic into Latin soon after the destruction of Carthage in 146 B.C. Columella frequently cited Mago in *De re rustica*; however, Mago's treatise is no longer extant. See Meyer, *Geschichte der Botanik* (1854–57), vol. 1, pp. 296–305.]

40. [Columella, *De re rustica* (1472; see n. 17). The following works provide a guide for studying Columella's treatise: Brown, Columella (1976); Burgess, *Asters* (1902), pp. 134–38; Duff, *Literary history . . . silver age* (1964), pp. 131–37; Fussell, *Classical tradition . . . farming* (1972), ch. 1; Kappelmacher, Columella (1919); Marshall, *L'horticulture . . . Columelle* (1918); Mayerhöfer *et al.*, *Lexikon* (1959–), vol. 1, pp. 674–75; Meyer, *Geschichte der Botanik* (1854–57), vol. 2, pp. 58–80; Sarton, *Introduction to history of science* (1927–48), vol. 1, pp. 256–57; Weinold, *Quellen . . . Columella* (1959); K. White, *Roman farming* (1970), pp. 26–27 *et passim.*]

41. Columella, *De re rustica*, bk. 11, ch. 3, secs. 52, 53. [The passage may refer to translucent stone rather than panes of glass: "Sed nihilo minus specularibus integi debebunt, ut etiam frigoribus serenis diebus tuto producantur ad solem. Hac ratione fere toto anno Tiberio Cæsari cucumis præbebatur." The crucial word here is "specularibus." This passage is translated by Forster and Heffner as follows: "In any case the vessels ought to be covered with slabs of transparent stone, so that even in cold weather, when the days are clear, they may safely be brought out into the sun. By this method Tiberius Cæsar was supplied with cucumbers during almost the whole year." Tiberius died in 37 A.D., so this practice predates that year. It was not just an isolated practice, for Martial (*ca.* 40 to *ca.* 120 A.D.) also alluded to it in his *Epigrams*, bk. 8, no. 14 (also using the word "specularia"). Seneca (4 B.C.?–65 A.D.)

stated that the development of closed windows that admitted light had occurred during his lifetime; *Epistolæ morales* (1917–25), no. 90, sec. 26. Alabaster, mica, and shells were sometimes used in windows, but the Romans eventually were able to produce windowpanes; Forbes, *Ancient technology* (1953–64), vol. 5, pp. 182–83.]

42. Columella, *De re rustica*, bk. 3, ch. 11. [Incorrect reference; on grafting, see bk. 4, ch. 29 (on grape vines); bk. 5, ch. 11 (on trees); bk. 11, ch. 2, secs. 11, 23, 26 (brief seasonal instructions); "De arboribus," ch. 8, secs. 1–4 (on grape vines), and ch. 26 (on trees). The last two discussions are very similar to the first two. See also Mendel, Transplantations (1953); Pease, Ancient grafting (1933).]

43. Sprengel, *Historia rei herbariæ* (1807–8), vol. 1, pp. 149–51 [(1817–18), vol. 1, pp. 126–34].

44. *Ibid.*, p. 151 [(1817–18), vol. 1, p. 134].

45. Meyer, *Geschichte der Botanik* (1854–57), vol. 2, pp. 96–100.

46. Sprengel, *Historia rei herbariæ* (1807–8), vol. 1, p. 152. [Tacitus, *Annals* (1931), bk. 15, ch. 33. This passage discusses Nero, without mention of Dioscorides.]

47. [On the life and work of Dioscorides, see Dönt, Dioskurides (1970); Meyer, *Geschichte der Botanik* (1854–57), vol. 2, pp. 96–134; Riddle, Dioscorides (1971); Sarton, *Introduction to history of science* (1927–48), vol. 1, pp. 258–60 *et passim*; Wellmann, Dioskurides (1903).]

48. [Since there were important treatises used by Dioscorides that have, except for very short fragments, subsequently disappeared, it is difficult to know what plants and remedies he discovered. However, it would be safest to assume a small number. Stadler compared Dioscorides' plant names with those of Theophrastus and found 236 used by both, 152 used only by Theophrastus, and 249 used only by Dioscorides; Theophrast und Dioscorides (1891a). Among the other predecessors of Dioscorides were Diocles of Carystos; Mithridates VI Eupator of Pontus (120–63 B.C.) and his rhizotomist physician Krateuas; Quintius Sextius Niger (fl. 25 B.C.); Pamphilos; Menecrates, physician to Tiberius (14–37 A.D.); Andromachos of Crete, physician to Nero (54–68 A.D.); and Servilius Damocrates, of the time of Nero. For descriptions and citations on these authors, see ch. 2, n. 19; Burgess, *Asters* (1902), pp. 118–23; Nissen, *Herbals* (1958), pp. 8–14; Sarton, *Introduction to history of science* (1927–48), vol. 1, pp. 121, 214, 241, 261; Singer, Herbal (1927), pp. 3–19; Wellmann, Sextius Niger (1889) and Krateuas (1899). The influence of the Hippocratic medical botany on Dioscorides is described in Mock, *Pflanzliche Arzneimittel* (1919). The Roman encyclopedist Aulus Aurelius Cornelius Celsus (fl. 10–37 A.D.) also wrote before Dioscorides; his *De medicina* was lost during the Middle Ages, but rediscovered in 1443 and published at Florence in 1478. The herbs mentioned by Celsus are listed in the Loeb edition (1935–38). See also Caturegli, *Farmacologia celsiana* (1966).

Aretæos of Cappadocia is of uncertain floruit. He was either a contemporary of Galen's or somewhat younger. He has earned a reputation among historians of medicine for his outstanding clinical descriptions and diagnoses. His works were published in Latin translation in 1552 and in the original Greek in 1554. Francis Adams published an English translation (1856), and Hude has edited the modern critical text (1958). The contributions of Aretæos to medical botany have been recently documented by Stannard, Materia medica . . . Aretæus (1964), who points out the importance of plant material to Aretæos both for medicines and for therapeutic diets. This article includes extensive citations to other accounts of Aretæos. Kudlien, Aretæus (1970c), places the floruit at about A.D. 50.]

49. [According to Regenbogen, Theophrastus (1940), col. 1467, Theophrastus mentioned about 550 kinds of plants. According to Stadler, Theophrast und Dioscorides (1891a), Dioscorides mentioned 537 kinds of plants. One could swell the

list by maintaining that some names refer to more than one species, but even with this qualification, the number of species would be closer to 550 than 600. See also Louis, Dénominations binominales . . . grecque (1971), pp. 13–26; Váczy, Origines . . . nomenclature binaire (1971), pp. 582–90.]

50. [Except, of course, that there is a treatise on medicinal plants in Theophrastus, *Historia plantarum*, bk. 9. For a discussion of the pharmacological aspects of Dioscorides' contribution, see Berendes, *Pharmacie* (1891), vol. 1, pp. 289–308. Berendes also annotated his translation of Dioscorides, *Materia medica*; see n. 62.]

51. [On the medieval tradition, see Riddle, Dioscorides (1971), Dioscorides (1980); see also Bologa, I sinonimi (1930); Bonnet, Essai d'identification (1903); Burgess, *Asters* (1902), pp. 149–52; Dubler, *Materia medica* (1953–59), vol. 1; Garrison, *History of medicine* (1929), pp. 109–10; Hofman, Auracher & Stadler, Dioskorides des Marcellus Virgilius (1883–1903); Mioni, *Un ignoto Dioscoride miniato* (1959); Penzig, *Storia della botanica* (1904, 1905), pp. 241–82; Piccinini, Dioscoride (1920), pp. 71–73; Sigerist, Materia medica in Middle Ages (1939); Singer, Greek biology (1921), pp. 64–66, and Herbal in antiquity (1927), pp. 22–29; Stannard, Mioni's *Miniato* (1962b) and Dioscorides and Renaissance (1966b), pp. 4–6; Thorndike, *Magic and experimental science* (1923–58), vol. 1, pp. 605–11.]

52. See Pritzel, *Thesaurus* (1871–77), p. 84. [This was bound with Nicander's *Theriaca* and *Alexipharmaca*, and it was the first book printed in Greek. The critical Greek text has been established by Wellmann in Dioscorides, *Materia medica* (1906–14).]

53. Pritzel, *Thesaurus* (1871–77), p. 84.

54. Said to bear the name of Petrus Paduanensis; see *ibid.*, p. 85. [*I.e.*, edited by Pietro d'Abano (*ca.* 1250 to *ca.* 1315). For further discussions of Pietro, see Part II, ch. 16, n. 8.]

55. [The Latin editions before this date would undoubtedly have been more numerous had there not been competition from medieval texts on materia medica compiled from Dioscorides and less important sources. See Stannard, Dioscorides and Renaissance (1966b). The terminology in the Ruel translation is discussed in Sprague, Technical terms in Ruellius' Dioscorides (1936), and Ruel is the subject of ch. 16, in Part II.]

56. [Anguillara, Mattioli, Dodœns, Cesalpino, and Colonna are discussed in Part II, chs. 19, 21, 24, 22, and 23. To Greene's list should be added Barbaro and Marcello Virgilio, discussed in chs. 13 and 14.]

57. [Piccinini, Dioscoride (1920), pp. 78–79, has argued persuasively that if Mattioli had had the courage to publish his observations independently of Dioscorides' text, Dioscorides' treatise would have become obsolete much sooner than it did.]

58. [Concerning Bauhin's *Pinax*, see Part II, Appendix C.]

59. [The Spanish translation by Don Andrés de Laguna (1499?–1560) is reprinted in facsimile, with critical apparatus, by Dubler as vol. 3 (1955) of his *Materia medica*, with indexes in vol. 6 (1959). His biography of Laguna is vol. 4 (1955) of the same work. See also Friedenwald, Andrés a Laguna (1939); Peset, Spanish . . . Dioscorides (1954); Riddle, Dioscorides (1980), pp. 89–92.]

60. Pritzel, *Thesaurus* (1871–77), pp. 86–87. [The more recent German translation by Berendes is cited in n. 62. When Greene wrote, the English translation by John Goodyer (1592–1664), made in 1652–55, had not been published. On the concordance of chapters in various translations, see n. 62. For further discussions of early editions and translations, see Burgess, *Asters* (1902), pp. 326–30; Piccinini, Dioscoride (1920), pp. 74–78, 110–13; Stannard, Mattioli . . . Dioscorides (1966c).]

61. [For a more detailed, statistical discussion of the contents, see Piccinini, Dioscoride (1920), pp. 79–80.]

62. Dioscorides, *Materia medica*, bk. 4, ch. 155. [Greene owned a copy of the 1543 Egenolph edition of Ruel's Latin translation of Dioscorides, and clearly his citations are based on that edition. Unfortunately, the chapter divisions are not constant among the various editions and translations. Therefore, the notes cite the standard book and chapter numbers to the Greek text established by Wellmann along with the page numbers to the Goodyer English and Berendes German translations. The above citation is in the Wellmann text (1906–14), bk. 4, ch. 173; Goodyer transl. (1934), pp. 574–76; Berendes transl. (1902), pp. 465–66. The Goodyer translation is not always precise. For comparisons with more precise translations by Jerry Stannard, see the references in the Editor's Introduction, n. 12. For a concordance of chapters in the translations by Ruel (1516), Sprengel (1829–30), and Goodyer (1934), see Sprague, Technical terms in Ruellius' Dioscorides (1936), pp. 150–52. For a concordance of chapters of the Wellmann text with Arabic, Latin, and Spanish translations, see Dubler, *Materia medica* (1953–59), vol. 1, pp. xxxiii–lxviii.]

63. K. Bauhin, *Pinax* (1671), pp. 455–56.

64. [Dioscorides, *Materia medica* (1906–14), bk. 3, ch. 137–39; (1934), pp. 379–82; (1902), pp. 352–54. According to Sibthorp and Lindley (Goodyer transl., p. 670), *βούφθαλμον* in *Codex Aniciæ Iulianæ* is *Chrysanthemum segetum* L.; Cecil Norman's identification, which Gunther accepted, is *Anthemis valentina* L. = *Anacyclus radiatus* Loisel. See also Burgess, *Asters* (1902), pp. 138–45.]

65. [Dioscorides, *Materia medica* (1906–14), bk. 4, chs. 23–27; (1934), pp. 422–24; (1902), pp. 378–79.]

66. [*Ibid*., bk. 4, ch. 70–75; (1934), pp. 467–74; (1902), pp. 404–11.]

67. Pliny the Younger, *Letters* (1915), bk. 3, epistle 5. [Among the recent general discussions of Pliny the Elder are Duff, *Literary history . . . silver age* (1964), pp. 281–310 *et passim*; *Eichholz*, Pliny (1975); Kroll, Plinius Secundus der Ältere (1951); Sarton, *Introduction to history of science* (1927–48), vol. 1, pp. 249–51 *et passim*; Stahl, *Roman science* (1962), ch. 7; Thorndike, *Magic and experimental science* (1923–58), vol. 1, ch. 2; K. White, *Roman farming* (1970), pp. 28–29 *et passim*.]

68. Pliny the Younger, *Letters* (1915), bk. 6, epistle 16. [See Zirkle, Gaius Plinius Secundus (1967).]

69. [Book 1 is especially noteworthy because it contains a list of the sources Pliny used for each subject.]

70. Meyer, *Geschichte der Botanik* (1854–57), vol. 2, p. 125. [Another factor was that Pliny's *Natural history* appeared to render obsolete many earlier treatises, which then ceased to be recopied or highly valued.]

71. Julius Sillig, quoted in Meyer, *Geschichte der Botanik* (1854–57), vol. 2, p. 125. [The earliest of these originated in the eighth or ninth century. None of them provides an authoritative text, and there are difficult textual problems that will probably remain. Sillig's discussion occurs as part of the introduction to his edition of Pliny, *Naturalis historiæ* (1831–36). Over 2,000 manuscripts are reported to be extant by Gudger, Pliny's *Historia naturalis* (1923); however, Gudger provides no documentation. For discussions of the most important manuscripts, see Ernout, Introduction, vol. 1, pp. 20–38, in Pliny, *Histoire naturelle* (1950–72); Finch, Excerpts from Pliny (1965); Rück, Plinius im Mittelalter (1898); Thorndike, Epitomes of Pliny (1936).]

72. [There were 55 editions published from 1469 to 1599, including translations into Italian (1476), German (1539), French (1562), English (1565), and Spanish

(1587). Labarre, L'historia naturalis de Pline (1973). A census of medieval and Renaissance commentaries is in Nauert, Plinius Secundus (1980).]

73. [Pliny's use of Theophrastus is discussed in Renjes, *Ratione . . . Plinii . . . Theophrasti* (1893); I. Sprengel, *Plinium et Theophrastum* (1880).]

74. [There is insufficient evidence to warrant the conclusion that Pliny used Dioscorides' work. Pliny did not list him as one of his sources, and the distinct similarities in some passages may be due to their reliance on common sources, probably Diocles of Carystos, Krateuas, and Sextius Niger. Jones, Introduction, in Pliny, *Natural history* (1938–62), vol. 6, pp. xix–xx; Wellmann, Sextius Niger (1889).]

75. [This belief was antiquated in the sense that some philosophers considered it naive, not in the sense that it was generally conceded to be false. One of the most influential defenses of it was by Cicero (106–43 B.C.), *De natura deorum* (1931), bk. 2.]

76. Pliny, *Naturalis historia* (1606), bk. 7, ch. 1, secs. 1, 2.

77. Adanson, *Familles des plantes* (1763), Preface, p. vii.

78. [Pliny, *Natural history* (1938–62), bks. 12–17.]

79. [Studies of Pliny's botany include André, Pline (1955); Brosig, *Botanik . . . Plinius* (1883); Burgess, *Asters* (1902), pp. 152–56; Meyer, *Geschichte der Botanik* (1854–57), vol. 2, pp. 118–34; Stannard, Plant called moly (1962a), pp. 270–75, Pliny and Roman botany (1965), and Squill (1974c), pp. 694–95; Váczy, Origines . . . nomenclature binaire (1971). Other studies on Pliny's botany, of restricted scope, are listed in Le Bonniec, Bibliographie . . . Pline (1945), pp. 233–39, or (1946), pp. 36–42.]

80. [Adanson received a copy of Pliny, *Historiæ mundi libri XXXVII* (1631) as a prize at school in 1739, when he was 12 years old (his copy is now in the Hunt Botanical Library's Adanson Collection, AD 103). He was influenced by his studies of Pliny, but he did not write a thesis on him. For a description of his copy of Pliny, see Margadant, Adanson collection (1963), p. 302. For his bibliography, see Nicolas, Adanson (1963), pp. 112–16. For Pliny's influence, see Stafleu, Adanson (1963), pp. 146, 148. For other studies on Adanson, see the Editor's Introduction, n. 3.]

81. Adanson, Histoire de la botanique, in *Familles des plantes* (1847), p. 94. [See also Sprague, Botanical terms in Pliny (1933a).]

82. ["Gemma" had been used previously in Cato, *De re rustica*, sec. 42. Both terms are included in André, *Termes de botanique en latin* (1956).]

83. [Plumier, *Nova plantarum americanarum* (1703), p. 9. Linnæus retained the name from Plumier, which is still valid. Manoel Blanco also named a genus *Plinia* (1837), since changed to *Kayea* Wall.]

84. [Through a series of misassumptions, "Claudius" was attached to Galenic works in the early Renaissance, but research has demonstrated that no name is known to precede "Galen." More probably, it would have been "Ælius Galen," since he was the son of Ælius Nicon, a famous architect of Roman Pergamon. Galen's life dates are uncertain, but a birth date of September 129 is convincingly defended by Nutton, Chronology of Galen (1973), pp. 159–61. Because of conflicting evidence, the details of Galen's career are hotly debated. Sarton provides a basic bibliography before 1926 in *Introduction to history of science* (1927–48), vol. 1, pp. 301–7. See also Kudlien & Wilson, Galen (1972); Mewaldt, Galenos (1910); Sarton, *Galen* (1954a); Scarborough, *Roman medicine* (1969), pp. 114–21, and Galenic question (1981); Temkin, *Galenism* (1973), pp. 3–9; Thorndike, *Magic and experimental science* [1923–58), vol. 1, pp. 117–81.]

85. [On Nicon, Galen's father, see Nutton, Chronology of Galen (1973), pp. 161–65; Scarborough, *Roman medicine* (1969), pp. 115, 214–15 with n. 41. Greene's "experimenting with grain sowing" is somewhat inaccurate, since Galen

speaks of his father's interest in the testing of seeds, and their possible changes from grains into the weedy grass with bristly flower clusters that we would term darnel. Galen, *De alimentorum facultatibus*, in *Opera omnia* (1821–33), vol. 6, bk. 1, sec. 37, pp. 551–53, esp. p. 552, and *De alimentorum facultatibus* (1923), pp. 261–62. Galen's passage bears great similarity to Theophrastus, *Enquiry into plants* (1916), bk. 8, ch. 8, sec. 4, and to *Geoponica* (1895), bk. 2, chs. 42, 43, pp. 78, 79. Galen's *De alimentorum facultatibus*, bk. 1, ch. 37, is an important document that reveals much about the Roman knowledge of grains; the consideration of ἀιγίλωψ (Havergrass, *Ægilops ovata* L.) introduces Galen's anecdote about his father. It is also apparent that Galen's account rests on a long tradition of herbal dietetics, since we find similar accounts in Dioscorides, *Materia medica* (1906–14), bk. 2, ch. 100, and bk. 4, ch. 137, and in Theophrastus, *De causis plantarum*, bk. 5, ch. 15, sec. 5, and *Enquiry into plants* (1916), bk. 1, ch. 5, sec. 2; bk. 2, ch. 4, sec. 1; bk. 4, ch. 4, sec. 10; and bk. 8, ch. 8, sec. 3. Havergrass appears as part of a multi-substance remedy in Nicander, *Theriaca*, line 857, in *Poems* (1953), pp. 84, 85, against scorpion stings, spider bites, shrew bites, and stingray poison.]

86. [Rather, his father had a dream that told him Galen should study medicine. Galen, *Method of medicine*, bk. 9, ch. 4, in *Opera omnia* (1821–33), vol. 10, pp. 608–16, esp. p. 609. If we have the chronology correct, Nicon's dream occurred when Galen was 16. He had begun studying the philosophy of the Epicureans, Stoics, and Platonists at 14. He then attended the anatomy lectures of Satyrus in Pergamon. Galen, *Anatomical procedures*, bk. 1, ch. 2, in *Opera omnia* (1821–33), vol. 2, pp. 218–27, esp. pp. 224–25; (1956), pp. 4–5. The chronology in Sarton, *Galen* (1954a), p. 94, is based on the articles of Walsh, but Walsh was less reliable than his main source: Ilberg, Schriftstellerei . . . Galenos (1889–97) and Galens Praxis (1905). Moreover, there is conflicting evidence concerning Galen's stint as gladiatorial physician in Pergamon, as seen in the differing dates offered by Scarborough, Galen and gladiators (1971), and Nutton, Chronology of Galen (1973). Sarton's chronology, as revised by Scarborough and Nutton, is as follows:

Sept. 129 A.D. Galen born at Pergamon
129–43. Educated at home, under supervision of his father, Nicon
143–45. Attended lectures on philosophy in Pergamon
145. Nicon's dream
145–48. Studied anatomy at Pergamon under Satyrus, Stratonicus, and Æschrion
ca. 147–48. Appointed a member of the staff at the shrine to Asclepius at Pergamon
148 or 149. Death of Nicon
149 or 150. Studied medicine and anatomy at Smyrna
151. Studied anatomy at Corinth
152–57. Studied anatomy, medicine, and philosophy at Alexandria
157–61. Physician to the gladiators at Pergamon (Nutton, 1973) *or*
159–61. Physician to the gladiators at Pergamon (Scarborough, 1971)
161. Outbreak of war with Parthia
161–62. Journey to Rome
162. First visit to Rome
166. Return to Pergamon from Rome
168. Summons from the emperors. Visit at Lemnos on the way to Aquilea
169–? Physician to Marcus Aurelius and Commodus, in Rome
199 or 200. Death, either in Rome or Pergamon.]

87. [Nutton, Chronology of Galen (1973), pp. 162–63; Scarborough, *Roman medicine* (1969), p. 115 with n. 43.]

88. [Greene has oversimplified the complexities of dating Galen's journeys to

Lemnos, Cyprus, Lycia, Palestine, and elsewhere. He probably visited Lemnos twice and Cyprus more than once. Nutton, Chronology of Galen (1973), pp. 166–70.]

89. [Walsh, Galen clashes . . . Rome (1928).]

90. [Nutton, Chronology of Galen (1973), pp. 166–67.]

91. [Quadi and Marcomanni are correct, but "Parthians" is not. Nutton, Chronology of Galen (1973), p. 167 with n. 2.]

92. [Ilberg, Schriftstellerei . . . Galenos (1889–97); Walsh, Galen's writings (1934–39).]

93. Edited by Cossæus, Fuchs, and Gemusæus. [The first collected edition in Latin was in 2 volumes (1490); the first Greek edition was in 5 volumes (1525). Concerning the manuscripts of Galen's writings, translations, and early editions, see Ackermann, Historia . . . Galeni (1821); Campbell, *Arabian medicine* (1926), vol. 2, pp. 13–220; Durling, Renaissance editions . . . Galen (1961); Harris, Latin translation . . . Galen (1919, 1921); MacKinney, Galen's *Opera omnia* (1950); Nutton, *Kühn and . . . Galen* (1976), pp. 15–16; Sudhoff, Liste von Galenschriften (1925); Thorndike, Translations . . . Galen (1942) and Translation . . . Galen (1946); Walzer, Djālīnūs (1962). Concerning later editions, see Sarton, *Galen* (1954a), pp. 26–27, 99–107.]

94. Galen, *Opera omnia* (1821–33). [Volumes 17 and 18 are in two parts each, so there are 20 volumes in 22. Volume 20 is the index, which is more detailed than the earlier one by Brasavola. The 1965 reprint of volume 20 includes a valuable Epilogum et notas bibliographicas by K. Schubring.]

95. Haller, *Bibliotheca botanica* (1771–72), vol. 1, p. 111.

96. [Galen, *Opera omnia* (1821–33), vol. 14, pp. 30–31. Galen's main work on materia medica is *De simplicium medicamentorum temperamentis ac facultatibus*, of which books 6–8 are relevant; *Opera omnia*, vol. 11, pp. 789–892, and vol. 12, pp. 1–158. Also of some botanical interest are his *De antidotis*, *De theriaca ad pamphilianum*, and *De theriaca ad pisonem*; all in *Opera omnia*, vol. 14. There is a recent edition of *De theriaca ad pisonem* in Latin with an Italian translation and introduction (1959). See also Berendes, *Pharmacie* (1891), vol. 2, pp. 56–76; Burgess, *Asters* (1902), pp. 88–91, 107–9, 163 *et passim*; Fabricius, *Galens Exzerpte* (1972); Israelson, *Materia medica . . . Galenos* (1894), pp. 21–149; Piccinini, Dioscoride (1920), pp. 106–7; Sarton, *Galen* (1954a), pp. 64–65; Stannard, Plant called moly (1962a), pp. 263–66, and Squill (1974c), p. 695. The recipe for theriac has been translated into English from *Opera omnia*, vol. 14, pp. 154–55, in Walsh, Galen's second sojourn in Italy (1930), pp. 491–94, and in Macht, History of opium (1915), p. 479.]

97. Meyer, *Geschichte der Botanik* (1854–57), vol. 2, p. 191. ["Galen states it is impossible to learn botany from books, and the doctor has to learn the plants by patiently hunting them, preferably under the direction of an experienced teacher"; Scarborough, *Roman medicine* (1969), p. 130.]

98. Galen, *Opera omnia* (1821–33), vol. 11, p. 796.

99. Adanson, *Familles des plantes* (1763), vol. 2, p. 572. [This genus, which includes the tomato, is indigenous to the Americas only, and was unknown in antiquity.]

100. *Ibid.*, p. 165.

Chapter 4: Sixteenth-Century German Fathers

1. Meyer alone, *Geschichte der Botanik* (1854–57), vols. 3 and 4, has treated the subject of botany in the Middle Ages with impartiality. [The situation has improved since Greene wrote this; see Appendix B.]

2. Emphatic examples of this kind of writing in the name of history are in Sprengel's *Historia rei herbariæ* (1807–8), vol. 1, particularly his chapter on "Monastic botany," pp. 222–28 [(1817–18), vol. 1, pp. 198–202], and on the "Latinobarbarous age," pp. 274–99 [(1817–18), vol. 1, pp. 228–48], wherein even concerning the botanical volume of Albertus Magnus he says, "Let him read it who has time to throw away"; though Meyer, only a half-century after Sprengel, and as much an antimonarchist as he, devoted 70 serious pages of his history to the merits of this same Albertus of the Middle Ages. [Sprengel (vol. 1, p. 280) quoted the statement from Gesner, but omitted it from his German ed. (1817–18), vol. 1, pp. 224–25. For references on Albertus Magnus, including Meyer, see Appendix B.]

3. Sprengel, *Historia rei herbariæ* (1807–8), vol. 1. [Of the four histories Greene includes in this list, only the one by Sachs can be fairly accused of neglecting the botanists before Brunfels, and Sachs should have a right to limit the scope of his book to the period 1530–1860. Greene omitted Jessen's *Botanik* (1864) from his list, evidently because he was unaware of it. Jessen discussed "Die Väter der deutschen Pflanzenkunde" on pp. 176–80.]

4. Meyer, *Geschichte der Botanik* (1854–57), vol. 4, p. 295.

5. Winckler, *Geschichte der Botanik* (1854), pp. 74–78.

6. Sachs, *Geschichte der Botanik* (1875), p. 16 [(1890), p. 15].

7. *Ibid.* (1875), p. 31 [(1890), p. 29].

8. Pliny, *Naturalis historia* (1606), bk. 25, ch. 2; see also Meyer, *Geschichte der Botanik* (1854–57), vol. 1, p. 250. The names of the ancient painters were Krateuas, Dionysius, and Metrodorus. [See Arber, *Herbals* (1938), ch. 7; Blunt, *Botanical illustration* (1950a), ch. 2; Nissen, *Botanische Buchillustration* (1966), vol. 1, pp. 15–19.]

9. A celebrated Greek manuscript of the *Materia medica* of Dioscorides, known as the *Codex Aniciæ Iulianæ* (1906), in which each plant is represented by a painting of natural size. The manuscript dates from the sixth century and was done at Constantinople. It has long been in the Imperial Library at Vienna. The whole has lately been reproduced photographically. The title page of the published work has the following: "Dioscurides. Codex Aniciæ Julianæ picturis illustratus nunc Vindobonensis Med. Gr. I. photographice editus. Moderante J. Karabacek. Prefati sunt A. de Premerstein, K. Wessely, J. Mantuani. Lugduni Batavorum, A. W. Sijthoff, 1906." [There is also a more recent facsimile edition, with color illustrations, Dioscorides (1965–70).]

10. For some accounts of these earliest specimens of printed books of popular medicine, chiefly botanical, the reader is referred to Pritzel's *Thesaurus* (1871–77), pp. 364–68; also Meyer, *Geschichte der Botanik* (1854–57), vol. 4, p. 189. [More recent studies include Abrahams & Savin, Herbarius latinus (1975); F. Anderson, *Illustrated history* (1977), chs. 11–14; Arber, *Herbals* (1938), ch. 2; Behling, Hausbuchmeister (1951); Burgess, *Asters* (1902), pp. 317–23; Choulant, *Graphische Incunabeln* (1858); Herm. Fischer, *Mittelalterliche Pflanzenkunde* (1929), pp. 79–104; Freeman, *Herbs . . . mediæval . . . cooking* (1943), which has illustrations from the 1485 and 1492 editions of *Hortus sanitatis*; R. Fuchs, Mainzer Frühdrucke (1960); Goff, Johann Petri (1973); Klebs, Incunabula scientifica et medica (1938) and Herbal facts (1925); Nissen, *Botanische Buchillustration* (1966), vol. 1, pp. 26–37, and *Herbals* (1958), pp. 56–60; Payne, On "Herbarius" and "Hortus sanitatis" (1901); Schreiber, *Kräuterbücher* (1924); Shaffer, *Garden of health* (1957).]

11. [Although "wretched" fairly describes the illustrations of some of the early herbals, such as those in the anonymous *Herbarium Apulei Platonici* (1481), the quality of Fig. 28, from the also anonymous *Gart der Gesundheit* (1485), shows

clearly that *Herbarum vivæ icones* (1530–36) was not the first printed herbal with adequate illustrations.]

12. [Brunfels and Fuchs probably deserve the credit for having the woodcuts made for their herbals and probably for supervising their production, but since they themselves did not produce the illustrations, it does not seem appropriate to style them "fathers of plant iconography." Brunfels's herbal might have been initiated by his friend Johann Schott, the publisher.]

13. See p. 232.

Chapter 5: Otto Brunfels

1. [Vol. 1 was printed in 1530–31 and reprinted in 1532, 1536, 1537, 1539. Vol. 2 was printed in 1531–32 and reprinted with additions in 1536, 1537, 1539. Vol. 3 was printed in 1536, and reprinted in 1537, 1539, 1540. The German version appeared under the name *Contrafayt Kreüterbuch* in 1532–37, with three later editions by 1546. Pagination is not the same for the different editions. Greene used his own copy of vols. 1 and 2 of the Latin version, both dated 1532, with the Appendix in vol. 2 numbered pp. 1–199. The editor has checked Greene's citations to the Appendix in Greene's own copy, but otherwise has used the edition at the Hunt Botanical Library, dated 1532–39, which is described in Quinby, *Catalogue of botanical books* (1958), no. 30, and which has the Appendix numbered continuously with the text. In his identification of the plants discussed by Brunfels, Sprague collated three Latin and three German editions; Brunfels (1928), pp. 91–113.

Illustrations from *Herbarum* were republished in 1543 in an anonymous work, *In Dioscoridis historiam herbarum certissima adaptatio*, without text but with plant names given in Greek, Latin, and German; Fulton, Notes and queries (1953). For other bibliographic information, see F. Anderson, *Illustrated history* (1977), ch. 16; Riddle, Dioscorides (1980), pp. 55–58; Roth, Brunfels (1900b), pp. 274–84; Stannard, Brunfels (1970d).]

2. Brunfels died of consumption at Berne, Switzerland, probably at the age of about 70. [Although Greene wrote this after publication of Roth's studies of Brunfels, he seems unaware of Roth's work. Brunfels was born about 1489, not in 1464 as Greene thought. The confusion has arisen because one of his relatives had the same name, and his birth was registered at Erfurt on Easter, 1465; Roth, Brunfels (1900a), p. 191. "Consumption" is an old term for what was usually pulmonary tuberculosis. Some recent studies indicate that Brunfels might not have died from this, but rather from diphtheria. Mayerhöfer *et al.*, *Lexikon* (1959–), vol. 1, p. 565; Miall, *Early naturalists* (1912), pp. 17–24; Stannard, Brunfels (1970d).]

3. Sprengel, *Historia rei herbariæ* (1807–8), vol. 1, p. 311 [(1817–18), vol. 1, pp. 258–59].

4. [On Brunfels's dates, see n. 2.]

5. [Mainz was the city in which Johann Gutenberg had lived and had printed the Bible in 1456.]

6. Brunsfels, rather than Brunfels, was the family name. In some of our author's earlier works he wrote it Brunsfelsius; but later he appears to have changed it to Brunfelsius.

7. Otto Brunfels was born in 1464, Martin Luther in 1483. [On Brunfels's birth date, see n. 2. Brunfels knew and corresponded with Luther and with Ulrich von Hutten, who assisted him in escaping from the monastery; Hartfelder, Brunfels (1893).]

8. [He received his master of arts degree in 1508 or 1509 from the University of Mainz. It seems likely that he entered the monastery because his father was unable to

support his education for a doctor of theology degree. The monastery was near Strasbourg, not Mainz. His first teacher there became its rector in 1515.]

9. Meyer, *Geschichte der Botanik* (1854–57), vol. 4, p. 296.

10. [As stated in n. 2, Greene's information was based on the confusion of Brunfels with an older relative having the same name.]

11. [Rather, he left Strasbourg late in July 1521.]

12. [Gesner, *Bibliotheca* (1545–55). Roth, Schriften . . . Brunfels (1900). On his theological writings, see Sanwald, *Brunfels* (1932).]

13. [For discussions of Weiditz and his botanical illustrations, see Arber, Draughtsman (1921) and Recent discovery (1936); Blunt, *Botanical illustration* (1950a), ch. 5; Church, Brunfels and Fuchs (1919); Nissen, *Botanische Buchillustration* (1966), vol. 1, pp. 40–44; Röttinger, *Weiditz* (1904); Sprague, Brunfels (1928), pp. 83–84. Walther Rytz discovered that some of Wieditz's original drawings had been preserved in the herbarium of Felix Platter (1536–1614); Rytz, Herbarium Felix Platters (1933) and *Pflanzenaquarelle . . . Weiditz* (1936).]

14. [If so, they apparently corresponded for several years before Brunfels's visit. This conclusion is based on Meyer's assertion, *Geschichte der Botanik* (1854–57), vol. 4, p. 300, that Brunfels's reference to "Hieronymus Herbarius" was a reference to Bock (Tragus). Since Brunfels makes the first such reference in vol. 1 of *Herbarum* (1532–39), it follows that he was in contact with Bock at least by 1530; the references are in vol. 1, pp. 55, 74, 81, 110, and vol. 3, pp. 15, 17. An essay by Bock on various plants appeared in vol. 2, pp. 272–81.]

15. [See Durling, *Catalogue of . . . books* (1967), nos. 721–32.]

16. [Plumier, *Nova plantarum americanarum* (1703), p. 12.]

17. Sachs, *Geschichte der Botanik* (1875), p. 4. ["As compared with the efforts of the philological commentators, who knew little of plants from their own observation, a great advance was made by the first German composers of herbals, who went straight to nature, described the wild plants growing around them and had figures of them carefully executed in wood"; (1890), p. 3.]

18. Brunfels, *Herbarum* (1532), Epistle dedicatory, to the Senate of Strasbourg, second page. [There was also a dedication of *Contrafayt Kreüterbuch* (1532–37) to Prince Ludwig, Elector of Palatine.]

19. [Brunfels, *Herbarum* (1532–39), vol. 1, p. 40. He gave a list of 47 sources, sig. A iii^{v}, but with few clues to which editions he used. Some of them can be identified as follows:

Theophrastus, *Enquiry into plants* (1916), bk. 9, ch. 13, sec. 1.

Dioscorides, *Materia medica* (1906–14), bk. 3, ch. 132; (1934), pp. 377–78; (1902), pp. 349–50.

Pliny, *Natural history* (1938–62), bk. 25, secs. 75, 76, 132; bk. 26, secs. 32, 44, 45, 57, 76, 81, 94, 135, 144, 147, 155, 163.

Apuleius Barbarus, *Herbarium* ([1481]), ch. 70, fol. 65^{r}. (Brunfels cited ch. 68, which is incorrect for ed. 1 but may apply to eds. 2 or 3, both of which appeared in 1528.)

Valla (1447?–1500?), *De simplicium natura* (1528). Simples are listed alphabetically; see "Nymphæa."

Ibn-Sīnā (Avicenna), *De viribus cordus*. This small treatise was often printed with *Canon medicinæ*, beginning in 1476. It is a good summary of some of his most important ideas, but it may have been written later by another Arabic physician; see Gruner, Interpretation of Avicenna (1921) and *Treatise on . . . Avicenna* (1930), p. 571. On ibn-Sīnā's writings of botanical relevance and further references, see Appendix B.

Ibn-Sarābī (Serapion the Younger), *Liber . . . medicinis simplicibus*, ch. 144. This work had been translated into Latin by Simon of Genoa and published in 1473. Brunfels edited the edition published in 1531.

Simon of Genoa, *Clavis sanationis* (1471–72). The simples are listed alphabetically. On Simon, see Part II, ch. 19, n. 24.

Al-Rāzī (Rhazes): Brunfels did not state which of his works he had in mind, but most likely "De simplicibus" in Brunfels, *In hoc volumine* (1531), pp. 373–97. For discussion and references, see Appendix B.

Giovanni de Vigo (1460?–1525?) wrote primarily on surgery and the healing of wounds; his mention of herbs must have been in this connection. For a list of his works, see Durling, *Catalogue of . . . books* (1967), nos. 4605–28.

Hieronymus Herbarius was probably a reference to Jerome Bock (Tragus), as Greene indicates on p. 304.

On Brunfels's use of his sources, see Roth, Brunfels (1900a), pp. 211–17; Sprague, Brunfels (1928), pp. 82–83.]

20. [Brunfels, *Herbarum* (1532–39), vol. 1, pp. 49–55. See also Dioscorides, *Materia medica* (1906–14), bk. 3, ch. 4; (1934), pp. 235–38; (1902), pp. 263–65. Pliny, *Natural history* (1938–62), bk. 25, secs. 95–98, 101, 119, 128, 134, 141, 143, 165, 166; bk. 26, secs. 32, 33, 41, 53, 75, 90, 111, 117, 126, 137, 142, 154; bk. 27, sec. 143; bk. 28, sec. 254.]

21. [The identification of both figures as *C. cava* (L.) Schweigg. & Koerte is suggested in Sprague, Brunfels (1928), p. 92.]

22. Fuchs, Epistle dedicatory, in *Stirpium* (1542). [For a list of Brunfels's plates incorrectly matched with descriptions, see Sprague, Brunfels (1928), p. 80. For Euricius Cordus's criticism of Brunfels, see p. 366.]

23. [Although they are isolated observations without discernible influence, al-Bīrunī's discussion of the number of parts of flowers provides the first suggestion that species might be identified by these numbers; the passage is quoted in Appendix B.]

24. Brunfels, *Herbarum* (1532), vol. 1, pp. 151–54. [*Urtica dioica* L., *Lamium album* L., and *L. maculatum* L., according to Sprague, Brunfels (1928), pp. 95–96. On the early use of these species, see Grieve, *Modern herbal* (1931), vol. 2, pp. 574–80.]

25. [See p. 185.]

26. [There is some support for this statement in Cicero, *De natura deorum* (1931), bk. 2.]

27. Brunfels, *Herbarum* (1532), vol. 1, p. 22. [Aristotle had defended the study of "ignoble" aspects of nature, such as fleas, in *De partibus animalium*, 645ª 1-25, in *Works* (1908-52).]

28. [Greene seems to be wrong about Brunfels's motive for including the wild species for which medicinal uses were unknown:

> By the irony of fate some of the more important additions of Brunfels to the iconography of the European flora—e.g., *Anemone pulsatilla*, *Cardamine pratensis*, and *Linaria vulgaris*—were included rather against his will, in order to keep the craftsmen employed, or to fill up space. The inclusion of *Isnardia palustris* was due to this species having been brought in to be drawn under the mistaken impression that it was Purslane (*Portulaca oleracea*). (Sprague, Brunfels (1928), p. 113.)]

29. Fuchs, Epistola nuncupatoria, in *Stirpium* (1542), p. viii.

30. Brunfels, *Herbarum* (1532), vol. 1, pp. 190–91. [These plants were called liverworts because their leaves have a shape vaguely similar to the mammalian liver, and according to the doctrine of signatures, this indicated that these plants were meant to cure liver diseases. For references on this doctrine, see ch. 1, n. 6.]

31. [Greene's analysis of generic names from Brunfels is extended in Bartlett, Generic concept (1940), pp. 354–56. On the other hand, Sprague observed: "To the

nomenclature of species it is evident Brunfels gave no thought"; Brunfels (1928), pp. 84–85.]

32. Brunfels, *Herbarum* (1532), vol. 1, pp. 119–23. [On the use of vervain (*Verbena officinalis* L.), see Grieve, *Modern herbal* (1931), vol. 2, pp. 831–32; Stannard, Magiferous plants (1977b).]

33. [Linnæus is considered the official author of these names, except that he named the last species *Potentilla tormentilla*, which is now accepted rather than *Tormentilla erecta*.]

34. [Brunfels, *Herbarum* (1532–39), vol. 1, pp. 77, 95, 80, 84, 85; vol. 3, pp. 6–8.]

35. [References to this literature are cited in Appendix B.]

36. [See n. 31.]

37. Brunfels, *Herbarum* (1532), vol. 1, pp. 151–57.

38. But this one was not figured, though by implication he included it, because he quoted from Ermolao Barbaro in Brunfels, *Herbarum* (1532), vol. 1, p. 154. [Barbaro is discussed in Part II, ch. 13.]

39. [See pp. 87–119.]

40. In modern botany the genus is written *Galeopsis*. Dioscorides wrote it *Galiopsis*, as did also Brunfels; Dioscorides, *Materia medica* (1543), bk. 4, ch. 80 [(1906–14), bk. 4, ch. 95; (1934), p. 491; (1902), pp. 421–22. Dioscorides, of course, wrote in Greek; some authors have transliterated the name with an "i" and others with an "e".]

41. Brunfels, *Herbarum* (1532), vol. 1, p. 155.

42. [Although Greene probably exaggerated Brunfels's concern for nomenclature, he was correct about Brunfels's discussing herbs with uneducated herb gatherers; Sprague, Brunfels (1928), p. 82.]

43. Brunfels, *Herbarum* (1532), vol. 1, pp. 103–10.

44. *Ibid*., pp. 181–82.

45. [*Ibid*., pp. 213–15. Linnæus is now considered the official author of these names, except that he gave *Ficaria* its currently accepted name *Ranunculus ficaria*. On the use of this species, see Grieve, *Modern herbal* (1931), vol. 1, pp. 179–82.]

46. Brunfels, *Herbarum* (1532), vol. 1, pp. 111–13. [*Echium vulgare* L., *Anchusa officinalis* L., and *Borago officinalis* L., according to Sprague, Brunfels (1928), p. 94. On the use of these species, see Grieve, *Modern herbal* (1931), vol. 1, pp. 18–19, 119–20, 142.]

47. Brunfels, *Herbarum* (1532), vol. 1, pp. 175–77. [*Cynoglossum officinale* L. and *Myosotis palustris* (L.) Lam., according to Sprague, Brunfels (1928), p. 96.]

48. [Brunfels, *Herbarum* (1532–39), vol. 1, pp. 99, 81, 107, 114, 120, 125, 215, 224.]

49. "Nomina generica ex duobus vocabulis integris, ac distinctis facta, e Republica Botanica releganda sunt"; Linnæus, *Philosophia botanica* (1751), sec. 221.

50. [But see n. 31; Burgess, *Asters* (1902), pp. 340–42.]

51. [Sprengel, *Linnæi Genera plantarum* (1830–31).]

52. Brunfels, *Herbarum* (1532), vol. 1, p. 218.

53. *Ibid*., vol. 2, p. 80. [Greene misread the page number; it should be p. 84. (The illustration in vol. 2, p. 80, is of *Mentha* sp.) Brunfels did refer the reader to vol. 1, p. 81, for a discussion of *Sanicula*. On the use of wood anemone, see Grieve, *Modern herbal* (1931), vol. 1, pp. 34–35.]

54. Brunfels, *Herbarum* (1532), vol. 2, p. 55.

55. The botanists of remote antiquity knew but one plant that they called *Trifolium*. It figures in modern nomenclature as *Psoralea bituminosa*. [Greene oversimplified the situation. See André, *Termes de botanique en latin* (1956), pp. 320–21.]

56. Brunfels, *Herbarum* (1532), vol. 2, p. 34.
57. [On the history of the Germanic names of plants, see the references in n. 63.]
58. Adanson, *Familles des plantes* (1763), vol. 2, p. 420.
59. Kuntze, *Revisio generum* (1891), vol. 1, p. 29. [This work is discussed on pp. 68, 76–78.]
60. Brunfels, *Herbarum* (1532), vol. 1, pp. 143–50. [*Ranunculus nemorosus* DC., *R. acris* L., and *R. bulbosus* L., according to Sprague, Brunfels (1928), p. 95.]
61. [Brunfels, *Herbarum* (1532–39), vol. 1, pp. 22–29. *P. media* L. and *P. lanceolata* L., according to Sprague, Brunfels (1928), p. 91. Brunfels pp. 24–25, also listed two other species (*Plantago aquatica* and *P. rubea*), which Sprague identified as *Alisma plantago-aquatica* L. and *P. major* L. There are many references in Pliny, *Natural history* (1938–62) to *Plantago*. Brunfels referred to those in bks. 25, 26.]
62. Brunfels, *Herbarum* (1532), vol. 1, p. 217. [On the use of this species, see Grieve, *Modern herbal* (1931), vol. 1, pp. 32–33.]
63. [A guide to medieval plant names in Latin and German is in Herm. Fischer, *Mittelalterliche Pflanzenkunde* (1929), pp. 257–318. For medieval German names, see also Björkman, Pflanzennamen (1902, 1905); Hegi, *Illustrierte Flora von Mittel-Europa* (1935–); Marzell, *Deutschen Pflanzennamen* (1943–58); Pritzel & Jessen, *Deutschen Volksnamen der Pflanzen* (1882). For medieval English plant names, see *Oxford English Dictionary*; Earle, *English plant names* (1880); Henslow, *Medical works* (1898).]

Chapter 6: Leonhard Fuchs

1. [Rather, he was born in the much smaller town of Wemding. On Fuchs's life, see Egerton, Fuchs (1978); Marzell, Fuchs (1938); Neumann, Fuchs (1917); Roth, Fuchs (1898b); Stübler, *Fuchs* (1928, which shows two sets of page numbers, those at the top corresponding to pagination in the *Münchener Beitr. Gesch. Lit. Naturwiss. Med.*, heft 13–14, and those, cited here, from the bottom of the pages).]
2. [His paternal grandfather, in whose house he lived, assisted in raising him.]
3. Adam, *Vitæ germanorum medicorum* [(1620), pp. 172–83; (1706), pp. 76–81].
4. [Not plague, but the English sweating sickness, which swept over Europe from 1485 to 1551. What it really was remains uncertain, but Roberts argues for influenza in English sweating sickness (1965).]
5. [Fuchs, *Compendiaria* (1531). Bibliographies are given in Roth, Fuchs (1898b), pp. 177–84; Stübler, *Fuchs* (1928), pp. 115–34. For a discussion of the importance of Fuchs's *Compendiaria*, see Wightman, *Science and Renaissance* (1962), vol. 1, pp. 213–21, and *Quid sit methodus?* (1964). Another of Fuchs's notable works was *Errata recentiorum medicorum* (1530), which is discussed in Samoggia, Leoniceniano . . . Fuchs (1964), pp. 18–39.]
6. [Hippocrates, *Epidemiorum* (1532).]
7. [*I.e.*, they fled with the Margrave to Culmbach.]
8. [Some of his correspondence during this period is discussed and quoted in Fichtner, Fuchs aus seinen Briefen (1968).]
9. Meyer, *Geschichte der Botanik* (1854–57), vol. 4, p. 311. [One reason Fuchs gave for refusing the offer of King Christian III was the difficulty of transporting his personal library to Denmark; another was salary. Stübler, *Fuchs* (1928), p. 8; Warburg, Salary . . . Fuchs (1970).]
10. [This "safe assumption" may be incorrect. Apparently, his grandfather had instilled in him a fondness for flowers at an early age.]
11. [Fuchs was elected rector of the university for 1536/37 and 1540/41, and he

wrote the statutes of the medical faculty that were issued in 1539. Among these statutes was the statement that medical students should collect and study plants from the country and the mountains during the summer. Tübingen Universität, *Geschichte der Universität* (1877), p. 312; in Latin, in Neumann, Fuchs (1917), p. 642.]

12. [Theophrastus, Dioscorides, and Galen are discussed in chs. 2, 3. Although there was a popular medieval treatise on materia medica attributed to Hippocrates, the Hippocratic corpus does not contain a treatise on the subject. Drugs were discussed throughout its treatises because they were recommended for various illnesses; Schöne, Hippokrates περι φαρμάκων (1920).]

13. [Harig, Fuchs (1966b). Fuchs also contributed to medical education by translating into Latin the following treatises by Galen: *De crisisbus*; *De curandi ratione per venæ sectionem*; *De differentiis febrium*; *De hirundinibus*; *De inæquali intemperie*; *De locis affectis*; *De morborum causis*; *De morborum differentiis*; *De symptomatum causis*; *De symptomatum differentiis*. See Durling, Renaissance editions . . . Galen (1961).]

14. "Leonardi Fuchsii annotationes aliquot herbarum et simplicium, a medicis hactenus non recte intellectorum," in Brunfels, *Herbarum* (1532), vol. 2, app., pp. 129–55 [(1532–39), vol. 2, pp. 245–71].

15. The portraits of these artists, with their names, Heinrich Füllmaurer, Albert Meyer, and Veit Rudolf Speckle, are appended to the first edition of Fuchs, *Stirpium* (1542). [For discussions of their illustrations, see Blunt, *Botanical illustration* (1950a), ch. 5; Church, Brunfels and Fuchs (1919); Eisendrath, Portraits of plants (1961) and Illustrations of Atropa (1963–64); Nissen, *Botanische Buchillustration* (1966), vol. 1, pp. 44–47; Sprague & Nelmes, Fuchs (1931), p. 550.]

16. [Fuchs, *Stirpium* (1542). The first German edition appeared in 1543 (facsimile eds.: 1938 and 1964). There were also translations into Dutch (1543), French (1550?), and Spanish (1557). Altogether, including editions without texts and the two facsimiles, there are 37 editions of his herbal. The illustrations also appeared in various later herbals. Arber, Fuchs (1928); Roth, Fuchs (1898b), pp. 177–84; Stübler, *Fuchs* (1928), pp. 128–33.]

17. [According to Sprague, Brunfels (1928), Brunfels provided a total of 285 plates, though all of these were not in the first edition; nor does each plate represent a different species. According to Sprague & Nelmes, Fuchs (1931), Fuchs published a total of 517 plates, though all of them were not in the first edition, and each plate did not always represent a different species.]

18. [In his preface, Epistola nuncupatoria, in *Stirpium* (1542), p. ix, Fuchs complained about the herbals' being published by the German printer and compiler Christian Egenolph (Egenolff) (1502–55). Ironically, Egenolph pirated Fuchs's plates when publishing some of his later herbals. To Egenolph's *Adversum . . . Fuchsii* ([1544]) Fuchs responded with *Adversus . . . Egenolphi* (1545). On Egenolph, see Grotefend, *Egenolff* (1881). Egenolph published Dorsten's *Botanicon* (1540). For lists of other herbals and medical books Egenolph published, see indexes in Durling, *Catalogue of . . . books* (1967), p. 650; Quinby, *Catalogue of botanical books* (1958), p. 477. For a discussion of the later use of Fuchs's plates, see Sprague & Nelmes, Fuchs (1931), p. 548.]

19. [Six volumes (1542–56, now bound together) that Fuchs owned during this period are now at the Royal College of Surgeons in London; Le Fanu, Volume . . . Fuchs (1956).]

20. Meyer, *Geschichte der Botanik* (1854–57), vol. 4, p. 313. [This account seems incorrect; the whole manuscript seems to have been completed before *Stirpium* (1542) was published, because *Stirpium* contains parts of all the volumes. The

manuscript, bound in nine folio volumes, has survived and is discussed in Ganzinger, Kräuterbuchmanuscript . . . Fuchs (1959). There is a microfilm copy of all nine volumes at the Hunt Botanical Library. The published illustrations are selected from several of each plant in the manuscript. Three of the illustrations that Fuchs did not publish are reproduced in Stübler, *Fuchs* (1928), between pp. 104 and 105. Ganzinger has shown that some of the drawings were made from specimens in Leonhard Rauwolf's herbarium; Rauwolf und Fuchs (1963).]

21. "Æquissimus majorum suorum judex"; Sprengel, *Historia rei herbariæ* (1807–8), vol. 1, p. 324.

22. [Fuchs issued complaints as well as praise, however. Those against Egenolph are mentioned in n. 18. Fuchs also accused Walther Hermann Ryff (Gualtherus Hermenius Rivius) of plagiarism of his works in 1542 and again in 1544, and Fuchs attempted to correct errors in Ryff's commentary on Dioscorides; Riddle, Dioscorides (1980), pp. 76–80; Stübler, *Fuchs* (1928), pp. 81–98; Thorndike, *Magic and experimental science* (1923–58), vol. 5, pp. 442–43. Nor was Fuchs the only contemporary to level such charges against Ryff; for Gesner's complaint, see Bock, *Stirpium* (1552), p. xii, and Gesner, *Bibliotheca* (1545–55), vol. 1, fols. 284ᵛ–285. On Ryff's life and work, see Benzing, *Ryff* (1959). Barbaro, Virgilio, and Ruel are discussed in Part II, chs. 13, 14, 16. Fuchs's use of his sources is summarized in Sprague & Nelmes, Fuchs (1931), pp. 554–55.]

23. [Fuchs, Epistola nuncupatoria, in *Stirpium* (1542), p. v. Similar complaints were made by Euricius Cordus and Antonio Musa Brasavola, whose works are discussed in ch. 8 and in Part II, ch. 17. Fuchs himself was not entirely free from such errors, which are listed in Sprague & Nelmes, Fuchs (1931), p. 556.]

24. [Plumier, *Nova plantarum americanarum* (1703), p. 14.]

25. [Fuchs, Explicatio quarundam vocum toto hoc opere passim occurrentium, in quibus assequendis non admodum peritus lector hærere posset, in *Stirpium* (1542), pp. xvi–xix. This glossary had, in fact, been discussed in Sachs, *Geschichte der Botanik* (1875), pp. 22–23; (1890), pp. 20–21. More recently, it has been translated and discussed in Choate, Earliest glossary (1917). However, its originality has been exaggerated. Four definitions from *Liber pandectarum medicinæ*, by Matthæus Sylvaticus (d. *ca.* 1342), are quoted in Sprague & Nelmes, Fuchs (1931), p. 553. (On Matthæus Sylvaticus, see Part II, Appendix C.) Sprague & Nelmes, *ibid.*, point out that many of the terms (but not the definitions) come from Pliny. There is also a glossary of six pages preceding the index in *Grete herball* (1526). See Larkey, Scientific glossaries (1937). After publishing this chapter on Fuchs, Greene concluded that Fuchs had "pillaged . . . his botanical vocabulary from Ruel"; Part II, ch. 16.]

26. [See pp. 147–48.]

27. Fuchs, *Stirpium* (1542), p. 366. [Sprague & Nelmes identified this species as *Lilium croceum* Chaix.; Fuchs (1931), p. 597. Marzell, *Fuchs* (1938), also provides modern identifications of the species discussed by Fuchs.]

28. Fuchs, *Stirpium* (1542), p. 115. [*Lilium martagon* L., according to Sprague & Nelmes, Fuchs (1931), p. 580. The confusion concerning the identity of the asphodel arose among the Romans. See Abbe, *Plants of Virgil* (1965), pp. 53–63; André, *Termes de botanique en latin* (1956), p. 44; LeClerc, Histoire de l'asphodèle (1929).]

29. [Fuchs, *Stirpium* (1542), pp. 356–57. Sprague & Nelmes identified both figures as *Colchicum autumnale* L.; Fuchs (1931), p. 596. On early uses of *Colchicum*, see Flückiger & Hanbury, *Pharmacographia* (1879), pp. 699–703; Grieve, *Modern herbal* (1931), vol. 2, p. 700; Hartung, History of . . . colchicum (1954); Rüegg, *Geschichte . . . Colchicum* (1936).]

30. [See pp. 150, 172.]

31. [See Columella, *De arboribus*, par. 3, sec. 4, in *On agriculture* (1941–55), vol. 3, p. 348; and Pliny, *Natural history* (1938–62), bk. 17, sec. 152; bk. 23, sec.

21; bk. 24, sec. 178. In connection with the following discussion, see André, *Termes de botanique en latin* (1956); Sprague, Botanical terms in Pliny (1933a), since Pliny used most of these terms.]

32. [See André, *Termes de botanique en latin* (1956), p. 21; Pliny, *Natural history* (1938–62), bk. 16, sec. 29; bk. 22, sec. 45; bk. 25, sec. 38.]

33. [André, *Termes de botanique en latin* (1956), p. 284; Columella, *On agriculture* (1941–55), bk. 9, ch. 4, par. 4; Pliny, *Natural history* (1938–62), bk. 18, sec. 95; bk. 20, sec. 198; bk. 21, sec. 26; Varro, *On agriculture* (1934), bk. 1, ch. 31, par. 5.]

34. ["A decussis was made by the ancients in the form of the letter X, which to the Latins meant ten. Hence to be decussate is to be arranged in the form of an X." Choate, Earliest glossary (1917), p. 192.]

35. ["A circle of leaves or flowers which crowns the stem or branch of an herb constitutes a verticillum, named from its resemblance to the woman's instrument which is wont to be fastened to the spindle to turn it"; *ibid.*, pp. 200–201.]

36. ["Crenae are a kind of incision on the extreme edge of leaves, giving them the name crenate, that is serrate, and cut on the edge"; *ibid.*, p. 192.]

37. ["Fimbriæ are the fringes of garments, and hence leaves are said to be fimbriate which end in a kind of fringe at the edge"; *ibid.*, p. 193. See also André, *Termes de botanique en latin* (1956), p. 139.]

38. ["Pediculus or petiolus is that part by which a leaf, flower, or fruit hangs"; Choate, Earliest glossary (1917), p. 197. See also André, *Termes de botanique en latin* (1956), p. 241; Columella, *On agriculture* (1941–55), bk. 12, ch. 15, sec. 3; Pliny, *Natural history* (1938–62), bk. 16, sec. 91, and bk. 22, secs. 96, 99.]

39. "Stipulæ sunt folia culmum ambientia"; Fuchs, *Stirpium* (1542), p. xix. ["Stipulæ are leaves surrounding the stem"; Choate, Earliest glossary (1917), p. 199. André, *Termes de botanique en latin* (1956), p. 304; Pliny, *Natural history* (1938–62), bk. 18, sec. 99; Varro, *On agriculture* (1934), bk. 1, ch. 49, sec. 1, 2.]

40. [See pp. 181–82.]

41. ["Frons is sometimes used for branch (ramus). Properly speaking, however, it is that which the branch (ramus) produces on all sides, having sometimes many leaves and some bark, like a little stem." Choate, Earliest glossary (1917), p. 193. As Greene mentioned on p. 277, Fuchs did not define the most basic terms like leaf (folium). On Theophrastus's differentiation of compound leaves, see p. 152.]

42. [Although "frons" is included in André, *Termes de botanique en latin* (1956), p. 141, he gives no indication of its having had much botanical usage in antiquity.]

43. [André, *Termes de botanique en latin* (1956), p. 316; Carnoy, *Noms grecs de plantes* (1959), p. 267; Pliny, *Natural history* (1938–62), bk. 19, sec. 146; bk. 24, sec. 150; bk. 25, sec. 81.]

44. Fuchs, *Stirpium* (1542), p. 840. [On early uses of hyssop, see Grieve, *Modern herbal* (1931), vol. 1, p. 426. See also Andrews, Hyssop (1961b); Ha-Reubeni, Biblical hyssop (1953).]

45. [André, *Termes de botanique en latin* (1956), p. 268; Pliny, *Natural history* (1938–62), bk. 14, sec. 41; bk. 15, sec. 115; bk. 16, secs. 112, 146, 151.]

46. Fuchs, *Stirpium* (1542), p. 772. [*I.e.*, *Sorghum vulgare* Pers. See André, *Termes de botanique en latin* (1956), p. 299; Cato, *On agriculture* (1934), secs. 51, 70; Choate, Earliest glossary (1917), p. 199; Pliny, *Natural history* (1938–62), bk. 12, sec. 42; bk. 21, sec. 47.]

47. Fuchs, *Stirpium* (1542), p. 824. [*I.e.*, *Zea mays* L. The first published description was in Bock, *Neu Kreüter Buch* (1539), pp. 21–22. On its history, see Finan, Maize . . . herbals (1948); Mangelsdorf, *Corn* (1974), p. 206; Simmonds, *Evolution of crop plants* (1976), pp. 128–36; Weatherwax, *Indian corn* (1954), ch. 6.]

48. Fuchs, *Stirpium* (1542), p. 777.

49. [André, *Termes de botanique en latin* (1956), p. 237; Pliny, *Natural history* (1938–62), bk. 16, secs. 49, 158; bk. 18, sec. 53; bk. 21, sec. 18; bk. 32, sec. 141.]

50. [André, *Termes de botanique en latin* (1956), p. 67; Pliny, *Natural history* (1938–62), bk. 15, secs. 86, 92; bk. 16, secs. 19, 20; bk. 20, sec. 198; bk. 21, secs. 14, 23; bk. 22, sec. 90; bk. 23, sec. 86; bk. 27, sec. 58.]

51. Fuchs, *Stirpium* (1542), p. 363.

52. [André, *Termes de botanique en latin* (1956), p. 335; Pliny, *Natural history* (1938–62), bk. 12, secs. 36, 70, and bk. 21, secs. 121, 122.]

53. "Stamina sunt qui in medio calycis erumpunt apices; sic dicta, quod veluti filamenta ab intimo floris sinu prosiliant"; Fuchs, *Stirpium* (1542), p. xix. ["Stamina are those knobs (apices) which spring up in the middle of the calyx; they are so called because they grow out like filaments from the lowest portion of the flower"; Choate, Earliest glossary (1917), p. 199.]

54. Fuchs, *Stirpium* (1542), p. 363.

55. [According to André, *Termes de botanique en latin* (1956), pp. 69, 139, these terms had rather different meanings in antiquity than "stamen."]

56. [Perhaps, but we do know most of the sources he used. See Sprague & Nelmes, Fuchs (1931), pp. 554–55.]

57. ["Nucamenta are those structures with compact scale-like coverings which hang from the branches of nut, oak and fir trees: so called because they seem attempts of nature to make pine nuts"; Choate, Earliest glossary (1917), p. 196. See also Pliny, *Natural history* (1938–62), bk. 16, sec. 49. On Theophrastus's discussion of nut-tree flowers, see pp. 156–60.]

58. [See pp. 165–67.]

59. [These four terms were commonly used in antiquity; André, *Termes de botanique en latin* (1956), pp. 69, 71, 293, 325.]

60. See p. 244.

61. See p. 186.

62. [On the contrary, Wightman has shown that Fuchs made important contributions to medical methodology; *Science and Renaissance* (1962), vol. 1, pp. 213–21, and *Quid sit methodus?* (1964). It seems that Greene established criteria too narrow to provide a fair assessment of Fuchs's capabilities.]

63. [Fuchs, *Stirpium* (1542), pp. 63, 40.]

64. [*Ibid.*, pp. 686–94. *S. nigrum* L., *P. alkekengi* L., *Cardiospermum halicacabum* L., *A. belladonna* L., and *D. metel* L., according to Sprague & Nelmes, Fuchs (1931), pp. 618–19.]

65. Fuchs, *Stirpium* (1542), p. 691. [On early uses of *Datura*, see Grieve, *Modern herbal* (1931), vol. 2, pp. 802–7.]

66. [Fuchs, *Stirpium* (1542), pp. 731–35. On the history of pepper, see Bois, *Plantes alimentaires* (1927–37), vol. 1, pp. 370–74, and vol. 3, pp. 68–75; Rosengarten, *Book of spices* (1969), pp. 128–47; Simmonds, *Evolution of crop plants* (1976), pp. 265–68; Xavier, Saffron and pepper (1944).]

67. Fuchs, *Stirpium* (1542), pp. 591–93. [*Sisymbrium officinale* Scop. and *Verbena officinalis* L., according to Sprague & Nelmes, Fuchs (1931), p. 612.]

68. [This name had already appeared in *Grete herball* (1526), ch. 356, though this may not have been the first time it had been published. Only one kind was mentioned.]

69. Fuchs, *Stirpium* (1542), pp. 604–7. [On early uses of these two species, see Grieve, *Modern herbal* (1931), vol. 1, pp. 175, 384–85.]

70. [*Artemisia absinthium* L. and *Sisymbrium sophia* L., according to Sprague & Nelmes, Fuchs (1931), p. 573. On the history of the uses of wormwood, see Coats, *Flowers* (1956), pp. 277–81; Fournier, *Plantes médicinales . . . France* (1947–48), vol. 1, pp. 115–120; Grieve, *Modern herbal* (1931), vol. 2, pp. 858–60.]

71. See p. 257.
72. Dioscorides, *Materia medica* (1543), bk. 3, ch. 131 [(1906–14), bk. 3, ch. 137; (1934), pp. 379–80; (1902), pp. 352–53].
73. See the fine plate of *Adonis* between two plates of anthemids in Mattioli, *Commentarii . . . Dioscoridis* (1565), pp. 904–6. [Mattioli is discussed in Part II, ch. 21.]
74. Fuchs, *Stirpium* (1542), p. 28.
75. Brunfels, *Herbarum* (1532), vol. 1, p. 84.
76. Fuchs, *Stirpium* (1542), pp. 105–9 on *Urtica*, three species; pp. 468–69 on *Lamium*, also three species.
77. [*Ibid.*, p. 628.]
78. *Ibid.*, pp. 892–94. [Since the vegetative parts of these two species are also similar, Greene has perhaps exaggerated Fuchs's concern for the flower. Fuchs's woodcuts were the first published illustrations of these two species. *D. purpurea* is the source of the heart drug, digitalis, which was first publicized by William Withering (1741–99), *Account of foxglove* (1785). On the history of digitalis, see also Bynum, Withering (1976); Grieve, *Modern herbal* (1931), vol. 1, pp. 322–26; Kreig, *Green medicine* (1964), ch. 7; LeClerc, Histoire de la digitale (1926); Roddis, *Withering* (1936).]
79. [For examples of Fuchs's accounts, see Bodenheimer, *History of biology* (1958), pp. 225–27, an abridged account of *Ruta graveolens* L., translated from *Stirpium* (1542), pp. 615–18; Burgess, *Asters* (1902), p. 352, for an account in Latin of *Aster amellus* L. from *Stirpium*, p. 133; Sachs, *Geschichte der Botanik* (1875), pp. 28–29, and (1890), pp. 26–27, for almost all of the account of *Convolvulus arvensis* L., quoted in Latin from *Stirpium*, p. 259.]
80. *Plantago* (of two species) is the topic of ch. 11, pp. 40–41; *Plantago aquatica* (monotypic) is of ch. 12, pp. 42–43. [For a more statistical list of Fuchs's plant names, see Sprague & Nelmes, Fuchs (1931), p. 560.]
81. [Fuchs, *Stirpium* (1542), pp. 83–86, 94–98. On the history of the wine grapes, see ch. 3, n. 20. On the history of the uses of bryony, see Grieve, *Modern herbal* (1931), vol. 1, pp. 131–33.]
82. [The "phrase names" developed as more and more species were discovered. In Fuchs's time, it was not yet a serious problem.]
83. Fuchs, *Stirpium* (1542), p. 892.
84. *Ibid.*, p. 895.
85. *Ibid.*, pp. 204–10.
86. [Greene means by cabalistic that these names had some mystical meaning. Perhaps so, but such names had been in use for so long that Fuchs was merely following a tradition. He might not have had the mystical ideas in mind. On an earlier use of *mas* and *fœmina*, see Sarton, *Introduction to history of science* (1927–48), vol. 2, p. 62, n. 45; Sprague, Brunfels (1928), p. 87.]

Chapter 7: Hieronymus Tragus (Jerome Bock)

1. Brunfels, *Herbarum* (1532), vol. 2, pp. 183–99 (in my copy).
2. [Plumier, *Nova plantarum americanarum* (1703), p. 14.]
3. [Greene was apparently unaware of Mayerhöfer, Bock (1896), and Roth, Bock (1898a) and Bock (1899). For more recent studies, see Hoppe, *Bock* (1969); Mayerhöfer *et al.*, *Lexikon* (1959–), vol. 1, p. 504; Stannard, Bock (1970c).]
4. Adam, *Vitæ germanorum medicorum* (1620), p. 68 [(1706), pp. 29–32].
5. Meyer, *Geschichte der Botanik* (1854–57), vol. 4, p. 303.
6. *Ibid.*
7. Adam, *Vitæ germanorum medicorum* (1620), p. 70 [(1706), p. 31. Greene

omitted from the above quotation the word "quondam," which comes after "animæ."]

8. [What Greene might have meant by this is not clear, since Bock was born before Fuchs and published his herbal before Fuchs. Perhaps Greene was thinking of the fact that both Brunfels and Fuchs had begun to publish books before Bock did.]

9. Both *Neu Kreüter Buch* (1539) and *Kreüter Buch* (1546) are in German; the former without illustrations. [There were 12 additional German editions and 2 Latin editions (Strasbourg, 1552 and 1553).]

10. Dioscorides, *Materia medica* (1543), bk. 2, ch. 156 [(1906–14), bk. 2, ch. 164; (1934), pp. 202–3; (1902), pp. 241–42].

11. Tragus, *Stirpium* (1552), bk. 2, ch. 134, p. 905. [The plants in Bock's *Stirpium* are identified by Hoppe, using Hegi, *Illustrierte Flora von Mittel-Europa* (1935–); Hoppe, *Bock* (1969), p. 330, identified the species on p. 905 as *Cyclamen europæum* L. On the history of its use, see Fournier, *Plantes médicinales* (1947–48), vol. 2, pp. 33–35.]

12. Dioscorides, *Materia medica* (1543), bk. 1, ch. 9 [(1906–14), bk. 1, ch. 10; (1934), pp. 12–13; (1902), pp. 32–33].

13. Tragus, *Stirpium* (1552), bk. 1, ch. 72, pp. 216, 217. [*Verbascum phlomoides* L., according to Hoppe, *Bock* (1969), pp. 155–56. On the history of the use of mullein, see Part II, ch. 24, n. 52.]

14. Tragus, *Stirpium* (1552), bk. 1, ch. 197, p. 572. [*Convallaria majalis* L., according to Hoppe, *Bock* (1969), p. 246. For other examples of Bock's descriptions quoted in English, see Burgess, *Asters* (1902), p. 343, for an account of *Aster*, which had been printed in Brunfels's *Herbarum* (1532–39), vol. 2, p. 273; Sachs, *Geschichte der Botanik* (1875), pp. 29–30, and (1890), pp. 27–28, for Bock's account of *Convolvulus arvensis* L., from *Stirpium*, bk. 2, ch. 89, p. 805. Greene later found that Manardo had described this species in 1532; see Part II, ch. 15. On the history of its use, see Grieve, *Modern herbal* (1931), vol. 2, pp. 480–82.]

15. Brunfels, *Herbarum* (1532), vol. 1, p. 212.

16. Tragus, *Stirpium* (1552), bk. 1, ch. 197, p. 573. [See Dioscorides, *Materia medica* (1906–14), bk. 3, ch. 122; (1934), pp. 367–68; (1902), pp. 344–45.]

17. Fuchs, *Stirpium* (1542), p. 397. [On Theophrastus, see pp. 159–60.]

18. Tragus, *Stirpium* (1552), bk. 3, ch. 67, p. 1095. [See Fuchs, *Stirpium* (1542), pp. 397–98. On Ruel, see Part II, ch. 16. On the history of the use of hazel, see Simmonds, *Evolution of crop plants* (1976), p. 305.]

19. Tragus, *Stirpium* (1552), bk. 3, ch. 63, p. 1085. [*Alnus glutinosa* G., according to Hoppe, *Bock* (1969), p. 370.]

20. Tragus, *Stirpium* (1552), bk. 3, ch. 75, p. 1114. [*Betula pendula* Roth., according to Hoppe, *Bock* (1969), p. 377.]

21. Tragus, *Stirpium* (1552), bk. 3, ch. 70, p. 1101.

22. *Ibid.*, bk. 3, ch. 60, p. 1078.

23. *Ibid.*, bk. 1, ch. 38, p. 119. [This species, corn poppy (*Papaver rhœas* L.), and also the opium poppy (*P. somniferum* L.) were known in antiquity. On the history of the use of the latter, see Andrews, Opium poppy (1952); Grieve, *Modern herbal* (1931), vol. 2, pp. 651–54; Kritikos & Papadaki, History of poppy (1967); Simmonds, *Evolution of crop plants* (1976), p. 318.]

24. [*I.e.*, this discovery could not be fitted into an existing paradigm, but discoveries are usually noticed when they have relevance for the kinds of questions that scientists of the time are asking. See Kuhn, *Structure of scientific revolutions* (1962), pp. 15–17.]

25. Tragus, *Stirpium* (1552), bk. 1, ch. 200, p. 582. [Although *Pæonia* is fairly closely related to the family Rosaceæ, it is more closely related to the Ranunculaceæ

(buttercups), in which some botanists place it. However, it is commonly considered as being in a monogeneric family, Pæoniaceæ. On the history of its use, see Coats, *Flowers* (1956); Grieve, *Modern herbal* (1931), vol. 2, pp. 606–7.]

26. [Hollyhocks are members of the Malvaceæ (mallows) and appear to be less closely related to roses than are peonies.]

27. Tragus, *Stirpium* (1552), bk. 1, chs. 194, 195, p. 568; bk. 2, ch. 132, p. 903.

28. *Ibid.*, bk. 3, ch. 25, p. 1002, under the name *Sambucus aquatica.* [On the history of the use of the guelder rose (*Viburnum opulus* L.), see Coats, *Garden shrubs* (1965), pp. 352–57; Grieve, *Modern herbal* (1931), vol. 1, pp. 381–82.]

29. Tragus, *Stirpium* (1552), bk. 1, ch. 32, p. 100. [On the history of the use of white mustard (*Sinapis alba* L.), see Grieve, *Modern herbal* (1931), vol. 2, pp. 566–67; Simmonds, *Evolution of crop plants* (1976), pp. 56–59.]

30. Tragus, *Stirpium* (1552), bk. 2, ch. 58, p. 718. [*Brassica oleracea* L. var. *capitata*, according to Hoppe, *Bock* (1969), p. 285. On the history of *B. oleracea*, see Simmonds, *Evolution of crop plants* (1976), pp. 49–52.]

31. Tragus, *Stirpium* (1552), bk. 2, ch. 61, p. 728. [On the history of the turnip (*Brassica campestris* L.), see Simmonds, *Evolution of crop plants* (1976), pp. 45–48.]

32. Tragus, *Stirpium* (1552), bk. 1, ch. 33, p. 106. [*Chelidonium maius* L., according to Hoppe, *Bock* (1969), p. 125.]

33. [Bock, *Stirpium* (1552), bk. 1, ch. 28, pp. 88–89. *Lepidium latifolium* L., according to Hoppe, *Bock* (1969), p. 119.]

34. [Bock, *Stirpium* (1552), bk. 1, ch. 71, pp. 214–16.]

35. Tragus, *Stirpium* (1552), bk. 2, ch. 87, p. 798. [*Glechoma hederaceum* L., according to Hoppe, *Bock* (1969), p. 306.]

36. Tragus, *Stirpium* (1552), bk. 2, ch. 95, p. 819. [*Bryonia dioica* Jacq., according to Hoppe, *Bock* (1969), p. 310. On the history of the uses of bryony, see Grieve, *Modern herbal* (1931), vol. 1, pp. 131–33.]

37. Tragus, *Stirpium* (1552), bk. 3, ch. 25, p. 1002.

38. *Ibid.*, bk. 1, ch. 37, p. 117.

39. Tournefort, *Élémens* (1694), vol. 1, p. 225.

40. Tragus, *Stirpium* (1552), bk. 1, ch. 41, p. 132. [On the history of the use of henbane (*Hyoscyamus niger* L.), see Grieve, *Modern herbal* (1931), vol. 1, pp. 397–404; Hocking, Henbane (1947).]

41. Tragus, *Stirpium* (1552), bk. 2, ch. 90, p. 724; bk. 2, ch. 147, p. 926.

42. *Ibid.*, bk. 2, ch. 125, p. 888.

43. *Ibid.*, bk. 3, ch. 13, p. 974.

44. [*Ibid.*, bk. 2, ch. 85, p. 794.]

45. [*Ibid.*, bk. 2, ch. 89, p. 805. Part of Bock's account of *Convolvulus arvensis* L. is quoted in Sachs, *Geschichte der Botanik* (1875), p. 29; (1890), pp. 27–28.]

46. [See n. 35.]

47. [Bock, *Stirpium* (1552), bk. 3, chs. 8, 9, pp. 960–65.]

48. [This term is discussed on p. 384 and in ch. 9, n. 60.]

49. Tragus, *Stirpium* (1552), bk. 2, ch. 85, p. 794. [*Lilium candidum* L., according to Hoppe, *Bock* (1969), p. 304. See also p. 286.]

50. Fuchs had said of gooseberry flowers only that they "are of a purplish green"; *Stirpium* (1542), p. 186.

51. Tragus, *Stirpium* (1552), bk. 3, ch. 14, p. 978. [*Ribes grossularia* L. (= *R. uva-crispa* L.), according to Hoppe, *Bock* (1969), p. 346. On the history of the use of gooseberry, see Grieve, *Modern herbal* (1931), vol. 1, pp. 364–65; Simmonds, *Evolution of crop plants* (1976), pp. 309–10.]

52. Tragus, *Stirpium* (1552), bk. 3, ch. 18, p. 984. [*Cratægus oxyacanthus* L.,

according to Hoppe, *Bock* (1969), p. 348. Unfortunately, Plantefol and Prévost, Notion d'étamine (1962), have taken notice of neither Bock's nor Greene's writings on this subject.]

53. Tragus, *Stirpium* (1552), bk. 3, ch. 33, p. 1016. [The blackthorn (*Prunus spinosa* L.) is thought to be one ancestor of the domestic plum. Simmonds, *Evolution of crop plants* (1976), pp. 244–45.]

54. The *Vaccinium* of recent botany.

55. Tragus, *Stirpium* (1552), bk. 3, ch. 13, p. 974.

56. [See pp. 358–59 on transmutations.]

57. Tragus, *Stirpium* (1552), bk. 3, ch. 80, pp. 1125–26. [There was little opposition to the concept of spontaneous generation until after 1650. For references on the early history of the concept, see Egerton, Guide to history of . . . ecology (1977), pp. 210–11.]

58. Tragus, *Stirpium* (1552), bk. 3, ch. 60, p. 1078.

59. [*Ibid.*, ch. 61, p. 1081.]

60. *Ibid.*, bk. 2, ch. 82, p. 784. [Orchid seeds are very small.]

61. *Ibid.*, bk. 1, ch. 186, p. 544. [On the history of the use of the royal fern (*Osmunda regalis* L.), see Grieve, *Modern herbal* (1931), vol. 1, p. 308.]

62. The twenty-fourth of June; otherwise called, at least in Old England, Midsummer Day.

63. Tragus, *Stirpium* (1552), bk. 1, ch. 186, p. 544.

64. *Ibid.*, bk. 2, ch. 25, p. 648. [On the history of buckwheat (*Fagopyrum esculentum* Moench), see Grieve, *Modern herbal* (1931), vol. 1, pp. 137–39; Simmonds, *Evolution of crop plants* (1976), pp. 235–37.]

65. [See pp. 168–69.]

66. [The separation of flowering plants into these two groups was made by John Ray; see Part II, Appendix C.]

67. Tragus, Præfatio, in *Stirpium* (1552), ch. 14.

68. Dioscorides, *Materia medica* (1543), bk. 4, ch. 155 [(1906–14), bk. 4, ch. 173; (1934), pp. 574–75; (1902), pp. 465–66].

69. Tragus, *Stirpium* (1552), bk. 2, ch. 86, p. 796; bk. 3, ch. 23, p. 996. [*Sambucus ebulus* L. and *S. nigra* L., according to Hoppe, *Bock* (1969), pp. 305, 351–52.]

70. [*I.e.*, there are exactly 1,200 numbered pages in Bock, *Stirpium* (1552). However, there are 67 unnumbered introductory pages and 63 unnumbered pages of indexes.]

71. [See ch. 3, n. 49; both Theophrastus and Dioscorides discussed about 550 kinds of plants, though not entirely the same kinds. Adanson estimated in *Familles des plantes* (1763), vol. 1, p. cxvi, that Pliny discussed about 800 kinds.]

72. [Bock, *Stirpium* (1552), bk. 1, ch. 1, pp. 1–7. *U. dioica* L., *U. urens* L., and *U. pilulifera* L., according to Hoppe, *Bock* (1969), p. 92.]

73. [Bock, *Stirpium* (1552), bk. 1, chs. 2–18, pp. 7–59. Greene is referring only to the illustrated species. There are actually 47 species mentioned on these pages (rather than the 36 Greene indicates), in 21 genera, according to Hoppe, *Bock* (1969), pp. 94–110.]

74. [Bock, *Stirpium* (1552), bk. 1, ch. 19, pp. 59–63. *V. phu* L., *V. officinalis* L., and *V. dioica* L., according to Hoppe, *Bock* (1969), pp. 111–12.]

75. See p. 256.

76. Tragus, *Stirpium* (1552), bk. 1, ch. 2, p. 10.

77. *Ibid.*, p. 8.

78. *Ibid.*, ch. 25, p. 79.

79. *Ibid.*, bk. 3, chs. 29–31, pp. 1008–11. [*Sorbus aucuparia* L., *S. torminalis*

Cr., and *S. domestica* L., according to Hoppe, *Bock* (1969), p. 354. The folklore of plants is prominent both in the text of Bock's herbal and also in some of the illustrations in it by David Kandel (as in Fig. 88). See Marzell, Buchsbaum-Bild (1954) (Buchsbaum = boxwood, that is, *Buxus*); Gérock, David Kandel (1923) and Illustrations de Kandel (1931); Nissen, *Botanische Buchillustration* (1966), vol. 1, pp. 51–52. Marzell has also written more generally on early folk beliefs concerning plants in Germany: *Pflanzen im Volksleben* (1925); *Unsere Heilpflanzen* (1930); *Volksbotanik* (1935); *Zauberpflanzen* ([1964]).]

80. Tragus, *Stirpium* (1552), bk. 1, chs. 159–62, pp. 474–83.

81. [*Alliaria officinalis* = *Sisymbrium alliaria* L.]

82. [Sachs, *Geschichte der Botanik* (1875), p. 15; (1890), p. 14.]

83. [See Andrews, Mints (1958a).]

84. [See Andrews, Hoary mustard (1942).]

85. [See Flückiger & Hanbury, *Pharmacographia* (1879), pp. 64–70; Grieve, *Modern herbal* (1931), vol. 2, pp. 566–70.]

86. [See Grieve, *Modern herbal* (1931), vol. 1, pp. 149–50, which makes similar remarks concerning *R. bulbosus* L.]

87. Tragus, *Stirpium* (1552), bk. 1, chs. 191–94, pp. 560–67. [*Cheiranthus cheiri* L., *Matthiola incana* R. Br., *M. annua* Sweet, and *Hesperis matronalis* L., according to Hoppe, *Bock* (1969), pp. 242–45.]

88. Tragus, *Stirpium* (1552), bk. 2, chs. 58–63, pp. 716–35.

89. The Latin for turnip being Rapum. [On the history of the turnip (*Brassica rapa* L. emend. Metzger [= *B. campestris* L.]), see Simmonds, *Evolution of crop plants* (1976), pp. 45–48.]

90. Tragus, *Stirpium* (1552), bk. 1, ch. 71, p. 214.

91. *Ibid.*, ch. 27, p. 82. [Evidently Greene meant to cite p. 83, where this species is illustrated; it is discussed on p. 85.]

92. Tragus, *Stirpium* (1552), bk. 2, ch. 28, p. 655. [*Camelina sativa* (L.) Cr., according to Hoppe, *Bock* (1969), p. 267.]

93. Tragus, *Stirpium* (1552), bk. 1, ch. 85, p. 255. [*Isatis tinctoria* L., according to Hoppe, *Bock* (1969), p. 165. On the history of the use of woad, see Grieve, *Modern herbal* (1931), vol. 2, pp. 852–53; Hurry, *Woad plant* (1930).]

94. Dioscorides, *Materia medica* (1543), bk. 2, ch. 180 [(1906–14), bk. 2, ch. 185; (1934), p. 230; (1902), pp. 258–59.]

95. Fuchs, *Stirpium* (1542), p. 330. [Fuchs did cite Dioscorides on that page, and therefore he may not have thought that he was plagiarizing. However, Greene is only one of many who accused Fuchs of it; see Part II, ch. 16, for his more serious accusation that Fuchs "pillaged . . . his botanical vocabulary from Ruel"; see also Egerton, Fuchs (1978).]

96. [See p. 292.]

97. Tragus, *Stirpium* (1552), bk. 1, ch. 32, p. 104.

98. [According to Sprague, Brunfels (1928), p. 115, Brunfels included 17 members of the family Cruciferæ in *Herbarum* (1532–39).]

99. See p. 290.

100. Tragus, *Stirpium* (1552), bk. 2, ch. 148, p. 928.

101. *Ibid.*, bk. 1, chs. 76–79, pp. 229–41. [*Cynoglossum officinale* L., *Anchusa officinalis* L., *A. azurea* Mill., *Lycopsis arvensis* L., *Borago officinalis* L., *Symphytum officinale* L., according to Hoppe, *Bock* (1969), pp. 159–61. On the history of the uses of alkanet, see Grieve, *Modern herbal* (1931), vol. 1, pp. 18–19.]

102. Tragus, *Stirpium* (1552), bk. 2, chs. 89–96, pp. 804–23. [*Convolvulus sepium* L., *C. arvensis* L., *Lysimachia nummularia* L., *Cuscuta epithymum* (L.) L.,

C. *europæa* L., *Humulus lupulus* L., *Smilax aspera* L., *Solanum dulcamara* L., *Clematis vitalba* L., *Bryonia dioica* Jacq., and *Lonicera periclymenum* L., according to Hoppe, *Bock* (1969), pp. 307–11. On the history of the use of honeysuckle (*Lonicera periclymenum* L.), see Coats, *Garden shrubs* (1965), pp. 200–207; Grieve, *Modern herbal* (1931), vol. 1, pp. 409–10.]

103. Tragus, *Stirpium* (1552), bk. 2, chs. 97–100, pp. 824–35. [*Lagenaria vulgaris* Ser., *Citrullus colocynthis* Schr., *Cucurbita pyxidaria* DC., *Cucumis melo* L., *C. sativus* L., *Citrullus vulgaris* Schr., *Ecballium elaterium* A. Rich., *Cucurbita pepo* L., according to Hoppe, *Bock* (1969), pp. 311–14. On the history of the cucurbits, see Eisendrath, Portraits of plants (1961); Grieve, *Modern herbal* (1931), vol. 1, pp. 49–50, and vol. 2, pp. 527–29; Simmonds, *Evolution of crop plants* (1976), pp. 64–69.]

104. Tragus, *Stirpium* (1552), bk. 3, chs. 47–50, pp. 1049–56.

105. [*Ibid.*, bk. 1, ch. 31, p. 99. See also Carnoy, *Noms grecs de plantes* (1959), p. 93. On the history of the use of buckhorn plantain (*Plantago coronopus* L.), see Grieve, *Modern herbal* (1931), vol. 2, p. 642.]

106. [Bock, *Stirpium* (1552), bk. 2, chs. 87, 88 pp. 798–804. Ground ivy is *Glechoma hederaceum* L. Greene further discussed Bock's grouping principles in Part II, ch. 20.]

107. [See pp. 183–86.]

108. Tragus, *Stirpium* (1552), bk. 2, ch. 87, p. 798.

109. *Ibid.*, chs. 101–4, pp. 836–44. [*Arctium lappa* L., *Xanthium strumarium* L., and *Trapa natans* L. are the first three species, according to Hoppe, *Bock* (1969), pp. 314–15. The next species in this series, which do not agree with Greene's identifications, are *Orlaya grandiflora* Hoffm. and *Circæa lutetiana* L., according to Hoppe, *Bock* (1969), p. 315. Bock's discussion of *Agrimonia* is in *Stirpium*, bk. 1, ch. 175, pp. 513–14.]

110. *Ibid.*, bk. 1, ch. 72, pp. 218–19. [The German name he gave it was Waldtkornblumen. It is *Centaurea montana* L., according to Hoppe, *Bock* (1969), p. 156. On the history of the use of knapweeds, see Grieve, *Modern herbal* (1931), vol. 2, pp. 456–57.]

111. Fuchs, *Stirpium* (1542), p. 43.

112. Tragus, *Stirpium* (1552), bk. 1, ch. 75, p. 227. [On the use of water plantain, see Grieve, *Modern herbal* (1931), vol. 2, pp. 645–46.]

113. Fuchs, *Stirpium* (1542), p. 892.

114. Tragus, *Stirpium* (1552), bk. 2, ch. 125, p. 889.

115. [With some justification, Greene already has listed the number of pages as 1,000 (see n. 70). However, only the first 1,127 pages are devoted to the discussion of individual species; the last 73 are devoted to various lists of plants. In this instance, Greene apparently arrived at his figure of 1,100 by rounding off 1,127.]

116. *Quinquefolium Tragi*; Tragus, *Stirpium* (1552), bk. 1, ch. 172, p. 507. [It is not established that Bock named this plant after himself. In the German edition he gave its name as Zweierlei Fünff finger kraut, and the other name appears only in Kyber's Latin translation. This species is now known as *Potentilla alba* L., according to Hoppe, *Bock* (1969), pp. 229–30.]

117. [Except for the second chapter, on mosses, which are not woody.]

118. [Again, except for mosses, which are neither saprophytes nor parasites; rather, they are autotrophs. On the early use of mosses and fungi, see Grieve, *Modern herbal* (1931), vol. 1, pp. 331–39; vol. 2, pp. 551–55. For a discussion of the accounts of fungi in herbals in the sixteenth century, see Ainsworth, *History of mycology* (1976), pp. 37–43. On the early use of mistletoe (*Viscum album* L.), see

Grieve, *Modern herbal* (1931), vol. 2, pp. 547–48; and Kanner, Mistletoe (1939).]

119. Tragus, *Stirpium* (1552), bk. 1, ch. 20, p. 65. [*Asarum europæum* L., according to Hoppe, *Bock* (1969), p. 112.]

120. Tragus, *Stirpium* (1552), bk. 1, ch. 27, p. 85. [*Alliaria officinalis* Andr., according to Hoppe, *Bock* (1969), p. 118.]

121. Tragus, *Stirpium* (1552), bk. 1, ch. 126, p. 384. [*Stellaria media* Vill., according to Hoppe, *Bock* (1969), p. 200.]

122. Tragus, *Stirpium* (1552), bk. 1, ch. 30, p. 93, under the name *Apium aquaticum.*

123. *Ibid.*, bk. 1, ch. 34, p. 111. [*Fumaria officinalis* L., according to Hoppe, *Bock* (1969), p. 126.]

124. Tragus, *Stirpium* (1552), bk. 1, ch. 128, p. 390.

125. *Ibid.*, ch. 42, p. 136. [*Aquilegia vulgaris* L., according to Hoppe, *Bock* (1969), p. 133.]

126. Tragus, *Stirpium* (1552), bk. 1, chs. 122–24, pp. 373–80. [*Sedum telephium* L., *S. album* L., *S. acre* L., *S. rupestre* L., *?S. villosum* L., *Portulaca oleracea* L., according to Hoppe, *Bock* (1969), pp. 196–99.]

127. Tragus, *Stirpium* (1552), bk. 1, ch. 61, p. 187. [On the use of brooklime (*Veronica beccabunga* L.), see Grieve, *Modern herbal* (1931), vol. 1, pp. 123–24.]

128. Tragus, *Stirpium* (1552), bk. 1, chs. 67, 68, pp. 203–9.

129. Dioscorides, *Materia medica* (1543), bk. 2, ch. 120 [(1906–14), bk. 2, ch. 127; (1934), p. 167; (1902), pp. 222–23].

130. The plant is *Sium angustifolium* L., type of the genus, though now called *Berula angustifolia.*

131. Tragus, *Stirpium* (1552), bk. 1, ch. 155, pp. 464–65. [*Apium graveolens* L. and *Sium erectum* Huds., according to Hoppe, *Bock* (1969), p. 219.]

132. Tragus, *Stirpium* (1552), bk. 2, ch. 35, p. 668. [Concerning Theophrastus's ideas on such transmutations, see pp. 205–9.]

133. Tragus, *Stirpium* (1552), bk. 1, ch. 32, pp. 101–2.

134. *Ibid.*, bk. 2, ch. 32, p. 662.

135. *Ibid.*, ch. 34, p. 666.

136. [It seems ironic that Greene should on the one hand defend the superstitions of the ancient rhizotomi (see ch. 1), and on the other hand condemn as childish Bock's belief in the transmutation of plants. The situation Bock describes seems to be one in which one kind of seed was planted and another kind of plant came up. This phenomenon may have a simple explanation (*e.g.*, blowing of foreign seeds or pollen), but his failure to discover it need not be considered childish.]

Chapter 8: Euricius Cordus

1. [On his life, see Dilg, Euricius Cordus (1971); Krause, *Euricius Cordus* (1863); *LGN*, vol. 1, pp. 684–85; Schmitz, Euricius Cordus (1971a).]

2. [On Hessus (1488–1540), see W. Grant, *Neo-Latin literature* (1965), pp. 164–69 *et passim*; Krause, *Hessus* (1879).]

3. "Præcepta bonæ valetudinis conservandæ" is the title of this treatise, according to Haller, *Bibliotheca botanica* (1771–72), vol. 1, p. 260. [Hessus's work was first published at Erfurt in 1524. Other editions appeared in 1533, 1539, 1550, 1551, and 1564. Durling, *Catalogue of . . . books* (1967), nos. 2283–87.]

4. [E. Cordus, *Elegie* ([1516]); *Epigramatū libri duo* (1517); *Bucolicum ludicrum* (1518); *Opera poetica* (1614). See Grant, *Neo-Latin literature* (1965), pp. 169–70, 276, 338, 358, 359, 380, 396 *et passim.*]

5. Winckler, *Geschichte der Botanik* (1854), p. 76.

6. [It seems likely that Nicolò Leoniceno (discussed in Part II, ch. 11) was an important stimulus for Cordus's critical attitude toward authorities.]

7. [These translations were reprinted in E. Cordus, *Opera poetica* (1614), pp. 68–131 (misnumbered as 66–129). On Nicander, see pp. 213–14.]

8. E. Cordus, *Botanologicon* (1534). [The index of plant names was reprinted in Dioscorides, *Medicinali materia* (1549).]

9. The *Botanologicon* abounds in expressions deploring the adverse influence of the religious dissensions of the time upon the universities of Germany.

10. [As has been indicated in this chapter and the three preceding, there was agreement among these four Protestant German scholars—Brunfels, Fuchs, Bock, and Euricius Cordus—that medical botany, like Christianity, needed a reformation. However, as Greene brings out later (Part II, ch. 17), this feeling was not limited to Germans. The Italian physician Antonio Musa Brasavola also wrote a critique of pharmacy in 1537 which, like Cordus's *Botanologicon*, was in the form of a dialogue. Before Cordus published his criticisms, the ebullient French physician Symphorien Champier (*ca.* 1471–1537) had attacked the errors of Arabic medical authors, while defending ancient Greek medical authors. See Thorndike, *Magic and experimental science* (1923–58), vol. 5, ch. 7. Even earlier, Roger Bacon (*ca.* 1214–92) had discussed the defects of pharmacy and medicine; Welborn, Doctors according to Friar Roger Bacon (1932).]

11. See p. 247.

12. E. Cordus, *Botanologicon* (1534), p. 96. [See Dioscorides, *Materia medica* (1906–14), bk. 3, ch. 4 (1934), p. 235; (1902), p. 263.]

13. E. Cordus, *Botanologicon* (1534), pp. 77–78.

14. [For other discussions of Cordus's *Botanologicon*, see Burgess, *Asters* (1902), p. 355; Roth, Euricius Cordus (1909a), Schulz, Euricius Cordus (1919). Schulz quotes extensively from the *Botanologicon* in Latin, and Greene quotes further from it in his English translation in Part II, ch. 17.]

15. [For a list of the plates incorrectly matched to descriptions, see Sprague, Brunfels (1928), p. 80.]

16. E. Cordus, *Botanologicon* (1534), pp. 26–27. [Ralla in the above discussion was E. Cordus's brother-in-law.]

Chapter 9: Valerius Cordus

1. Sprengel, *Historia rei herbariæ* (1807–8) vol. 1, p. 346 [(1817–18), vol. 1, p. 273].

2. Meyer, *Geschichte der Botanik* (1854–57), vol. 4, p. 317.

3. Sachs, *Geschichte der Botanik* (1875), p. 31 [(1890), p. 29. *I.e.*, unimportant for the history of classification. The fourth German historian to whom Greene referred must be Winckler, *Geschichte der Botanik* (1854), p. 77, where Valerius Cordus is mentioned only briefly. Greene was apparently unaware of Jessen, *Botanik* (1864), which also briefly discussed V. Cordus, pp. 177–78.]

4. Sachs, *Geschichte der Botanik* (1875), p. 579 [(1890), p. 536. On the early history of plant sensitivity, see ch. 2, n. 189.]

5. Walter Riffius, Preface to V. Cordus, *Annotationes*, in Dioscorides, *Medicinali materia* (1549); also to Cordus, *Annotationes*, in Gesner, *Collecta* (1561). [V. Cordus was born at Erfurt on 18 February 1515, according to Dann, Leben . . . Valerius Cordus (1968b), p. 3; Schmitz, Valerius Cordus (1971b).]

6. [He received his baccalaureate degree from the University of Marburg in 1531. There is no known record of his having received an M.D. degree. Dann, Leben . . .

Valerius Cordus (1968b), p. 5. See also Dann, Familienkreis Valerius Cordus (1967), Cordus-Bildnisse (1968a), Neuere Ergebnisse der Cordus (1968c); *LGN*, vol. 1, p. 685.]

7. Meyer, *Geschichte der Botanik* (1854–57), vol. 4, p. 317. [Dannenfeldt, Wittenberg botanists (1972).]

8. Meyer, *Geschichte der Botanik* (1854–57), vol. 4, p. 318.

9. Haller, *Bibliotheca botanica* (1771–72), vol. 1, p. 282. [Although Greene's citation of the date is the one given by Haller, it is now believed that the first edition of V. Cordus, *Pharmacorum omnium*, appeared posthumously in 1546.This date is supported by the official *Nurenberger Ratsbuche*, which has several contemporary entries concerning Valerius Cordus; these are quoted in H. Peters, *Aus pharmazeutische Vorzeit* (1910), vol. 1, pp. 193–98, and in Leake, Valerius Cordus (1924), p. 20. See also Ferguson, *Bibliotheca chemica* (1906), vol. 1, pp. 177–78; A. Lutz, Nürnberger Dispensatorium des Valerius Cordus (1949); H. Peters, Ancient pharmacy (1889), pp. 113–20; Schmitz, Bibliographie . . . Valerii Cordi (1958).]

10. Meyer, *Geschichte der Botanik* (1854–57), vol. 4, p. 318. [Dioscorides, *Medicinali materia* (1549). For an example of V. Cordus's *Annotationes* from this edition of Dioscorides, see Burgess, *Asters* (1902), p. 356 (quoted in Latin). A good account of the writing and editing of Cordus's *Annotationes* is in Riddle, Dioscorides (1980), pp. 67–72.]

11. [Leake, Cordus (1924), p. 23, credited him with having written the first clear account of the synthesis of ether, from sulfuric acid and alcohol. The account was first published posthumously in Gesner, *Collecta* (1561). For further discussion, quotes from the Latin text, and an English translation, see Tallmadge, *Extractione* of Cordus (1925). Robinson provides another translation, with a close study of Cordus's procedure and results; Sweet oil of vitriol (1959). An English translation apparently unknown to Tallmadge and Robinson is by George Baker: Gesner, *Newe jewell* (1576), bk. 3, ch. 39, fols. 196^{v}–197^{r}. See also Partington, *History of chemistry* (1961–62), vol. 2, pp. 82, 164–67. Priority claims for Paracelsus are made in Sudhoff, Valerius Cordus, der Äther und Theophrast von Hohenheim (1929).]

12. Haller, *Bibliotheca botanica* (1771–72), vol. 1, p. 281. [Of the more than 500 species described or illustrated in Cordus's *Historia plantarum*, in Gesner, *Collecta* (1561), about 66 are new, according to Sprague & Sprague, Cordus (1939), p. 1.]

13. [Sprengel, *Historia rei herbariæ* (1807–8), vol. 1, pp. 347–49; (1817–18), vol. 1, pp. 273–74. See also Sprague & Sprague, Cordus (1939), pp. 11–18.]

14. [According to Sprague, Brunfels (1928), pp. 114–17, Brunfels illustrated 47 species that had not been previously recorded. According to Sprague & Nelmes, Fuchs (1931), p. 565, Fuchs first illustrated 103 species, though some of them had been described earlier by Bock.]

15. [See pp. 364–65.]

16. Riffius, Preface to V. Cordus, *Annotationes*, in Gesner, *Collecta* (1561).

17. [According to Sprague & Sprague, Cordus (1939), p. 101, there were 477 plants described in the first four books; in a few cases there was more than one description of a species.]

18. [Gesner, the editor, published a bibliographic nightmare. Although V. Cordus's herbal occupies most of the volume (fols. 86–212^{r}), there are 26 parts, which are listed in Sprague & Sprague, Cordus (1939), pp. 5–7. The bibliographic awkwardness of this volume arises mainly from two facts: the titles given internally to the 26 parts often differ from their listing on the volume's title page, and the volume title is very long, with no satisfactory short title. Important for the present chapter on V. Cordus is the fact that his herbal (books 1–4) is entitled *Historiæ stirpium* on

the volume's title page, but on fol. 86ʳ the title is *Historiæ plantarum*. Greene consistently used *Historia plantarum*, and his example is followed here and in Sprague & Sprague. (Other authors, such as Schmitz, Valerius Cordus (1971), cite the former title.) The whole volume is cited here as Gesner, *Collecta* (1561).]

19. [According to Sprague & Sprague, Cordus (1931), p. 21, the species are *Drosera rotundifolia* L. (but the illustration supplied by Gesner is *D. longifolia* L.), *Gratiola officinalis* L., *Sagittaria sagittifolia* L., and *Polygonum bistorta* L. On the use of bistort, see Grieve, *Modern herbal* (1931), vol. 1, pp. 105–7.]

20. [As indicated in the legend of Fig. 99, Bock had published illustrations of *Drosera rotundifolia* L. in his *Stirpium* (1552), bk. 1, ch. 181, and bk. 3, ch. 2, though confusing it with moss. These figures were published after Cordus wrote, but before his work was published.]

21. [Brunfels, *Herbarum* (1532–39), vol. 1, pp. 61–62. See also Dioscorides, *Materia medica* (1906–14), bk. 2, ch. 166; (1934), pp. 206–8; (1902), pp. 244–45. The name "bistorta" was used during the Middle Ages, according to Herm. Fischer, *Mittelalterliche Pflanzenkunde* (1929), p. 279.]

22. [On Ghini, see Part II, ch. 18. On his German and Italian journeys Cordus was accompanied by his former pupil, the French naturalist Pierre Belon, on whom see Part II, ch. 18, n. 45.]

23. Gesner, *Opera botanica* (1751–71), vol. 1, p. 20. [Tallmadge translated the epitaph as follows (in Leake, Cordus (1924), p. 18):

> Valerius Cordus of Siemershausen, Hessia, the son of Euricius, unequalled in virtue, brilliance, and fidelity, worthy of the admiration of all teachers, who while still a youth, explained to men the working of Nature and the powers of plants. When he could not satiate his desire for knowledge in cultured Germany, he came to Italy. He was received with honour by the Venetians; he had hardly entered Rome when he was suddenly stricken, amidst the tears of his friends;—an irreparable loss to Science, his life was extinguished, in its fairest season, on the twenty-fifth day of September in the fifteen hundred and forty-fourth year of salvation. Cordus lives still in his genius, his mind itself is received into Heaven: mighty Rome guards what is left to the earth.]

24. [V. Cordus, *Stirpium . . . Italia* (1563).]

25. [See Bessler, Valerius Cordus (1952); Schulz, Valerius Cordus (1916).]

26. Gesner, letter to Hieronymus Herold, in Gesner, *Collecta* (1561), fol. 85.

27. Tournefort, *Institutiones* (1700), vol. 1, p. 26.

28. Haller, *Bibliotheca botanica* (1771–72), vol. 1, p. 282. [Though agreeing with Greene and the earlier historians cited here on the importance of Cordus's work, Sprague & Sprague nevertheless concluded (Cordus (1939), pp. 2–3) that Greene "gives the impression, gained from a superficial examination of the 'Historia,' that Cordus had paid considerable attention to morphology. Actually, Cordus was a keen observer, who described the parts of plants in empirical language, paying more attention to their function than to their morphological nature."]

29. [Gesner correctly matched about 220 figures with Cordus's descriptions and incorrectly matched about 40 figures and descriptions, according to Sprague & Sprague, Cordus (1939), p. 9. The species involved are indicated in the systematic list, pp. 21–101. Gesner got 192 of the illustrations from Bock, *Kreüter Buch* (1546), which themselves were largely copied from Fuchs, *Stirpium* (1542). Sprague & Sprague indicate these by the abbreviation "Tr." followed by the page number and an exclamation point (*e.g.*, Tr. 93!).]

30. Tournefort, *Institutiones* (1700), vol. 1, p. 26.

31. [Herbaria may have been formed before the sixteenth century, but the extant evidence indicates that the modern tradition of making herbaria began with Luca Ghini, who is discussed in Part II, ch. 18. On the early history of herbaria, see Arber,

Herbals (1938), pp. 138–43; Camus, Premiers herbiers (1895); Meyer, *Geschichte der Botanik* (1854–57), vol. 4, pp. 266–73, and On . . . herbaria (1863), a translation of his discussion without the footnotes; Rytz, Herbarium Felix Platters (1933); Saint-Lager, *Histoire des herbiers* (1885); Schuster, Herbarium (1919).]

32. [See p. 168.]

33. [Dioscorides, *Materia medica* (1906–14), bk. 2, ch. 166; (1934), p. 208; (1902), p. 245.]

34. Cordus's Latin word here is *involucrum.*

35. [For other accounts by Cordus in English translation, see Bodenheimer, *History of biology* (1958), pp. 224–25, on the chaste tree (*Vitex agnus-castus* L.), from V. Cordus, *Stirpium . . . Italia* (1563), fol. 4^{v}; Stearn, *Botanical Latin* (1966), pp. 26, 28, on *Butomus umbellatus* L., from V. Cordus, *Historia plantarum*, in Gesner, *Collecta* (1561), fol. 124.]

36. V. Cordus, *Historia plantarum*, in Gesner, *Collecta* (1561), fol. 95. [On the ancient use of the term "cauliculus," see André, *Termes de botanique en latin* (1956), p. 77; Sprague & Sprague, Cordus (1939), p. 3.]

37. [V. Cordus, *Historia plantarum*, in Gesner, *Collecta* (1561), fols. 132^{v}–134^{r}, 203^{r}.]

38. V. Cordus, *Historia plantarum*, in Gesner, *Collecta* (1561), fol. 143^{r}. [*Sorghum bicolor* (L.) Moench, according to Sprague & Sprague, Cordus (1939), p. 52. On the history of sorghum, see ch. 6, n. 46, and Simmonds, *Evolution of crop plants* (1976), pp. 112–17.]

39. V. Cordus, *Historia plantarum*, in Gesner, *Collecta* (1561), fol. 112^{r}.

40. *Ibid.*, fol. 103^{v}. [*I.e.*, ch. 52; this folio is incorrectly numbered 104. This term as used by Fuchs is discussed on p. 287.]

41. *Ibid.*, fol. 89^{r}. [This was the first published account of *Lathræa squamaria* L., according to Sprague & Sprague, Cordus (1931), p. 22.]

42. V. Cordus, *Stirpium . . . Italia* (1563, 2nd ed.), p. 5. [*Cytinus hypocistis* L., according to Sprague & Sprague, Cordus (1939), p. 101.]

43. Meyer, *Geschichte der Botanik* (1854–57), vol. 1, p. 166. [On Theophrastus, see p. 165. Heinrich Friedrich Link (1767–1851) was a professor of botany in Berlin, and Robert Brown (1773–1858) a prominent British botanist. Snelders, Link (1973); Stearn, Brown (1970).]

44. V. Cordus, *Historia plantarum*, in Gesner, *Collecta* (1561), fol. 87^{r}.

45. *Ibid.*, fol. 139^{r}. [*Achillea millefolium* L., according to Sprague & Sprague, Cordus (1939), p. 50. On the use of yarrow, see Grieve, *Modern herbal* (1931), vol. 2, pp. 863–65.]

46. [On the history of the understanding of the Compositæ, see Burgess, *Asters* (1902); Jas. Small, *Compositæ* (1919).]

47. V. Cordus, *Historia plantarum*, in Gesner, *Collecta* (1561), fol. 90^{r}. [*Ammi visnaga* (L.) Lam., according to Sprague and Sprague, Cordus (1939), p. 22.]

48. Tournefort, *Institutiones* (1700), vol. 1, p. 158.

49. V. Cordus, *Historia plantarum*, in Gesner, *Collecta* (1561), fol. 102^{v}. [*Arum maculatum* L., according to Sprague & Sprague, Cordus (1939), p. 29.]

50. [See p. 284.]

51. Tragus, *Stirpium* (1552), bk. 1, ch. 41, p. 132; Fuchs, *Stirpium* (1542), p. 832.

52. Tragus, *Stirpium* (1552), bk. 2, ch. 118, p. 896.

53. Fuchs, *Stirpium* (1542), p. 881. [This page number is incorrect (it refers to *Carlina acaulis* L., which is in the family Compositæ, not the Solanaceæ). *Physalis alkekengi* L. is described in *Stirpium* on p. 691 and illustrated on p. 687. The identifications are from Sprague & Nelmes, Fuchs (1931), pp. 618, 632.]

54. Tragus, *Stirpium* (1552), bk. 1, ch. 101, p. 304.
55. [See André, *Termes de botanique en latin* (1956), p. 67.]
56. V. Cordus, *Historia plantarum*, in Gesner, *Collecta* (1561), fol. 99^{r}.
57. *Ibid.*, fol. 153^{r}.
58. [On the history of the understanding of the Compositæ, see Burgess, *Asters* (1902); Jas. Small, *Compositæ* (1919).]
59. Fuchs, *Stirpium* (1542), p. 628, under *Pisum*.
60. Gesner, *Historia plantarum* (1541), p. 192. [Later Greene discovered that the term had been used in Ruel, *Stirpium* (1536), bk. 2, ch. 79. See Part II, chs. 16, 20.]
61. "Flores parvi . . . papilionum figura." V. Cordus, *Historia plantarum*, in Gesner, *Collecta* (1561), fol. 99^{v}.

"Flores . . . figura papilionibus similes." *Ibid.*, fol. 100^{r}.

"Flores . . . papilionum figura." *Ibid.*, fol. 101^{r}.

"Flores papilionibus similes." *Ibid.*, fol. 127^{v}.

"Flores . . . figura papilionum." *Ibid.*, fol. 137^{r}.

"Pediculus . . . parvis oblongis papilionaceis floribus circumdatus." *Ibid.*, fol. 164^{v}.

"Flores forma papilionacei et oblongi." *Ibid.*, fol. 166^{v}.

"Flores . . . producit . . . papilionaceos, quales in omnibus leguminibus est videre (!)." *Ibid.*, fol. 187^{v}.

The reader must not fail to note that, when once the term *papilionacea* has presented itself to Cordus's mind he thenceforward employs it constantly.

62. *Ibid.*, fol. 133.
63. *Ibid.*, fol. 90^{r}. [*D. metel* L., according to Sprague & Sprague, Cordus (1939), p. 22. See Fig. 60, which is a copy from Fuchs, *Stirpium* (1542), p. 690. The illustration had also been used in Bock, *Stirpium* (1552), bk. 2, ch. 128, p. 896.]
64. [*I.e.*, Fabio Colonna, who is discussed in Part II, ch. 23.]
65. V. Cordus, *Historia plantarum*, in Gesner, *Collecta* (1561), fol. 93^{r}. [*Tussilago farfara* L., according to Sprague & Sprague, Cordus (1939), p. 23. On the use of coltsfoot, see Grieve, *Modern herbal* (1931), vol. 1, pp. 212–14.]
66. V. Cordus, *Historia plantarum*, in Gesner, *Collecta* (1561), fol. 121. [*Anemone pulsatilla* L., according to Sprague & Sprague, Cordus (1939), p. 39. The figure is based on one in Brunfels, *Herbarum* (1532–39), vol. 1, p. 217, though Gesner took it from the copy in Bock, *Stirpium* (1552), bk. 1, ch. 137, p. 413.]
67. V. Cordus, *Historia plantarum*, in Gesner, *Collecta* (1561), fol. 97^{v}. [*Thalictrum flavum* L., according to Sprague & Sprague, Cordus (1939), p. 25.]
68. V. Cordus, *Historia plantarum*, in Gesner, *Collecta* (1561), fol. 90^{v}. [*Solanum melongena* L., according to Sprague & Sprague, Cordus (1939), p. 22.]
69. V. Cordus, *Historia plantarum*, in Gesner, *Collecta* (1561), fol. 139^{v}. Also in the flowers of *Paris*, fol. 152^{r}, he calls the stigmas, not the anthers, apices. [*Lilium martagon* L. and *Paris quadrifolia* L., according to Sprague & Sprague, Cordus (1939), pp. 51, 60.]
70. V. Cordus, *Historia plantarum*, in Gesner, *Collecta* (1561), fol. 152^{r}. [See n. 69. On its use, see Grieve, *Modern herbal* (1931), vol. 2, pp. 610–11.]
71. [On an earlier use of this term, see p. 287.]
72. V. Cordus, *Historia plantarum*, in Gesner, *Collecta* (1561), fol. 135^{v}. [*Pæonia officinalis* L. ex Willd., according to Sprague & Sprague, Cordus (1939), p. 49.]
73. [See pp. 156–60.]
74. Tragus, *Stirpium* (1552), bk. 3, ch. 19, p. 988.
75. [Greene bases this judgment on more or less implicit advances. Although such advances can be important in the progress of science, it is necessary to show their influence rather than merely assume it.]

76. V. Cordus, *Historia plantarum*, in Gesner, *Collecta* (1561), fol. 184r. In my copy of Cordus, and presumably in the whole edition, a word has been omitted by the printer. As printed, the first line about fig flowers reads: "Flores, ut omnes tradiderunt, fert, sed statim parva fructuum rudimenta." This is wholly unintelligible, as a sentence, until you supply the negative "non" before "fert," so that the reading shall be, "non fert." Cordus's manuscript was not printed until 17 years after his death. [Greene omitted "hactenus" before "tradiderunt."]

77. [*I.e.*, in Linnæus, *Systema naturæ* (1735). However, Linnæus, *Species plantarum* (1753), vol. 2, p. 1059, listed *Ficus* under "Classis xxiii. Polygamia. Polyœcia."]

78. [See p. 165.]

79. [See Fuchs, *Stirpium* (1542), pp. 688, 691; Bock, *Stirpium* (1552), bk. 2, ch. 129.]

80. V. Cordus, *Historia plantarum*, in Gesner, *Collecta* (1561), fol. 89r. [*Cardiospermum halicacabum* L., according to Sprague & Sprague, Cordus (1939), p. 22. On the subsequent history of the understanding of the parts of the fruit, see Lorch, Naked seed controversy (1959) and Carpel (1963).]

81. See p. 326.

82. Tragus states the following on the reproduction of junipers: "Maio mense tenuissimus ac luteus pulvis è juniperis in auras avolare conspicitur, quod semen illius esse animadverti. Post hunc quem diximus pulverem baccæ prorumpunt exiguæ, virides, quæ altero demum anno autumno appetente, quod illi tempur maturitatis est, cœruleo tinguntur colore, etc., etc. È nucleis lapidosis, qui in hisce baccis continentur, novæ Juniperi plantulæ fruticant." Bock, *Stirpium* (1552), p. 1073. If in this passage "semen" is used botanically, the meaning can be no other than that junipers may grow from pollen. If, what is improbable, he employs it in a zoological sense, as meaning that the dust which he says sails away upon the air is needful to the fertilization of the seed within the juniper berry, then he is the first to proclaim the modern doctrine of bisexuality in plants. He who knows Tragus's belief in the reproductibility of even seed plants apart from any kind of seed or germ will not very readily accept this latter interpretation of his language.

83. V. Cordus, *Historia plantarum*, in Gesner, *Collecta* (1561), fol. 113r. [*Phyllitis solopendrium* (L.) Newman, according to Sprague & Sprague, Cordus (1939), p. 34. On the use of hart's-tongue fern, see Grieve, *Modern herbal* (1931), vol. 1, p. 304.]

84. V. Cordus, *Historia plantarum*, in Gesner, *Collecta* (1561), fol. 169v. [*Dryopteris filix-mas* (L.) Schott, according to Sprague & Sprague, Cordus (1939), p. 71. They suggest that he had, however, confused other ferns with it when describing the range of habitats. On the use of male fern, see Grieve, *Modern herbal* (1931), vol. 1, pp. 300–302.]

85. V. Cordus, *Historia plantarum*, in Gesner, *Collecta* (1561), fol. 170r. [*Asplenium trichomanes* L., according to Sprague & Sprague, Cordus (1939), p. 71.]

86. The individual spores of ferns were first seen some 70 years after Cordus's demise.

87. [Greene is probably overly optimistic here. An understanding of the prothallus was not achieved until early in the nineteenth century. See Goebel, *Hofmeister* (1926), pp. 47–48.]

88. V. Cordus, *Historia plantarum*, in Gesner, *Collecta* (1561), fol. 137r. [*Lupinus albus* L., according to Sprague & Sprague, Cordus (1939), p. 50. On the use of lupine, see Grieve, *Modern herbal* (1931), vol. 2, pp. 502–3; Simmonds, *Evolution of crop plants* (1976), pp. 312–13.]

89. V. Cordus, *Historia plantarum*, in Gesner, *Collecta* (1561), fol. 166v. [*Vicia*

faba L., according to Sprague & Sprague, Cordus (1939), p. 69. Cordus described three cultivated forms. For a modern discussion, see Simmonds, *Evolution of crop plants* (1976), pp. 179–82.]

90. V. Cordus, *Historia plantarum*, in Gesner, *Collecta* (1561), fol. 99ᵛ. [Chickpea (*Cicer arietinum* L.), according to Sprague & Sprague, Cordus (1939), p. 27. On the history of this species, see Simmonds, *Evolution of crop plants* (1976), pp. 157–59.]

91. V. Cordus, *Historia plantarum*, in Gesner, *Collecta* (1561), fol. 127ᵛ. [Cowpea (*Vigna unguiculata* (L.) Walp.), according to Sprague & Sprague, Cordus (1939), p. 44. On the history of this species, see Simmonds, *Evolution of crop plants* (1976), pp. 183–85.]

92. [See ch. 9, n. 60, for a later conclusion.]

93. [See Carnoy, *Noms grecs de plantes* (1959), p. 132.]

94. V. Cordus, *Historia plantarum*, in Gesner, *Collecta* (1561), fol. 164ᵛ. [*Glycyrrhiza glabra* L., according to Sprague & Sprague, Cordus (1939), p. 68. On the use of licorice, see Grieve, *Modern herbal* (1931), vol. 2, pp. 487–92; Simmonds, *Evolution of crop plants* (1976), p. 312.]

95. See Tournefort, *Élémens* (1694); Linnæus, *Philosophia botanica* (1751).

96. V. Cordus, *Historia plantarum*, in Gesner, *Collecta* (1561), fols. 114ᵛ–119ʳ. [For references on the usage of cucurbits, see ch. 7, n. 103.]

97. *Ibid.*, fol. 94ʳ. [*Lithospermum officinale* L., according to Sprague & Sprague, Cordus (1939), p. 24.]

98. [V. Cordus, *Historia plantarum*, in Gesner, *Collecta* (1561), fols. 119ᵛ–121ᵛ. *Ranunculus sceleratus* L., *R. sardous* Crantz, *Anemone ranunculoides* L., *A. nemorosa* L., *R. arvensis* L., ?*R. polyanthemos* L., *R. repens* L., ?*R. acris* L., *R. bulbosus* L., *A. pulsatilla* L., *A. sylvestris* L., and *R. flammula* L., according to Sprague & Sprague, Cordus (1939), pp. 38–39.]

99. Cesalpino was but four years younger than Cordus, but lived to old age. [On Cesalpino, see Part II, ch. 22.]

100. V. Cordus, *Historia plantarum*, in Gesner, *Collecta* (1561), fol. 122ʳ.

101. *Ibid.*, fol. 120ᵛ. [*I.e., septima.*]

102. *Ibid.*, fols. 174ᵛ–175ʳ

103. On the history of the *Dipsacus* controversy, and its nomenclature, see Greene, Fullers' teasel (1896). [On its usage, see Grieve, *Modern herbal* (1931), vol. 2, pp. 793–94.]

104. [V. Cordus, *Historia plantarum*, in Gesner, *Collecta* (1561), fols. 105ᵛ–106ʳ. These species are *Dipsacus sylvestris* Huds. and *D. fullonum* L. emend. Huds., according to Sprague & Sprague, Cordus (1939), p. 31.]

105. V. Cordus, *Annotationes*, bk. 4, ch. 165, fol. 74ᵛ, in Gesner, *Collecta* (1561).

106. [*Opulus* Mill. = *Viburnum* L.; *Pneumonanthe* Gled. = *Gentiana* L.]

107. [*Balsamilla* was a name for *Impatiens*, and *Millegrana* for *Herniaria* during the Middle Ages, according to Herm. Fischer, *Mittelalterliche Pflanzenkunde* (1929), p. 271.]

108. V. Cordus, *Historia plantarum*, in Gesner, *Collecta* (1561), fol. 161ʳ.

109. *Ibid.*, fol. 173ʳ.

110. [By Plumier, in *Nova plantarum americanarum* (1703).]

111. V. Cordus, *Historia plantarum*, in Gesner, *Collecta* (1561), fol. 169ʳ.

112. See pp. 258, 263.

113. V. Cordus, *Historia plantarum*, in Gesner, *Collecta* (1561), fol. 122ʳ.

114. *Ibid.*, fol. 86ᵛ. ["Sagitta pediculos habet triquetros, intus raros & spongiosos, bicubitales, interdum longiores, breuiorésque, secundum lacuum in quibus nascitur altitudinem iis in summo triplici acumine folia alatæ sagittæ modo prefixa

sunt. . . ." *Sagittaria sagittifolia* L., according to Sprague & Sprague, Cordus (1939), p. 21.]

115. V. Cordus, *Historia plantarum*, in Gesner, *Collecta* (1561), fol. 99.

116. *Ibid.*, fol. 105^{v}. [See nn. 103, 104.]

117. *Ibid.*, fol. 164^{v}. [*Glycyrrhiza glabra* L., according to Sprague & Sprague, Cordus (1939), p. 68. For references on the history of studies of plant movements, see ch. 2, n. 189. For references on licorice, see n. 94.]

118. V. Cordus, *Historia plantarum*, in Gesner, *Collecta* (1561), fol. 100^{r}. [*Trigonella fœnum-græcum* L., according to Sprague & Sprague, Cordus (1939), p. 27. On the use of fenugreek, see Grieve, *Modern herbal* (1931), vol. 1, p. 299; Rosengarten, *Book of spices* (1969), pp. 248–53; Wein, *Trigonella fœnum-græcum* L. (1955).]

119. V. Cordus, *Historia plantarum*, in Gesner, *Collecta* (1561), fol. 86^{r}. [*Drosera rotundifolia* L., according to Sprague & Sprague, Cordus (1939), p. 21.]

120. Tragus had known *Drosera rotundifolia* and has it figured as a species of *Polytrichum*, *P. minus* [Bock, *Stirpium* (1552), p. 528]. He said its habitat was "dewy rocks," and had no idea that the "dew" on its leaves was an exudation. [See Fig. 99.]

121. V. Cordus, *Historia plantarum*, in Gesner, *Collecta* (1561), fol. 137^{r}. [*Lupinus alba* L., according to Sprague & Sprague, Cordus (1939), p. 50. For references on lupines, see n. 88.]

122. V. Cordus, *Historia plantarum*, in Gesner, *Collecta* (1561), fol. 113^{r}. [*Phyllitis scolopendrium* (L.) Newm., as indicated in n. 83.]

123. *Ibid.*, fol. 92^{r}.

124. *Ibid.*, fol. 144^{r}. [*Sanguisorba officinalis* L. and *Poterium sanguisorba* L., according to Sprague & Sprague, Cordus (1939), p. 53. On the use of the great burnet, see Grieve, *Modern herbal* (1931), vol. 1, pp. 145–46.]

125. V. Cordus, *Historia plantarum*, in Gesner, *Collecta* (1561), fol. 140^{v}. [*Oxycoccus palustris* Pers., according to Sprague & Sprague, Cordus (1939), p. 51.]

126. V. Cordus, *Historia plantarum*, in Gesner, *Collecta* (1561), fol. 86^{r}.

127. *Ibid.*, fol. 172^{v}. [*Adoxa moschatellina* L., according to Sprague & Sprague, Cordus (1939), p. 73.]

128. V. Cordus, *Historia plantarum*, in Gesner, *Collecta* (1561), fol. 189, as *Genista angulosa*; where the editor, Gesner, made the inexcusable error of inserting Tragus's woodcut of the extremely different *Genista sagittalis*. [This species is now called *Sarothamnus scoparius* (L.) Koch., according to Sprague & Sprague, Cordus (1939), p. 78.]

129. [See the references at ch. 3, n. 20.]

130. V. Cordus, *Historia plantarum*, in Gesner, *Collecta* (1561), fols. 176^{v}–182^{r}. [Later, Greene discovered that Ruel had published descriptions of French varieties of pears and apples in 1536 (see Part II, ch. 16). See also Bois, *Plantes alimentaires* (1927–37), vol. 2, pp. 281–307, on pears (*Pyrus* L.), and vol. 2, pp. 259–79, vol. 4, pp. 97–129, on apples (*Malus* L.); LeClerc, *Fruits de France* (1948), pp. 110–21; Simmonds, *Evolution of crop plants* (1976), pp. 247–50.]

131. That is to say, a handbell, with its handle.

132. [Plumier, *Nova plantarum americanarum* (1703), p. 13.]